DISCRETE INVERSE AND STATE ESTIMATION PROBLEMS

With Geophysical Fluid Applications

The problems of making inferences about the natural world from noisy observations and imperfect theories occur in almost all scientific disciplines. This book addresses these problems using examples taken from geophysical fluid dynamics. It focuses on discrete formulations, both static and time-varying, known variously as inverse, state estimation or data assimilation problems. Starting with fundamental algebraic and statistical ideas, the book guides the reader through a range of inference tools including the singular value decomposition, Gauss–Markov and minimum variance estimates, Kalman filters and related smoothers, and adjoint (Lagrange multiplier) methods. The final chapters discuss a variety of practical applications to geophysical flow problems.

Discrete Inverse and State Estimation Problems: With Geophysical Fluid Applications is an ideal introduction to the topic for graduate students and researchers in oceanography, meteorology, climate dynamics, geophysical fluid dynamics, and any field in which models are used to interpret observations. It is accessible to a wide scientific audience, as the only prerequisite is an understanding of linear algebra.

CARL WUNSCH is Cecil and Ida Green Professor of Physical Oceanography at the Department of Earth, Atmospheric and Planetary Sciences, Massachusetts Institute of Technology. After gaining his Ph.D. in geophysics in 1966 at MIT, he has risen through the department, becoming its head for the period between 1977–81. He subsequently served as Secretary of the Navy Research Professor and has held senior visiting positions at many prestigious universities and institutes across the world. His previous books include *Ocean Acoustic Tomography* (Cambridge University Press, 1995) with W. Munk and P. Worcester, and *The Ocean Circulation Inverse Problem* (Cambridge University Press, 1996).

DISCRETE INVERSE AND STATE ESTIMATION PROBLEMS

With Geophysical Fluid Applications

CARL WUNSCH

Department of Earth, Atmospheric and Planetary Sciences
Massachusetts Institute of Technology

CAMBRIDGE
UNIVERSITY PRESS

CAMBRIDGE UNIVERSITY PRESS
Cambridge, New York, Melbourne, Madrid, Cape Town,
Singapore, São Paulo, Delhi, Mexico City

Cambridge University Press
The Edinburgh Building, Cambridge CB2 8RU, UK

Published in the United States of America by Cambridge University Press, New York

www.cambridge.org
Information on this title: www.cambridge.org/9781107406063

First published 2006
First paperback edition 2011

A catalogue record for this publication is available from the British Library

ISBN 978-0-521-85424-5 Hardback
ISBN 978-1-107-40606-3 Paperback

Additional resources for this publication at www.cambridge.org/9781107406063

To Walter Munk for decades of friendship and exciting collaboration.

Colour plate section appears between pages 182 and 183, and is also available for download in colour from www. cambridge.org/9781107406063

Contents

Colour plates between pp. 182 and 183.

Preface

This book is to a large extent the second edition of *The Ocean Circulation Inverse Problem*, but it differs from the original version in a number of ways. While teaching the basic material at MIT and elsewhere over the past ten years, it became clear that it was of interest to many students outside of physical oceanography – the audience for whom the book had been written. The oceanographic material, instead of being a motivating factor, was in practice an obstacle to understanding for students with no oceanic background. In the revision, therefore, I have tried to make the examples more generic and understandable, I hope, to anyone with even rudimentary experience with simple fluid flows.

Also many of the oceanographic applications of the methods, which were still novel and controversial at the time of writing, have become familiar and almost commonplace. The oceanography, now confined to the two last chapters, is thus focussed less on explaining why and how the calculations were done, and more on summarizing what has been accomplished. Furthermore, the time-dependent problem (here called "state estimation" to distinguish it from meteorological practice) has evolved rapidly in the oceanographic community from a hypothetical methodology to one that is clearly practical and in ever-growing use.

The focus is, however, on the basic concepts and not on the practical numerical engineering required to use the ideas on the very large problems encountered with real fluids. Anyone attempting to model the global ocean or atmosphere or equivalent large scale system must confront issues of data storage, code parallelization, truncation errors, grid refinement, and the like. Almost none of these important problems are taken up here. Before constructive approaches to the practical problems can be found, one must understand the fundamental ideas. An analogy is the need to understand the implications of Maxwell's equations for electromagnetic phenomena before one undertakes to build a high fidelity receiver. The effective engineering of an electronic instrument can only be helped by good understanding of how one works in principle, albeit the details of making one work in practice can be quite different.

In the interests of keeping the book as short as possible, I have, however, omitted some of the more interesting theoretical material of the original version, but which readers can find in the wider literature on control theory. It is assumed that the reader has a familiarity at the introductory level with matrices and vectors, although everything is ultimately defined in Chapter 2.

Finally, I have tried to correct the dismaying number of typographical and other errors in the previous book, but have surely introduced others. Reports of errors of any type will be gratefully received.

I thank the students and colleagues who over the years have suggested corrections, modifications, and clarifications. My time and energies have been supported financially by the National Aeronautics and Space Administration, and the National Science Foundation through grants and contracts, as well as by the Massachusetts Institute of Technology through the Cecil and Ida Green Professorship.

Acknowledgements

The following figures are reproduced by permission of the American Geophysical Union: 4.8, 6.16–6.20, 6.23–6.26, 6.31, 7.3–7.5 and 7.7–7.10. Figures 2.16, 6.21, 6.27, 6.32, 7.1 and 7.11 are reproduced by permission of the American Meteorological Society. Figure 6.22 is reproduced by permission of Kluwer.

Errata

A correction list compiled by the author can be found here: http://ocean.mit.edu/~cwunsch

Part I

Fundamental machinery

1

Introduction

The most powerful insights into the behavior of the physical world are obtained when observations are well described by a theoretical framework that is then available for predicting new phenomena or new observations. An example is the observed behavior of radio signals and their extremely accurate description by the Maxwell equations of electromagnetic radiation. Other such examples include planetary motions through Newtonian mechanics, or the movement of the atmosphere and ocean as described by the equations of fluid mechanics, or the propagation of seismic waves as described by the elastic wave equations. To the degree that the theoretical framework supports, and is supported by, the observations one develops sufficient confidence to calculate similar phenomena in previously unexplored domains or to make predictions of future behavior (e.g., the position of the moon in 1000 years, or the climate state of the earth in 100 years).

Developing a coherent view of the physical world requires some mastery, therefore, of both a framework, and of the meaning and interpretation of real data. Conventional scientific education, at least in the physical sciences, puts a heavy emphasis on learning how to solve appropriate differential and partial differential equations (Maxwell, Schrödinger, Navier–Stokes, etc.). One learns which problems are "well-posed," how to construct solutions either exactly or approximately, and how to interpret the results. Much less emphasis is placed on the problems of understanding the implications of data, which are inevitably imperfect – containing noise of various types, often incomplete, and possibly inconsistent and thus considered mathematically "ill-posed" or "ill-conditioned." When working with observations, ill-posedness is the norm, not the exception.

Many interesting problems arise in using observations in conjunction with theory. In particular, one is driven to conclude that there are no well-posed problems outside of textbooks, that stochastic elements are inevitably present and must be confronted, and that more generally, one must make inferences about the world from data that are necessarily always incomplete. The main purpose of this introductory chapter is to provide some comparatively simple examples of the type of problems one confronts in practice, and for which many interesting and useful tools exist for their solution. In an older context, this subject was called the "calculus of observations."[1] Here we refer to "inverse methods," although many different approaches are so labeled.

1.1 Differential equations

Differential equations are often used to describe natural processes. Consider the elementary problem of finding the temperature in a bar where one end, at $r = r_A$, is held at constant temperature T_A, and at the other end, $r = r_B$, it is held at temperature T_B. The only mechanism for heat transfer within the bar is by molecular diffusion, so that the governing equation is

$$\kappa \frac{\mathrm{d}^2 T}{\mathrm{d}r^2} = 0, \tag{1.1}$$

subject to the boundary conditions

$$T(r_A) = T_A, \quad T(r_B) = T_B. \tag{1.2}$$

Equation (1.1) is so simple we can write its solution in a number of different ways. One form is

$$T(r) = a + br, \tag{1.3}$$

where a, b are unknown parameters, until some additional information is provided. Here the additional information is contained in the boundary conditions (1.2), and, with two parameters to be found, there is just sufficient information, and

$$T(r) = \frac{r_B T_A - r_A T_B}{r_B - r_A} + \left(\frac{T_B - T_A}{r_B - r_A} \right) r, \tag{1.4}$$

which is a straight line. Such problems, or analogues for much more complicated systems, are sometimes called "forward" or "direct" and they are "well-posed": exactly enough information is available to produce a unique solution insensitive to perturbations in any element (easily proved here, not so easily in other cases). The solution is both stable and differentiable. This sort of problem and its solution is what is generally taught in elementary science courses.

On the other hand, the problems one encounters in actually doing science differ significantly – both in the questions being asked, and in the information available.

For example:

1. One or both of the boundary values T_A, T_B is known from measurements; they are thus given as $T_A = T_A^{(c)} \pm \Delta T_A$, $T_B = T_B^{(c)} \pm \Delta T_B$, where the $\Delta T_{A,B}$ are an estimate of the possible inaccuracies in the theoretical values $T_i^{(c)}$. (Exactly what that might mean is taken up later.)
2. One or both of the positions, $r_{A,B}$ is also the result of measurement and are of the form $r_{A,B}^{(c)} \pm \Delta r_{A,B}$.
3. T_B is missing altogether, but is known to be positive, $T_B > 0$.
4. One of the boundary values, e.g., T_B, is unknown, but an interior value $T_{int} = T_{int}^{(c)} \pm \Delta T_{int}$ is provided instead. Perhaps many interior values are known, but none of them perfectly.

Other possibilities exist. But even this short list raises a number of interesting, practical problems. One of the themes of this book is that almost nothing in reality is known perfectly. It is possible that ΔT_A, ΔT_B are very small; but as long as they are not actually zero, there is no longer any possibility of finding a unique solution.

Many variations on this model and theme arise in practice. Suppose the problem is made slightly more interesting by introducing a "source" $S_T(r)$, so that the temperature field is thought to satisfy the equation

$$\frac{\mathrm{d}^2 T(r)}{\mathrm{d}r^2} = S_T(r), \tag{1.5}$$

along with its boundary conditions, producing another conventional forward problem. One can convert (1.5) into a different problem by supposing that one knows $T(r)$, and seeks $S_T(r)$. Such a problem is even easier to solve than the conventional one: differentiate T twice. Because convention dictates that the "forward" or "direct" problem involves the determination of $T(r)$ from a known $S_T(r)$ and boundary data, this latter problem might be labeled as an "inverse" one – simply because it contrasts with the conventional formulation.

In practice, a whole series of new problems can be raised: suppose $S_T(r)$ is imperfectly known. How should one proceed? If one knows $S_T(r)$ and $T(r)$ at a series of positions $r_i \neq r_A, r_B$, could one nonetheless deduce the boundary conditions? Could one deduce $S_T(r)$ if it were not known at these interior values?

$T(r)$ has been supposed to satisfy the differential equation (1.1). For many purposes, it is helpful to reduce the problem to one that is intrinsically discrete. One way to do this would be to expand the solution in a system of polynomials,

$$T(r) = \alpha_0 r^0 + \alpha_1 r^1 + \cdots + \alpha_m r^m, \tag{1.6}$$

and

$$S_T(r) = \beta_0 r^0 + \beta_1 r^1 + \cdots + \beta_n r^n, \tag{1.7}$$

where the β_i would conventionally be known, and the problem has been reduced from the need to find a function $T(r)$ defined for all values of r, to one in which only the finite number of parameters α_i, $i = 0, 1, \ldots, m$, must be found.

An alternative discretization is obtained by using the coordinate r. Divide the interval $r_A = 0 \leq r \leq r_B$ into $N - 1$ intervals of length Δr, so that $r_B = (N - 1)\Delta r$. Then, taking a simple one-sided difference:

$$T(2\Delta r) - 2T(\Delta r) + T(0) = (\Delta r)^2 S_T(\Delta r),$$
$$T(3\Delta r) - 2T(2\Delta r) + T(1\Delta r) = (\Delta r)^2 S_T(2\Delta r), \tag{1.8}$$

$$\vdots$$

$$T((N-1)\Delta r) - 2T((N-2)\Delta r) + T((N-3)\Delta r) = (\Delta r)^2 S_T((N-2)\Delta r).$$

If one counts the number of equations in (1.8) it is readily found that there are $N - 2$, but with a total of N unknown $T(p\Delta r)$. The two missing pieces of information are provided by the two boundary conditions $T(0\Delta r) = T_0$, $T((N-1)\Delta r) = T_{N-1}$. Thus the problem of solving the differential equation has been reduced to finding the solution of a set of ordinary linear simultaneous algebraic equations, which we will write, in the notation of Chapter 2, as

$$\mathbf{Ax} = \mathbf{b}, \tag{1.9}$$

where $\mathbf{A}$ is a square matrix, $\mathbf{x}$ is the vector of unknowns $T(p\Delta r)$, and $\mathbf{b}$ is the vector of values $\mathbf{q}(p\Delta t)$, and of boundary values. The list above, of variations, e.g., where a boundary condition is missing, or where interior values are provided instead of boundary conditions, then becomes statements about having too few, or possibly too many, equations for the number of unknowns. Uncertainties in the T_i or in the $q(p\Delta r)$ become statements about having to solve simultaneous equations with uncertainties in some elements. That models, even non-linear ones, can be reduced to sets of simultaneous equations, is the unifying theme of this book. One might need truly vast numbers of grid points, $p\Delta r$, or polynomial terms, and ingenuity in the formulation to obtain adequate accuracy, but as long as the number of parameters $N < \infty$, one has achieved a great, unifying simplification.

Consider a little more interesting ordinary differential equation, that for the simple mass–spring oscillator:

$$m\frac{\mathrm{d}^2\xi(t)}{\mathrm{d}t^2} + \varepsilon\frac{\mathrm{d}\xi(t)}{\mathrm{d}t} + k_0\xi(t) = S_\xi(t), \tag{1.10}$$

where m is mass, k_0 is a spring constant, and ε is a dissipation parameter. Although the equation is slightly more complicated than (1.5), and we have relabeled the independent variable as t (to suggest time), rather than as r, there really is no fundamental difference. This differential equation can also be solved in any number of ways. As a second-order equation, it is well-known that one must provide two extra conditions to have enough information to have a unique solution. Typically, there are *initial* conditions, $\xi(0)$, $d\xi(0)/dt$ – a position and velocity, but there is nothing to prevent us from assigning two end conditions, $\xi(0)$, $\xi(t = t_f)$, or even two velocity conditions $d\xi(0)/dt$, $d\xi(t_f)/dt$, etc.

If we naively discretize (1.10) as we did the straight-line equation, we have

$$\xi(p\Delta t + \Delta t) - \left(2 - \frac{\varepsilon\Delta t}{m} - \frac{k_0(\Delta t)^2}{m}\right)\xi(p\Delta t) - \left(\frac{\varepsilon\Delta t}{m} - 1\right)\xi(p\Delta t - \Delta t)$$

$$= (\Delta t)^2 \frac{S_\xi\left((p - 1)\,\Delta t\right)}{m}, \quad 2 \le p \le N - 1, \tag{1.11}$$

which is another set of simultaneous equations as in (1.9) in the unknown $\xi(p\Delta t)$; an equation count again would show that there are two fewer equations than unknowns – corresponding to the two boundary or two initial conditions. In Chapter 2, several methods will be developed for solving sets of simultaneous linear equations, even when there are apparently too few or too many of them. In the present case, if one were given $\xi(0)$, $\xi(1\Delta t)$, Eq. (1.11) could be stepped forward in time, generating $\xi(3\Delta t)$, $\xi(4\Delta t)$, $\dots$, $\xi((N - 1)\Delta t)$. The result would be identical to the solution of the simultaneous equations – but with far less computation.

But if one were given $\xi((N - 1)\Delta t)$ instead of $\xi(1\Delta t)$, such a simple time-stepping rule could no longer be used. A similar difficulty would arise if $q(j\Delta t)$ were missing for some j, but instead one had knowledge of $\xi(p\Delta t)$, for some p. Looked at as a set of simultaneous equations, there is no conceptual problem: one simply solves it, all at once, by Gaussian elimination or equivalent. There *is* a problem only if one sought to time-step the equation forward, but without the required second condition at the starting point – there would be inadequate information to go forward in time. Many of the so-called inverse methods explored in this book are ways to solve simultaneous equations while avoiding the need for all-at-once brute-force solution. Nonetheless, one is urged to always recall that most of the interesting algorithms are just clever ways of solving large sets of such equations.

1.2 Partial differential equations

Finding the solutions of linear differential equations is equivalent, when discretized, to solving sets of simultaneous linear algebraic equations. Unsurprisingly, the same is true of partial differential equations. As an example, consider a very familiar problem:

Solve

$$\nabla^2 \phi = \rho, \tag{1.12}$$

for ϕ, given ρ, in the domain $\mathbf{r} \in D$, subject to the boundary conditions $\phi = \phi_0$ on the boundary ∂D, where $\mathbf{r}$ is a spatial coordinate of dimension greater than 1.

This statement is the Dirichlet problem for the Laplace–Poisson equation, whose solution is well-behaved, unique, and stable to perturbations in the boundary data, ϕ_0, and the source or forcing, ρ. Because it is the familiar boundary value problem, it is by convention again labeled a forward or direct problem. Now consider a different version of the above:

Solve (1.12) for ρ given ϕ in the domain D.

This latter problem is again easier to solve than the forward one: differentiate ϕ twice to obtain the Laplacian, and ρ is obtained from (1.12). Because the problem as stated is inverse to the conventional forward one, it is labeled, as with the ordinary differential equation, an *inverse problem*. It is inverse to a more familiar boundary value problem in the sense that the usual unknowns ϕ have been inverted or interchanged with (some of) the usual knowns ρ. Notice that both forward and inverse problems, as posed, are well-behaved and produce uniquely determined answers (ruling out mathematical pathologies in any of ρ, ϕ_0, ∂D, or ϕ). Again, there are many variations possible: one could, for example, demand computation of the boundary conditions, ϕ_0, from given information about some or all of ϕ, ρ.

Write the Laplace–Poisson equation in finite difference form for two Cartesian dimensions:

$$\phi_{i+1,j} - 2\phi_{i,j} + \phi_{i-1,j} + \phi_{i,j+1} - 2\phi_{i,j} + \phi_{i,j-1} = (\Delta x)^2 \rho_{ij}, \quad i,\, j \in D, \tag{1.13}$$

with square grid elements of dimension Δx. To make the bookkeeping as simple as possible, suppose the domain D is the square $N \times N$ grid displayed in Fig. 1.1, so that ∂D is the four line segments shown. There are $(N-2) \times (N-2)$ interior grid points, and Eqs. (1.13) are then $(N-2) \times (N-2)$ equations in N^2 of the ϕ_{ij}. If this is the forward problem with ρ_{ij} specified, there are fewer equations than unknowns. But appending the set of boundary conditions to (1.13):

$$\phi_{ij} = \phi_{0ij}, \quad i,\, j \in \partial D, \tag{1.14}$$

there are precisely $4N - 4$ of these conditions, and thus the combined set (1.13) plus (1.14), written as (1.9) with,

$$\mathbf{x} = \text{vec}\{\phi_{ij}\} = [\phi_{11}\ \phi_{12} \cdot \phi_{NN}]^{\text{T}},$$

$$\mathbf{b} = \text{vec}\{\rho_{ij}, \phi_{ij}^0\} = [\rho_{22}\ \rho_{23} \cdot \rho_{N-1,N-1}\ \phi_{011} \cdot \phi_{0N,N}]^{\text{T}},$$

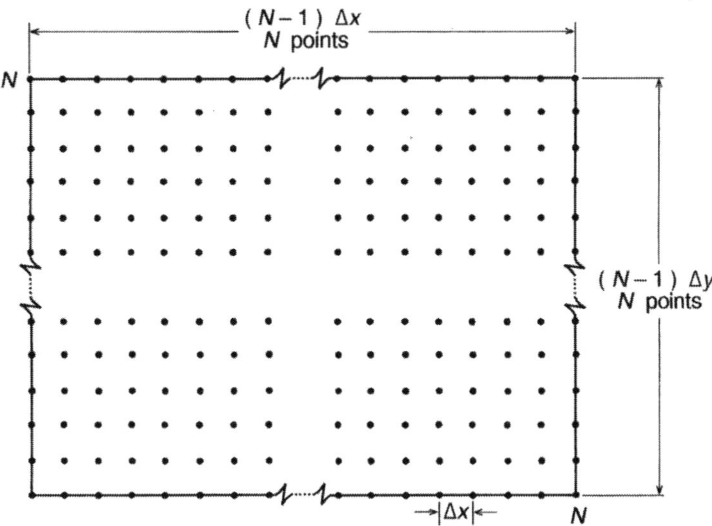

Figure 1.1 Square, homogeneous grid used for discretizing the Laplacian, thus reducing the partial differential equation to a set of linear simultaneous equations.

which is a set of $M = N^2$ equations in $M = N^2$ unknowns. (The operator, vec, forms a column vector out of the two-dimensional array ϕ_{ij}; the superscript T is the vector transpose, defined in Chapter 2.) The nice properties of the Dirichlet problem can be deduced from the well-behaved character of the matrix $\mathbf{A}$. Thus the forward problem corresponds directly with the solution of an ordinary set of simultaneous algebraic equations.[2] One complementary inverse problem says: "Using (1.9) compute ρ_{ij} and the boundary conditions, given ϕ_{ij}," which is an even simpler computation – it involves just multiplying the known $\mathbf{x}$ by the known matrix $\mathbf{A}$.

But now let us make one small change in the forward problem, making it the Neumann one:

Solve

$$\nabla^2 \phi = \rho, \tag{1.15}$$

for ϕ, given ρ, in the domain $\mathbf{r} \in D$ subject to the boundary conditions $\partial\phi/\partial\hat{\mathbf{m}} = \phi'_0$ on the boundary ∂D, where $\mathbf{r}$ is the spatial coordinate and $\hat{\mathbf{m}}$ is the unit normal to the boundary.

This new problem is another classical, much analyzed forward problem. It is, however, well-known that the solution is indeterminate up to an additive constant. This indeterminacy is clear in the discrete form: Eqs. (1.14) are now replaced by

$$\phi_{i+1,j} - \phi_{i,j} = \phi'_{0ij}, \quad i, j \in \partial D' \tag{1.16}$$

etc., where $\partial D'$ represents the set of boundary indices necessary to compute the local normal derivative. There is a new combined set:

$$\mathbf{A}\mathbf{x} = \mathbf{b}_1, \quad \mathbf{x} = \mathrm{vec}\{\phi_{ij}\}, \quad \mathbf{b}_1 = \mathrm{vec}\{\rho_{ij}, \phi'_{0ij}\}. \tag{1.17}$$

Because only *differences* of the ϕ_{ij} are specified, there is no information concerning the absolute value of $\mathbf{x}$. When some machinery is obtained in Chapter 2, we will be able to demonstrate automatically that even though (1.17) appears to be M equations in M unknowns, in fact only $M - 1$ of the equations are independent, and thus the Neumann problem is an underdetermined one. This property of the Neumann problem is well-known, and there are many ways of handling it, either in the continuous or discrete forms. In the discrete form, a simple way is to add one equation setting the value at any point to zero (or anything else). A further complication with the Neumann problem is that it can be set up as a contradiction – even while underdetermined – if the flux boundary conditions do not balance the interior sources. The simultaneous presence of underdetermination and contradiction is commonplace in real problems.

1.3 More examples

A tracer box model

In scientific practice, one often has observations of elements of the solution of the differential system or other model. Such situations vary enormously in the complexity and sophistication of both the data and the model. A useful and interesting example of a simple system, with applications in many fields, is one in which there is a large reservoir (Fig. 1.2) connected to a number of source regions which provide fluid to the reservoir. One would like to determine the rate of mass transfer from each source region to the reservoir.

Suppose that some chemical tracer or dye, C_0, is measured in the reservoir, and that the concentrations of the dye, C_i, in each source region are known. Let the unknown transfer rates be J_{i0} (transfer from source i to reservoir 0). Then we must have

$$C_1 J_{10} + C_2 J_{20} + \cdots + C_N J_{N0} = C_0 J_{0\infty}, \tag{1.18}$$

which says that, for a steady state, the rate of transfer in must equal the rate of transfer out (written $J_{0\infty}$). To conserve mass,

$$J_{10} + J_{20} + \cdots + J_{N0} = J_{0\infty}. \tag{1.19}$$

This model has produced two equations in $N + 1$ unknowns, $[J_{10}, J_{20}, \ldots, J_{N0}, J_{0\infty}]$, which evidently is insufficient information if $N > 1$. The equations

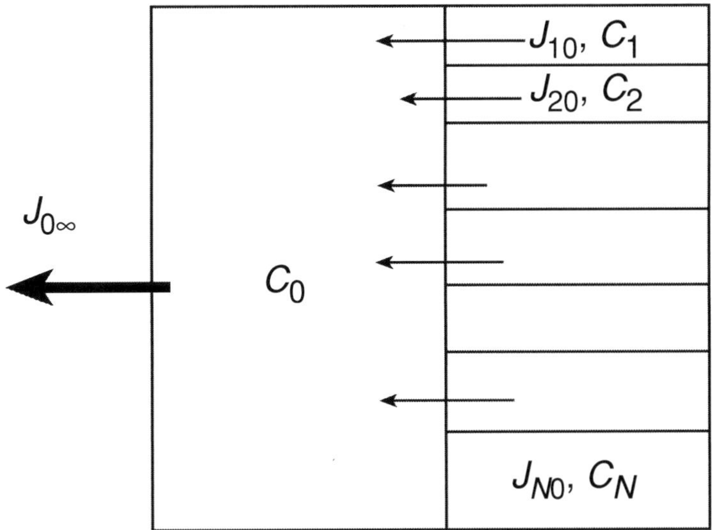

Figure 1.2 A simple reservoir problem in which there are multiple sources of flow, at rates J_{i0}, each carrying an identifiable property C_i, perhaps a chemical concentration. In the forward problem, given J_{i0}, C_i one could calculate C_0. One form of inverse problem provides C_0 and the C_i and seeks the values of J_{i0}.

have also been written as though everything were perfect. If, for example, the tracer concentrations C_i were measured with finite precision and accuracy (they always are), the resulting inaccuracy might be accommodated as

$$C_1 J_{10} + C_2 J_{20} + \cdots + C_N J_{N0} + n = C_0 J_{0\infty}, \qquad (1.20)$$

where n represents the resulting error in the equation. Its introduction produces another unknown. If the reservoir were capable of some degree of storage or fluctuation in level, an error term could be introduced into (1.19) as well. One should also notice that, as formulated, one of the apparently infinite number of solutions to Eqs. (6.1, 1.19) includes $J_{i0} = J_{0\infty} = 0$ – no flow at all. More information is required if this null solution is to be excluded.

To make the problem slightly more interesting, suppose that the tracer C is radioactive, and diminishes with a decay constant λ. Equation (6.1) becomes

$$C_1 J_{10} + C_2 J_{20} + \cdots + C_N J_{N0} - C_0 J_{0\infty} = -\lambda C_0. \qquad (1.21)$$

If $C_0 > 0$, $J_{ij} = 0$ is no longer a possible solution, but there remain many more unknowns than equations. These equations are once again in the canonical linear form $\mathbf{Ax} = \mathbf{b}$.

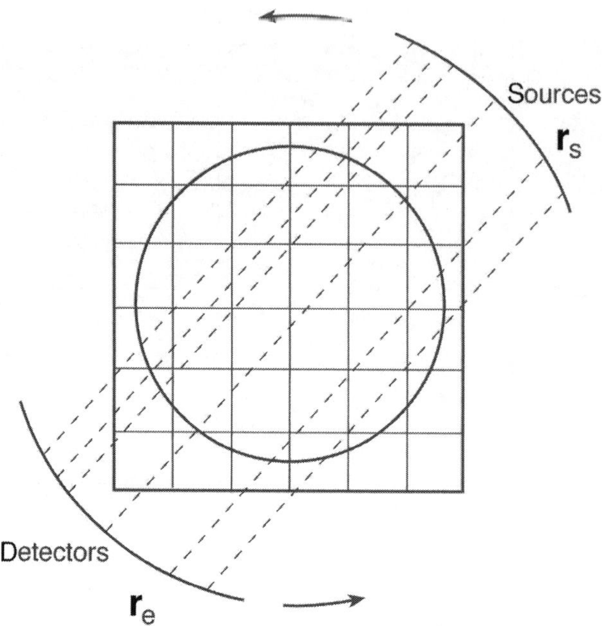

Figure 1.3 Generic tomographic problem in two dimensions. Measurements are made by integrating through an otherwise impenetrable solid between the transmitting sources and receivers using x-rays, sound, radio waves, etc. Properties can be anything measurable, including travel times, intensities, group velocities, etc., as long as they are functions of the parameters sought (such as the density or sound speed). The tomographic problem is to reconstruct the interior from these integrals. In the particular configuration shown, the source and receiver are supposed to revolve so that a very large number of paths can be built up. It is also supposed that the division into small rectangles is an adequate representation. In principle, one can have many more integrals than the number of squares defining the unknowns.

A tomographic problem

So-called tomographic problems occur in many fields, most notably in medicine, but also in materials testing, oceanography, meteorology, and geophysics. Generically, they arise when one is faced with the problem of inferring the distribution of properties inside an area or volume based upon a series of integrals through the region. Consider Fig. 1.3, where, to be specific, suppose we are looking at the top of the head of a patient lying supine in a so-called CAT-scanner. The two external shell sectors represent a source of x-rays, and a set of x-ray detectors. X-rays are emitted from the source and travel through the patient along the indicated lines where the intensity of the received beam is measured. Let the absorptivity/unit length within the patient be a function, $c(\mathbf{r})$, where $\mathbf{r}$ is the vector position within the patient's head. Consider one source at $\mathbf{r}_s$ and a receptor at $\mathbf{r}_e$ connected by the

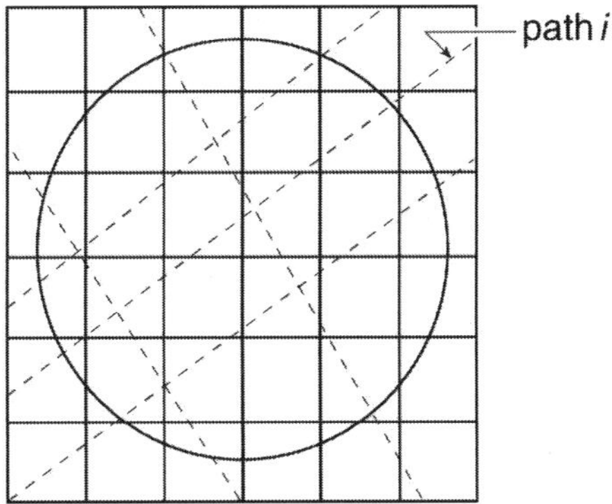

Figure 1.4 Simplified geometry for defining a tomographic problem. Some squares may have no integrals passing through them; others may be multiply-covered. Boxes outside the physical body can be handled in a number of ways, including the addition of constraints setting the corresponding $c_j = 0$.

path as indicated. Then the intensity measured at the receptor is

$$I\left(\mathbf{r}_s, \mathbf{r}_e\right) = \int_{\mathbf{r}_s}^{\mathbf{r}_e} c\left(\mathbf{r}\left(s\right)\right) ds, \tag{1.22}$$

where s is the arc length along the path. The basic tomographic problem is to determine $c(\mathbf{r})$ for all $\mathbf{r}$ in the patient, from measurements of I. c can be a function of both position and the physical parameters of interest. In the medical problem, the shell sectors rotate around the patient, and an enormous number of integrals along (almost) all possible paths are obtained. An analytical solution to this problem, as the number of paths becomes infinite, is produced by the Radon transform.[3] Given that tumors and the like have a different absorptivity to normal tissue, the reconstructed image of $c(\mathbf{r})$ permits physicians to "see" inside the patient. In most other situations, however, the number of paths tends to be much smaller than the formal number of unknowns and other solution methods must be found.

Note first, however, that Eq. (1.22) should be modified to reflect the inability of any system to produce a perfect measurement of the integral, and so, more realistically,

$$I\left(\mathbf{r}_s, \mathbf{r}_e\right) = \int_{\mathbf{r}_s}^{\mathbf{r}_e} c\left(\mathbf{r}\left(s\right)\right) ds + n(\mathbf{r}_s, \mathbf{r}_e), \tag{1.23}$$

where n is the measurement noise.

To proceed, surround the patient with a bounding square (Fig. 1.4) – to produce a simple geometry – and divide the area into sub-squares as indicated, each numbered in sequence, $j = 1, 2, \ldots, N$. These squares are supposed sufficiently small that $c(\mathbf{r})$ is effectively constant within them. Also number the paths, $i = 1, 2, \ldots, M$.

Then Eq. (1.23) can be approximated with arbitrary accuracy (by letting the sub-square dimensions become arbitrarily small) as

$$I_i = \sum_{j=1}^{N} c_j \Delta r_{ij} + n_i. \tag{1.24}$$

Here Δr_{ij} is the arc length of path i within square j (most of them will vanish for any particular path). These last equations are of the form

$$\mathbf{Ex} + \mathbf{n} = \mathbf{y}, \tag{1.25}$$

where $\mathbf{E} = \{\Delta r_{ij}\}$, $\mathbf{x} = [c_j]$, $\mathbf{y} = [I_i]$, $\mathbf{n} = [n_i]$. Quite commonly there are many more unknown c_j than there are integrals I_i. (In the present context, there is no distinction made between using matrices $\mathbf{A}$, $\mathbf{E}$. $\mathbf{E}$ will generally be used where noise elements are present, and $\mathbf{A}$ where none are intended.)

Tomographic measurements do not always consist of x-ray intensities. In seismology or oceanography, for example, c_j is commonly $1/v_j$, where v_j is the speed of sound or seismic waves within the area; I is then a travel time rather than an intensity. The equations remain the same, however. This methodology also works in three dimensions, the paths need not be straight lines, and there are many generalizations.[4] A problem of great practical importance is determining what one can say about the solutions to Eqs. (1.25) even where many more unknowns exist than formal pieces of information y_i.

As with all these problems, many other forms of discretization are possible. For example, the continuous function $c(\mathbf{r})$ can be expanded:

$$c(\mathbf{r}) = \sum_n \sum_m a_{nm} T_n(r_x) T_m(r_y), \tag{1.26}$$

where $\mathbf{r} = (r_x, r_y)$, and the T_n are any suitable expansion functions (sines and cosines, Chebyschev polynomials, etc.). The linear equations (4.35) then represent constraints leading to the determination of the a_{nm}.

A second tracer problem

Consider the closed volume in Fig. 1.5 enclosed by four boundaries as shown. There are steady flows, $v_i(z)$, $i = 1, \ldots, 4$, either into or out of the volume, each carrying a corresponding fluid of constant density $\rho_0 \cdot z$ is the vertical coordinate. If the width of each boundary is l_i, the statement that mass is conserved within the volume is simply

$$\sum_{i=1}^{r} l_i \rho_0 \int_{-h}^{0} v_i(z)\, dz = 0, \tag{1.27}$$

where the convention is made that flows into the box are positive, and flows out are negative. $z = -h$ is the lower boundary of the volume and $z = 0$ is the top

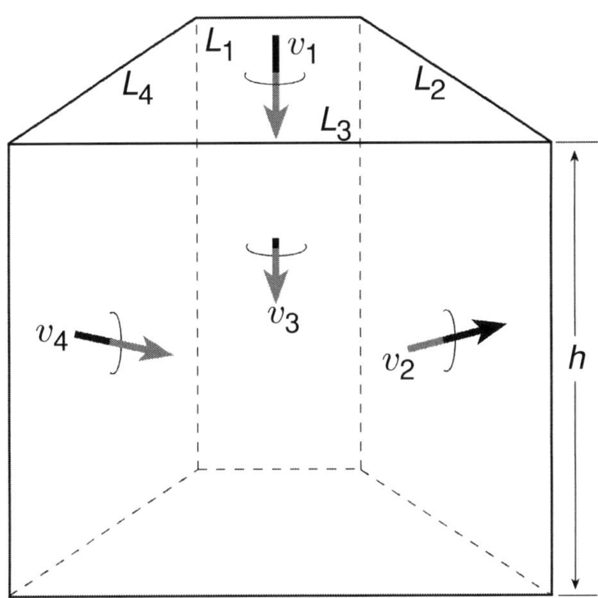

Figure 1.5 Volume of fluid bounded on four open vertical and two horizontal sides across which fluid is supposed to flow. Mass is conserved, giving one relationship among the fluid transports v_i; conservation of one or more other tracers C_i leads to additional useful relationships.

one. If the v_i are unknown, Eq. (1.27) represents one equation (constraint) in four unknowns:

$$\int_{-h}^{0} v_i(z)\,dz, \quad 1 \le i \le 4. \tag{1.28}$$

One possible, if boring, solution is $v_i(z) = 0$. To make the problem somewhat more interesting, suppose that, for some mysterious reason, the vertical derivatives, $v_i'(z) = dv_i(z)/dz$, are known so that

$$v_i(z) = \int_{z_0}^{z} v_i'(z)\,dz + b_i(z_0), \tag{1.29}$$

where z_0 is a convenient place to start the integration (but can be any value). b_i are integration constants ($b_i = v_i(z_0)$) that remain unknown. Constraint (1.27) becomes

$$\sum_{i=1}^{4} l_i \rho_0 \int_{-h}^{0} \left[\int_{z_0}^{z} v_i'(z')dz' + b_i(z_0) \right] dz = 0, \tag{1.30}$$

or

$$\sum_{i=1}^{4} hl_i b_i (z_0) = -\sum_{i=1}^{4} l_i \int_{-h}^{0} dz \int_{z_0}^{z} v_i'(z')dz', \tag{1.31}$$

where the right-hand side is known. Equation (1.31) is still one equation in four unknown b_i, but the zero-solution is no longer possible, unless the right-hand side vanishes. Equation (1.31) is a statement that the weighted average of the b_i on the left-hand side is known. If one seeks to obtain estimates of the b_i separately, more information is required.

Suppose that information pertains to a tracer, perhaps a red dye, known to be conservative, and that the box concentration of red dye, C, is known to be in a steady state. Then conservation of C becomes

$$\sum_{i=1}^{4} \left[hl_i \int_{-h}^{0} C_i (z)\, dz \right] b_i = -\sum_{i=1}^{4} l_i \int_{-h}^{0} dz \int_{-z_0}^{z} C_i(z')v_i'(z')dz', \tag{1.32}$$

where $C_i (z)$ is the concentration of red dye on each boundary. Equation (1.32) provides a second relationship for the four unknown b_i. One might try to measure another dye concentration, perhaps green dye, and write an equation for this second tracer, exactly analogous to (1.32). With enough such dye measurements, there might be more constraint equations than unknown b_i. In any case, no matter how many dyes are measured, the resulting equation set is of the form (1.9). The number of boundaries is not limited to four, but can be either fewer, or many more.[5]

Vibrating string

Consider a uniform vibrating string anchored at its ends $r_x = 0$, $r_x = L$. The free motion of the string is governed by the wave equation

$$\frac{\partial^2 \eta}{\partial r_x^2} - \frac{1}{c^2}\frac{\partial^2 \eta}{\partial t^2} = 0, \quad c^2 = T/\rho, \tag{1.33}$$

where T is the tension and ρ the density. Free modes of vibration (eigen-frequencies) are found to exist at discrete frequencies, s_q,

$$2\pi s_q = \frac{q\pi c}{L}, \quad q = 1, 2, 3, \ldots, \tag{1.34}$$

which is the solution to a classical forward problem. A number of interesting and useful inverse problems can be formulated. For example, given $s_q \pm \Delta s_q$, $q = 1, 2, \ldots, M$, to determine L or c. These are particularly simple problems, because there is only one parameter, either c or L, to determine. More generally, it is obvious from Eq. (1.34) that one has information only about the ratio c/L – they could not be determined separately.

Suppose, however, that the density varies along the string, $\rho = \rho(r_x)$, so that $c = c(r_x)$. Then, it may be confirmed that the observed frequencies are no longer given by Eq. (1.34), but by expressions involving the integral of c over the length of the string. An important problem is then to infer $c(r_x)$, and hence $\rho(r_x)$. One might wonder whether, under these new circumstances, L can be determined independently of c?

A host of such problems exist, in which the observed frequencies of free modes are used to infer properties of media in one to three dimensions. The most elaborate applications are in geophysics and solar physics, where the normal mode frequencies of the vibrating whole Earth or Sun are used to infer the interior properties (density, elastic parameters, magnetic field strength, etc.).[6] A good exercise is to render the spatially variable string problem into discrete form.

1.4 Importance of the forward model

Inference about the physical world from data requires assertions about the structure of the data and its internal relationships. One sometimes hears claims from people who are expert in measurements that "I don't use models." Such a claim is almost always vacuous. What the speaker usually means is that he doesn't use equations, but is manipulating his data in some simple way (e.g., forming an average) that seems to be so unsophisticated that no model is present. Consider, however, a simple problem faced by someone trying to determine the average temperature in a room. A thermometer is successively placed at different three-dimensional locations, $\mathbf{r}_i$, at times t_i. Let the measurements be y_i and the value of interest be

$$\tilde{m} = \frac{1}{M} \sum_{i=1}^{M} y_i. \tag{1.35}$$

In deciding to compute, and use, $\tilde{m}$ the observer has probably made a long list of very sophisticated, but implicit, model assumptions. Among them we might suggest: (1) Thermometers actually measure the length of a fluid, or an oscillator frequency, or a voltage and require knowledge of the relation to temperature as well as potentially elaborate calibration methods. (2) That the temperature in the room is sufficiently slowly changing that all of the t_i can be regarded as effectively identical. A different observer might suggest that the temperature in the room is governed by shock waves bouncing between the walls at intervals of seconds or less. Should that be true, $\tilde{m}$ constructed from the available samples might prove completely meaningless. It might be objected that such an hypothesis is far-fetched. But the assumption that the room temperature is governed, e.g., by a slowly evolving diffusion process, is a specific, and perhaps incorrect model. (3) That the errors in the thermometer are such that the best estimate of the room mean temperature is obtained by the simple sum in Eq. (1.35). There are many measurement devices for which this assumption is a very poor one (perhaps the instrument is drifting, or has a calibration that varies with temperature), and we will discuss how to determine averages in Chapter 2. But the assumption that property $\tilde{m}$ is useful, is a strong model assumption concerning both the instrument being used and the physical process it is measuring.

This list can be extended, but more generally, the inverse problems listed earlier in this chapter only make sense to the degree that the underlying forward model is likely to be an adequate physical description of the observations. For example, if one is attempting to determine ρ in Eq. (1.15) by taking the Laplacian $\nabla^2\phi$, (analytically or numerically), the solution to the inverse problem is only sensible if this equation really represents the correct governing physics. If the correct equation to use were, instead,

$$\frac{\partial^2\phi}{\partial r_x^2} + \frac{1}{2}\frac{\partial\phi}{\partial r_y} = \rho, \tag{1.36}$$

where r_y is another coordinate, the calculated value of ρ would be incorrect. One might, however, have good reason to use Eq. (1.15) as the most likely hypothesis, but nonetheless remain open to the possibility that it is not an adequate descriptor of the required field, ρ. A good methodology, of the type to be developed in subsequent chapters, permits posing the question: is my model consistent with the data? If the answer to the question is "yes," a careful investigator would *never* claim that the resulting answer is the correct one and that the model has been "validated" or "verified." One claims only that the answer and the model are consistent with the observations, and remains open to the possibility that some new piece of information will be obtained that completely *invalidates* the model (e.g., some direct measurements of ρ showing that the inferred value is simply wrong). One can never validate or verify a model, one can only show consistency with existing observations.[7]

Notes

1 Whittaker and Robinson (1944).
2 Lanczos (1961) has a much fuller discussion of this correspondence.
3 Herman (1980).
4 Herman (1980); Munk *et al.* (1995).
5 Oceanographers will recognize this apparently highly artificial problem as being a slightly simplified version of the so-called geostrophic inverse problem, and which is of great practical importance. It is a central subject in Chapter 6.
6 Aki and Richards (1980). A famous two-dimensional version of the problem is described by Kač (1966); see also Gordon and Webb (1996).
7 Oreskes *et al.* (1994).

2

Basic machinery

2.1 Background

The purpose of this chapter is to record a number of results that are useful in finding and understanding the solutions to sets of usually noisy simultaneous linear equations and in which formally there may be too much or too little information. A lot of the material is elementary; good textbooks exist, to which the reader will be referred. Some of what follows is discussed primarily so as to produce a consistent notation for later use. But some topics are given what may be an unfamiliar interpretation, and I urge everyone to at least skim the chapter.

Our basic tools are those of matrix and vector algebra as they relate to the solution of linear simultaneous equations, and some elementary statistical ideas – mainly concerning covariance, correlation, and dispersion. Least-squares is reviewed, with an emphasis placed upon the arbitrariness of the distinction between knowns, unknowns, and noise. The singular-value decomposition is a central building block, producing the clearest understanding of least-squares and related formulations. Minimum variance estimation is introduced through the Gauss–Markov theorem as an alternative method for obtaining solutions to simultaneous equations, and its relation to and distinction from least-squares is discussed. The chapter ends with a brief discussion of recursive least-squares and estimation; this part is essential background for the study of time-dependent problems in Chapter 4.

2.2 Matrix and vector algebra

This subject is very large and well-developed and it is not my intention to repeat material better found elsewhere.[1] Only a brief survey of essential results is provided.

A matrix is an $M \times N$ array of elements of the form

$$\mathbf{A} = \{A_{ij}\}, \quad i = 1, 2, \ldots, M, \quad j = 1, 2, \ldots, N.$$

Normally a matrix is denoted by a bold-faced capital letter. A vector is a special case of an $M \times 1$ matrix, written as a bold-face lower case letter, for example, $\mathbf{q}$. Corresponding capital or lower case letters for Greek symbols are also indicated in bold-face. Unless otherwise stipulated, vectors are understood to be columnar. The transpose of a matrix $\mathbf{A}$ is written $\mathbf{A}^T$ and is defined as $\{A^T\}_{ij} = A_{ji}$, an interchange of the rows and columns of $\mathbf{A}$. Thus $(\mathbf{A}^T)^T = \mathbf{A}$. Transposition applied to vectors is sometimes used to save space in printing, for example, $\mathbf{q} = [q_1, q_2, \ldots, q_N]^T$ is the same as

$$\mathbf{q} = \begin{bmatrix} q_1 \\ q_2 \\ \vdots \\ q_N \end{bmatrix}.$$

Matrices and vectors

A conventional measure of length of a vector is $\sqrt{\mathbf{a}^T\mathbf{a}} = \sqrt{\sum_i^N a_i^2} = \|\mathbf{a}\|$. The inner, or dot, product between two $L \times 1$ vectors $\mathbf{a}, \mathbf{b}$ is written $\mathbf{a}^T\mathbf{b} \equiv \mathbf{a} \cdot \mathbf{b} = \sum_{i=1}^L a_i b_i$ and is a scalar. Such an inner product is the "projection" of $\mathbf{a}$ onto $\mathbf{b}$ (or vice versa). It is readily shown that $|\mathbf{a}^T\mathbf{b}| = \|\mathbf{a}\| \, \|\mathbf{b}\| \, |\cos\phi| \leq \|\mathbf{a}\| \, \|\mathbf{b}\|$, where the magnitude of $\cos\phi$ ranges between zero, when the vectors are orthogonal, and one, when they are parallel.

Suppose we have a collection of N vectors, $\mathbf{e}_i$, each of dimension N. If it is possible to represent perfectly an arbitrary N-dimensional vector $\mathbf{f}$ as the linear sum

$$\mathbf{f} = \sum_{i=1}^N \alpha_i \mathbf{e}_i, \tag{2.1}$$

then $\mathbf{e}_i$ are said to be a "basis." A necessary and sufficient condition for them to have that property is that they should be "independent," that is, no one of them should be perfectly representable by the others:

$$\mathbf{e}_j - \sum_{i=1, i \neq j}^N \beta_i \mathbf{e}_i \neq 0, \quad j = 1, 2, \ldots, N. \tag{2.2}$$

A subset of the $\mathbf{e}_j$ are said to span a subspace (all vectors perfectly representable by the subset). For example, $[1, -1, 0]^T$, $[1, 1, 0]^T$ span the subspace of all vectors $[v_1, v_2, 0]^T$. A "spanning set" completely describes the subspace too, but might have additional, redundant vectors. Thus the vectors $[1, -1, 0]^T$, $[1, 1, 0]^T$, $[1, 1/2, 0]$ span the subspace but are not a basis for it.

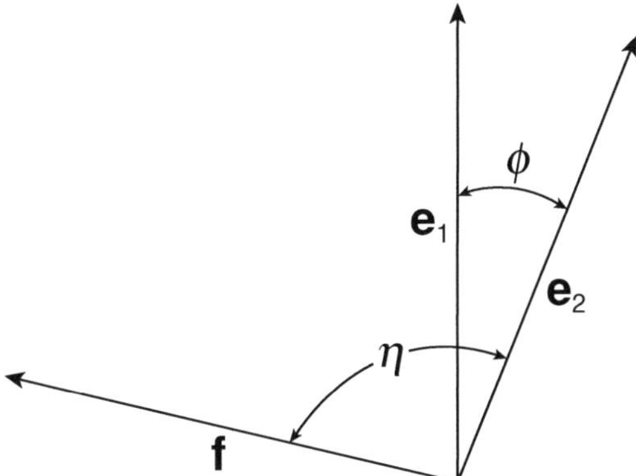

Figure 2.1 Schematic of expansion of an arbitrary vector $\mathbf{f}$ in two vectors $\mathbf{e}_1, \mathbf{e}_2$ which may nearly coincide in direction.

The expansion coefficients α_i in (2.1) are obtained by taking the dot product of (2.1) with each of the vectors in turn:

$$\sum_{i=1}^{N} \alpha_i \mathbf{e}_k^{\mathrm{T}} \mathbf{e}_i = \mathbf{e}_k^{\mathrm{T}} \mathbf{f}, \quad k = 1, 2, \ldots, N, \tag{2.3}$$

which is a system of N equations in N unknowns. The α_i are most readily found if the $\mathbf{e}_i$ are a mutually orthonormal set, that is, if

$$\mathbf{e}_i^{\mathrm{T}} \mathbf{e}_j = \delta_{ij},$$

but this requirement is not a necessary one. With a basis, the information contained in the set of projections, $\mathbf{e}_i^{\mathrm{T}} \mathbf{f} = \mathbf{f}^{\mathrm{T}} \mathbf{e}_i$, is adequate then to determine the α_i and thus all the information required to reconstruct $\mathbf{f}$ is contained in the dot products.

The concept of "nearly dependent" vectors is helpful and can be understood heuristically. Consider Fig. 2.1, in which the space is two-dimensional. Then the two vectors $\mathbf{e}_1, \mathbf{e}_2$, as depicted there, are independent and can be used to expand an arbitrary two-dimensional vector $\mathbf{f}$ in the plane. The simultaneous equations become

$$\alpha_1 \mathbf{e}_1^{\mathrm{T}} \mathbf{e}_1 + \alpha_2 \mathbf{e}_1^{\mathrm{T}} \mathbf{e}_2 = \mathbf{e}_1^{\mathrm{T}} \mathbf{f}, \tag{2.4}$$
$$\alpha_1 \mathbf{e}_2^{\mathrm{T}} \mathbf{e}_1 + \alpha_2 \mathbf{e}_2^{\mathrm{T}} \mathbf{e}_2 = \mathbf{e}_2^{\mathrm{T}} \mathbf{f}.$$

The vectors become nearly parallel as the angle ϕ in Fig. 2.1 goes to zero; as long as they are not identically parallel, they can still be used mathematically to represent $\mathbf{f}$ perfectly. An important feature is that even if the lengths of $\mathbf{e}_1, \mathbf{e}_2, \mathbf{f}$ are all order-one, the expansion coefficients α_1, α_2 can have unbounded magnitudes when the angle ϕ becomes small and $\mathbf{f}$ is nearly orthogonal to both (measured by angle η).

That is to say, we find readily from (2.4) that

$$\alpha_1 = \frac{\left(\mathbf{e}_1^{\mathrm{T}}\mathbf{f}\right)\left(\mathbf{e}_2^{\mathrm{T}}\mathbf{e}_2\right) - \left(\mathbf{e}_2^{\mathrm{T}}\mathbf{f}\right)\left(\mathbf{e}_1^{\mathrm{T}}\mathbf{e}_2\right)}{\left(\mathbf{e}_1^{\mathrm{T}}\mathbf{e}_1\right)\left(\mathbf{e}_2^{\mathrm{T}}\mathbf{e}_2\right) - \left(\mathbf{e}_1^{\mathrm{T}}\mathbf{e}_2\right)^2}, \tag{2.5}$$

$$\alpha_2 = \frac{\left(\mathbf{e}_2^{\mathrm{T}}\mathbf{f}\right)\left(\mathbf{e}_1^{\mathrm{T}}\mathbf{e}_1\right) - \left(\mathbf{e}_1^{\mathrm{T}}\mathbf{f}\right)\left(\mathbf{e}_2^{\mathrm{T}}\mathbf{e}_1\right)}{\left(\mathbf{e}_1^{\mathrm{T}}\mathbf{e}_1\right)\left(\mathbf{e}_2^{\mathrm{T}}\mathbf{e}_2\right) - \left(\mathbf{e}_1^{\mathrm{T}}\mathbf{e}_2\right)^2}. \tag{2.6}$$

Suppose for simplicity that $\mathbf{f}$ has unit length, and that the $\mathbf{e}_i$ have also been normalized to unit length as shown in Fig. 2.1. Then,

$$\alpha_1 = \frac{\cos(\eta - \phi) - \cos\phi\cos\eta}{1 - \cos^2\phi} = \frac{\sin\eta}{\sin\phi}, \tag{2.7}$$

$$\alpha_2 = \cos\eta - \sin\eta\cot\phi \tag{2.8}$$

and whose magnitudes can become arbitrarily large as $\phi \to 0$. One can imagine a situation in which $\alpha_1\mathbf{e}_1$ and $\alpha_2\mathbf{e}_2$ were separately measured and found to be very large. One could then erroneously infer that the sum vector, $\mathbf{f}$, was equally large. This property of the expansion in non-orthogonal vectors potentially producing large coefficients becomes important later (Chapter 5) as a way of gaining insight into the behavior of so-called non-normal operators. The generalization to higher dimensions is left to the reader's intuition. One anticipates that as ϕ becomes very small, numerical problems can arise in using these "almost parallel" vectors.

Gram–Schmidt process

One often has a set of p independent, but non-orthonormal vectors, $\mathbf{h}_i$, and it is convenient to find a new set $\mathbf{g}_i$, which are orthonormal. The "Gram–Schmidt process" operates by induction. Suppose the first k of the $\mathbf{h}_i$ have been orthonormalized to a new set, $\mathbf{g}_i$. To generate vector $k + 1$, let

$$\mathbf{g}_{k+1} = \mathbf{h}_{k+1} - \sum_j^k \gamma_j\mathbf{g}_j. \tag{2.9}$$

Because $\mathbf{g}_{k+1}$ must be orthogonal to the preceding $\mathbf{g}_i$, $i = 1, \ldots, k$, take the dot products of (2.9) with each of these vectors, producing a set of simultaneous equations for determining the unknown γ_j. The resulting $\mathbf{g}_{k+1}$ is easily given unit norm by dividing by its length.

Given the first k of N necessary vectors, an additional $N - k$ independent vectors, $\mathbf{h}_i$, are needed. There are several possibilities. The necessary extra vectors might be generated by filling their elements with random numbers. Or a very simple trial set like $\mathbf{h}_{k+1} = [1, 0, 0, \ldots, 0]^{\mathrm{T}}$, $\mathbf{h}_{k+2} = [0, 1, 0, \ldots 0]$, $\ldots$ might be adequate. If one is unlucky, the set chosen might prove not to be independent of the existing $\mathbf{g}_i$.

But a simple numerical perturbation usually suffices to render them so. In practice, the algorithm is changed to what is usually called the "modified Gram–Schmidt process" for purposes of numerical stability.[2]

2.2.1 Matrix multiplication and identities

It has been found convenient and fruitful to usually define multiplication of two matrices $\mathbf{A}$, $\mathbf{B}$, written as $\mathbf{C} = \mathbf{AB}$, such that

$$C_{ij} = \sum_{p=1}^{P} A_{ip} B_{pj}. \tag{2.10}$$

For the definition (2.10) to make sense, $\mathbf{A}$ must be an $M \times P$ matrix and $\mathbf{B}$ must be $P \times N$ (including the special case of $P \times 1$, a column vector). That is, the two matrices must be "conformable." If two matrices are multiplied, or a matrix and a vector are multiplied, conformability is implied – otherwise one can be assured that an error has been made. Note that $\mathbf{AB} \neq \mathbf{BA}$ even where both products exist, except under special circumstances. Define $\mathbf{A}^2 = \mathbf{AA}$, etc. Other definitions of matrix multiplication exist, and are useful, but are not needed here.

The mathematical operation in (2.10) may appear arbitrary, but a physical interpretation is available: Matrix multiplication is the dot product of all of the rows of $\mathbf{A}$ with all of the columns of $\mathbf{B}$. Thus multiplication of a vector by a matrix represents the projections of the rows of the matrix onto the vector.

Define a matrix, $\mathbf{E}$, each of whose columns is the corresponding vector $\mathbf{e}_i$, and a vector, $\boldsymbol{\alpha} = \{\alpha_i\}$, in the same order. Then the expansion (2.1) can be written compactly as

$$\mathbf{f} = \mathbf{E}\boldsymbol{\alpha}. \tag{2.11}$$

A "symmetric matrix" is one for which $\mathbf{A}^{\mathrm{T}} = \mathbf{A}$. The product $\mathbf{A}^{\mathrm{T}}\mathbf{A}$ represents the array of all the dot products of the columns of $\mathbf{A}$ with themselves, and similarly, $\mathbf{AA}^{\mathrm{T}}$ represents the set of all dot products of all the rows of $\mathbf{A}$ with themselves. It follows that $(\mathbf{AB})^{\mathrm{T}} = \mathbf{B}^{\mathrm{T}}\mathbf{A}^{\mathrm{T}}$. Because we have $(\mathbf{AA}^{\mathrm{T}})^{\mathrm{T}} = \mathbf{AA}^{\mathrm{T}}$, $(\mathbf{A}^{\mathrm{T}}\mathbf{A})^{\mathrm{T}} = \mathbf{A}^{\mathrm{T}}\mathbf{A}$, both of these matrices are symmetric.

The "trace" of a square $M \times M$ matrix $\mathbf{A}$ is defined as $\mathrm{trace}(\mathbf{A}) = \sum_i^M A_{ii}$. A "diagonal matrix" is square and zero except for the terms along the main diagonal, although we will later generalize this definition. The operator $\mathrm{diag}(\mathbf{q})$ forms a square diagonal matrix with $\mathbf{q}$ along the main diagonal.

The special $L \times L$ diagonal matrix $\mathbf{I}_L$, with $I_{ii} = 1$, is the "identity." Usually, when the dimension of $\mathbf{I}_L$ is clear from the context, the subscript is omitted. $\mathbf{IA} = \mathbf{A}$, $\mathbf{AI} = \mathbf{I}$, for any $\mathbf{A}$ for which the products make sense. If there is a matrix $\mathbf{B}$, such that $\mathbf{BE} = \mathbf{I}$, then $\mathbf{B}$ is the "left inverse" of $\mathbf{E}$. If $\mathbf{B}$ is the left inverse of $\mathbf{E}$ and $\mathbf{E}$ is square, a standard result is that it must also be a right inverse: $\mathbf{EB} = \mathbf{I}$, $\mathbf{B}$ is then called "the inverse of $\mathbf{E}$" and is usually written $\mathbf{E}^{-1}$. Square matrices with inverses are "non-singular." Analytical expressions exist for a few inverses; more generally, linear algebra books explain how to find them numerically when they exist. If $\mathbf{E}$ is not square, one must distinguish left and right inverses, sometimes written, $\mathbf{E}^{+}$, and referred to as "generalized inverses." Some of them will be encountered later. A useful result is that $(\mathbf{AB})^{-1} = \mathbf{B}^{-1}\mathbf{A}^{-1}$, if the inverses exist. A notational shorthand is $(\mathbf{A}^{-1})^{\mathrm{T}} = (\mathbf{A}^{\mathrm{T}})^{-1} \equiv \mathbf{A}^{-\mathrm{T}}$.

The "length," or norm, of a vector has already been introduced. But several choices are possible; for present purposes, the conventional l_2 norm already defined,

$$\|\mathbf{f}\|_2 \equiv (\mathbf{f}^{\mathrm{T}}\mathbf{f})^{1/2} = \left(\sum_{i=1}^{N} f_i^2\right)^{1/2}, \tag{2.12}$$

is most useful; often the subscript is omitted. Equation (2.12) leads in turn to the measure of distance between two vectors, $\mathbf{a}$, $\mathbf{b}$, as

$$\|\mathbf{a} - \mathbf{b}\|_2 = \sqrt{(\mathbf{a} - \mathbf{b})^{\mathrm{T}} (\mathbf{a} - \mathbf{b})}, \tag{2.13}$$

which is the familiar Cartesian distance. Distances can also be measured in such a way that deviations of certain elements of $\mathbf{c} = \mathbf{a} - \mathbf{b}$ count for more than others – that is, a metric, or set of weights can be introduced with a definition,

$$\|\mathbf{c}\|_W = \sqrt{\sum_i c_i W_{ii} c_i}, \tag{2.14}$$

depending upon the importance to be attached to magnitudes of different elements, stretching and shrinking various coordinates. Finally, in the most general form, distance can be measured in a coordinate system both stretched and rotated relative to the original one

$$\|\mathbf{c}\|_W = \sqrt{\mathbf{c}^{\mathrm{T}}\mathbf{W}\mathbf{c}}, \tag{2.15}$$

where $\mathbf{W}$ is an arbitrary matrix (but usually, for physical reasons, symmetric and positive definite,[3] implying that $\mathbf{c}^{\mathrm{T}}\mathbf{W}\mathbf{c} \geq 0$).

2.2.2 Linear simultaneous equations

Consider a set of M-linear equations in N-unknowns,

$$\mathbf{Ex} = \mathbf{y}. \tag{2.16}$$

Because of the appearance of simultaneous equations in situations in which the y_i are observed, and where $\mathbf{x}$ are parameters whose values are sought, it is often convenient to refer to (2.16) as a set of measurements of $\mathbf{x}$ that produced the observations or data, $\mathbf{y}$. If $M > N$, the system is said to be "formally overdetermined." If $M < N$, it is "underdetermined," and if $M = N$, it is "formally just-determined." The use of the word "formally" has a purpose we will come to. Knowledge of the matrix inverse to $\mathbf{E}$ would make it easy to solve a set of L equations in L unknowns, by left-multiplying (2.16) by $\mathbf{E}^{-1}$:

$$\mathbf{E}^{-1}\mathbf{E}\mathbf{x} = \mathbf{I}\mathbf{x} = \mathbf{x} = \mathbf{E}^{-1}\mathbf{y}.$$

The reader is cautioned that although matrix inverses are a very powerful theoretical tool, one is usually ill-advised to solve large sets of simultaneous equations by employing $\mathbf{E}^{-1}$; better numerical methods are available for the purpose.[4]

There are several ways to view the meaning of any set of linear simultaneous equations. If the columns of $\mathbf{E}$ continue to be denoted $\mathbf{e}_i$, then (2.16) is

$$x_1\mathbf{e}_1 + x_2\mathbf{e}_2 + \cdots + x_N\mathbf{e}_N = \mathbf{y}. \tag{2.17}$$

The ability to so describe an arbitrary $\mathbf{y}$, or to solve the equations, would thus depend upon whether the $M \times 1$ vector $\mathbf{y}$ can be specified by a sum of N-column vectors, $\mathbf{e}_i$. That is, it would depend upon their being a spanning set. In this view, the elements of $\mathbf{x}$ are simply the corresponding expansion coefficients. Depending upon the ratio of M to N, that is, the number of equations compared to the number of unknown elements, one faces the possibility that there are fewer expansion vectors $\mathbf{e}_i$ than elements of $\mathbf{y}$ ($M > N$), or that there are more expansion vectors available than elements of $\mathbf{y}$ ($M < N$). Thus the overdetermined case corresponds to having *fewer* expansion vectors, and the underdetermined case corresponds to having *more* expansion vectors, than the dimension of $\mathbf{y}$. It is possible that in the overdetermined case, the too-few expansion vectors are not actually independent, so that there are even fewer vectors available than is first apparent. Similarly, in the underdetermined case, there is the possibility that although it appears we have more expansion vectors than required, fewer may be independent than the number of elements of $\mathbf{y}$, and the consequences of that case need to be understood as well.

An alternative interpretation of simultaneous linear equations denotes the rows of $\mathbf{E}$ as $\mathbf{r}_i^{\mathrm{T}}$, $i = 1, 2, \ldots, M$. Then Eq. (2.16) is a set of M-inner products,

$$\mathbf{r}_i^{\mathrm{T}}\mathbf{x} = y_i, \quad i = 1, 2, \ldots, M. \tag{2.18}$$

That is, the set of simultaneous equations is also equivalent to being provided with the value of M-dot products of the N-dimensional unknown vector, $\mathbf{x}$, with M known vectors, $\mathbf{r}_i$. Whether that is sufficient information to determine $\mathbf{x}$ depends upon whether the $\mathbf{r}_i$ are a spanning set. In this view, in the overdetermined case, one has *more* dot products available than unknown elements x_i, and, in the underdetermined case, there are *fewer* such values than unknowns.

A special set of simultaneous equations for square matrices, $\mathbf{A}$, is labelled the "eigenvalue/eigenvector problem,"

$$\mathbf{A}\mathbf{e} = \lambda \mathbf{e}. \tag{2.19}$$

In this set of linear simultaneous equations one seeks a special vector, $\mathbf{e}$, such that for some as yet unknown scalar eigenvalue, λ, there is a solution. An $N \times N$ matrix will have up to N solutions $(\lambda_i, \mathbf{e}_i)$, but the nature of these elements and their relations require considerable effort to deduce. We will look at this problem more later; for the moment, it again suffices to say that numerical methods for solving Eq. (2.19) are well-known.

2.2.3 Matrix norms

A number of useful definitions of a matrix size, or norm, exist. The so-called "spectral norm" or "2-norm" defined as

$$\|\mathbf{A}\|_2 = \sqrt{\text{maximum eigenvalue of } (\mathbf{A}^{\mathrm{T}}\mathbf{A})} \tag{2.20}$$

is usually adequate. Without difficulty, it may be seen that this definition is equivalent to

$$\|\mathbf{A}\|_2 = \max \frac{\mathbf{x}^{\mathrm{T}}\mathbf{A}^{\mathrm{T}}\mathbf{A}\mathbf{x}}{\mathbf{x}^{\mathrm{T}}\mathbf{x}} = \max \frac{\|\mathbf{A}\mathbf{x}\|_2}{\|\mathbf{x}\|_2} \tag{2.21}$$

where the maximum is defined over all vectors $\mathbf{x}$.[5] Another useful measure is the "Frobenius norm,"

$$\|\mathbf{A}\|_{\mathrm{F}} = \sqrt{\sum_{i=1}^{M} \sum_{j=1}^{N} A_{ij}^2} = \sqrt{\text{trace}(\mathbf{A}^{\mathrm{T}}\mathbf{A})}. \tag{2.22}$$

Neither norm requires $\mathbf{A}$ to be square. These norms permit one to derive various useful results. Consider the following illustration. Suppose $\mathbf{Q}$ is square, and $\|\mathbf{Q}\| < 1$, then

$$(\mathbf{I} + \mathbf{Q})^{-1} = \mathbf{I} - \mathbf{Q} + \mathbf{Q}^2 - \cdots, \tag{2.23}$$

which may be verified by multiplying both sides by $\mathbf{I} + \mathbf{Q}$, doing term-by-term multiplication and measuring the remainders with either norm.

Nothing has been said about actually finding the numerical values of either the matrix inverse or the eigenvectors and eigenvalues. Computational algorithms for obtaining them have been developed by experts, and are discussed in many good textbooks.[6] Software systems like MATLAB, Maple, IDL, and Mathematica implement them in easy-to-use form. For purposes of this book, we assume the reader has at least a rudimentary knowledge of these techniques and access to a good software implementation.

2.2.4 Identities: differentiation

There are some identities and matrix/vector definitions that prove useful.

A square "positive definite" matrix $\mathbf{A}$, is one for which the scalar "quadratic form,"

$$J = \mathbf{x}^{\mathrm{T}}\mathbf{A}\mathbf{x},$$

is positive for all possible vectors $\mathbf{x}$. (It suffices to consider only symmetric $\mathbf{A}$ because for a general matrix, $\mathbf{x}^{\mathrm{T}}\mathbf{A}\mathbf{x} = \mathbf{x}^{\mathrm{T}}[(\mathbf{A} + \mathbf{A}^{\mathrm{T}})/2]\mathbf{x}$, which follows from the scalar property of the quadratic form.) If $J \geq 0$ for all $\mathbf{x}$, $\mathbf{A}$ is "positive semi-definite," or "non-negative definite." Linear algebra books show that a necessary and sufficient requirement for positive definiteness is that $\mathbf{A}$ has only positive eigenvalues (Eq. 2.19) and a semi-definite one must have all non-negative eigenvalues.

We end up doing a certain amount of differentiation and other operations with respect to matrices and vectors. A number of formulas are very helpful, and save a lot of writing. They are all demonstrated by doing the derivatives term-by-term. Let $\mathbf{q}, \mathbf{r}$ be $N \times 1$ column vectors, and $\mathbf{A}, \mathbf{B}, \mathbf{C}$ be matrices. The derivative of a matrix by a scalar is just the matrix of element by element derivatives. Alternatively, if s is any scalar, its derivative by a vector,

$$\frac{\partial s}{\partial \mathbf{q}} = \left[\frac{\partial s}{\partial q_1} \cdots \frac{\partial s}{\partial q_N} \right]^{\mathrm{T}}, \tag{2.24}$$

is a column vector (the gradient; some authors define it to be a row vector). The derivative of one vector by another is defined as a matrix:

$$\frac{\partial \mathbf{r}}{\partial \mathbf{q}} = \left\{ \frac{\partial r_i}{\partial q_j} \right\} = \begin{Bmatrix} \frac{\partial r_1}{\partial q_1} & \frac{\partial r_2}{\partial q_1} & . & \frac{\partial r_M}{\partial q_1} \\ \frac{\partial r_1}{\partial q_2} & . & . & \frac{\partial r_M}{\partial q_2} \\ . & . & . & . \\ \frac{\partial r_1}{\partial q_N} & . & . & \frac{\partial r_M}{\partial q_N} \end{Bmatrix} \equiv \mathbf{B}. \tag{2.25}$$

If $\mathbf{r}, \mathbf{q}$ are of the same dimension, the determinant of $\mathbf{B} = \det(\mathbf{B})$ is the "Jacobian" of $\mathbf{r}$.[7]

The second derivative of a scalar,

$$\frac{\partial^2 s}{\partial \mathbf{q}^2} = \left\{ \frac{\partial}{\partial \mathbf{q}_i} \frac{\partial s}{\partial \mathbf{q}_j} \right\} = \begin{Bmatrix} \frac{\partial^2 s}{\partial q_1^2} & \frac{\partial^2 s}{\partial q_1 q_2} & . & . & \frac{\partial^2 s}{\partial q_1 q_N} \\ . & . & . & . & . \\ \frac{\partial^2 s}{\partial q_N \partial q_1} & . & . & . & \frac{\partial^2 s}{\partial q_N^2} \end{Bmatrix}, \tag{2.26}$$

is the "Hessian" of s and is the derivative of the gradient of s.

Assuming conformability, the inner product, $J = \mathbf{r}^\mathrm{T}\mathbf{q} = \mathbf{q}^\mathrm{T}\mathbf{r}$, is a scalar. The differential of J is

$$dJ = d\mathbf{r}^\mathrm{T}\mathbf{q} + \mathbf{r}^\mathrm{T}d\mathbf{q} = d\mathbf{q}^\mathrm{T}\mathbf{r} + \mathbf{q}^\mathrm{T}d\mathbf{r}, \tag{2.27}$$

and hence the partial derivatives are

$$\frac{\partial(\mathbf{q}^\mathrm{T}\mathbf{r})}{\partial\mathbf{q}} = \frac{\partial(\mathbf{r}^\mathrm{T}\mathbf{q})}{\partial\mathbf{q}} = \mathbf{r}, \tag{2.28}$$

$$\frac{\partial(\mathbf{q}^\mathrm{T}\mathbf{q})}{\partial\mathbf{q}} = 2\mathbf{q}. \tag{2.29}$$

It follows immediately that, for matrix/vector products,

$$\frac{\partial}{\partial\mathbf{q}}(\mathbf{B}\mathbf{q}) = \mathbf{B}^\mathrm{T}, \qquad \frac{\partial}{\partial\mathbf{q}}(\mathbf{q}^\mathrm{T}\mathbf{B}) = \mathbf{B}. \tag{2.30}$$

The first of these is used repeatedly, and attention is called to the apparently trivial fact that differentiation of $\mathbf{B}\mathbf{q}$ with respect to $\mathbf{q}$ produces the transpose of $\mathbf{B}$ – the origin, as seen later, of so-called adjoint models. For a quadratic form,

$$J = \mathbf{q}^\mathrm{T}\mathbf{A}\mathbf{q}$$
$$\frac{\partial J}{\partial\mathbf{q}} = (\mathbf{A} + \mathbf{A}^\mathrm{T})\mathbf{q}, \tag{2.31}$$

and the Hessian of the quadratic form is $2\mathbf{A}$ if $\mathbf{A} = \mathbf{A}^\mathrm{T}$.

Differentiation of a scalar function (e.g., J in Eq. 2.31) or a vector by a matrix, $\mathbf{A}$, is readily defined.[8] Differentiation of a matrix by another matrix results in a third, very large, matrix. One special case of the *differential* of a matrix function proves useful later on. It can be shown[9] that

$$d\mathbf{A}^n = (d\mathbf{A})\,\mathbf{A}^{n-1} + \mathbf{A}\,(d\mathbf{A})\,\mathbf{A}^{n-2} + \cdots + \mathbf{A}^{n-1}(d\mathbf{A}), \tag{2.32}$$

where $\mathbf{A}$ is square. Thus the derivative with respect to some scalar, k, is

$$\frac{d\mathbf{A}^n}{dk} = \frac{(d\mathbf{A})}{dk}\mathbf{A}^{n-1} + \mathbf{A}^{n-2}\frac{(d\mathbf{A})}{dk}\mathbf{A} + \cdots + \mathbf{A}^{n-1}\left(\frac{d\mathbf{A}}{dk}\right). \tag{2.33}$$

There are a few, unfortunately unintuitive, matrix inversion identities that are essential later. They are derived by considering the square, partitioned matrix,

$$\begin{Bmatrix} \mathbf{A} & \mathbf{B} \\ \mathbf{B}^\mathrm{T} & \mathbf{C} \end{Bmatrix}, \tag{2.34}$$

where $\mathbf{A}^\mathrm{T} = \mathbf{A}$, $\mathbf{C}^\mathrm{T} = \mathbf{C}$, but $\mathbf{B}$ can be rectangular of conformable dimensions in (2.34).[10] The most important of the identities, sometimes called the "matrix inversion lemma" is, in one form,

$$\{C - B^T A^{-1} B\}^{-1} = \{I - C^{-1} B^T A^{-1} B\}^{-1} C^{-1}$$
$$= C^{-1} - C^{-1} B^T (BC^{-1} B^T - A)^{-1} BC^{-1}, \tag{2.35}$$

where it is assumed that the inverses exist.[11] A variant is

$$AB^T(C + BAB^T)^{-1} = (A^{-1} + B^T C^{-1} B)^{-1} B^T C^{-1}. \tag{2.36}$$

Equation (2.36) is readily confirmed by left-multiplying both sides by $(A^{-1} + B^T C^{-1} B)$, and right-multiplying by $(C + BAB^T)$ and showing that the two sides of the resulting equation are equal.

Another identity, found by "completing the square," is demonstrated by directly multiplying it out, and requires $C = C^T$ (A is unrestricted, but the matrices must be conformable as shown):

$$ACA^T - BA^T - AB^T = (A - BC^{-1})C(A - BC^{-1})^T - BC^{-1}B^T. \tag{2.37}$$

2.3 Simple statistics: regression

2.3.1 Probability densities, moments

Some statistical ideas are required, but the discussion is confined to stating some basic notions and to developing a notation.[12] We require the idea of a probability density for a random variable x. This subject is a very deep one, but our approach is heuristic.[13] Suppose that an arbitrarily large number of experiments can be conducted for the determination of the values of x, denoted $X_i, i = 1, 2, \ldots, M$, and a histogram of the experimental values found. The frequency function, or probability density, will be defined as the limit, supposing it exists, of the histogram of an arbitrarily large number of experiments, $M \to \infty$, divided into bins of arbitrarily small value ranges, and normalized by M, to produce the fraction of the total appearing in the ranges. Let the corresponding limiting frequency function be denoted $p_x(X)dX$, interpreted as the fraction (probability) of values of x lying in the range, $X \le x \le X + dX$. As a consequence of the definition, $p_x(X) \ge 0$ and

$$\int_{\text{all } X} p_x(X)\, dX = \int_{-\infty}^{\infty} p_x(X)\, dX = 1. \tag{2.38}$$

The infinite integral is a convenient way of representing an integral over "all X," as p_x simply vanishes for impossible values of X. (It should be noted that this so-called frequentist approach has fallen out of favor, with Bayesian assumptions being regarded as ultimately more rigorous and fruitful. For introductory purposes, however, empirical frequency functions appear to provide an adequate intuitive basis for proceeding.)

The "average," or "mean," or "expected value" is denoted $\langle x \rangle$ and defined as

$$\langle x \rangle \equiv \int_{\text{all } X} X p_x(X) \mathrm{d}X = m_1. \tag{2.39}$$

The mean is the center of mass of the probability density. Knowledge of the true mean value of a random variable is commonly all that we are willing to assume known. If forced to "forecast" the numerical value of x under such circumstances, often the best we can do is to employ $\langle x \rangle$. If the deviation from the true mean is denoted x' so that $x = \langle x \rangle + x'$, such a forecast has the virtue that we are assured the average forecast error, $\langle x' \rangle$, would be zero if many such forecasts are made. The bracket operation is very important throughout this book; it has the property that if a is a non-random quantity, $\langle ax \rangle = a \langle x \rangle$ and $\langle ax + y \rangle = a \langle x \rangle + \langle y \rangle$.

Quantity $\langle x \rangle$ is the "first-moment" of the probability density. Higher order moments are defined as

$$m_n = \langle x^n \rangle = \int_{-\infty}^{\infty} X^n p_x(X) \mathrm{d}X,$$

where n are the non-negative integers. A useful theoretical result is that a knowledge of all the moments is usually enough to completely define the probability density themselves. (There are troublesome situations with, e.g., non-existent moments, as with the so-called Cauchy distribution, $p_x(X) = (2/\pi)(1/(1 + X^2))$ $X \geq 0$, whose mean is infinite.) For many important probability densities, including the Gaussian, a knowledge of the first two moments $n = 1, 2$ is sufficient to define all the others, and hence the full probability density. It is common to define the moments for $n > 1$ about the mean, so that one has

$$\mu_n = \langle (x - \langle x \rangle)^n \rangle = \int_{-\infty}^{\infty} (X - \langle X \rangle)^n \, p_x(X) \mathrm{d}X.$$

μ_2 is the variance and often written $\mu_2 = \sigma^2$, where σ is the "standard deviation."

2.3.2 Sample estimates: bias

In observational sciences, one normally must estimate the values defining the probability density from the data itself. Thus the first moment, the mean, is often computed as the "sample average,"

$$\tilde{m}_1 = \langle x \rangle_M \equiv \frac{1}{M} \sum_{i=1}^{M} X_i. \tag{2.40}$$

The notation $\tilde{m}_1$ is used to distinguish the sample estimate from the true value, m_1. On the other hand, if the experiment of computing $\tilde{m}_1$ from M samples could be repeated many times, the mean of the sample estimates would be the true mean. This conclusion is readily seen by considering the expected value of the difference from the true mean:

$$\langle \langle x \rangle_M - \langle x \rangle \rangle = \left\langle \frac{1}{M} \sum_{i=1}^{M} X_i - \langle x \rangle \right\rangle$$

$$= \frac{1}{M} \sum_{i=1}^{M} \langle X_i \rangle - \langle x \rangle = \frac{M}{M} \langle x \rangle - \langle x \rangle = 0.$$

Such an estimate is said to be "unbiassed": its expected value is the quantity one seeks.

The interpretation is that, for finite M, we do not expect that the sample mean will equal the true mean, but that if we could produce sample averages from distinct groups of observations, the sample averages would themselves have an average that will fluctuate about the true mean, with equal probability of being higher or lower. There are many sample estimates, however, some of which we encounter, where the expected value of the sample estimate is not equal to the true estimate. Such an estimator is said to be "biassed." A simple example of a biassed estimator is the "sample variance," defined as

$$s^2 \equiv \frac{1}{M} \sum_{i}^{M} (X_i - \langle x \rangle_M)^2. \tag{2.41}$$

For reasons explained later in this chapter (p. 42), one finds that

$$\langle s^2 \rangle = \frac{M-1}{M} \sigma^2 \neq \sigma^2,$$

and thus the expected value is not the true variance. (This particular estimate is "asymptotically unbiassed," as the bias vanishes as $M \to \infty$.)

We are assured that the sample mean is unbiassed. But the probability that $\langle x \rangle_M = \langle x \rangle$, that is that we obtain exactly the true value, is very small. It helps to have a measure of the extent to which $\langle x \rangle_M$ is likely to be very far from $\langle x \rangle$. To do so, we need the idea of dispersion – the expected or average squared value of some quantity about some interesting value, like its mean. The most familiar measure of dispersion is the variance, already used above, the expected fluctuation of a random variable about its mean:

$$\sigma^2 = \langle (x - \langle x \rangle)^2 \rangle.$$

More generally, define the dispersion of any random variable, z, as

$$D^2(z) = \langle z^2 \rangle.$$

Thus, the variance of x is $D^2(x - \langle x \rangle)$.

The variance of $\langle x \rangle_M$ about the correct value is obtained by a little algebra using the bracket notation,

$$D^2(((\langle x \rangle_M - x)^2) = \frac{\sigma^2}{M}. \tag{2.42}$$

This expression shows the well-known result that as M becomes large, any tendency of the sample mean to lie far from the true value will diminish. It does not prove that some particular value will not, by accident, be far away, merely that it becomes increasingly unlikely as M grows. (In statistics textbooks, the Chebyschev inequality is used to formalize this statement.)

An estimate that is unbiassed and whose expected dispersion about the true value goes to zero with M is evidently desirable. In more interesting estimators, a bias is often present. Then for a fixed number of samples, M, there would be two distinct sources of deviation (error) from the true value: (1) the bias – how far, on average, it is expected to be from the true value, and (2) the tendency – from purely random events – for the value to differ from the true value (the random error). In numerous cases, one discovers that tolerating a small bias error can greatly reduce the random error – and thus the bias may well be worth accepting for that reason. In some cases therefore, a bias is deliberately introduced.

2.3.3 Functions and sums of random variables

If the probability density of x is $p_x(x)$, then the mean of a function of x, $g(x)$, is just

$$\langle g(x) \rangle = \int_{-\infty}^{\infty} g(X) p_x(X) \mathrm{d}X, \tag{2.43}$$

which follows from the definition of the probability density as the limit of the outcome of a number of trials.

The probability density for g regarded as a new random variable is obtained from

$$p_g(G) = p_x(X(G)) \frac{\mathrm{d}x}{\mathrm{d}g} \, \mathrm{d}G, \tag{2.44}$$

where $\mathrm{d}x/\mathrm{d}g$ is the ratio of the differential intervals occupied by x and g and can be understood by reverting to the original definition of probability densities from histograms.

The Gaussian, or normal, probability density is one that is mathematically handy (but is potentially dangerous as a general model of the behavior of natural

processes – many geophysical and fluid processes are demonstrably non-Gaussian).
For a single random variable x, it is defined as

$$p_x(X) = \frac{1}{\sqrt{2\pi}\,\sigma_x} \exp\left[-\frac{(X-m_x)^2}{2\sigma_x^2}\right]$$

(sometimes abbreviated as $G(m_x, \sigma_x)$). It is readily confirmed that $\langle x \rangle = m_x$, $\langle (x - \langle x \rangle)^2 \rangle = \sigma_x^2$.

One important special case is the transformation of the Gaussian to another Gaussian of zero-mean and unit standard deviation, $G(0, 1)$,

$$z = \frac{x-m}{\sigma_x},$$

which can always be done, and thus,

$$p_z(Z) = \frac{1}{\sqrt{2\pi}} \exp\left[-\frac{Z^2}{2}\right].$$

A second important special case of a change of variable is $g = z^2$, where z is Gaussian of zero mean and unit variance. Then the probability density of g is

$$p_g(G) = \frac{1}{G^{1/2}\sqrt{2\pi}} \exp(-G/2), \tag{2.45}$$

a special probability density usually denoted χ_1^2 ("chi-square-sub-1"), the probability density for the square of a Gaussian. One finds $\langle g \rangle = 1$, $\langle (g - \langle g \rangle)^2 \rangle = 2$.

2.3.4 Multivariable probability densities: correlation

The idea of a frequency function generalizes easily to two or more random variables, x, y. We can, in concept, do an arbitrarily large number of experiments in which we count the occurrences of differing pair values, (X_i, Y_i), of x, y and make a histogram normalized by the total number of samples, taking the limit as the number of samples goes to infinity, and the bin sizes go to zero, to produce a joint probability density $p_{xy}(X, Y)$. $p_{xy}(X, Y)\,dX\,dY$ is then the fraction of occurrences such that $X \le x \le X + dX, Y \le y \le Y + dY$. A simple example would be the probability density for the simultaneous measurement of the two components of horizontal velocity at a point in a fluid. Again, from the definition, $p_{xy}(X, Y) \ge 0$ and

$$\int_{-\infty}^{\infty} p_{xy}(X, Y)\,dY = p_x(X), \tag{2.46}$$

$$\int_{-\infty}^{\infty}\int_{-\infty}^{\infty} p_{xy}(X, Y)\,dX\,dY = 1. \tag{2.47}$$

An important use of joint probability densities is in what is known as "conditional probability." Suppose that the joint probability density for x, y is known and, furthermore, $y = Y$, that is, information is available concerning the actual value of y. What then is the probability density for x given that a particular value for y is known to have occurred? This new frequency function is usually written as $p_{x|y}(X|Y)$ and read as "the probability of x, given that y has occurred," or, "the probability of x conditioned on y." It follows immediately from the definition of the probability density that

$$p_{x|y}(X|Y) = \frac{p_{xy}(X, Y)}{p_y(Y)} \tag{2.48}$$

(This equation can be interpreted by going back to the original experimental concept, and understanding the restriction on x, given that y is known to lie within a strip paralleling the X axis).

Using the joint frequency function, define the average product as

$$\langle xy \rangle = \int \int_{\text{all} X, Y} XY p_{xy}(X, Y) dX \, dY. \tag{2.49}$$

Suppose that upon examining the joint frequency function, one finds that $p_{xy}(X, Y) = p_x(X)p_y(Y)$, that is, it factors into two distinct functions. In that case, x, y are said to be "independent." Many important results follow, including

$$\langle xy \rangle = \langle x \rangle \langle y \rangle.$$

Non-zero mean values are often primarily a nuisance. One can always define modified variables, e.g., $x' = x - \langle x \rangle$, such that the new variables have zero mean. Alternatively, one computes statistics centered on the mean. Should the centered product $\langle (x - \langle x \rangle)(y - \langle y \rangle) \rangle$ be non-zero, x, y are said to "co-vary" or to be "correlated." If $\langle (x - \langle x \rangle)(y - \langle y \rangle) \rangle = 0$, then the two variables are "uncorrelated." If x, y are independent, they are uncorrelated. Independence thus implies lack of correlation, but the reverse is not necessarily true. (These are theoretical relationships, and if $\langle x \rangle$, $\langle y \rangle$ are determined from observation, as described below, one must carefully distinguish estimated behavior from that expected theoretically.)

If the two variables are independent, then (2.48) is

$$p_{x|y}(X|Y) = p_x(X), \tag{2.50}$$

that is, the probability of x given y does not depend upon Y, and thus

$$p_{xy}(X, Y) = p_x(X) p_y(Y)$$

– and *there is then no predictive power for one variable given knowledge of the other.*

Suppose there are two random variables x, y between which there is anticipated to be some linear relationship,

$$x = ay + n, \tag{2.51}$$

where n represents any contributions to x that remain unknown despite knowledge of y, and a is a constant. Then,

$$\langle x \rangle = a\langle y \rangle + \langle n \rangle, \tag{2.52}$$

and (2.51) can be re-written as

$$x - \langle x \rangle = a(y - \langle y \rangle) + (n - \langle n \rangle),$$

or

$$x' = ay' + n', \quad \text{where } x' = x - \langle x \rangle, \quad \text{etc.} \tag{2.53}$$

From this last equation,

$$a = \frac{\langle x'y' \rangle}{\langle y'^2 \rangle} = \frac{\langle x'y' \rangle}{(\langle y'^2 \rangle \langle x'^2 \rangle)^{1/2}} \frac{\langle x'^2 \rangle^{1/2}}{\langle y'^2 \rangle^{1/2}} = \rho \frac{\langle x'^2 \rangle^{1/2}}{\langle y'^2 \rangle^{1/2}}, \tag{2.54}$$

where it was supposed that $\langle y'n' \rangle = 0$, thus defining n'. The quantity

$$\rho \equiv \frac{\langle x'y' \rangle}{\langle y'^2 \rangle^{1/2} \langle x'^2 \rangle^{1/2}} \tag{2.55}$$

is the "correlation coefficient" and has the property,[14] $|\rho| \le 1$. If ρ should vanish, then so does a. If a vanishes, then knowledge of y' carries no information about the value of x'. If $\rho = \pm 1$, then it follows from the definitions that $n = 0$ and knowledge of a permits perfect prediction of x' from knowledge of y'. (Because probabilities are being used, rigorous usage would state "perfect prediction almost always," but this distinction will be ignored.)

A measure of how well the prediction of x' from y' will work can be obtained in terms of the variance of x'. We have

$$\langle x'^2 \rangle = a^2 \langle y'^2 \rangle + \langle n'^2 \rangle = \rho^2 \langle x'^2 \rangle + \langle n'^2 \rangle,$$

or

$$(1 - \rho^2)\langle x'^2 \rangle = \langle n'^2 \rangle. \tag{2.56}$$

That is, $(1 - \rho^2)\langle x'^2 \rangle$ is the fraction of the variance in x' that is *unpredictable* from knowledge of y' and is the "unpredictable power." Conversely, $\rho^2 \langle x'^2 \rangle$ is the "predictable" power in x' given knowledge of y'. The limits as $\rho \to 0$, 1 are readily apparent.

Thus we interpret the statement that two variables x', y' "are correlated" or "co-vary" to mean that knowledge of one permits at least a partial prediction of the other, the expected success of the prediction depending upon the magnitude of ρ. If ρ is not zero, the variables cannot be independent, and the conditional probability $p_{x|y}(X|Y) \neq p_x(X)$. This result represents an implementation of the statement that if two variables are not independent, then knowledge of one permits some skill in the prediction of the other. If two variables do not co-vary, but are known not to be independent, a linear model like (2.51) would not be useful – a non-linear one would be required. Such non-linear methods are possible, and are touched on briefly later. The idea that correlation or covariance between various physical quantities carries useful predictive skill between them is an essential ingredient of many of the methods taken up in this book.

In most cases, quantities like ρ, $\langle x'^2 \rangle$ are determined from the available measurements, e.g., of the form

$$ay_i + n_i = x_i, \tag{2.57}$$

and are not known exactly. They are thus sample values, are not equal to the true values, and must be interpreted carefully in terms of their inevitable biasses and variances. This large subject of regression analysis is left to the references.[15]

2.3.5 Change of variables

Suppose we have two random variables x, y with joint probability density $p_{xy}(X, Y)$. They are known as functions of two new variables $x = x(\xi_1, \xi_2)$, $y = y(\xi_1, \xi_2)$ and an inverse mapping $\xi_1 = \xi_1(x, y)$, $\xi_2 = \xi_2(x, y)$. What is the probability density for these new variables? The general rule for changes of variable in probability densities follows from area conservation in mapping from the x, y space to the ξ_1, ξ_2 space, that is,

$$p_{\xi_1 \xi_2}(\Xi_1, \Xi_2) = p_{xy}(X(\Xi_1, \Xi_2), Y(\Xi_1, \Xi_2)) \frac{\partial(X, Y)}{\partial(\Xi_1, \Xi_2)}, \tag{2.58}$$

where $\partial(X, Y)/\partial(\Xi_1, \Xi_2)$ is the Jacobian of the transformation between the two variable sets. As in any such transformation, one must be alert for zeros or infinities in the Jacobian, indicative of multiple valuedness in the mapping. Texts on multivariable calculus discuss such issues in detail.

Example *Suppose x_1, x_2 are independent Gaussian random variables of zero mean and variance σ^2. Then*

$$p_{\mathbf{x}}(\mathbf{X}) = \frac{1}{2\pi\sigma^2} \exp\left(-\frac{(X_1^2 + X_2^2)}{2\sigma^2}\right).$$

Define new random variables

$$r = (x_1^2 + x_2^2)^{1/2}, \qquad \phi = \tan^{-1}(x_2/x_1), \tag{2.59}$$

whose mapping in the inverse direction is

$$x_1 = r \cos \phi, \qquad y_1 = r \sin \phi, \tag{2.60}$$

that is, the relationship between polar and Cartesian coordinates. The Jacobian of the transformation is $J_a = r$. Thus

$$p_{r,\phi}(R, \Phi) = \frac{R}{2\pi} \frac{2}{\sigma^2} \exp(-R^2/\sigma^2), \quad 0 \le r, \ -\pi \le \phi \le \pi \tag{2.61}$$

The probability density for r alone is obtained by integrating

$$p_r(R) = \int_{-\pi}^{\pi} p_{r,\phi} d\phi = \frac{R}{\sigma^2} \exp[-R^2/(2\sigma^2)], \tag{2.62}$$

which is known as a Rayleigh distribution. By inspection then,

$$p_\phi(\Phi) = \frac{1}{2\pi},$$

which is the uniform distribution, independent of Φ. (These results are very important in signal processing.)

To generalize to n dimensions, let there be N variables, x_i, $i = 1, 2, \ldots, N$, with known joint probability density $p_{x_1 \cdots x_N}$. Let there be N new variables, ξ_i, that are known functions of the x_i, and an inverse mapping between them. Then the joint probability density for the new variables is just

$$p_{\xi_1 \cdots \xi_N}(\Xi_1, \ldots, \Xi_N)$$
$$= p_{x_1 \cdots x_N}(\Xi_1(X_1, \ldots, X_N) \ldots, \Xi_N(X_1, \ldots, X_N)) \frac{\partial(X_1, \ldots, X_N)}{\partial(\Xi_1, \ldots, \Xi_N)}. \tag{2.63}$$

Suppose that x, y are *independent* Gaussian variables $G(m_x, \sigma_x)$, $G(m_y, \sigma_y)$. Then their joint probability density is just the product of the two individual densities,

$$p_{x,y}(X, Y) = \frac{1}{2\pi \sigma_x \sigma_y} \exp\left(-\frac{(X - m_x)^2}{2\sigma_x^2} - \frac{(Y - m_y)^2}{2\sigma_y^2}\right). \tag{2.64}$$

Let two new random variables, ξ_1, ξ_2, be defined as a linear combination of x, y,

$$\xi_1 = a_{11}(x - m_x) + a_{12}(y - m_y) + m_{\xi_1}$$
$$\xi_2 = a_{21}(x - m_x) + a_{22}(y - m_y) + m_{\xi_2}, \tag{2.65}$$

or, in vector form,

$$\boldsymbol{\xi} = \mathbf{A}(\mathbf{x} - \mathbf{m}_x) + \mathbf{m}_\xi,$$

where $\mathbf{x} = [x, y]^{\mathrm{T}}$, $\mathbf{m}_x = [m_x, m_y]^{\mathrm{T}}$, $\mathbf{m}_y = [m_{\xi_1}, m_{\xi_2}]^{\mathrm{T}}$, and the numerical values satisfy the corresponding functional relations,

$$\Xi_1 = a_{11}(X - m_x) + a_{12}(Y - m_y) + m_{\xi_1},$$

etc. Suppose that the relationship (2.65) is invertible, that is, we can solve for

$$x = b_{11}(\xi_1 - m_{\xi_1}) + b_{12}(\xi_2 - m_{\xi_2}) + m_x$$
$$y = b_{21}(\xi_1 - m_{\xi_1}) + b_{22}(\xi_2 - m_{\xi_2}) + m_y,$$

or

$$\mathbf{x} = \mathbf{B}(\boldsymbol{\xi} - \mathbf{m}_\xi) + \mathbf{m}_x. \tag{2.66}$$

Then the Jacobian of the transformation is

$$\frac{\partial(X, Y)}{\partial(\Xi_1, \Xi_2)} = b_{11}b_{22} - b_{12}b_{21} = \det(\mathbf{B}) \tag{2.67}$$

($\det(\mathbf{B})$ is the determinant). Equation (2.65) produces

$$\langle \xi_1 \rangle = m_{\xi_1}$$
$$\langle \xi_2 \rangle = m_{\xi_2}$$
$$\langle (\xi_1 - \langle \xi_1 \rangle)^2 \rangle = a_{11}^2 \sigma_x^2 + a_{12}^2 \sigma_y^2, \quad \langle (\xi_2 - \langle \xi_2 \rangle)^2 \rangle = a_{21}^2 \sigma_x^2 + a_{22}^2 \sigma_y^2$$
$$\langle (\xi_1 - \langle \xi_1 \rangle)(\xi_2 - \langle \xi_2 \rangle) \rangle = a_{11}a_{21}\sigma_x^2 + a_{12}a_{22}\sigma_y^2 \neq 0. \tag{2.68}$$

In the special case,

$$\mathbf{A} = \left\{ \begin{array}{cc} \cos\phi & \sin\phi \\ -\sin\phi & \cos\phi \end{array} \right\}, \quad \mathbf{B} = \left\{ \begin{array}{cc} \cos\phi & -\sin\phi \\ \sin\phi & \cos\phi \end{array} \right\}, \tag{2.69}$$

the transformation (2.69) is a simple coordinate rotation through angle ϕ, and the Jacobian is 1. The new second-order moments are

$$\langle (\xi_1 - \langle \xi_1 \rangle)^2 \rangle = \sigma_{\xi_1}^2 = \cos^2\phi \, \sigma_x^2 + \sin^2\phi \, \sigma_y^2, \tag{2.70}$$
$$\langle (\xi_2 - \langle \xi_2 \rangle)^2 \rangle = \sigma_{\xi_2}^2 = \sin^2\phi \, \sigma_x^2 + \cos^2\phi \, \sigma_y^2, \tag{2.71}$$
$$\langle (\xi_1 - \langle \xi_1 \rangle)(\xi_2 - \langle \xi_2 \rangle) \rangle \equiv \mu_{\xi_1\xi_2} = \left(\sigma_y^2 - \sigma_x^2\right) \cos\phi \, \sin\phi. \tag{2.72}$$

The new probability density is

$$p_{\xi_1\xi_2}(\Xi_1, \Xi_2) = \frac{1}{2\pi\sigma_{\xi_1}\sigma_{\xi_2}\sqrt{1-\rho_\xi^2}} \tag{2.73}$$

$$\exp\left\{-\frac{1}{2\sqrt{1-\rho_\xi^2}}\left[\frac{(\Xi_1-m_{\xi_1})^2}{\sigma_{\xi_1}^2} - \frac{2\rho_\xi(\Xi_1-m_{\xi_1})(\Xi_2-m_{\xi_2})}{\sigma_{\xi_1}\sigma_{\xi_2}} + \frac{(\Xi_2-m_{\xi_2})^2}{\sigma_{\xi_2}^2}\right]\right\},$$

where $\rho_\xi = (\sigma_y^2 - \sigma_x^2)\sin\phi\cos\phi/(\sigma_x^2+\sigma_y^2)^{1/2} = \mu_{\xi_1\xi_2}/\sigma_{\xi_1}\sigma_{\xi_2}$ is the correlation coefficient of the new variables. A probability density derived through a linear transformation from two independent variables that are Gaussian will be said to be "jointly Gaussian" and (2.73) is a canonical form. Because a coordinate rotation is invertible, it is important to note that if we had two random variables ξ_1, ξ_2 that were jointly Gaussian with $\rho \neq 1$, then we could find a pure rotation (2.69), which produces two other variables x, y that are uncorrelated, and therefore *independent*. Two such uncorrelated variables x, y will necessarily have different variances, otherwise ξ_1, ξ_2 would have zero correlation, too, by Eq. (2.72).

As an important by-product, it is concluded that two jointly Gaussian random variables that are uncorrelated, are also independent. This property is one of the reasons Gaussians are so nice to work with; but it is not generally true of uncorrelated variables.

Vector random processes

Simultaneous discussion of two random processes, x, y, can regarded as discussion of a vector random process $[x, y]^T$, and suggests a generalization to N dimensions. Label N random processes as x_i and define them as the elements of a vector $\mathbf{x} = [x_1, x_2, \ldots, x_N]^T$. Then the mean is a vector: $\langle\mathbf{x}\rangle = \mathbf{m}_x$, and the covariance is a matrix:

$$\mathbf{C}_{xx} = D^2(\mathbf{x} - \langle\mathbf{x}\rangle) = \langle(\mathbf{x}-\langle\mathbf{x}\rangle)(\mathbf{x}-\langle\mathbf{x}\rangle)^T\rangle, \tag{2.74}$$

which is necessarily symmetric and positive semi-definite. The cross-covariance of two vector processes $\mathbf{x}$, $\mathbf{y}$ is

$$\mathbf{C}_{xy} = \langle(\mathbf{x}-\langle\mathbf{x}\rangle)(\mathbf{y}-\langle\mathbf{y}\rangle)^T\rangle, \tag{2.75}$$

and $\mathbf{C}_{xy} = \mathbf{C}_{yx}^T$.

It proves convenient to introduce two further moment matrices in addition to the covariance matrices. The "second moment" matrices will be defined as

$$\mathbf{R}_{xx} \equiv D^2(\mathbf{x}) = \langle\mathbf{xx}^T\rangle, \qquad \mathbf{R}_{xy} = \langle\mathbf{xy}^T\rangle,$$

that is, not taken about the means. Note $\mathbf{R}_{xy} = \mathbf{R}_{yx}^{\mathrm{T}}$, etc. Let $\tilde{\mathbf{x}}$ be an "estimate" of the true value, $\mathbf{x}$. Then the dispersion of $\tilde{\mathbf{x}}$ about the true value will be called the "uncertainty" (it is sometimes called the "error covariance") and is

$$\mathbf{P} \equiv D^2(\tilde{\mathbf{x}} - \mathbf{x}) = \langle(\tilde{\mathbf{x}} - \mathbf{x})(\tilde{\mathbf{x}} - \mathbf{x})^{\mathrm{T}}\rangle.$$

$\mathbf{P}$ is similar to $\mathbf{C}$, but differs in being taken about the true value, rather than about the mean value; the distinction can be very important.

If there are N variables, ξ_i, $i = 1, 2, \ldots, N$, they will be said to have an "N-dimensional jointly normal probability density" if it is of the form

$$p_{\xi_1,\ldots,\xi_N}(\Xi_1, \ldots, \Xi_N) = \frac{\exp\left[-\frac{1}{2}(\Xi - \mathbf{m})^{\mathrm{T}}\mathbf{C}_{\xi\xi}^{-1}(\Xi - \mathbf{m})\right]}{(2\pi)^{N/2}\sqrt{\det(\mathbf{C}_{\xi\xi})}}. \tag{2.76}$$

One finds $\langle\xi\rangle = \mathbf{m}$, $\langle(\xi - \mathbf{m})(\xi - \mathbf{m})^{\mathrm{T}}\rangle = \mathbf{C}_{\xi\xi}$. Equation 2.73 is a special case for $N = 2$, and so the earlier forms are consistent with this general definition.

Positive definite symmetric matrices can be factored as

$$\mathbf{C}_{\xi\xi} = \mathbf{C}_{\xi\xi}^{\mathrm{T}/2}\mathbf{C}_{\xi\xi}^{1/2}, \tag{2.77}$$

which is called the "Cholesky decomposition," where $\mathbf{C}_{\xi\xi}^{1/2}$ is an upper triangular matrix (all zeros below the main diagonal) and non-singular.[16] It follows that the transformation (a rotation and stretching)

$$\mathbf{x} = \mathbf{C}_{\xi\xi}^{-\mathrm{T}/2}(\xi - \mathbf{m}) \tag{2.78}$$

produces new variables $\mathbf{x}$ of zero mean, and diagonal covariance, that is, a probability density

$$\begin{aligned}
p_{x_1,\ldots,x_N}(X_1, \ldots, X_N) &= \frac{\exp -\frac{1}{2}\left(X_1^2 + \cdots + X_N^2\right)}{(2\pi)^{N/2}} \\
&= \frac{\exp\left(-\frac{1}{2}X_1^2\right)}{(2\pi)^{1/2}} \cdots \frac{\exp\left(-\frac{1}{2}X_N^2\right)}{(2\pi)^{1/2}},
\end{aligned} \tag{2.79}$$

which factors into N-independent, normal variates of zero mean and unit variance ($\mathbf{C}_{xx} = \mathbf{R}_{xx} = \mathbf{I}$). Such a process is often called Gaussian "white noise," and has the property $\langle x_i x_j \rangle = 0$, $i \neq j$.[17]

2.3.6 Sums of random variables

It is often helpful to be able to compute the probability density of sums of independent random variables. The procedure for doing so is based upon (2.43). Let x be a random variable and consider the expected value of the function e^{ixt}:

$$\langle e^{ixt} \rangle = \int_{-\infty}^{\infty} p_x(X)\, e^{iXt}\, dX \equiv \phi_x(t), \tag{2.80}$$

which is the Fourier transform of $p_x(X)$; $\phi_x(t)$ is the "characteristic function" of x. Now consider the sum of two independent random variables x, y with probability densities p_x, p_y, respectively, and define a new random variable $z = x + y$. What is the probability density of z? One starts by first determining the characteristic function, $\phi_z(t)$, for z and then using the Fourier inversion theorem to obtain $p_x(Z)$. To obtain ϕ_z,

$$\phi_z(t) = \langle e^{izt} \rangle = \langle e^{i(x+y)t} \rangle = \langle e^{ixt} \rangle \langle e^{iyt} \rangle,$$

where the last step depends upon the independence assumption. This last equation shows

$$\phi_z(t) = \phi_x(t)\phi_y(t). \tag{2.81}$$

That is, the characteristic function for a sum of two independent variables is the product of the characteristic functions. The "convolution theorem"[18] asserts that the Fourier transform (forward or inverse) of a product of two functions is the convolution of the Fourier transforms of the two functions. That is,

$$p_z(Z) = \int_{-\infty}^{\infty} p_x(r)\, p_y(Z - r)\, dr. \tag{2.82}$$

We will not explore this relation in any detail, leaving the reader to pursue the subject in the references.[19] But it follows immediately that the multiplication of the characteristic functions of a sum of independent Gaussian variables produces a new variable, which is also Gaussian, with a mean equal to the sum of the means and a variance that is the sum of the variances ("sums of Gaussians are Gaussian"). It also follows immediately from Eq. (2.81) that if a variable ξ is defined as

$$\xi = x_1^2 + x_2^2 + \cdots + x_\nu^2, \tag{2.83}$$

where each x_i is Gaussian of zero mean and unit variance, then the probability density for ξ is

$$p_\xi(\Xi) = \frac{\Xi^{\nu/2-1}}{2^{\nu/2}\Gamma\left(\frac{\nu}{2}\right)} \exp(-\Xi/2), \tag{2.84}$$

known as χ_ν^2 – "chi-square sub-ν." The chi-square probability density is central to the discussion of the expected sizes of vectors, such as $\tilde{\mathbf{n}}$, measured as $\tilde{\mathbf{n}}^{\mathrm{T}}\tilde{\mathbf{n}} = \|\tilde{\mathbf{n}}\|^2 = \sum_i \tilde{n}_i^2$ if the elements of $\tilde{\mathbf{n}}$ can be assumed to be independent and Gaussian. One has $\langle \xi \rangle = \nu$, $\langle (\xi - \langle \xi \rangle)^2 \rangle = 2\nu$. Equation (2.45) is the special case $\nu = 1$.

Degrees-of-Freedom

The number of independent variables described by a probability density is usually called the "number of degrees-of-freedom." Thus the densities in (2.76) and (2.79) have N degrees-of-freedom and (2.84) has ν of them. If a sample average (2.40) is formed, it is said to have N degrees-of-freedom if each of the x_j is independent. But what if the x_j have a covariance $\mathbf{C}_{xx}$ that is non-diagonal? This question of how to interpret averages of correlated variables will be explicitly discussed later (p. 133).

Consider the special case of the sample variance Eq. (2.41) – which we claimed was biassed. The reason is that even if the sample values, x_i, are independent, the presence of the sample average in the sample variance means that there are only $N-1$ independent terms in the sum. That this is so is most readily seen by examining the two-term case. Two samples produce a sample mean, $\langle x \rangle_2 = (x_1 + x_2)/2$. The two-term sample variance is

$$s^2 = \tfrac{1}{2}[(x_1 - \langle x \rangle_2)^2 + (x_2 - \langle x \rangle_2)^2],$$

but knowledge of x_1, and of the sample average, permits perfect prediction of $x_2 = 2\langle x \rangle_2 - x_1$. The second term in the sample variance as written is not independent of the first term, and thus there is just one independent piece of information in the two-term sample variance. To show it in general, assume without loss of generality that $\langle x \rangle = 0$, so that $\sigma^2 = \langle x^2 \rangle$. The sample variance about the sample mean (which will not vanish) of independent samples is given by Eq. (2.41), and so

$$
\begin{aligned}
\langle s^2 \rangle &= \frac{1}{M}\sum_{i=1}^{M}\left\langle \left(x_i - \frac{1}{M}\sum_{j=1}^{M}x_j\right)\left(x_i - \frac{1}{M}\sum_{p=1}^{M}x_p\right)\right\rangle \\
&= \frac{1}{M}\sum_{i=1}^{M}\left\{\langle x_i^2\rangle - \frac{1}{M}\sum_{j=1}^{M}\langle x_i x_j\rangle - \frac{1}{M}\sum_{p=1}^{M}\langle x_i x_p\rangle + \frac{1}{M^2}\sum_{j=1}^{M}\sum_{p=1}^{M}\langle x_j x_p\rangle\right\} \\
&= \frac{1}{M}\sum_{i=1}^{M}\left\{\sigma^2 - \frac{\sigma^2}{M}\sum_{j}\delta_{ij} - \frac{\sigma^2}{M}\sum_{p}\delta_{ip} + \frac{\sigma^2}{M^2}\sum_{j}\sum_{p}\delta_{jp}\right\} \\
&= \frac{(M-1)}{M}\sigma^2 \neq \sigma^2.
\end{aligned}
$$

Stationarity

Consider a vector random variable, with element x_i where the subscript i denotes a position in time or space. Then x_i, x_j are two different random variables – for example, the temperature at two different positions in a moving fluid, or the temperature at two different times at the same position. If the physics governing these two different random variables are independent of the parameter i (i.e., independent of time or space), then x_i is said to be "stationary" – meaning that all the underlying statistics are independent of i.[20] Specifically, $\langle x_i \rangle = \langle x_j \rangle \equiv \langle x \rangle$, $D^2(x_i) = D^2(x_j) = D^2(x)$, etc. Furthermore, x_i, x_j have a covariance

$$C_{xx}(i, j) = \langle (x_i - \langle x_i \rangle)(x_j - \langle x_j \rangle) \rangle, \tag{2.85}$$

that is, independent of i, j, and might as well be written $C_{xx}(|i - j|)$, depending only upon the difference $|i - j|$. The distance, $|i - j|$, is often called the "lag." $C_{xx}(|i - j|)$ is called the "autocovariance" of $\mathbf{x}$ or just the covariance, because x_i, x_j are now regarded as intrinsically the same process.[21] If C_{xx} does not vanish, then by the discussion above, knowledge of the numerical value of x_i implies some predictive skill for x_j and vice-versa – a result of great importance for map-making and objective analysis. For stationary processes, all moments having the same $|i - j|$ are identical; it is seen that all diagonals of such a matrix $\{C_{xx}(i, j)\}$, are constant, for example, $\mathbf{C}_{\xi\xi}$ in Eq. (2.76). Matrices with constant diagonals are thus defined by the vector $C_{xx}(|i - j|)$, and are said to have a "Toeplitz form."

2.4 Least-squares

Much of what follows in this book can be described using very elegant and power-ful mathematical tools. On the other hand, by restricting the discussion to discrete models and finite numbers of measurements (all that ever goes into a digital com-puter), almost everything can also be viewed as a form of ordinary least-squares, providing a much more intuitive approach than one through functional analysis. It is thus useful to go back and review what "everyone knows" about this most-familiar of all approximation methods.

2.4.1 Basic formulation

Consider the elementary problem motivated by the "data" shown in Fig. 2.2. t is supposed to be an independent variable, which could be time, a spatial coordinate, or just an index. Some physical variable, call it $\theta(t)$, perhaps temperature at a point in a laboratory tank, has been measured at coordinates $t = t_i$, $i = 1, 2, \ldots, M$, as depicted in the figure.

We have reason to believe that there is a linear relationship between $\theta(t)$ and t in the form $\theta(t) = a + bt$, so that the measurements are

$$y(t_i) = \theta(t_i) + n(t_i) = a + bt_i + n(t_i), \tag{2.86}$$

where $n(t)$ is the inevitable measurement noise. The straight-line relationship might as well be referred to as a "model," as it represents the present conception of the data structure. We want to determine a, b.

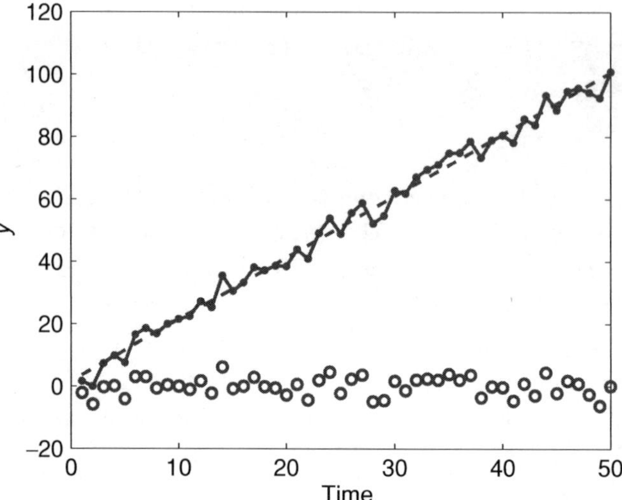

Figure 2.2 "Data" generated through the rule $y = 1 + 2t + n_t$, where $\langle n_t \rangle = 0$, $\langle n_i n_j \rangle = 9\delta_{ij}$, shown as solid dots connected by the solid line. The dashed line is the simple least-squares fit, $\tilde{y} = 1.69 \pm 0.83 + (1.98 \pm 0.03)t$. Residuals are plotted as open circles, and at least visually, show no obvious structure. Note that the fit is correct within its estimated standard errors. The sample variance of the estimated noise, not the theoretical value, was used for calculating the uncertainty.

The set of observations can be written in the general standard form,

$$\mathbf{E}\mathbf{x} + \mathbf{n} = \mathbf{y}, \tag{2.87}$$

where

$$\mathbf{E} = \left\{ \begin{matrix} 1 & t_1 \\ 1 & t_2 \\ \cdot & \cdot \\ \cdot & \cdot \\ 1 & t_M \end{matrix} \right\}, \quad \mathbf{x} = \begin{bmatrix} a \\ b \end{bmatrix}, \quad \mathbf{y} = \begin{bmatrix} y(t_1) \\ y(t_2) \\ \cdot \\ \cdot \\ y(t_M) \end{bmatrix}, \quad \mathbf{n} = \begin{bmatrix} n(t_1) \\ n(t_2) \\ \cdot \\ \cdot \\ n(t_M) \end{bmatrix}. \tag{2.88}$$

Equation sets like (2.87) are seen in many practical situations, including the ones described in Chapter 1. The matrix $\mathbf{E}$ in general represents arbitrarily complicated linear relations between the parameters $\mathbf{x}$, and the observations $\mathbf{y}$. In some real cases, it has many thousands of rows and columns. Its construction involves specifying what those relations are, and, in a very general sense, it requires a "model" of the data set. Unfortunately, the term "model" is used in a variety of other ways in this context, including statistical assumptions, and often for auxiliary relationships among the elements of $\mathbf{x}$ that are independent of those contained in $\mathbf{E}$. To separate these difference usages, we will sometimes append various adjectives to the use ("statistical model," "exact relationships," etc.).

One sometimes sees (2.87) written as

$$\mathbf{Ex} \sim \mathbf{y},$$

or even

$$\mathbf{Ex} = \mathbf{y}.$$

But Eq. (2.87) is preferable, because it explicitly recognizes that $\mathbf{n} = \mathbf{0}$ is exceptional. Sometimes, by happenstance or arrangement, one finds $M = N$ and that $\mathbf{E}$ has an inverse. But the obvious solution, $\mathbf{x} = \mathbf{E}^{-1}\mathbf{y}$, leads to the conclusion that $\mathbf{n} = \mathbf{0}$, which should be unacceptable if the $\mathbf{y}$ are the result of measurements. We will need to return to this case, but, for now, consider the commonplace problem where $M > N$.

Then, one often sees a "best possible" solution – defined as producing the smallest possible value of $\mathbf{n}^{\mathrm{T}}\mathbf{n}$, that is, the minimum of

$$J = \sum_{i=1}^{M} n_i^2 = \mathbf{n}^{\mathrm{T}}\mathbf{n} = (\mathbf{y} - \mathbf{Ex})^{\mathrm{T}}(\mathbf{y} - \mathbf{Ex}). \tag{2.89}$$

(Whether the smallest noise solution really is the best one is considered later.) In the special case of the straight-line model,

$$J = \sum_{i=1}^{M} (y_i - a - bt_i)^2. \tag{2.90}$$

J is an example of what is called an "objective," "cost" or "misfit" function.[22]

Taking the differential of (2.90) with respect to a, b or $\mathbf{x}$ (using (2.27)), and setting it to zero produces

$$dJ = \sum_i \frac{\partial J}{\partial x_i} dx_i = \left(\frac{\partial J}{\partial \mathbf{x}}\right)^{\mathrm{T}} d\mathbf{x}$$

$$= 2d\mathbf{x}^{\mathrm{T}}(\mathbf{E}^{\mathrm{T}}\mathbf{y} - \mathbf{E}^{\mathrm{T}}\mathbf{Ex}) = 0. \tag{2.91}$$

This equation is of the form

$$dJ = \sum a_i dx_i = 0. \tag{2.92}$$

It is an elementary result of multivariable calculus that an extreme value (here a minimum) of J is found where $dJ = 0$. Because the x_i are free to vary independently, dJ will vanish only if the coefficients of the dx_i are separately zero, or

$$\mathbf{E}^{\mathrm{T}}\mathbf{y} - \mathbf{E}^{\mathrm{T}}\mathbf{Ex} = \mathbf{0}. \tag{2.93}$$

That is,

$$\mathbf{E}^{\mathrm{T}}\mathbf{Ex} = \mathbf{E}^{\mathrm{T}}\mathbf{y}, \tag{2.94}$$

which are called the "normal equations." Note that Eq. (2.93) asserts that the columns of $\mathbf{E}$ are orthogonal (that is "normal") to $\mathbf{n} = \mathbf{y} - \mathbf{Ex}$. Making the sometimes-valid assumption that $(\mathbf{E}^{\mathrm{T}}\mathbf{E})^{-1}$ exists,

$$\tilde{\mathbf{x}} = (\mathbf{E}^{\mathrm{T}}\mathbf{E})^{-1}\mathbf{E}^{\mathrm{T}}\mathbf{y}. \tag{2.95}$$

Note on notation: Solutions to equations involving data will be denoted $\tilde{\mathbf{x}}$, to show that they are an estimate of the solution and not necessarily identical to the "true" one in a mathematical sense.

Second derivatives of J with respect to $\mathbf{x}$, make clear that we have a minimum and not a maximum. The relationship between 2.95 and the "correct" value is obscure. $\tilde{\mathbf{x}}$ can be substituted everywhere for $\mathbf{x}$ in Eq. (2.89), but usually the context makes clear the distinction between the calculated and true values. Figure 2.2 displays the fit along with the residuals,

$$\tilde{\mathbf{n}} = \mathbf{y} - \mathbf{E}\tilde{\mathbf{x}} = [\mathbf{I} - \mathbf{E}(\mathbf{E}^{\mathrm{T}}\mathbf{E})^{-1}\mathbf{E}^{\mathrm{T}}]\mathbf{y}. \tag{2.96}$$

That is, the M equations have been used to estimate N values, $\tilde{x}_i$, and M values $\tilde{n}_i$, or $M + N$ altogether. The combination

$$\mathbf{H} = \mathbf{E}(\mathbf{E}^{\mathrm{T}}\mathbf{E})^{-1}\mathbf{E}^{\mathrm{T}} \tag{2.97}$$

occurs sufficiently often that it is worth a special symbol. Note the "idempotent" property $\mathbf{H}^2 = \mathbf{H}$. If the solution, $\tilde{\mathbf{x}}$, is substituted into the original equations, the result is

$$\mathbf{E}\tilde{\mathbf{x}} = \mathbf{H}\mathbf{y} = \tilde{\mathbf{y}}, \tag{2.98}$$

and

$$\tilde{\mathbf{n}}^{\mathrm{T}}\tilde{\mathbf{y}} = [(\mathbf{I} - \mathbf{H})\,\mathbf{y}]^{\mathrm{T}}\mathbf{H}\mathbf{y} = \mathbf{0}. \tag{2.99}$$

The residuals are orthogonal (normal) to the inferred noise-free "data" $\tilde{\mathbf{y}}$.

All of this is easy and familiar and applies to any set of simultaneous linear equations, not just the straight-line example. Before proceeding, let us apply some of the statistical machinery to understanding (2.95). Notice that no statistics were used in obtaining (2.95), but we can nonetheless ask the extent to which this value for $\tilde{\mathbf{x}}$ is affected by the random elements: the noise in $\mathbf{y}$. Let $\mathbf{y}_0$ be the value of $\mathbf{y}$ that would be obtained in the hypothetical situation for which $\mathbf{n} = \mathbf{0}$. Assume further that $\langle \mathbf{n} \rangle = \mathbf{0}$ and that $\mathbf{R}_{nn} = \mathbf{C}_{nn} = \langle \mathbf{n}\mathbf{n}^{\mathrm{T}} \rangle$ is known. Then the expected value of $\tilde{\mathbf{x}}$ is

$$\langle \tilde{\mathbf{x}} \rangle = (\mathbf{E}^{\mathrm{T}}\mathbf{E})^{-1}\mathbf{E}^{\mathrm{T}}\mathbf{y}_0. \tag{2.100}$$

If the matrix inverse exists, then in many situations, including the problem of fitting a straight line to data, perfect observations would produce the correct answer,

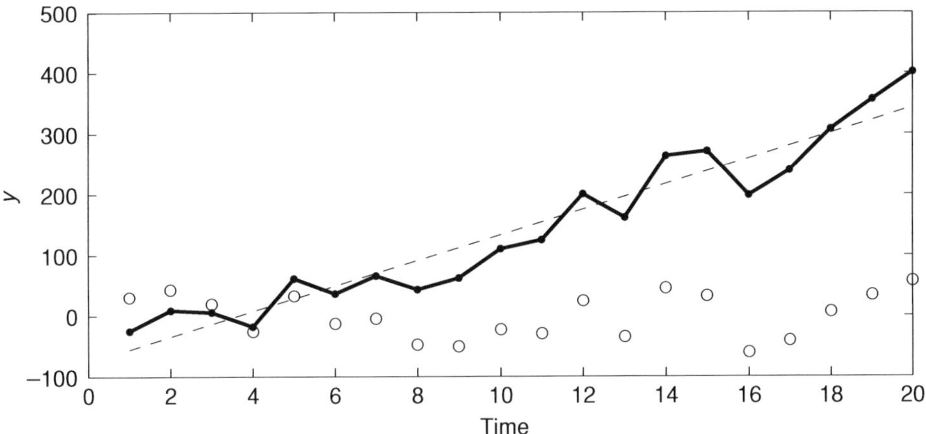

Figure 2.3 Here the "data" were generated from a quadratic rule, $y = 1 + t^2 + n(t)$, $\langle n^2 \rangle = 900$. Note that only the first 20 data points are used. An incorrect straight line fit was used resulting in $\tilde{y} = (-76.3 \pm 17.3) + (20.98 \pm 1.4)t$, which is incorrect, but the residuals, at least visually, do not appear unacceptable. At this point some might be inclined to claim the model has been "verified," or "validated."

and Eq. (2.95) provides an unbiassed estimate of the true solution, $\langle \tilde{x} \rangle = x$. A more transparent demonstration of this result will be given later in this chapter (see p. 103).

On the other hand, if the data were actually produced from physics governed, for example, by a quadratic rule, $\theta(t) = a + ct^2$, then fitting the linear rule to such observations, even if they are perfect, could never produce the right answer and the solution would be biassed. An example of such a fit is shown in Figs. 2.3 and 2.4. Such errors are conceptually distinguishable from the noise of observation, and are properly labeled "model errors."

Assume, however, that the correct model is being used, and therefore that $\langle \tilde{x} \rangle = x$. Then the uncertainty of the solution is

$$
\begin{aligned}
\mathbf{P} = \mathbf{C}_{\tilde{x}\tilde{x}} &= \langle (\tilde{x} - x)(\tilde{x} - x)^{\mathrm{T}} \rangle \\
&= (\mathbf{E}^{\mathrm{T}}\mathbf{E})^{-1}\mathbf{E}^{\mathrm{T}}\langle \mathbf{nn}^{\mathrm{T}} \rangle \mathbf{E}(\mathbf{E}^{\mathrm{T}}\mathbf{E})^{-1} \\
&= (\mathbf{E}^{\mathrm{T}}\mathbf{E})^{-1}\mathbf{E}^{\mathrm{T}}\mathbf{R}_{nn}\mathbf{E}(\mathbf{E}^{\mathrm{T}}\mathbf{E})^{-1}.
\end{aligned}
\tag{2.101}
$$

In the special case, $\mathbf{R}_{nn} = \sigma_n^2\mathbf{I}$, that is, no correlation between the noise in different equations (white noise), Eq. (2.101) simplifies to

$$
\mathbf{P} = \sigma_n^2(\mathbf{E}^{\mathrm{T}}\mathbf{E})^{-1}.
\tag{2.102}
$$

If we are not confident that $\langle \tilde{x} \rangle = x$, perhaps because of doubts about the straight-line model, Eqs. (2.101) and (2.102) are still interpretable, but as $C_{\tilde{x}\tilde{x}} =$

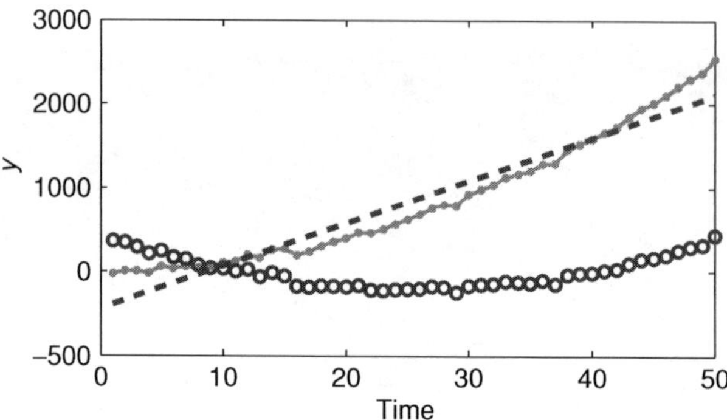

Figure 2.4 The same situation as in Fig. 2.3, except the series was extended to
50 points. Now the residuals ($^\circ$) are visually structured, and one would have a
powerful suggestion that some hypothesis (something about the model or data) is
not correct. This straight-line fit should be rejected as being inconsistent with the
assumption that the residuals are unstructured: the model has been "invalidated."

$D^2(\tilde{\mathbf{x}} - \langle\tilde{\mathbf{x}}\rangle)$, the covariance of $\tilde{\mathbf{x}}$. The "standard error" of $\tilde{x}_i$ is usually defined
to be $\pm\sqrt{C_{\tilde{\mathbf{x}}\tilde{\mathbf{x}}_{ii}}}$ and is used to understand the adequacy of data for distinguishing different possible estimates of $\tilde{\mathbf{x}}$. If applied to the straight-line fit of Fig. 2.2,
$\tilde{\mathbf{x}}^{\mathrm{T}} = [\tilde{a}, \tilde{b}] = [1.69 \pm 0.83, 1.98 \pm 0.03]$, which are within one standard deviation
of the true values, $[a, b] = [1, 2]$. If the noise in $\mathbf{y}$ is Gaussian, it follows that the
probability density of $\tilde{\mathbf{x}}$ is also Gaussian, with mean $\langle\tilde{\mathbf{x}}\rangle$ and covariance $\mathbf{C}_{\tilde{\mathbf{x}}\tilde{\mathbf{x}}}$. Of
course, if $\mathbf{n}$ is not Gaussian, then the estimate won't be either, and one must be
wary of the utility of the standard errors. A Gaussian, or other, assumption should
be regarded as part of the model definition. The uncertainty of the residuals is

$$\mathbf{C}_{nn} = \langle(\tilde{\mathbf{n}} - \langle\tilde{\mathbf{n}}\rangle)(\tilde{\mathbf{n}} - \langle\tilde{\mathbf{n}}\rangle)^{\mathrm{T}}\rangle = (\mathbf{I} - \mathbf{H})\,\mathbf{R}_{nn}\,(\mathbf{I} - \mathbf{H})^{\mathrm{T}} \qquad (2.103)$$
$$= \sigma_n^2\,(\mathbf{I} - \mathbf{H})^2 = \sigma_n^2\,(\mathbf{I} - \mathbf{H})\,,$$

where zero-mean white noise was assumed, and $\mathbf{H}$ was defined in Eq. (2.97). The
true noise, $\mathbf{n}$, was assumed to be white, but the estimated noise, $\tilde{\mathbf{n}}$, has a non-
diagonal covariance and so in a formal sense does not have the expected structure.
We return to this point below.

The fit of a straight line to observations demonstrates many of the issues involved
in making inferences from real, noisy data that appear in more complex situations.
In Fig. 2.5, the correct model used to generate the data was the same as in Fig. 2.2,
but the noise level is very high. The parameters $[\tilde{a}, \tilde{b}]$ are numerically inexact, but
consistent within one standard error with the correct values, which is all one can
hope for.

In Fig. 2.3, a quadratic model $y = 1 + t^2 + n(t)$ was used to generate the numbers, with $\langle n^2 \rangle = 900$. Using only the first 20 points, and fitting an incorrect model,

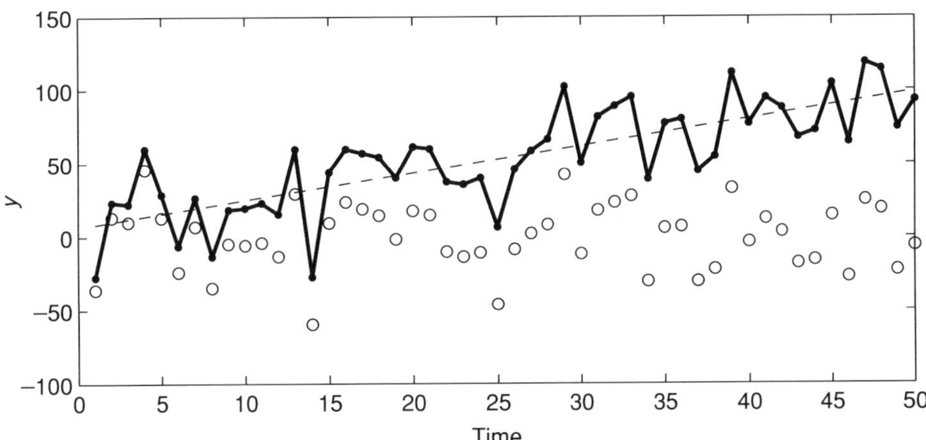

Figure 2.5 The same situation as in Fig. 2.2, $y = 1 + 2t$, except $\langle n^2 \rangle = 900$ to give very noisy data. Now the best-fitting straight line is $y = (6.62 \pm 6.50) + (1.85 \pm 0.22)\,t$, which includes the correct answer within one standard error. Note that the intercept value is indistinguishable from zero.

produces a reasonable straight-line fit to the data as shown. Modeling a quadratic field with a linear model produces a systematic or "model" error, which is not easy to detect here. One sometimes hears it said that "least-squares failed" in situations such as this one. But this conclusion shows a fundamental misunderstanding: least-squares did exactly what it was asked to do – to produce the best-fitting straight line to the data. Here, one might conclude that "the straight-line fit *is* consistent with the data." Such a conclusion is completely different from asserting that one has proven a straight-line fit correctly "explains" the data or, in modeler's jargon, that the model has been "verified" or "validated." If the outcome of the fit were sufficiently important, one might try more powerful tests on the $\tilde{n}_i$ than a mere visual comparison. Such tests might lead to rejection of the straight-line hypothesis; but even if the tests are passed, the model has *never* been verified: it has only been shown to be consistent with the available data.

If the situation remains unsatisfactory (perhaps one suspects the model is inadequate, but there are not enough data to produce sufficiently powerful tests), it can be very frustrating. But sometimes the only remedy is to obtain more data. So, in Fig. 2.4, the number of observations was extended to 50 points. Now, even visually, the $\tilde{n}_i$ are obviously structured, and one would almost surely reject any hypothesis that a straight line was an adequate representation of the data. *The model has been invalidated.* A quadratic rule, $y = a + bt + ct^2$, produces an acceptable solution (see Fig. 2.6).

One must always confirm, after the fact, that J, which is a direct function of the residuals, behaves as expected when the solution is substituted. In particular, its

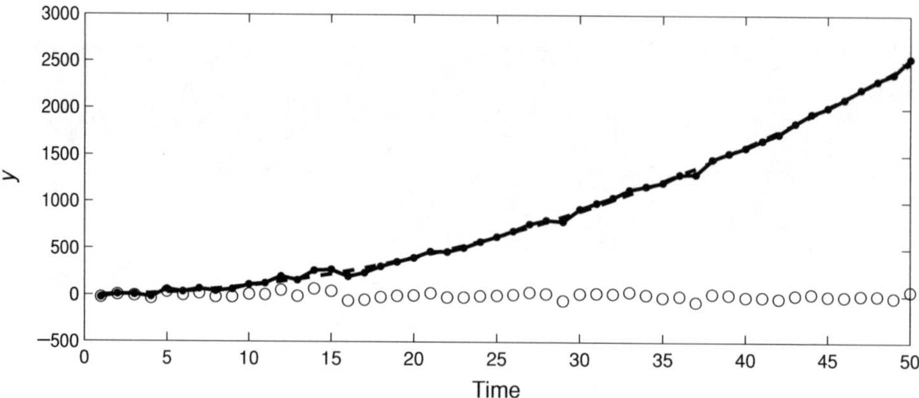

Figure 2.6 Same as Fig. 2.4, except a more complete model, $y = a + bt + ct^2$, was used, and which gives acceptable residuals.

expected value,

$$\langle J \rangle = \sum_i^M \langle n_i^2 \rangle = M - N, \qquad (2.104)$$

assuming that the n_i have been scaled so that each has an expected value $\langle n_i^2 \rangle = 1$. That there are only $M - N$ independent terms in (2.104) follows from the N supposed-independent constraints linking the variables. For any particular solution, $\tilde{\mathbf{x}}, \tilde{\mathbf{n}}, J$ will be a random variable, whose expectation is (2.104). Assuming the n_i are at least approximately Gaussian, J itself is the sum of $M - N$ independent χ_1^2 variables, and is therefore distributed in χ_{M-N}^2. One can and should make histograms of the individual n_i^2 to check them against the expected χ_1^2 probability density. This type of argument leads to the large literature on hypothesis testing.

As an illustration of the random behavior of residuals, 30 equations, $\mathbf{Ex} + \mathbf{n} = \mathbf{y}$, in 15 unknowns were constructed, such that $\mathbf{E}^\mathrm{T}\mathbf{E}$ was non-singular. Fifty different values of $\mathbf{y}$ were then constructed by generating 50 separate $\mathbf{n}$ using a pseudo-random number generator. An ensemble of 50 different solutions were calculated using (2.95), producing $50 \times 30 = 1500$ separate values of $\tilde{n}_i^2$. These are plotted in Fig. 2.7 and compared to χ_1^2. The corresponding value, $\tilde{J}^{(p)} = \sum_1^{30} \tilde{n}_i^2$, was found for each set of equations, and also plotted. A corresponding frequency function for $\tilde{J}^{(p)}$ is compared in Fig. 2.7 to χ_{15}^2, with reasonably good results. The empirical mean value of all $\tilde{J}_i$ is 14.3. Any particular solution may, completely correctly, produce individual residuals $\tilde{n}_i^2$ differing considerably from the mean of $\langle \chi_1^2 \rangle = 1$, and, similarly, their sums, $J^{(p)}$, may differ greatly from $\langle \chi_{15}^2 \rangle = 15$. But one can readily calculate the probability of finding a much larger or smaller value, and employ it to help evaluate the possibility that one has used an incorrect model.

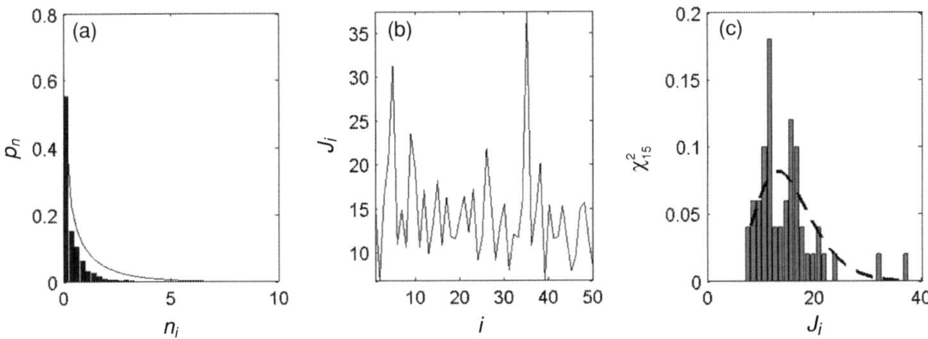

Figure 2.7 (a) χ_1^2 probability density and the empirical frequency function of *all* residuals, $\tilde{n}_i^2$, from 50 separate experiments for the simple least-squares solution of $\mathbf{Ex} + \mathbf{n} = \mathbf{y}$. There is at least rough agreement between the theoretical and calculated frequency functions. (b) The 50 values of J_i computed from the same experiments in (a). (c) The empirical frequency function for J_i as compared to the theoretical value of χ_{15}^2 (dashed line). Tests exist (not discussed here) of the hypothesis that the calculated J_i are consistent with the theoretical distribution.

Visual tests for randomness of residuals have obvious limitations, and elaborate statistical tests in addition to the comparison with χ^2 exist to help determine objectively whether one should accept or reject the hypothesis that no significant structure remains in a sequence of numbers. Books on regression analysis[23] should be consulted for general methodologies. As an indication of what can be done, Fig. 2.8 shows the "sample autocorrelation"

$$\tilde{\phi}_{nn}(\tau) = \frac{1/M \sum_{i=1}^{M-|\tau|} \tilde{n}_i \tilde{n}_{i+\tau}}{1/M \sum_{i=1}^{M} \tilde{n}_i^2} \tag{2.105}$$

for the residuals of the fits shown in Figs. 2.4 and 2.6. For white noise,

$$\langle \tilde{\phi}(\tau) \rangle = \delta_{0\tau}, \tag{2.106}$$

and deviations of the estimated $\tilde{\phi}(t)$ from Eq. (2.106) can be used in simple tests. The adequate fit (Fig. 2.6) produces an autocorrelation of the residuals indistinguishable from a delta function at the origin, while the inadequate fit shows a great deal of structure that would lead to the conclusion that the residuals are too different from white noise to be acceptable. (Not all cases are this obvious.)

As already pointed out, the residuals of the least-squares fit cannot be expected to be precisely white noise. Because there are M relationships among the parameters of the problem (M equations), and the number of $\tilde{\mathbf{x}}$ elements determined is N, there are $M - N$ degrees of freedom in the determination of $\tilde{\mathbf{n}}$ and structures are imposed upon them. The failure, for this reason, of $\tilde{\mathbf{n}}$ strictly to be white noise, is generally only an issue in practice when $M - N$ becomes small compared to M.[24]

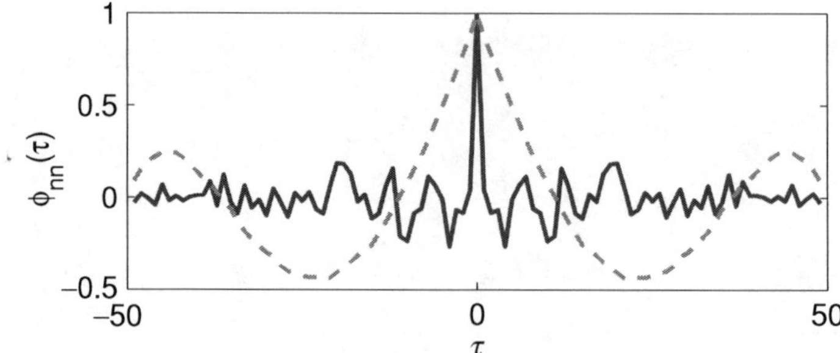

Figure 2.8 Autocorrelations of the estimated residuals in Figs. 2.4 (dashed line) and 2.6 (solid). The latter is indistinguishable, by statistical test, from a delta function at the origin, and so, with this test, the residuals are not distinguishable from white noise.

2.4.2 Weighted and tapered least-squares

The least-squares solution (Eqs. (2.95) and (2.96)) was derived by minimizing the objective function (2.89), in which each residual element is given equal weight. An important feature of least-squares is that we can give whatever emphasis we please to minimizing individual equation residuals, for example, by introducing an objective function,

$$J = \sum_i W_{ii}^{-1} n_i^2, \qquad (2.107)$$

where W_{ii} are any numbers desired. The choice $W_{ii} = 1$, as used above, might be reasonable, but it is an arbitrary one that without further justification does not produce a solution with any special claim to significance. In the least-squares context, we are free to make any other reasonable choice, including demanding that some residuals should be much smaller than others – perhaps just to see if it is possible.

A general formalism is obtained by defining a diagonal weight matrix, $W = \text{diag}(W_{ii})$. Dividing each equation by $\sqrt{W_{ii}}$,

$$W_{ii}^{-\text{T}/2} \sum_i E_{ij} x_j + W_{ii}^{-\text{T}/2} n_i = W_{ii}^{-\text{T}/2} y_i, \qquad (2.108)$$

or

$$\mathbf{E}'\mathbf{x} + \mathbf{n}' = \mathbf{y}'$$
$$\mathbf{E}' = \mathbf{W}^{-\text{T}/2}\mathbf{E}, \quad \mathbf{n}' = \mathbf{W}^{-\text{T}/2}\mathbf{n}, \quad \mathbf{y}' = \mathbf{W}^{-\text{T}/2}\mathbf{y} \qquad (2.109)$$

where we used the fact that the square root of a diagonal matrix is the diagonal matrix of element-by-element square roots. The operation in (2.108) or (2.109) is usually called "row scaling" because it operates on the rows of $\mathbf{E}$ (as well as on $\mathbf{n}, \mathbf{y}$).

For the new equations (2.109), the objective function,

$$J = \mathbf{n}'^{\mathrm{T}}\mathbf{n}' = (\mathbf{y}' - \mathbf{E}'\mathbf{x})^{\mathrm{T}}(\mathbf{y}' - \mathbf{E}'\mathbf{x}) \tag{2.110}$$
$$= \mathbf{n}^{\mathrm{T}}\mathbf{W}^{-1}\mathbf{n} = (\mathbf{y} - \mathbf{E}\mathbf{x})^{\mathrm{T}}\mathbf{W}^{-1}(\mathbf{y} - \mathbf{E}\mathbf{x}),$$

weights the residuals as desired. If, for some reason, $\mathbf{W}$ is non-diagonal, but symmetric and positive-definite, then it has a Cholesky decomposition (see p. 40) and

$$\mathbf{W} = \mathbf{W}^{\mathrm{T}/2}\mathbf{W}^{1/2},$$

and (2.109) remains valid more generally.

The values $\tilde{\mathbf{x}}$, $\tilde{\mathbf{n}}$, minimizing (2.110), are

$$\tilde{\mathbf{x}} = (\mathbf{E}'^{\mathrm{T}}\mathbf{E}')^{-1}\mathbf{E}'^{\mathrm{T}}\mathbf{y}' = (\mathbf{E}^{\mathrm{T}}\mathbf{W}^{-1}\mathbf{E})^{-1}\mathbf{E}^{\mathrm{T}}\mathbf{W}^{-1}\mathbf{y}, \tag{2.111}$$
$$\tilde{\mathbf{n}} = \mathbf{W}^{\mathrm{T}/2}\mathbf{n}' = [\mathbf{I} - \mathbf{E}(\mathbf{E}^{\mathrm{T}}\mathbf{W}^{-1}\mathbf{E})^{-1}\mathbf{E}^{\mathrm{T}}\mathbf{W}^{-1}]\mathbf{y},$$
$$\mathbf{C}_{xx} = (\mathbf{E}^{\mathrm{T}}\mathbf{W}^{-1}\mathbf{E})^{-1}\mathbf{E}^{\mathrm{T}}\mathbf{W}^{-1}\mathbf{R}_{nn}\mathbf{W}^{-1}\mathbf{E}(\mathbf{E}^{\mathrm{T}}\mathbf{W}^{-1}\mathbf{E})^{-1}. \tag{2.112}$$

Uniform diagonal weights are a special case. The rationale for choosing differing diagonal weights, or a non-diagonal $\mathbf{W}$, is probably not very obvious to the reader. Often one chooses $\mathbf{W} = \mathbf{R}_{nn} = \{\langle n_i n_j \rangle\}$, that is, the weight matrix is chosen to be the expected second moment matrix of the residuals. Then

$$\langle \mathbf{n}'\mathbf{n}'^{\mathrm{T}} \rangle = \mathbf{I},$$

and Eq. (2.112) simplifies to

$$\mathbf{C}_{xx} = \left(\mathbf{E}^{\mathrm{T}}\mathbf{R}_{nn}^{-1}\mathbf{E}\right)^{-1}. \tag{2.113}$$

In this special case, the weighting (2.109) has a ready interpretation: The equations (and hence the residuals) are rotated and stretched so that in the new coordinate system of n_i', the covariances are all diagonal and the variances are all unity. Under these circumstances, an objective function

$$J = \sum_i n_i'^2,$$

as used in the original form of least-squares (Eq. (2.89)), is a reasonable choice.

Example *Consider the system*

$$x_1 + x_2 + n_1 = 1$$
$$x_1 - x_2 + n_2 = 2$$
$$x_1 - 2x_2 + n_3 = 4.$$

Then if $\langle n_i \rangle = 0$, $\langle n_i^2 \rangle = \sigma^2$, the least-squares solution is $\tilde{\mathbf{x}} = [2.0, 0.5]^{\mathrm{T}}$. Now suppose that

$$\langle n_i n_j \rangle = \begin{Bmatrix} 1 & 0.99 & 0.98 \\ 0.99 & 1 & 0.99 \\ 0.98 & 0.99 & 4 \end{Bmatrix}.$$

Then from Eq. (2.112), $\tilde{\mathbf{x}} = [1.51, -0.48]^{\mathrm{T}}$. Calculation of the two different solution uncertainties is left to the reader.

We emphasize that this choice of $\mathbf{W}$ is a very special one and has confused many users of inverse methods. To emphasize again: Least-squares is an approximation procedure in which $\mathbf{W}$ is a set of weights wholly at the disposal of the investigator; setting $\mathbf{W} = \mathbf{R}_{nn}$ is a special case whose significance is best understood after we examine a different, statistical, estimation procedure.

Whether the equations are scaled or not, the previous limitations of the simple least-squares solutions remain. In particular, we still have the problem that the solution may produce elements in $\tilde{\mathbf{x}}$, $\tilde{\mathbf{n}}$ whose relative values are not in accord with expected or reasonable behavior, and the solution uncertainty or variances could be unusably large, as the solution is determined, mechanically, and automatically, from combinations such as $(\mathbf{E}^{\mathrm{T}}\mathbf{W}^{-1}\mathbf{E})^{-1}$. Operators like these are neither controllable nor very easy to understand; if any of the matrices are singular, they will not even exist.

It was long ago recognized that some control over the magnitudes of $\tilde{\mathbf{x}}$, $\tilde{\mathbf{n}}$, $\mathbf{C}_{xx}$ could be obtained in the simple least-squares context by modifying the objective function (2.107) to have an additional term:

$$J' = \mathbf{n}^{\mathrm{T}}\mathbf{W}^{-1}\mathbf{n} + \gamma^2 \mathbf{x}^{\mathrm{T}}\mathbf{x} \tag{2.114}$$

$$= (\mathbf{y} - \mathbf{E}\mathbf{x})^{\mathrm{T}}\mathbf{W}^{-1}(\mathbf{y} - \mathbf{E}\mathbf{x}) + \gamma^2 \mathbf{x}^{\mathrm{T}}\mathbf{x}, \tag{2.115}$$

in which γ^2 is a positive constant.

If the minimum of (2.114) is sought by setting the derivatives with respect to $\mathbf{x}$ to zero, then we arrive at the following:

$$\tilde{\mathbf{x}} = (\mathbf{E}^{\mathrm{T}}\mathbf{W}^{-1}\mathbf{E} + \gamma^2 \mathbf{I})^{-1}\mathbf{E}^{\mathrm{T}}\mathbf{W}^{-1}\mathbf{y} \tag{2.116}$$

$$\tilde{\mathbf{n}} = \mathbf{y} - \mathbf{E}\tilde{\mathbf{x}} \tag{2.117}$$

$$\mathbf{C}_{xx} = (\mathbf{E}^{\mathrm{T}}\mathbf{W}^{-1}\mathbf{E} + \gamma^2 \mathbf{I})^{-1}\mathbf{E}^{\mathrm{T}}\mathbf{W}^{-1}\mathbf{R}_{nn}\mathbf{W}^{-1}\mathbf{E}(\mathbf{E}^{\mathrm{T}}\mathbf{W}^{-1}\mathbf{E} + \gamma^2 \mathbf{I})^{-1}. \tag{2.118}$$

By letting $\gamma^2 \to 0$, the solution to (2.111) and (2.112) is recovered, and if $\gamma^2 \to \infty$, $\|\tilde{\mathbf{x}}\|_2 \to 0$, $\tilde{\mathbf{n}} \to \mathbf{y}$; γ^2 is called a " trade-off parameter," because it trades the magnitude of $\tilde{\mathbf{x}}$ against that of $\tilde{\mathbf{n}}$. By varying the size of γ^2 we gain some influence over the norm of the residuals relative to that of $\tilde{\mathbf{x}}$. The expected value of $\tilde{\mathbf{x}}$ is now,

$$\langle \tilde{\mathbf{x}} \rangle = [\mathbf{E}^{\mathrm{T}}\mathbf{W}^{-1}\mathbf{E} + \gamma^2 \mathbf{I}]^{-1}\mathbf{E}^{\mathrm{T}}\mathbf{W}^{-1}\mathbf{y}_0. \tag{2.119}$$

If the true solution is believed to be (2.100), then this new solution is biassed. But the variance of $\tilde{\mathbf{x}}$ has been reduced, (2.118), by introduction of $\gamma^2 > 0$ – that is, the acceptance of a bias reduces the variance, possibly very greatly. Equations (2.116) and (2.117) are sometimes known as the "tapered least-squares" solution, a label whose implication becomes clear later. $\mathbf{C}_{nn}$, which is not displayed, is readily found by direct computation as in Eq. (2.103).

The most basic, and commonly seen, form of this solution assumes that $\mathbf{W} = \mathbf{R}_{nn} = \mathbf{I}$, so that

$$\tilde{\mathbf{x}} = (\mathbf{E}^{\mathrm{T}}\mathbf{E} + \gamma^2 \mathbf{I})^{-1}\mathbf{E}^{\mathrm{T}}\mathbf{y}, \tag{2.120}$$

$$\mathbf{C}_{xx} = (\mathbf{E}^{\mathrm{T}}\mathbf{E} + \gamma^2 \mathbf{I})^{-1}\mathbf{E}^{\mathrm{T}}\mathbf{E}(\mathbf{E}^{\mathrm{T}}\mathbf{E} + \gamma^2 \mathbf{I})^{-1}. \tag{2.121}$$

A physical motivation for the modified objective function (2.114) is obtained by noticing that a preference for a bounded $\|\mathbf{x}\|$ is easily produced by adding an equation set, $\mathbf{x} + \mathbf{n}_1 = \mathbf{0}$, so that the combined set is

$$\mathbf{E}\mathbf{x} + \mathbf{n} = \mathbf{y}, \tag{2.122}$$

$$\mathbf{x} + \mathbf{n}_1 = \mathbf{0}, \tag{2.123}$$

or

$$\mathbf{E}_1 \mathbf{x} + \mathbf{n}_2 = \mathbf{y}_2,$$

$$\mathbf{E}_1 = \left\{ \begin{matrix} \mathbf{E} \\ \gamma \mathbf{I} \end{matrix} \right\}, \quad \mathbf{n}_2^{\mathrm{T}} = \left[\mathbf{n}^{\mathrm{T}} \ \gamma \mathbf{n}_1^{\mathrm{T}} \right], \quad \mathbf{y}_2^{\mathrm{T}} = [\mathbf{y}^{\mathrm{T}} \ \mathbf{0}^{\mathrm{T}}], \tag{2.124}$$

in which $\gamma > 0$ expresses a preference for fitting the first or second sets more closely. Then J in Eq. (2.114) becomes the natural objective function to use. A preference that $\mathbf{x} \approx \mathbf{x}_0$ is readily imposed instead, with an obvious change in (2.114) or (2.123).

Note the important points, to be shown later, that the matrix inverses in Eqs. (2.116) and (2.117) will *always* exist, as long as $\gamma^2 > 0$, and that the expressions remain valid even if $M < N$. Tapered least-squares produces some control over the sum of squares of the relative norms of $\tilde{\mathbf{x}}$, $\tilde{\mathbf{n}}$, but still does not produce control over the individual elements $\tilde{x}_i$.

To gain some of that control, we can further generalize the objective function by introducing another non-singular $N \times N$ weight matrix, $\mathbf{S}$ (which is usually symmetric), and

$$J = \mathbf{n}^T \mathbf{W}^{-1} \mathbf{n} + \mathbf{x}^T \mathbf{S}^{-1} \mathbf{x} \tag{2.125}$$

$$= (\mathbf{y} - \mathbf{E}\mathbf{x})^T \mathbf{W}^{-1} (\mathbf{y} - \mathbf{E}\mathbf{x}) + \mathbf{x}^T \mathbf{S}^{-1} \mathbf{x}, \tag{2.126}$$

for which Eq. (2.114) is a special case. Setting the derivatives with respect to $\mathbf{x}$ to zero results in the following:

$$\tilde{\mathbf{x}} = (\mathbf{E}^T \mathbf{W}^{-1} \mathbf{E} + \mathbf{S}^{-1})^{-1} \mathbf{E}^T \mathbf{W}^{-1} \mathbf{y}, \tag{2.127}$$

$$\tilde{\mathbf{n}} = \mathbf{y} - \mathbf{E}\tilde{\mathbf{x}}, \tag{2.128}$$

$$\mathbf{C}_{xx} = (\mathbf{E}^T \mathbf{W}^{-1} \mathbf{E} + \mathbf{S}^{-1})^{-1} \mathbf{E}^T \mathbf{W}^{-1} \mathbf{R}_{nn} \mathbf{W}^{-1} \mathbf{E} (\mathbf{E}^T \mathbf{W}^{-1} \mathbf{E} + \mathbf{S}^{-1})^{-1}, \tag{2.129}$$

which are identical to Eqs. (2.116)–(2.118) with $\mathbf{S}^{-1} = \gamma^2 \mathbf{I}$. $\mathbf{C}_{xx}$ simplifies if $\mathbf{R}_{nn} = \mathbf{W}$.

Suppose that $\mathbf{S}$, $\mathbf{W}$ are positive definite and symmetric and thus have Cholesky decompositions. Then employing both matrices directly on the equations, $\mathbf{E}\mathbf{x} + \mathbf{n} = \mathbf{y}$,

$$\mathbf{W}^{-T/2} \mathbf{E} \mathbf{S}^{-T/2} \mathbf{S}^{T/2} \mathbf{x} + \mathbf{W}^{-T/2} \mathbf{n} = \mathbf{W}^{-T/2} \mathbf{y} \tag{2.130}$$

$$\mathbf{E}' \mathbf{x}' + \mathbf{n}' = \mathbf{y}' \tag{2.131}$$

$$\mathbf{E}' = \mathbf{W}^{-T/2} \mathbf{E} \mathbf{S}^{T/2}, \ \mathbf{x}' = \mathbf{S}^{-T/2} \mathbf{x}, \ \mathbf{n}' = \mathbf{W}^{-T/2} \mathbf{n}, \ \mathbf{y}' = \mathbf{W}^{-T/2} \mathbf{y} \tag{2.132}$$

The use of $\mathbf{S}$ in this way is called "column scaling" because it weights the columns of $\mathbf{E}$. With Eqs. (2.131) the obvious objective function is

$$J = \mathbf{n}'^T \mathbf{n}' + \mathbf{x}'^T \mathbf{x}', \tag{2.133}$$

which is identical to Eq. (2.125) in the original variables, and the solution must be that in Eqs. (2.127)–(2.129).

Like $\mathbf{W}$, one is completely free to choose $\mathbf{S}$ as one pleases. A common example is to write, where $\mathbf{F}$ is $N \times N$, that

$$\mathbf{S} = \mathbf{F}^T \mathbf{F}$$

$$\mathbf{F} = \gamma^2 \begin{Bmatrix} 1 & -1 & 0 & \cdots & 0 \\ 0 & 1 & -1 & \cdots & 0 \\ \vdots & \vdots & \vdots & \vdots & \vdots \\ 0 & \cdots & & 0 & 1 \end{Bmatrix}, \tag{2.134}$$

The effect of (2.134) is to minimize a term $\gamma^2 \sum_i (x_i - x_{i+1})^2$, which can be regarded as a "smoothest" solution, and, using γ^2 to trade smoothness against the size of $\|\tilde{\mathbf{n}}\|_2$, $\mathbf{F}$ is obtained from the Cholesky decomposition of $\mathbf{S}$.

By invoking the matrix inversion lemma, an alternate form for Eqs. (2.127)–(2.129) is found:

$$\tilde{\mathbf{x}} = \mathbf{S}\mathbf{E}^{\mathrm{T}}(\mathbf{E}\mathbf{S}\mathbf{E}^{\mathrm{T}} + \mathbf{W})^{-1}\mathbf{y}, \tag{2.135}$$

$$\tilde{\mathbf{n}} = \mathbf{y} - \mathbf{E}\tilde{\mathbf{x}}, \tag{2.136}$$

$$\mathbf{C}_{xx} = \mathbf{S}\mathbf{E}^{\mathrm{T}}(\mathbf{E}\mathbf{S}\mathbf{E}^{\mathrm{T}} + \mathbf{W})^{-1}\mathbf{R}_{nn}(\mathbf{E}\mathbf{S}\mathbf{E}^{\mathrm{T}} + \mathbf{W})^{-1}\mathbf{E}\mathbf{S}. \tag{2.137}$$

A choice of which form to use is sometimes made on the basis of the dimensions of the matrices being inverted. Note again that $\mathbf{W} = \mathbf{R}_{nn}$ is a special case.

So far, all of this is conventional. But we have made a special point of displaying explicitly not only the elements $\tilde{\mathbf{x}}$, but those of the residuals, $\tilde{\mathbf{n}}$. Notice that although we have considered only the formally over determined system, $M > N$, we *always* determine not only the N elements of $\tilde{\mathbf{x}}$, but also the M elements of $\tilde{\mathbf{n}}$, for a total of $M + N$ values – extracted from the M equations. It is apparent that any change in any element $\tilde{n}_i$ forces changes in $\tilde{\mathbf{x}}$. In this view, to which we adhere, systems of equations involving observations *always* contain more unknowns than equations. Another way to make the point is to re-write Eq. (2.87) without distinction between $\mathbf{x}, \mathbf{n}$ as

$$\mathbf{E}_1\boldsymbol{\xi} = \mathbf{y}, \tag{2.138}$$

$$\mathbf{E}_1 = \{\mathbf{E}, \mathbf{I}_M\}, \quad \boldsymbol{\xi}^{\mathrm{T}} = [\mathbf{x}, \mathbf{n}]^{\mathrm{T}}. \tag{2.139}$$

A combined weight matrix,

$$\mathbf{S}_1 = \begin{Bmatrix} \mathbf{S} & \mathbf{0} \\ \mathbf{0} & \mathbf{W} \end{Bmatrix}, \tag{2.140}$$

would be used, and any distinction between the $\mathbf{x}, \mathbf{n}$ solution elements is suppressed. Equation (2.138) describes a formally underdetermined system, derived from the formally over determined observed one. This identity leads us to the problem of formal underdetermination in the next section.

In general with least-squares problems, the solution sought can be regarded as any of the following equivalents:

1. The $\tilde{\mathbf{x}}, \tilde{\mathbf{n}}$ satisfying

$$\mathbf{E}\mathbf{x} + \mathbf{n} = \mathbf{y}. \tag{2.141}$$

2. $\tilde{\mathbf{x}}, \tilde{\mathbf{n}}$ satisfying the normal equations arising from J (2.125).
3. $\tilde{\mathbf{x}}, \tilde{\mathbf{n}}$ producing the minimum of J in (2.125).

The point of this list lies with item (3): algorithms exist to find minima of functions by deterministic methods ("go downhill" from an initial guess),[25] or stochastic search methods (Monte Carlo) or even, conceivably, through a shrewd guess by the investigator. If an acceptable minimum of J is found, by whatever means, it is an acceptable solution (subject to further testing, and the possibility that there is more than one such solution). Search methods become essential for the non-linear problems taken up later.

2.4.3 Underdetermined systems and Lagrange multipliers

What does one do when the number, M, of equations is less than the number, N, of unknowns and no more observations are possible? We have seen that the claim that a problem involving observations is ever over determined is misleading – because each equation or observation always has a noise unknown, but to motivate some of what follows, it is helpful to first pursue the conventional approach.

One often attempts when $M < N$ to reduce the number of unknowns so that the formal overdeterminism is restored. Such a parameter reduction procedure may be sensible; but there are pitfalls. For example, let $p_i\,(t)$, $i = 0, 1, \ldots$, be a set of polynomials, e.g., Chebyschev or Laguerre, etc. Consider data produced from the formula

$$y\,(t) = 1 + a_M\,p_M\,(t) + n(t), \tag{2.142}$$

which might be deduced by fitting a parameter set $[a_0, \ldots, a_M]$ and finding $\tilde{a}_M$. If there are fewer than M observations, an attempt to fit with fewer parameters,

$$y = \sum_{j=0}^{Q} a_j p_j\,(t), \quad Q < M \tag{2.143}$$

may give a good, even perfect fit; but it would be incorrect. The reduction in model parameters in such a case biasses the result, perhaps hopelessly so. One is better off retaining the underdetermined system and making inferences concerning the possible values of a_i rather than using the form (2.143), in which any possibility of learning something about a_M has been eliminated.

Example *Consider a tracer problem, not unlike those encountered in medicine, hydrology, oceanography, etc. A box (Fig. 1.2) is observed to contain a steady tracer concentration C_0, and is believed to be fed at the rates J_1, J_2 from two reservoirs each with tracer concentration of C_1, C_2 respectively. One seeks to determine J_1, J_2. Tracer balance is*

$$J_1 C_1 + J_2 C_2 - J_0 C_0 = 0, \tag{2.144}$$

where J_0 is the rate at which fluid is removed. Mass balance requires that

$$J_1 + J_2 - J_0 = 0. \tag{2.145}$$

Evidently, there are but two equations in three unknowns (and a perfectly good solution would be $J_1 = J_2 = J_3 = 0$); but, as many have noticed, we can nonetheless, determine the relative fraction of the fluid coming from each reservoir. Divide both equations through by J_0,

$$\frac{J_1}{J_0}C_1 + \frac{J_2}{J_0}C_2 = C_0$$

$$\frac{J_1}{J_0} + \frac{J_2}{J_0} = 1,$$

producing two equations in two unknowns, J_1/J_0, J_2/J_0, which has a unique stable solution (noise is being ignored). Many examples can be given of such calculations in the literature – determining the flux ratios – apparently definitively. But suppose the investigator is suspicious that there might be a third reservoir with tracer concentration C_3. Then the equations become

$$\frac{J_1}{J_0}C_1 + \frac{J_2}{J_0}C_2 + \frac{J_3}{J_0}C_3 = C_0$$

$$\frac{J_1}{J_0} + \frac{J_2}{J_0} + \frac{J_3}{J_0} = 1,$$

which are now underdetermined with two equations in three unknowns. If it is obvious that no such third reservoir exists, then the reduction to two equations in two unknowns is the right thing to do. But if there is even a suspicion of a third reservoir (or more), one should solve these equations with one of the methods we will develop – permitting construction and understanding of all possible solutions.

In general terms, parameter reduction can lead to model errors, that is, bias errors, which can produce wholly illusory results.[26] A common situation, particularly in problems involving tracer movements in groundwater, ocean, or atmosphere, is fitting a one or two-dimensional model to data that represent a fully three-dimensional field. The result may be apparently pleasing, but possibly completely erroneous.

A general approach to solving underdetermined problems is to render the answer apparently unique by minimizing an objective function, subject to satisfaction of the linear constraints. To see how this can work, suppose that $\mathbf{A}\mathbf{x} = \mathbf{b}$, exactly and formally underdetermined, $M < N$, and seek the solution that exactly satisfies the equations and simultaneously renders an objective function, $J = \mathbf{x}^{\mathrm{T}}\mathbf{x}$, as small as possible. Direct minimization of J leads to

$$\mathrm{d}J = \mathrm{d}\mathbf{x}^{\mathrm{T}}\frac{\partial J}{\partial \mathbf{x}} = 2\mathbf{x}^{\mathrm{T}}\mathrm{d}\mathbf{x} = 0, \tag{2.146}$$

but, unlike the case in Eq. (2.91), the coefficients of the individual $\mathrm{d}x_i$ can no longer be separately set to zero (i.e., $\mathbf{x} = 0$ is an incorrect solution) because the $\mathrm{d}x_i$ no longer vary independently, but are restricted to values satisfying $\mathbf{Ax} = \mathbf{b}$. One approach is to use the known dependencies to reduce the problem to a new one in which the differentials are independent. For example, suppose that there are general functional relationships

$$\begin{bmatrix} x_1 \\ \vdots \\ x_M \end{bmatrix} = \begin{bmatrix} \xi_1(x_{M+1}, \ldots, x_N) \\ \vdots \\ \xi_M(x_{M+1}, \ldots, x_N) \end{bmatrix}.$$

Then the first M elements of x_i may be eliminated, and the objective function becomes

$$J = [\xi_1(x_{M+1}, \ldots, x_N)^2 + \cdots + \xi_M(x_{M+1}, \ldots, x_N)^2] + [x_{M+1}^2 + \cdots + x_N^2],$$

in which the remaining x_i, $M + i = 1, 2, \ldots, N$ are independently varying. In the present case, one can choose (arbitrarily) the first M unknowns, $\mathbf{q} = [x_i]$, and define the last $N - M$ unknowns $\mathbf{r} = [x_i]$, $i = N - M + 1, \ldots, N$, and rewrite the equations as

$$\{\mathbf{A}_1\ \mathbf{A}_2\} \begin{bmatrix} \mathbf{q} \\ \mathbf{r} \end{bmatrix} = \mathbf{b} \tag{2.147}$$

where $\mathbf{A}_1$ is $M \times M$, $\mathbf{A}_2$ is $M \times (N - M)$. Then solving the first set for $\mathbf{q}$,

$$\mathbf{q} = \mathbf{b} - \mathbf{A}_2\mathbf{r}. \tag{2.148}$$

$\mathbf{q}$ can be eliminated from J leaving an unconstrained minimization problem in the independent variables, $\mathbf{r}$. If $\mathbf{A}_1^{-1}$ does not exist, one can try any other subset of the x_i to eliminate until a suitable group is found. This approach is completely correct, but finding an explicit solution for L elements of $\mathbf{x}$ in terms of the remaining ones may be difficult or inconvenient.

Example *Solve*

$$x_1 - x_2 + x_3 = 1,$$

for the solution of minimum norm. The objective function is $J = x_1^2 + x_2^2 + x_3^2$. With one equation, one variable can be eliminated. Arbitrarily choosing $x_1 = 1 +$

$x_2 - x_3, J = (1 + x_2 - x_3)^2 + x_2^2 + x_3^2$. x_2, x_3 *are now independent variables, and the corresponding derivatives of J can be independently set to zero.*

Example *A somewhat more interesting example involves two equations in three unknowns:*

$$x_1 + x_2 + x_3 = 1,$$
$$x_1 - x_2 + x_3 = 2,$$

and we choose to find a solution minimizing,

$$J = x_1^2 + x_2^2 + x_3^2.$$

Solving for two unknowns x_1, x_2 from

$$x_1 + x_2 = 1 - x_3,$$
$$x_1 - x_2 = 2 - x_3,$$

produces $x_2 = -1/2$, $x_1 = 3/2 - x_3$, and then

$$J = (3/2 - x_3)^2 + 1/4 + x_3^2,$$

whose minimum with respect to x_3 (the only remaining variable) is $x_3 = 3/4$, and the full solution is

$$x_1 = \frac{3}{4}, \quad x_2 = -\frac{1}{2}, \quad x_3 = \frac{3}{4}.$$

Lagrange multipliers and adjoints

When it is inconvenient to find such an explicit representation by eliminating some variables in favor of others, a standard procedure for finding the constrained minimum is to introduce a new vector "Lagrange multiplier," $\boldsymbol{\mu}$, of M unknown elements, to make a new objective function

$$J' = J - 2\boldsymbol{\mu}^{\mathrm{T}}(\mathbf{Ax} - \mathbf{b}) \tag{2.149}$$
$$= \mathbf{x}^{\mathrm{T}}\mathbf{x} - 2\boldsymbol{\mu}^{\mathrm{T}}(\mathbf{Ax} - \mathbf{b}),$$

and ask for its stationary point – treating both $\boldsymbol{\mu}$ and $\mathbf{x}$ as independently varying unknowns. The numerical 2 is introduced solely for notational tidiness.

The rationale for this procedure is straightforward.[27] Consider first a very simple example of one equation in two unknowns,

$$x_1 - x_2 = 1. \tag{2.150}$$

We seek the minimum norm solution,

$$J = x_1^2 + x_2^2, \tag{2.151}$$

subject to Eq. (2.150). The differential

$$dJ = 2x_1 dx_1 + 2x_2 dx_2 = 0 \tag{2.152}$$

leads to the unacceptable solution $x_1 = x_2 = 0$ if we incorrectly set the coefficients of dx_1, dx_2 to zero. Consider instead a modified objective function

$$J' = J - 2\mu (x_1 - x_2 - 1), \tag{2.153}$$

where μ is unknown. The differential of J' is

$$dJ' = 2x_1 dx_1 + 2x_2 dx_2 - 2\mu (dx_1 - dx_2) - 2 (x_1 - x_2 - 1) d\mu = 0, \tag{2.154}$$

or

$$dJ'/2 = dx_1 (x_1 - \mu) + dx_2 (x_2 + \mu) - d\mu (x_1 - x_2 - 1) = 0. \tag{2.155}$$

We are free to choose $x_1 = \mu$, which kills off the differential involving dx_1. But then only the differentials dx_2, $d\mu$ remain; as they can vary independently, their coefficients must vanish separately, and we have

$$x_2 = -\mu \tag{2.156}$$
$$x_1 - x_2 = 1. \tag{2.157}$$

Note that the second of these recovers the original equation. Substituting $x_1 = \mu$, we have $2\mu = 1$, or $\mu = 1/2$, and $x_1 = 1/2, x_2 = -1/2, J = 0.5$, and one can confirm that this is indeed the "constrained" minimum. (A "stationary" value of J' was found, not an absolute minimum value, because J' is no longer necessarily positive; it has a saddle point, which we have found.)

Before writing out the general case, note the following question: Suppose the constraint equation was changed to

$$x_1 - x_2 = \Delta. \tag{2.158}$$

How much would J change as Δ is varied? With variable Δ, (2.154) becomes

$$dJ' = 2dx_1 (x_1 - \mu) + 2dx_2 (x_2 + \mu) - 2d\mu (x_1 - x_2 - \Delta) + 2\mu d\Delta. \tag{2.159}$$

But the first three terms on the right vanish, and hence

$$\frac{\partial J'}{\partial \Delta} = 2\mu = \frac{\partial J}{\partial \Delta}, \tag{2.160}$$

because $J = J'$ at the stationary point (from (2.158)). *Thus 2μ is the sensitivity of the objective function J to perturbations in the right-hand side of the* *constraint equation.* If Δ is changed from 1 to 1.2, it can be confirmed that the approximate change in the value of J is 0.2, as one deduces immediately from Eq. (2.160). Keep in mind, however, that this sensitivity corresponds to infinitesimal perturbations.

We now develop this method generally. Reverting to Eq. (2.149),

$$
\begin{aligned}
\mathrm{d}J' &= \mathrm{d}J - 2\boldsymbol{\mu}^{\mathrm{T}}\mathbf{A}\mathrm{d}\mathbf{x} - 2\left(\mathbf{A}\mathbf{x} - \mathbf{b}\right)^{\mathrm{T}}\mathrm{d}\boldsymbol{\mu} \\
&= \left(\frac{\partial J}{\partial x_1} - 2\boldsymbol{\mu}^{\mathrm{T}}\mathbf{a}_1\right)\mathrm{d}x_1 + \left(\frac{\partial J}{\partial x_2} - 2\boldsymbol{\mu}^{\mathrm{T}}\mathbf{a}_2\right)\mathrm{d}x_2 + \cdots \quad (2.161) \\
&\quad + \left(\frac{\partial J}{\partial x_N} - 2\boldsymbol{\mu}^{\mathrm{T}}\mathbf{a}_N\right)\mathrm{d}x_N - 2(\mathbf{A}\mathbf{x} - \mathbf{b})^{\mathrm{T}}\mathrm{d}\boldsymbol{\mu} \\
&= (2x_1 - 2\boldsymbol{\mu}^{\mathrm{T}}\mathbf{a}_1)\mathrm{d}x_1 + (2x_2 - 2\boldsymbol{\mu}^{\mathrm{T}}\mathbf{a}_2)\mathrm{d}x_2 + \cdots \quad (2.162) \\
&\quad + (2x_N - 2\boldsymbol{\mu}^{\mathrm{T}}\mathbf{a}_N)\mathrm{d}x_N - 2(\mathbf{A}\mathbf{x} - \mathbf{b})^{\mathrm{T}}\mathrm{d}\boldsymbol{\mu} = 0
\end{aligned}
$$

Here the $\mathbf{a}_i$ are the corresponding columns of $\mathbf{A}$. The coefficients of the first M differentials $\mathrm{d}x_i$ can be set to zero by assigning $x_i = \boldsymbol{\mu}^{\mathrm{T}}\mathbf{a}_i$, leaving $N - M$ differentials $\mathrm{d}x_i$ whose coefficients must separately vanish (hence they *all* vanish, but for two separate reasons), plus the coefficient of the $M - \mathrm{d}\mu_i$, which must also vanish separately. This recipe produces, from Eq. (2.162),

$$
\frac{1}{2}\frac{\partial J'}{\partial \mathbf{x}} = \mathbf{x} - \mathbf{A}^{\mathrm{T}}\boldsymbol{\mu} = 0, \tag{2.163}
$$

$$
\frac{1}{2}\frac{\partial J'}{\partial \boldsymbol{\mu}} = \mathbf{A}\mathbf{x} - \mathbf{b} = \mathbf{0}, \tag{2.164}
$$

where the first equation set is the result of the vanishing of the coefficients of $\mathrm{d}x_i$ and the second, which is the original set of equations, arises from the vanishing of the coefficients of the $\mathrm{d}\mu_i$. The convenience of being able to treat all the x_i as independently varying is offset by the increase in problem dimensions by the introduction of the M unknown μ_i. The first set is N equations for $\boldsymbol{\mu}$ in terms of $\mathbf{x}$, and the second set is M equations in $\mathbf{x}$ in terms of $\mathbf{y}$. Taken together, these are $M + N$ equations in $M + N$ unknowns, and hence just-determined no matter what the ratio of M to N.

Equation (2.163) is

$$
\mathbf{A}^{\mathrm{T}}\boldsymbol{\mu} = \mathbf{x}, \tag{2.165}
$$

and, substituting for $\mathbf{x}$ into (2.164),

$$
\mathbf{A}\mathbf{A}^{\mathrm{T}}\boldsymbol{\mu} = \mathbf{b},
$$
$$
\tilde{\boldsymbol{\mu}} = (\mathbf{A}\mathbf{A}^{\mathrm{T}})^{-1}\mathbf{b}, \tag{2.166}
$$

assuming the inverse exists, and that

$$\tilde{\mathbf{x}} = \mathbf{A}^{\mathrm{T}}(\mathbf{A}\mathbf{A}^{\mathrm{T}})^{-1}\mathbf{b} \tag{2.167}$$

$$\tilde{\mathbf{n}} = \mathbf{0} \tag{2.168}$$

$$\mathbf{C}_{xx} = 0. \tag{2.169}$$

($\mathbf{C}_{xx} = 0$ because formally $\tilde{\mathbf{n}} = \mathbf{0}$.)

Equations (2.167)–(2.169) are the classical solution, satisfying the constraints exactly while minimizing the solution length. That a minimum is achieved can be verified by evaluating the second derivatives of J' at the solution point. The minimum occurs at a saddle point in $\mathbf{x}$, $\boldsymbol{\mu}$ space[28] and where the term proportional to $\boldsymbol{\mu}$ necessarily vanishes. The operator $\mathbf{A}^{\mathrm{T}}(\mathbf{A}\mathbf{A}^{\mathrm{T}})^{-1}$ is sometimes called a "Moore–Penrose inverse."

Equation (2.165) for $\boldsymbol{\mu}$ in terms of $\mathbf{x}$ involves the coefficient matrix $\mathbf{A}^{\mathrm{T}}$. An intimate connection exists between matrix transposes and adjoints of differential equations (see the appendix to this chapter), and thus $\boldsymbol{\mu}$ is sometimes called the "adjoint solution," with $\mathbf{A}^{\mathrm{T}}$ defining the "adjoint model"[29] in Eq. (2.165), and $\mathbf{x}$ acting as a forcing term. The original $\mathbf{A}\mathbf{x} = \mathbf{b}$ were assumed formally underdetermined, and thus the adjoint model equations in (2.165) are necessarily formally over determined.

Example *The last example now using matrix vector notation is*

$$\mathbf{A} = \begin{Bmatrix} 1 & 1 & 1 \\ 1 & -1 & 1 \end{Bmatrix}, \mathbf{b} = \begin{bmatrix} 1 \\ 2 \end{bmatrix}$$

$$J = \mathbf{x}^{\mathrm{T}}\mathbf{x} - 2\boldsymbol{\mu}^{\mathrm{T}}(\mathbf{A}\mathbf{x} - \mathbf{b})$$

$$\frac{\mathrm{d}}{\mathrm{d}\mathbf{x}}(\mathbf{x}^{\mathrm{T}}\mathbf{x} - 2\boldsymbol{\mu}^{\mathrm{T}}(\mathbf{A}\mathbf{x} - \mathbf{b})) = 2\mathbf{x} - 2\mathbf{A}^{\mathrm{T}}\boldsymbol{\mu} = \mathbf{0}$$

$$\mathbf{A}\mathbf{x} = \mathbf{b}$$

$$\mathbf{x} = \mathbf{A}^{\mathrm{T}}(\mathbf{A}\mathbf{A}^{\mathrm{T}})^{-1}\mathbf{b}$$

$$\mathbf{x} = [3/4, -1/2, 3/4]^{\mathrm{T}}.$$

Example *Write out J':*

$$J' = x_1^2 + x_2^2 + x_3^2 - 2\mu_1(x_1 + x_2 + x_3 - 1) - 2\mu_2(x_1 - x_2 + x_3 - 2)$$

$$\mathrm{d}J' = (2x_1 - 2\mu_1 - 2\mu_2)\mathrm{d}x_1 + (2x_2 - 2\mu_1 + 2\mu_2)\mathrm{d}x_2 + (2x_3 - 2\mu_1 - 2\mu_2)\mathrm{d}x_3$$
$$+ (-2x_1 - 2x_2 + 2 - 2x_3)\,\mathrm{d}\mu_1 + (-2x_1 + 2x_2 - 2x_3 + 4)\,\mathrm{d}\mu_2$$
$$= 0.$$

Set $x_1 = \mu_1 + \mu_2$, $x_2 = \mu_1 - \mu_2$ so that the first two terms vanish, and set the coefficients of the differentials of the remaining, independent terms to zero:

$$\frac{\mathrm{d}J'}{\mathrm{d}x_1} = 2x_1 - 2\mu_1 - 2\mu_2 = 0,$$

$$\frac{\mathrm{d}J'}{\mathrm{d}x_2} = 2x_2 - 2\mu_1 + 2\mu_2 = 0,$$

$$\frac{\mathrm{d}J'}{\mathrm{d}x_3} = 2x_3 - 2\mu_1 - 2\mu_2 = 0,$$

$$\frac{\mathrm{d}J'}{\mathrm{d}\mu_1} = -2x_1 - 2x_2 + 2 - 2x_3 = 0,$$

$$\frac{\mathrm{d}J'}{\mathrm{d}\mu_2} = -2x_1 + 2x_2 - 2x_3 + 4 = 0.$$

Then,

$$\mathrm{d}J' = (2x_3 - 2\mu_1 - 2\mu_2)\mathrm{d}x_3 + (-2x_1 - 2x_2 + 2 - 2x_3)\mathrm{d}\mu_1$$
$$+ (-2x_1 + 2x_2 - 2x_3 + 4)\mathrm{d}\mu_2$$
$$= 0,$$

or

$$x_1 = \mu_1 + \mu_2$$
$$x_2 = \mu_1 - \mu_2$$
$$x_3 - \mu_1 - \mu_2 = 0$$
$$-x_1 - x_2 + 1 - x_3 = 0$$
$$-x_1 + x_2 - x_3 + 2 = 0.$$

That is,

$$\mathbf{x} = \mathbf{A}^\mathrm{T}\boldsymbol{\mu}$$
$$\mathbf{A}\mathbf{x} = \mathbf{b},$$

or

$$\begin{Bmatrix} \mathbf{I} & -\mathbf{A}^\mathrm{T} \\ \mathbf{A} & \mathbf{0} \end{Bmatrix} \begin{bmatrix} \mathbf{x} \\ \boldsymbol{\mu} \end{bmatrix} = \begin{bmatrix} \mathbf{0} \\ \mathbf{b} \end{bmatrix}.$$

In this particular case, the first set can be solved for $\mathbf{x} = \mathbf{A}^\mathrm{T}\boldsymbol{\mu}$:

$$\boldsymbol{\mu} = (\mathbf{A}\mathbf{A}^\mathrm{T})^{-1}\mathbf{b} = [\,1/8 \quad 5/8\,]^\mathrm{T},$$

$$\mathbf{x} = \mathbf{A}^\mathrm{T}\begin{bmatrix} \frac{1}{8} \\ \frac{5}{8} \end{bmatrix} = [\,3/4 \quad -1/2 \quad 3/4\,]^\mathrm{T}.$$

Example *Suppose, instead, we wanted to minimize*

$$J = (x_1 - x_2)^2 + (x_2 - x_3)^2 = \mathbf{x}^{\mathrm{T}}\mathbf{F}^{\mathrm{T}}\mathbf{F}\mathbf{x},$$

where

$$\mathbf{F} = \begin{Bmatrix} 1 & -1 & 0 \\ 0 & 1 & -1 \end{Bmatrix}$$

$$\mathbf{F}^{\mathrm{T}}\mathbf{F} = \begin{Bmatrix} 1 & -1 & 0 \\ 0 & 1 & -1 \end{Bmatrix}^{\mathrm{T}} \begin{Bmatrix} 1 & -1 & 0 \\ 0 & 1 & -1 \end{Bmatrix} = \begin{Bmatrix} 1 & -1 & 0 \\ -1 & 2 & -1 \\ 0 & -1 & 1 \end{Bmatrix}.$$

Such an objective function might be used to find a "smooth" solution. One confirms that

$$\begin{bmatrix} x_1 & x_2 & x_3 \end{bmatrix} \begin{Bmatrix} 1 & -1 & 0 \\ -1 & 2 & -1 \\ 0 & -1 & 1 \end{Bmatrix} \begin{bmatrix} x_1 \\ x_2 \\ x_3 \end{bmatrix} = x_1^2 - 2x_1 x_2 + 2x_2^2 - 2x_2 x_3 + x_3^2$$

$$= (x_1 - x_2)^2 + (x_2 - x_3)^2.$$

The stationary point of

$$J' = \mathbf{x}^{\mathrm{T}}\mathbf{F}^{\mathrm{T}}\mathbf{F}\mathbf{x} - 2\mu^{\mathrm{T}}(\mathbf{A}\mathbf{x} - \mathbf{b})$$

leads to

$$\mathbf{F}^{\mathrm{T}}\mathbf{F}\mathbf{x} = \mathbf{A}^{\mathrm{T}}\mu$$

$$\mathbf{A}\mathbf{x} = \mathbf{b}.$$

But,

$$\mathbf{x} \neq (\mathbf{F}^{\mathrm{T}}\mathbf{F})^{-1}\mathbf{A}^{\mathrm{T}}\mu,$$

because there is no inverse (guaranteed). But the coupled set

$$\begin{Bmatrix} \mathbf{F}^{\mathrm{T}}\mathbf{F} & -\mathbf{A}^{\mathrm{T}} \\ \mathbf{A} & 0 \end{Bmatrix} \begin{bmatrix} \mathbf{x} \\ \mu \end{bmatrix} = \begin{bmatrix} \mathbf{0} \\ \mathbf{b} \end{bmatrix}$$

does have a solution.

The physical interpretation of μ can be obtained as above by considering the way in which J would vary with infinitesimal changes in $\mathbf{b}$. As in the special case above, $J = J'$ at the stationary point. Hence,

$$\mathrm{d}J' = \mathrm{d}J - 2\mu^{\mathrm{T}}\mathbf{A}\,\mathrm{d}\mathbf{x} - 2(\mathbf{A}\mathbf{x} - \mathbf{b})^{\mathrm{T}}\,\mathrm{d}\mu + 2\mu^{\mathrm{T}}\mathrm{d}\mathbf{b} = 0, \tag{2.170}$$

or, since the first three terms on the right vanish at the stationary point,

$$\frac{\partial J'}{\partial \mathbf{b}} = \frac{\partial J}{\partial \mathbf{b}} = 2\mu. \tag{2.171}$$

Thus, as inferred previously, the Lagrange multipliers are the sensitivity of J, at the stationary point, to perturbations in the parameters $\mathbf{b}$. This conclusion leads, in Chapter 4, to the scrutiny of the Lagrange multipliers as a means of understanding the sensitivity of models and the flow of information within them.

Now revert to $\mathbf{E}\mathbf{x} + \mathbf{n} = \mathbf{y}$, that is, equations containing noise. If these are first column scaled using $\mathbf{S}^{-T/2}$, Eqs. (2.167)–(2.169) are in the primed variables, and the solution in the original variables is

$$\tilde{\mathbf{x}} = \mathbf{S}\mathbf{E}^{T}(\mathbf{E}\mathbf{S}\mathbf{E}^{T})^{-1}\mathbf{y}, \tag{2.172}$$

$$\tilde{\mathbf{n}} = \mathbf{0}, \tag{2.173}$$

$$\mathbf{C}_{xx} = \mathbf{0}, \tag{2.174}$$

and the result depends directly upon $\mathbf{S}$. If a row scaling with $\mathbf{W}^{-T/2}$ is used, it is readily shown that $\mathbf{W}$ disappears from the solution and has no effect on it (see p. 107).

Equations (2.172)–(2.174) are a solution, but there is the same fatal defect as in Eq. (2.173) – $\tilde{\mathbf{n}} = \mathbf{0}$ is usually unacceptable when $\mathbf{y}$ are observations. Furthermore, $\|\tilde{\mathbf{x}}\|$ is again uncontrolled, and $\mathbf{E}\mathbf{S}\mathbf{E}^{T}$ may not have an inverse.

Noise vector $\mathbf{n}$ must be regarded as fully an element of the solution, as much as $\mathbf{x}$. Equations representing observations can always be written as in (2.138), and can be solved exactly. Therefore, we now use a modified objective function, allowing for general $\mathbf{S}, \mathbf{W}$,

$$J = \mathbf{x}^{T}\mathbf{S}^{-1}\mathbf{x} + \mathbf{n}^{T}\mathbf{W}^{-1}\mathbf{n} - 2\mu^{T}(\mathbf{E}\mathbf{x} + \mathbf{n} - \mathbf{y}), \tag{2.175}$$

with both $\mathbf{x}, \mathbf{n}$ appearing in the objective function. Setting the derivatives of (2.175) with respect to $\mathbf{x}, \mathbf{n}, \mu$ to zero, and solving the resulting normal equations produces the following:

$$\tilde{\mathbf{x}} = \mathbf{S}\mathbf{E}^{T}(\mathbf{E}\mathbf{S}\mathbf{E}^{T} + \mathbf{W})^{-1}\mathbf{y}, \tag{2.176}$$

$$\tilde{\mathbf{n}} = \mathbf{y} - \mathbf{E}\tilde{\mathbf{x}}, \tag{2.177}$$

$$\mathbf{C}_{xx} = \mathbf{S}\mathbf{E}^{T}(\mathbf{E}\mathbf{S}\mathbf{E}^{T} + \mathbf{I})^{-1}\mathbf{R}_{nn}(\mathbf{E}\mathbf{S}\mathbf{E}^{T} + \mathbf{I})^{-1}\mathbf{E}\mathbf{S}, \tag{2.178}$$

$$\tilde{\mu} = W^{-1}\tilde{\mathbf{n}}, \tag{2.179}$$

Equations (2.176)–(2.179) are identical to Eqs. (2.135)–(2.137) or to the alternate form Eqs. (2.127) − (2.129) derived from an objective function without Lagrange multipliers.

Equations (2.135)–(2.137) and (2.176)–(2.178) result from two very different appearing objective functions – one in which the equations are imposed in the mean-square, and one in which they are imposed exactly, using Lagrange multipliers. Constraints in the mean-square will be termed "soft," and those imposed exactly are "hard."[30] The distinction is, however, largely illusory: although (2.87) are being imposed exactly, it is only the presence of the error term, $\mathbf{n}$, which permits the equations to be written as equalities and thus as hard constraints. The hard and soft constraints here produce an identical solution. In some (rare) circumstances, which we will discuss briefly below, one may wish to impose exact constraints upon the elements of $\tilde{x}_i$. The solution in (2.167)–(2.169) was derived from the noise-free hard constraint, $\mathbf{A}\mathbf{x} = \mathbf{b}$, but we ended by rejecting it as generally inapplicable.

Once again, $\mathbf{n}$ is only by convention discussed separately from $\mathbf{x}$, and is fully a part of the solution. The combined form (2.138), which literally treats $\mathbf{x}$, $\mathbf{n}$ as the solution, are imposed through a hard constraint on the objective function,

$$ J = \xi^{\mathrm{T}}\xi - 2\mu^{\mathrm{T}}(\mathbf{E}_1\xi - \mathbf{y}), \qquad (2.180) $$

where $\xi = [(\mathbf{S}^{-\mathrm{T}/2}\mathbf{x})^{\mathrm{T}}, (\mathbf{W}^{-\mathrm{T}/2}\mathbf{n})^{\mathrm{T}}]^{\mathrm{T}}$, which is Eq. (2.175). (There are numerical advantages, however, in working with objects in two spaces of dimensions M and N, rather than a single space of dimension $M + N$.)

2.4.4 Interpretation of discrete adjoints

When the operators are matrices, as they are in discrete formulations, then the adjoint is just the transposed matrix. Sometimes the adjoint has a simple physical interpretation. Suppose, e.g., that scalar y was calculated from a sum,

$$ y = \mathbf{A}\mathbf{x}, \quad \mathbf{A} = \{1\ 1\ .\ 1\ 1\}. \qquad (2.181) $$

Then the adjoint operator applied to y is evidently

$$ \mathbf{r} = \mathbf{A}^{\mathrm{T}}y = \{1\ 1\ 1\ .\ 1\}^{\mathrm{T}}y = \mathbf{x}. \qquad (2.182) $$

Thus the adjoint operator "sprays" the average back out onto the originating vector, and might be thought of as an inverse operator.

A more interesting case is a first-difference forward operator,

$$ \mathbf{A} = \left\{ \begin{array}{ccccc} -1 & 1 & & & \\ & -1 & 1 & & \\ & & -1 & 1 & \\ & & & . & . & . \\ & & & & -1 & 1 \\ & & & & & -1 \end{array} \right\}, \qquad (2.183) $$

that is,

$$y_i = x_{i+1} - x_i, \tag{2.184}$$

(with the exception of the last element, $y_N = -x_N$).

Then its adjoint is

$$\mathbf{A}^{\mathrm{T}} = \left\{ \begin{array}{ccccccc} -1 & & & & & & \\ 1 & -1 & & & & & \\ & 1 & -1 & & & & \\ & & & \cdot & \cdot & & \\ & & & & 1 & -1 & \\ & & & & & 1 & -1 \end{array} \right\}, \tag{2.185}$$

which is a first-difference *backward* operator with $\mathbf{z} = \mathbf{A}^{\mathrm{T}}\mathbf{y}$, producing $z_i = y_{i-1} - y_i$, again with the exception of the end point, which is now z_1.

In general, the transpose matrix, or adjoint operator is *not* simply interpretable as an inverse operation as the summation/spray-out case might have suggested.[31] A more general understanding of the relationship between adjoints and inverses will be obtained in the next section.

2.5 The singular vector expansion

Least-squares is a very powerful, very useful method for finding solutions of linear simultaneous equations of any dimensionality and one might wonder why it is necessary to discuss any other form of solution. But even in the simplest form of least-squares, the solution is dependent upon the inverses of $\mathbf{E}^{\mathrm{T}}\mathbf{E}$, or $\mathbf{E}\mathbf{E}^{\mathrm{T}}$. In practice, their existence cannot be guaranteed, and we need to understand what that means, the extent to which solutions can be found when the inverses do not exist, and the effect of introducing weight matrices $\mathbf{W}, \mathbf{S}$. This problem is intimately related to the issue of controlling solution and residual norms. Second, the relationship between the equations and the solutions is somewhat impenetrable, in the sense that structures in the solutions are not easily relatable to particular elements of the data y_i. For many purposes, particularly physical insight, understanding the structure of the solution is essential. We will return to examine the least-squares solutions using some extra machinery.

2.5.1 Simple vector expansions

Consider again the elementary problem (2.1) of representing an L-dimensional vector $\mathbf{f}$ as a sum of a basis of L-orthonormal vectors $\mathbf{g}_i, i = 1, 2, \ldots, L, \mathbf{g}_i^{\mathrm{T}}\mathbf{g}_j = \delta_{ij}$.

Without error,

$$\mathbf{f} = \sum_{j=1}^{L} a_j \mathbf{g}_j, \quad a_j = \mathbf{g}_j^{\mathrm{T}} \mathbf{f}. \tag{2.186}$$

But if for some reason only the first K coefficients a_j are known, we can only approximate $\mathbf{f}$ by its first K terms:

$$\tilde{\mathbf{f}} = \sum_{j=1}^{K} b_j \mathbf{g}_j \tag{2.187}$$
$$= \mathbf{f} + \delta \mathbf{f}_1,$$

and there is an error, $\delta \mathbf{f}_1$. From the orthogonality of the $\mathbf{g}_i$, it follows that $\delta \mathbf{f}_1$ will have minimum l_2 norm only if it is orthogonal to the K vectors retained in the approximation, and then only if $b_j = a_j$ as given by (2.186). The only way the error could be reduced further is by increasing K.

Define an $L \times K$ matrix, $\mathbf{G}_K$, whose columns are the first K of the $\mathbf{g}_j$. Then $\mathbf{b} = \mathbf{a} = \mathbf{G}_K^{\mathrm{T}} \mathbf{f}$ is the vector of coefficients $a_j = \mathbf{g}_j^{\mathrm{T}} \mathbf{f}$, $j = 1, 2, \ldots, K$, and the finite representation (2.187) is (one should write it out)

$$\tilde{\mathbf{f}} = \mathbf{G}_K \mathbf{a} = \mathbf{G}_K \left(\mathbf{G}_K^{\mathrm{T}} \mathbf{f} \right) = \left(\mathbf{G}_K \mathbf{G}_K^{\mathrm{T}} \right) \mathbf{f}, \quad \mathbf{a} = \{a_i\}, \tag{2.188}$$

where the third equality follows from the associative properties of matrix multiplication. This expression shows that a *representation of a vector in an incomplete orthonormal set produces a resulting approximation that is a simple linear combination of the elements of the correct values* (i.e., a weighted average, or "filtered" version of them). Column i of $\mathbf{G}_K \mathbf{G}_K^{\mathrm{T}}$ produces the weighted linear combination of the true elements of $\mathbf{f}$ that will appear as $\tilde{f}_i$.

Because the columns of $\mathbf{G}_K$ are orthonormal, $\mathbf{G}_K^{\mathrm{T}} \mathbf{G}_K = \mathbf{I}_K$, that is, the $K \times K$ identity matrix; but $\mathbf{G}_K \mathbf{G}_K^{\mathrm{T}} \neq \mathbf{I}_L$ unless $K = L$. (That $\mathbf{G}_L \mathbf{G}_L^{\mathrm{T}} = \mathbf{I}_L$ for $K = L$ follows from the theorem for *square* matrices that shows a left inverse is also a right inverse.) If $K < L$, $\mathbf{G}_K$ is "semi-orthogonal." If $K = L$, it is "orthogonal"; in this case, $\mathbf{G}_L^{-1} = \mathbf{G}_L^{\mathrm{T}}$. If it is only semi-orthogonal, $\mathbf{G}_K^{\mathrm{T}}$ is a left inverse, but not a right inverse. Any orthogonal matrix has the property that its transpose is identical to its inverse.

The matrix $\mathbf{G}_K \mathbf{G}_K^{\mathrm{T}}$ is known as a "resolution matrix," with a simple interpretation. Suppose the true value of $\mathbf{f}$ were $\mathbf{f}_{j_0} = [0\,0\,0\,\ldots\,0\,1\,0\,.\,0\,.\,.\,0]^{\mathrm{T}}$, that is, a Kronecker delta $\delta_{j j_0}$, with unity in element j_0 and zero otherwise. Then the incomplete expansion (2.187) or (2.188) would not reproduce the delta function, but

$$\tilde{\mathbf{f}}_{j_0} = \mathbf{G}_K \mathbf{G}_K^{\mathrm{T}} \mathbf{f}_{j_0}, \tag{2.189}$$

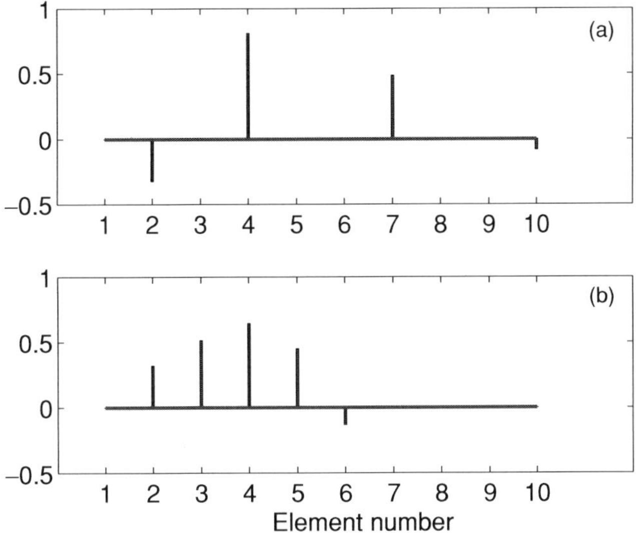

Figure 2.9 (a) Example of a row, j_0, of a 10×10 resolution matrix, perhaps the fourth one, showing widely distributed averaging in forming $\mathbf{f}_{j0}$. (b) The so-called compact resolution, in which the solution is readily interpreted as representing a local average of the true solution. Such situations are not common.

which is column j_0 of $\mathbf{G}_K \mathbf{G}_K^{\mathrm{T}}$. Each column (or row) of the resolution matrix tells one what the corresponding form of the approximating vector would be, if its true form were a Kronecker delta.

To form a Kronecker delta requires a spanning set of vectors. An analogous elementary result of Fourier analysis shows that a Dirac delta function demands contributions from all frequencies to represent a narrow, very high pulse. Removal of some of the requisite vectors (sinusoids) produces peak broadening and side-lobes. Here, depending upon the precise structure of the $\mathbf{g}_i$, the broadening and sidelobes can be complicated. If one is lucky, the effect could be a simple broadening (schematically shown in Fig. 2.9) without distant sidelobes), leading to the tidy interpretation of the result as a local average of the true values, called "compact resolution."[32]

A resolution matrix has the property

$$\text{trace}\left(\mathbf{G}_K \mathbf{G}_K^{\mathrm{T}}\right) = K, \tag{2.190}$$

which follows from noting that

$$\text{trace}\left(\mathbf{G}_K \mathbf{G}_K^{\mathrm{T}}\right) = \text{trace}\left(\mathbf{G}_K^{\mathrm{T}} \mathbf{G}_K\right) = \text{trace}(\mathbf{I}_K) = K.$$

2.5.2 *Square-symmetric problem: eigenvalues/eigenvectors*

Orthogonal vector expansions are particularly simple to use and interpret, but might seem irrelevant when dealing with simultaneous equations where neither the row nor column vectors of the coefficient matrix are so simply related. What we will show, however, is that we can always find sets of orthonormal vectors to greatly simplify the job of solving simultaneous equations. To do so, we digress to recall the basic elements of the "eigenvector/eigenvalue problem" mentioned in passing on p. 26.

Consider a *square, $M \times M$* matrix $\mathbf{E}$ and the simultaneous equations

$$\mathbf{E}\mathbf{g}_i = \lambda_i \mathbf{g}_i, \quad i = 1, 2, \ldots, M, \tag{2.191}$$

that is, the problem of finding a set of vectors $\mathbf{g}_i$ whose dot products with the rows of $\mathbf{E}$ are proportional to themselves. Such vectors are "eigenvectors," and the constants of proportionality are the "eigenvalues." Under special circumstances, the eigenvectors form an orthonormal spanning set: *Textbooks show that if $\mathbf{E}$ is square and symmetric, such a result is guaranteed.* It is easy to see that if two λ_j, λ_k are distinct, then the corresponding eigenvectors are orthogonal:

$$\mathbf{E}\mathbf{g}_j = \lambda_j \mathbf{g}_j, \tag{2.192}$$

$$\mathbf{E}\mathbf{g}_k = \lambda_k \mathbf{g}_k. \tag{2.193}$$

Left multiply the first of these by $\mathbf{g}_k^{\mathrm{T}}$, and the second by $\mathbf{g}_j^{\mathrm{T}}$, and subtract:

$$\mathbf{g}_k^{\mathrm{T}}\mathbf{E}\mathbf{g}_j - \mathbf{g}_j^{\mathrm{T}}\mathbf{E}\mathbf{g}_k = (\lambda_j - \lambda_k)\mathbf{g}_k^{\mathrm{T}}\mathbf{g}_j. \tag{2.194}$$

But because $\mathbf{E} = \mathbf{E}^{\mathrm{T}}$, the left-hand side vanishes, and hence $\mathbf{g}_k^{\mathrm{T}}\mathbf{g}_j$ by the assumption $\lambda_j \neq \lambda_k$. A similar construction proves that the λ_i are all real, and an elaboration shows that for coincident λ_i, the corresponding eigenvectors can always be made orthogonal.

Example *To contrast with the above result, consider the non-symmetric, square matrix*

$$\begin{Bmatrix} 1 & 2 & 3 \\ 0 & 1 & 4 \\ 0 & 0 & 1 \end{Bmatrix}.$$

Solution to the eigenvector/eigenvalue problem produces $\lambda_i = 1$, and $\mathbf{u}_i = [1, 0, 0]^{\mathrm{T}}$, $i = 1, 2, 3$. The eigenvectors are not orthogonal, and are certainly not a spanning set. On the other hand, the eigenvector/eigenvalues of

$$\begin{Bmatrix} 1 & -1 & -2 \\ -1 & 2 & -1 \\ 1.5 & 2 & -2.5 \end{Bmatrix}$$

are

$$\mathbf{g}_1 = \begin{bmatrix} -0.29 + 0.47i \\ -0.17 + 0.25i \\ 0.19 + 0.61i \end{bmatrix}, \mathbf{g}_2 = \begin{bmatrix} -0.29 - 0.47i \\ -0.17 - 0.25i \\ 0.19 - 0.61i \end{bmatrix}, \mathbf{g}_3 = \begin{bmatrix} -0.72 \\ 0.90 \\ 0.14 \end{bmatrix},$$

$$\lambda_j = [-1.07 + 1.74i, -1.07 - 1.74i, 2.64]$$

and (rounded) are not orthogonal, but are a basis. The eigenvalues/eigenvectors appear in complex conjugate pairs and in some contexts are called "principal oscillation patterns" (POPs).

Suppose for the moment that we have the square, symmetric, special case, and recall how eigenvectors can be used to solve (2.16). By convention, the pairs $(\lambda_i, \mathbf{g}_i)$ are ordered in the sense of decreasing λ_i. If some λ_i are repeated, an arbitrary order choice is made.

With an orthonormal, spanning set, both the known $\mathbf{y}$ and the unknown $\mathbf{x}$ can be written as

$$\mathbf{x} = \sum_{i=1}^{M} \alpha_i \mathbf{g}_i, \quad \alpha_i = \mathbf{g}_i^{\mathrm{T}} \mathbf{x}, \tag{2.195}$$

$$\mathbf{y} = \sum_{i=1}^{M} \beta_i \mathbf{g}_i, \quad \beta_i = \mathbf{g}_i^{\mathrm{T}} \mathbf{y}. \tag{2.196}$$

By convention, $\mathbf{y}$ is known, and therefore β_i can be regarded as given. If the α_i could be found, $\mathbf{x}$ would be known.

Substitute (2.195) into (2.16), to give

$$\mathbf{E} \sum_{i=1}^{M} \alpha_i \mathbf{g}_i = \sum_{i=1}^{M} \left(\mathbf{g}_i^{\mathrm{T}} \mathbf{y} \right) \mathbf{g}_i, \tag{2.197}$$

or, using the eigenvector property,

$$\sum_{i=1}^{M} \alpha_i \lambda_i \mathbf{g}_i = \sum_{i} \left(\mathbf{g}_i^{\mathrm{T}} \mathbf{y} \right) \mathbf{g}_i. \tag{2.198}$$

But the expansion vectors are orthonormal and so

$$\lambda_i \alpha_i = \mathbf{g}_i^{\mathrm{T}} \mathbf{y}, \tag{2.199}$$

$$\alpha_i = \frac{\mathbf{g}_i^{\mathrm{T}} \mathbf{y}}{\lambda_i}, \tag{2.200}$$

$$\mathbf{x} = \sum_{i=1}^{M} \frac{\mathbf{g}_i^{\mathrm{T}} \mathbf{y}}{\lambda_i} \mathbf{g}_i. \tag{2.201}$$

Apart from the obvious difficulty if an eigenvalue vanishes, the problem is now completely solved. Define a diagonal matrix, $\mathbf{\Lambda}$, with elements, λ_i, in descending numerical value, and the matrix $\mathbf{G}$, whose columns are the corresponding $\mathbf{g}_i$ in the same order, the solution to (2.16) can be written, from (2.195), (2.199)–(2.201) as

$$\boldsymbol{\alpha} = \mathbf{\Lambda}^{-1}\mathbf{G}^{\mathrm{T}}\mathbf{y}, \tag{2.202}$$

$$\mathbf{x} = \mathbf{G}\mathbf{\Lambda}^{-1}\mathbf{G}^{\mathrm{T}}\mathbf{y}, \tag{2.203}$$

where $\mathbf{\Lambda}^{-1} = \mathrm{diag}(1/\lambda_i)$.

Vanishing eigenvalues, $i = i_0$, cause trouble and must be considered. Let the corresponding eigenvectors be $\mathbf{g}_{i_0}$. Then any part of the solution which is proportional to such an eigenvector is "annihilated" by $\mathbf{E}$, that is, $\mathbf{g}_{i_0}$ is orthogonal to all the rows of $\mathbf{E}$. Such a result means that there is no possibility that anything in $\mathbf{y}$ could provide any information about the coefficient α_{i_0}. If $\mathbf{y}$ corresponds to a set of observations (data), then $\mathbf{E}$ represents the connection ("mapping") between system unknowns and observations. The existence of zero eigenvalues shows that the act of observation of $\mathbf{x}$ removes certain structures in the solution which are then indeterminate. Vectors $\mathbf{g}_{i_0}$ (and there may be many of them) are said to lie in the "nullspace" of $\mathbf{E}$. Eigenvectors corresponding to non-zero eigenvalues lie in its "range." The simplest example is given by the "observations"

$$x_1 + x_2 = 3,$$
$$x_1 + x_2 = 3.$$

Any structure in $\mathbf{x}$ such that $x_1 = -x_2$ is destroyed by this observation, and, by inspection, the nullspace vector must be $\mathbf{g}_2 = [1, -1]^{\mathrm{T}}/\sqrt{2}$. (The purpose of showing the observation twice is to produce an $\mathbf{E}$ that is square.)

Suppose there are $K < M$ non-zero λ_i. Then for $i > K$, Eq. (2.199) is

$$0\alpha_i = \mathbf{g}_i^{\mathrm{T}}\mathbf{y}, \quad K + i = 1, 2, \dots, M, \tag{2.204}$$

and two cases must be distinguished.

Case (1)

$$\mathbf{g}_i^{\mathrm{T}}\mathbf{y} = 0, \quad K + i = 1, 2, \dots, M. \tag{2.205}$$

We could then put $\alpha_i = 0$, $K + i = 1, 2, \dots, M$, and the solution can be written

$$\tilde{\mathbf{x}} = \sum_{i=1}^{K} \frac{\mathbf{g}_i^{\mathrm{T}}\mathbf{y}}{\lambda_i} \mathbf{g}_i, \tag{2.206}$$

and $\mathbf{E}\tilde{\mathbf{x}} = \mathbf{y}$, *exactly*. Equation (2.205) is often known as a "solvability condition." A tilde has been placed over $\mathbf{x}$ because a solution of the form

$$\tilde{\mathbf{x}} = \sum_{i=1}^{K} \frac{\mathbf{g}_i^{\mathrm{T}}\mathbf{y}}{\lambda_i}\mathbf{g}_i + \sum_{i=K+1}^{M} \alpha_i\mathbf{g}_i, \tag{2.207}$$

with the remaining α_i taking on arbitrary values, also satisfies the equations exactly. That is, the true value of $\mathbf{x}$ *could* contain structures proportional to the nullspace vectors of $\mathbf{E}$, but the equations (2.16) neither require their presence, nor provide information necessary to determine their amplitudes. We thus have a situation with a "solution nullspace." Define the matrix $\mathbf{G}_K$ to be $M \times K$, carrying only the first K of the $\mathbf{g}_i$, that is, the range vectors, $\mathbf{\Lambda}_K$, to be $K \times K$ with only the first K, non-zero eigenvalues, and the columns of $\mathbf{Q}_G$ are the $M - K$ nullspace vectors (it is $M \times (M - K)$), then the solutions (2.206) and (2.207) are

$$\tilde{\mathbf{x}} = \mathbf{G}_K \mathbf{\Lambda}_K^{-1} \mathbf{G}_K^{\mathrm{T}} \mathbf{y}, \tag{2.208}$$

$$\tilde{\mathbf{x}} = \mathbf{G}_K \mathbf{\Lambda}_K^{-1} \mathbf{G}_K^{\mathrm{T}} \mathbf{y} + \mathbf{Q}_G \boldsymbol{\alpha}_G, \tag{2.209}$$

where $\boldsymbol{\alpha}_G$ is the vector of unknown nullspace coefficients. The solution in (2.208), with no nullspace contribution, will be called the "particular" solution. If $\mathbf{y} = \mathbf{0}$, however, Eq. (2.16) is a homogeneous set of equations, then the nullspace represents the only possible solution.

If $\mathbf{G}$ is written as a partitioned matrix,

$$\mathbf{G} = \{\mathbf{G}_K \quad \mathbf{Q}_G\},$$

it follows from the column orthonormality that

$$\mathbf{G}\mathbf{G}^{\mathrm{T}} = \mathbf{I} = \mathbf{G}_K \mathbf{G}_K^{\mathrm{T}} + \mathbf{Q}_G \mathbf{Q}_G^{\mathrm{T}}, \tag{2.210}$$

or

$$\mathbf{Q}_G \mathbf{Q}_G^{\mathrm{T}} = \mathbf{I} - \mathbf{G}_K \mathbf{G}_K^{\mathrm{T}}. \tag{2.211}$$

Vectors $\mathbf{Q}_G$ span the nullspace of $\mathbf{G}$.

Case (2)

$$\mathbf{g}_i^{\mathrm{T}}\mathbf{y} \neq 0, \quad i > K + 1, \tag{2.212}$$

for one or more of the nullspace vectors. In this case, Eq. (2.199) is the contradiction

$$0\alpha_i \neq 0,$$

and Eq. (2.198) is actually

$$\sum_{i=1}^{K} \lambda_i \alpha_i \mathbf{g}_i = \sum_{i=1}^{M} \left(\mathbf{g}_i^{\mathrm{T}} \mathbf{y}\right) \mathbf{g}_i, \quad K < M, \tag{2.213}$$

that is, with differing upper limits on the sums. Therefore, the solvability condition is not satisfied. Owing to the orthonormality of the $\mathbf{g}_i$, there is no choice of α_i, $i = 1, \ldots, K$ on the left that can match the last $M - K$ terms on the right. Evidently there is no solution in the conventional sense unless (2.205) is satisfied, hence the name "solvability condition." What is the best we might do? Define "best" to mean that the solution $\tilde{\mathbf{x}}$ should be chosen such that

$$\mathbf{E}\tilde{\mathbf{x}} = \tilde{\mathbf{y}},$$

where the difference, $\tilde{\mathbf{n}} = \mathbf{y} - \tilde{\mathbf{y}}$, which we call the "residual," should be as small as possible (in the l_2 norm). If this choice is made, then the orthogonality of the $\mathbf{g}_i$ shows immediately that the best choice is still (2.200), $i = 1, 2, \ldots, K$. No choice of nullspace vector coefficients, nor any other value of the coefficients of the range vectors, can reduce the norm of $\tilde{\mathbf{n}}$. The best solution is then also (2.206) or (2.208).

In this situation, we are no longer solving the equations (2.16), but rather are dealing with a set that could be written

$$\mathbf{E}\mathbf{x} \sim \mathbf{y}, \tag{2.214}$$

where the demand is for a solution that is the "best possible," in the sense just defined. Such statements of approximation are awkward, so as before rewrite (2.214) as

$$\mathbf{E}\mathbf{x} + \mathbf{n} = \mathbf{y}, \tag{2.215}$$

where $\mathbf{n}$ is the residual. If $\tilde{\mathbf{x}}$ is given by (2.207), then by (2.213),

$$\tilde{\mathbf{n}} = \sum_{i=K+1}^{M} \left(\mathbf{g}_i^{\mathrm{T}} \mathbf{y}\right) \mathbf{g}_i. \tag{2.216}$$

Notice that $\tilde{\mathbf{n}}^{\mathrm{T}} \tilde{\mathbf{y}} = \mathbf{0} : \tilde{\mathbf{y}}$ is orthogonal to the residuals.

Example *Let*

$$x_1 + x_2 = 1,$$
$$x_1 + x_2 = 3.$$

Then using $\lambda_1 = 2$, $\mathbf{g}_1 = [1, 1]^{\mathrm{T}}/\sqrt{2}$, $\lambda_2 = 0$, $\mathbf{g}_2 = [1, -1]^{\mathrm{T}}/\sqrt{2}$, *one has* $\tilde{\mathbf{x}} = [1/2, 1/2]^{\mathrm{T}} \propto \mathbf{g}_1$, $\tilde{\mathbf{y}} = [2, 2]^{\mathrm{T}} \propto \mathbf{g}_1$, $\tilde{\mathbf{n}} = [-1, 1]^{\mathrm{T}} \propto \mathbf{g}_2$.

This outcome, where M equations in M unknowns were found in practice not to be able to determine some solution structures, is labeled "formally just-determined." The expression "formally" alludes to the fact that the appearance of a just-determined system did not mean that the characterization was true in practice. One or more vanishing eigenvalues mean that neither the rows nor columns of $\mathbf{E}$ are spanning sets.

Some decision has to be made about the coefficients of the nullspace vectors in (2.209). The form could be used as it stands, regarding it as the "general solution." The analogy with the solution of differential equations should be apparent – typically, there is a particular solution and a homogeneous solution – here the nullspace vectors. When solving a differential equation, determination of the magnitude of the homogeneous solution requires additional information, often provided by boundary or initial conditions; here additional information is also necessary, but missing.

Despite the presence of indeterminate elements in the solution, a great deal is known about them: They are proportional to the nullspace vectors. Depending upon the specific situation, we might conceivably be in a position to obtain more observations, and would seriously consider observational strategies directed at observing these missing structures. The reader is also reminded of the discussion of the Neumann problem in Chapter 1.

Another approach is to define a "simplest" solution, appealing to what is usually known as "Ockham's razor," or the "principle of parsimony," that in choosing between multiple explanations of a given phenomenon, the simplest one is usually the best. What is "simplest" can be debated, but here there is a compelling choice: The solution (2.208), which is without any nullspace contributions, is less structured than any other solution. (It is often, but not always, true that the nullspace vectors are more "wiggly" than those in the range. The nullspace of the Neumann problem is a counter example. In any case, including any vector not required by the data is arguably producing more structure than is required.) Setting all the unknown α_i to zero is thus one plausible choice. It follows from the orthogonality of the $\mathbf{g}_i$ that this particular solution is also the one of minimum solution norm. Later, other choices for the nullspace vectors will be made. If $\mathbf{y} = \mathbf{0}$, then the nullspace *is* the solution.

If the nullspace vector contributions are set to zero, the true solution has been expanded in an incomplete set of orthonormal vectors. Thus, $\mathbf{G}_K \mathbf{G}_K^{\mathrm{T}}$ is the resolution matrix, and the relationship between the true solution and the minimal one is just

$$\tilde{\mathbf{x}} = \mathbf{G}_K \mathbf{G}_K^{\mathrm{T}} \mathbf{x} = \mathbf{x} - \mathbf{Q}_G \boldsymbol{\alpha}_G, \quad \tilde{\mathbf{y}} = \mathbf{G}_K \mathbf{G}_K^{\mathrm{T}} \mathbf{y}, \quad \tilde{\mathbf{n}} = \mathbf{Q}_G \mathbf{Q}_G^{\mathrm{T}} \mathbf{y}. \qquad (2.217)$$

These results are very important, so we recapitulate them: (2.207) or (2.209) is the general solution. There are three vectors involved, one of them, $\mathbf{y}$, known, and two of them, $\mathbf{x}$, $\mathbf{n}$, unknown. Because of the assumption that $\mathbf{E}$ has an orthonormal basis of eigenvectors, all three of these vectors can be expanded exactly as

$$\mathbf{x} = \sum_{i=1}^{M} \alpha_i \mathbf{g}_i, \quad \mathbf{n} = \sum_{i=1}^{M} \gamma_i \mathbf{g}_i, \quad \mathbf{y} = \sum_{i=1}^{M} (\mathbf{y}^{\mathrm{T}} \mathbf{g}_i) \mathbf{g}_i. \tag{2.218}$$

Substituting into (2.215),

$$\sum_{i=1}^{K} \lambda_i \alpha_i \mathbf{g}_i + \sum_{i=1}^{M} \gamma_i \mathbf{g}_i = \sum_{i=1}^{M} (\mathbf{y}^{\mathrm{T}} \mathbf{g}_i) \mathbf{g}_i.$$

From the orthogonality property,

$$\lambda_i \alpha_i + \gamma_i = \mathbf{y}^{\mathrm{T}} \mathbf{g}_i, \quad i = 1, 2, \ldots, K, \tag{2.219}$$

$$\gamma_i = \mathbf{y}^{\mathrm{T}} \mathbf{g}_i, \quad K + i = 1, 2, \ldots, M. \tag{2.220}$$

In dealing with the first relationship, a choice is required. If we set

$$\gamma_i = \mathbf{g}_i^{\mathrm{T}} \mathbf{n} = 0, \quad i = 1, 2, \ldots, K, \tag{2.221}$$

the residual norm is made as small as possible, by completely eliminating the range vectors from the residual. This choice is motivated by the attempt to satisfy the equations as well as possible, but is seen to have elements of arbitrariness. A decision about other possibilities depends upon knowing more about the system and will be the focus of attention later.

The relative contributions of any structure in $\mathbf{y}$, determined by the projection, $\mathbf{g}_i^{\mathrm{T}} \mathbf{y}$, will depend upon the ratio $\mathbf{g}_i^{\mathrm{T}} \mathbf{y}/\lambda_i$. Comparatively weak values of $\mathbf{g}_i^{\mathrm{T}} \mathbf{y}$ may well be amplified by small, but non-zero, elements of λ_i. One must keep track of both $\mathbf{g}_i^{\mathrm{T}} \mathbf{y}$, and $\mathbf{g}_i^{\mathrm{T}} \mathbf{y}/\lambda_i$.

Before leaving this special case, note one more useful property of the eigenvector/ eigenvalues. For the moment, let $\mathbf{G}$ have all its columns, containing both the range and nullspace vectors, with the nullspace vectors being last in arbitrary order. It is thus an $M \times M$ matrix. Correspondingly, let Λ contain all the eigenvalues on its diagonal, including the zero ones; it too, is $M \times M$. Then the eigenvector definition (2.191) produces

$$\mathbf{EG} = \mathbf{G}\Lambda. \tag{2.222}$$

Multiply both sides of (2.222) by $\mathbf{G}^{\mathrm{T}}$:

$$\mathbf{G}^{\mathrm{T}} \mathbf{EG} = \mathbf{G}^{\mathrm{T}} \mathbf{G}\Lambda = \Lambda. \tag{2.223}$$

$\mathbf{G}$ is said to "diagonalize" $\mathbf{E}$. Now multiply both sides of (2.223) on the left by $\mathbf{G}$ and on the right by $\mathbf{G}^{\mathrm{T}}$:

$$\mathbf{GG}^{\mathrm{T}} \mathbf{EGG}^{\mathrm{T}} = \mathbf{G}\Lambda\mathbf{G}^{\mathrm{T}}. \tag{2.224}$$

Using the orthogonality of $\mathbf{G}$,

$$\mathbf{E} = \mathbf{G}\boldsymbol{\Lambda}\mathbf{G}^{\mathrm{T}}. \qquad (2.225)$$

This is a useful representation of $\mathbf{E}$, consistent with its symmetry.

Recall that $\boldsymbol{\Lambda}$ has zeros on the diagonal corresponding to the zero eigenvalues, and the corresponding rows and columns are entirely zero. Writing out (2.225), these zero rows and columns multiply all the nullspace vector columns of $\mathbf{G}$ by zero, and it is found that the nullspace columns of $\mathbf{G}$ can be eliminated, $\boldsymbol{\Lambda}$ can be reduced to its $K \times K$ form, and the decomposition (2.225) is still exact – in the form

$$\mathbf{E} = \mathbf{G}_K \boldsymbol{\Lambda}_K \mathbf{G}_K^{\mathrm{T}}. \qquad (2.226)$$

The representation (decomposition) in either Eq. (2.225) or (2.226) is identical to

$$\mathbf{E} = \lambda_1 \mathbf{g}_1 \mathbf{g}_1^{\mathrm{T}} + \lambda_2 \mathbf{g}_2 \mathbf{g}_2^{\mathrm{T}} + \cdots + \lambda_K \mathbf{g}_K \mathbf{g}_K^{\mathrm{T}}. \qquad (2.227)$$

That is, a square symmetric matrix can be exactly represented by a sum of products of orthonormal vectors $\mathbf{g}_i \mathbf{g}_i^{\mathrm{T}}$ multiplied by a scalar, λ_i.

Example *Consider the matrix from the last example,*

$$\mathbf{E} = \left\{ \begin{matrix} 1 & 1 \\ 1 & 1 \end{matrix} \right\}.$$

We have

$$\mathbf{E} = \frac{2}{\sqrt{2}} \begin{bmatrix} 1 \\ 1 \end{bmatrix} \begin{bmatrix} 1 & 1 \end{bmatrix} \frac{1}{\sqrt{2}} + \frac{0}{\sqrt{2}} \begin{bmatrix} 1 \\ -1 \end{bmatrix} \begin{bmatrix} 1 & -1 \end{bmatrix} \frac{1}{\sqrt{2}}.$$

The simultaneous equations (2.215) are

$$\mathbf{G}_K \boldsymbol{\Lambda}_K \mathbf{G}_K^{\mathrm{T}} \mathbf{x} + \mathbf{n} = \mathbf{y}. \qquad (2.228)$$

Left multiply both sides by $\boldsymbol{\Lambda}_K^{-1} \mathbf{G}_K^{\mathrm{T}}$ to get

$$\mathbf{G}_K^{\mathrm{T}} \mathbf{x} + \boldsymbol{\Lambda}_K^{-1} \mathbf{G}_K^{\mathrm{T}} \mathbf{n} = \boldsymbol{\Lambda}_K^{-1} \mathbf{G}_K^{\mathrm{T}} \mathbf{y}. \qquad (2.229)$$

But $\mathbf{G}_K^{\mathrm{T}} \mathbf{x}$ are the projection of $\mathbf{x}$ onto the range vectors of $\mathbf{E}$, and $\mathbf{G}_K^{\mathrm{T}} \mathbf{n}$ is the projection of the noise. We have agreed to set the latter to zero, and obtain

$$\mathbf{G}_K^{\mathrm{T}} \mathbf{x} = \boldsymbol{\Lambda}_K^{-1} \mathbf{G}_K^{\mathrm{T}} \mathbf{y},$$

the dot products of the range of $\mathbf{E}$ with the solution. Hence, it must be true, since the range vectors are orthonormal, that

$$\tilde{\mathbf{x}} \equiv \mathbf{G}_K \mathbf{G}_K^{\mathrm{T}} \mathbf{x} \equiv \mathbf{G}_K \boldsymbol{\Lambda}_K^{-1} \mathbf{G}_K^{\mathrm{T}} \mathbf{y}, \qquad (2.230)$$

$$\tilde{\mathbf{y}} = \mathbf{E}\tilde{\mathbf{x}} = \mathbf{G}_K \mathbf{G}_K^{\mathrm{T}} \mathbf{y}, \qquad (2.231)$$

which is identical to the particular solution (2.206). The residuals are

$$\tilde{\mathbf{n}} = \mathbf{y} - \tilde{\mathbf{y}} = \mathbf{y} - \mathbf{E}\tilde{\mathbf{x}} = (\mathbf{I}_M - \mathbf{G}_K\mathbf{G}_K^T)\mathbf{y} = \mathbf{Q}_G\mathbf{Q}_G^T\mathbf{y}, \tag{2.232}$$

with $\tilde{\mathbf{n}}^T\tilde{\mathbf{y}} = 0$. Notice that matrix $\mathbf{H}$ of Eq. (2.97) is just $\mathbf{G}_K\mathbf{G}_K^T$, and hence $(\mathbf{I} - \mathbf{H})$ is the projector of $\mathbf{y}$ onto the nullspace vectors.

The expected value of the solution (2.206) or (2.230) is

$$\langle \tilde{\mathbf{x}} - \mathbf{x} \rangle = \mathbf{G}_K\mathbf{\Lambda}_K^{-1}\mathbf{G}_K^T\langle\mathbf{y}\rangle - \sum_{i=1}^{N}\alpha_i\mathbf{g}_i = -\mathbf{Q}_G\alpha_G, \tag{2.233}$$

and so the solution is biassed unless $\alpha_G = 0$.

The uncertainty is given by

$$\begin{aligned}
\mathbf{P} = D^2(\tilde{\mathbf{x}} - \mathbf{x}) &= \langle \mathbf{G}_K\mathbf{\Lambda}_K^{-1}\mathbf{G}_K^T(\mathbf{y}_0 + \mathbf{n} - \mathbf{y}_0)(\mathbf{y}_0 + \mathbf{n} - \mathbf{y}_0)^T\mathbf{G}_K\mathbf{\Lambda}_K^{-1}\mathbf{G}_K^T \rangle \\
&\quad + \langle \mathbf{Q}_G\alpha_G\alpha_G^T\mathbf{Q}_G^T \rangle \\
&= \mathbf{G}_K\mathbf{\Lambda}_K^{-1}\mathbf{G}_K^T\langle\mathbf{n}\mathbf{n}^T\rangle\mathbf{G}_K\mathbf{\Lambda}_K^{-1}\mathbf{G}_K^T + \mathbf{Q}_G\langle\alpha_G\alpha_G^T\rangle\mathbf{Q}_G^T \tag{2.234} \\
&= \mathbf{G}_K\mathbf{\Lambda}_K^{-1}\mathbf{G}_K^T\mathbf{R}_{nn}\mathbf{G}_K\mathbf{\Lambda}_K^{-1}\mathbf{G}_K^T + \mathbf{Q}_G\mathbf{R}_{\alpha\alpha}\mathbf{Q}_G^T \\
&= \mathbf{C}_{xx} + \mathbf{Q}_G\mathbf{R}_{\alpha\alpha}\mathbf{Q}_G^T,
\end{aligned}$$

defining the second moments, $\mathbf{R}_{\alpha\alpha}$, of the coefficients of the nullspace vectors. Under the special circumstances that the residuals, $\mathbf{n}$, are white noise, with $\mathbf{R}_{nn} = \sigma_n^2\mathbf{I}$, (2.234) reduces to

$$\mathbf{P} = \sigma_n^2\mathbf{G}_K\mathbf{\Lambda}_K^{-2}\mathbf{G}_K^T + \mathbf{Q}_G\mathbf{R}_{\alpha\alpha}\mathbf{Q}_G^T. \tag{2.235}$$

Either case shows that the uncertainty of the minimal solution is made up of two distinct parts. The first part, the solution covariance, $\mathbf{C}_{xx}$, arises owing to the noise present in the observations, and generates uncertainty in the coefficients of the range vectors; the second contribution arises from the missing nullspace vector contribution. Either term can dominate. The magnitude of the noise term depends largely upon the ratio of the noise variance, σ_n^2, to the smallest non-zero eigenvalue, λ_K^2.

Example *Suppose that*

$$\mathbf{E}\mathbf{x} = \mathbf{y},$$

$$\begin{Bmatrix} 1 & 1 \\ 1 & 1 \end{Bmatrix}\begin{bmatrix} x_1 \\ x_2 \end{bmatrix} = \mathbf{y} = \begin{bmatrix} 1 \\ 3 \end{bmatrix}, \tag{2.236}$$

which is inconsistent and has no solution in the conventional sense. Solving,

$$\mathbf{E}\mathbf{g}_i = \lambda_i\mathbf{g}_i, \tag{2.237}$$

or

$$\left\{ \begin{bmatrix} 1-\lambda & 1 \\ 1 & 1-\lambda \end{bmatrix} \right\} \begin{bmatrix} g_{i1} \\ g_{i2} \end{bmatrix} = \begin{bmatrix} 0 \\ 0 \end{bmatrix}. \tag{2.238}$$

This equation requires that

$$g_{i1} \begin{bmatrix} 1-\lambda \\ 1 \end{bmatrix} + g_{i2} \begin{bmatrix} 1 \\ 1-\lambda \end{bmatrix} = \begin{bmatrix} 0 \\ 0 \end{bmatrix},$$

or

$$\begin{bmatrix} 1-\lambda \\ 1 \end{bmatrix} + \frac{g_{i2}}{g_{i1}} \begin{bmatrix} 1 \\ 1-\lambda \end{bmatrix} = \begin{bmatrix} 0 \\ 0 \end{bmatrix},$$

which is

$$\frac{g_{i2}}{g_{i1}} = -(1-\lambda)$$

$$\frac{g_{i2}}{g_{i1}} = -\frac{1}{1-\lambda}.$$

Both equations are satisfied only if $\lambda = 2, 0$. This method, which can be generalized, in effect derives the usual statement that for Eq. (2.238) to have a solution, the determinant,

$$\begin{vmatrix} 1-\lambda & 1 \\ 1 & 1-\lambda \end{vmatrix},$$

must vanish. The first solution is labeled $\lambda_1 = 2$, and substituting back in produces $\mathbf{g}_1 = \frac{1}{\sqrt{2}} [1, 1]^{\mathrm{T}}$, when given unit length. Also $\mathbf{g}_2 = \frac{1}{\sqrt{2}}[-1, 1]^{\mathrm{T}}$, $\lambda_2 = 0$. Hence,

$$\mathbf{E} = \frac{1}{\sqrt{2}} \begin{bmatrix} 1 \\ 1 \end{bmatrix} 2 \frac{1}{\sqrt{2}} \begin{bmatrix} 1 \\ 1 \end{bmatrix}^{\mathrm{T}} = \begin{bmatrix} 1 \\ 1 \end{bmatrix} \begin{bmatrix} 1 & 1 \end{bmatrix}. \tag{2.239}$$

The equations have no solution in the conventional sense. There is, however, a sensible "best" solution:

$$\tilde{\mathbf{x}} = \frac{\mathbf{g}_1^{\mathrm{T}} \mathbf{y}}{\lambda_1} \mathbf{g}_1 + \alpha_2 \mathbf{g}_2, \tag{2.240}$$

$$= \left(\frac{4}{2\sqrt{2}} \right) \frac{1}{\sqrt{2}} \begin{bmatrix} 1 \\ 1 \end{bmatrix} + \alpha_2 \frac{1}{\sqrt{2}} \begin{bmatrix} -1 \\ 1 \end{bmatrix} \tag{2.241}$$

$$= \begin{bmatrix} 1 \\ 1 \end{bmatrix} + \alpha_2 \frac{1}{\sqrt{2}} \begin{bmatrix} -1 \\ 1 \end{bmatrix}. \tag{2.242}$$

Notice that

$$\mathbf{E}\tilde{\mathbf{x}} = \begin{bmatrix} 2 \\ 2 \end{bmatrix} + 0 \neq \begin{bmatrix} 1 \\ 3 \end{bmatrix}. \tag{2.243}$$

The solution has compromised the inconsistency. No choice of α_2 can reduce the residual norm. The equations would more sensibly have been written

$$\mathbf{E}\mathbf{x} + \mathbf{n} = \mathbf{y},$$

and the difference, $\mathbf{n} = \mathbf{y} - \mathbf{E}\tilde{\mathbf{x}}$ is proportional to $\mathbf{g}_2$. A system like (2.236) would most likely arise from measurements (if both equations are divided by 2, they represent two measurements of the average of (x_1, x_2), and $\mathbf{n}$ would be best regarded as the noise of observation).

Example *Suppose the same problem as in the last example is solved using Lagrange multipliers, that is, minimizing*

$$J = \mathbf{n}^{\mathrm{T}}\mathbf{n} + \gamma^2\mathbf{x}^{\mathrm{T}}\mathbf{x} - 2\mu^{\mathrm{T}}\left(\mathbf{y} - \mathbf{E}\mathbf{x} - \mathbf{n}\right).$$

Then, the normal equations are

$$\frac{1}{2}\frac{\partial J}{\partial \mathbf{x}} = \gamma^2\mathbf{x} + \mathbf{E}^{\mathrm{T}}\mu = 0,$$

$$\frac{1}{2}\frac{\partial J}{\partial \mathbf{n}} = \mathbf{n} + \mu = 0,$$

$$\frac{1}{2}\frac{\partial J}{\partial \mu} = \mathbf{y} - \mathbf{E}\mathbf{x} - \mathbf{n} = 0,$$

which produces

$$\tilde{\mathbf{x}} = \mathbf{E}^{\mathrm{T}}(\mathbf{E}\mathbf{E}^{\mathrm{T}} + \gamma^2\mathbf{I})^{-1}\mathbf{y}$$

$$= \begin{Bmatrix} 1 & 1 \\ 1 & 1 \end{Bmatrix}\left\{\begin{Bmatrix} 2 & 2 \\ 2 & 2 \end{Bmatrix} + \gamma^2\begin{Bmatrix} 1 & 0 \\ 0 & 1 \end{Bmatrix}\right\}^{-1}\begin{bmatrix} 1 \\ 3 \end{bmatrix}.$$

The limit $\gamma^2 \to \infty$ is readily evaluated. Letting $\gamma^2 \to 0$ involves inverting a singular matrix. To understand what is going on, use

$$\mathbf{E} = \mathbf{G}\mathbf{\Lambda}\mathbf{G}^{\mathrm{T}} = \mathbf{g}_1\lambda_1\mathbf{g}_1^{\mathrm{T}} + 0 = \frac{1}{\sqrt{2}}\begin{bmatrix} 1 \\ 1 \end{bmatrix}2\frac{1}{\sqrt{2}}\begin{bmatrix} 1 \\ 1 \end{bmatrix}^{\mathrm{T}} + 0 \qquad (2.244)$$

Hence,

$$\mathbf{E}\mathbf{E}^{\mathrm{T}} = \mathbf{G}\mathbf{\Lambda}^2\mathbf{G}^{\mathrm{T}}.$$

Note that the full $\mathbf{G}$, $\mathbf{\Lambda}$ are being used, and $\mathbf{I} = \mathbf{G}\mathbf{G}^{\mathrm{T}}$. Thus,

$$(\mathbf{E}\mathbf{E}^{\mathrm{T}} + \gamma^2\mathbf{I}) = (\mathbf{G}\mathbf{\Lambda}^2\mathbf{G}^{\mathrm{T}} + \mathbf{G}(\gamma^2)\mathbf{G}^{\mathrm{T}}) = \mathbf{G}(\mathbf{\Lambda}^2 + \gamma^2\mathbf{I})\mathbf{G}^{\mathrm{T}}.$$

By inspection, the inverse of this last matrix is

$$(\mathbf{E}\mathbf{E}^{\mathrm{T}} + \mathbf{I}/\gamma^2)^{-1} = \mathbf{G}(\mathbf{\Lambda}^2 + \gamma^2\mathbf{I})^{-1}\mathbf{G}^{\mathrm{T}}.$$

But $(\Lambda^2 + \gamma^2 \mathbf{I})^{-1}$ is the inverse of a diagonal matrix,

$$(\Lambda^2 + \gamma^2 \mathbf{I})^{-1} = \operatorname{diag}\left\{1/\left(\lambda_i^2 + \gamma^2\right)\right\}.$$

Then

$$\tilde{\mathbf{x}} = \mathbf{E}^{\mathrm{T}}(\mathbf{E}\mathbf{E}^{\mathrm{T}} + \gamma^2 \mathbf{I})^{-1}\mathbf{y} = \mathbf{G}\Lambda\mathbf{G}^{\mathrm{T}}\left(\mathbf{G}\operatorname{diag}\left\{1/\left(\lambda_i^2 + \gamma^2\right)\right\}\mathbf{G}^{\mathrm{T}}\right)\mathbf{y}$$
$$= \mathbf{G}\operatorname{diag}\left\{\lambda_i/\left(\lambda_i^2 + \gamma^2\right)\right\}\mathbf{G}^{\mathrm{T}}\mathbf{y}$$
$$= \sum_{i=1}^{K}\mathbf{g}_i\frac{\lambda_i}{\lambda_i^2 + \gamma^2}\mathbf{g}_i^{\mathrm{T}}\mathbf{y} = \frac{1}{\sqrt{2}}\begin{bmatrix}1\\1\end{bmatrix}\frac{2}{2+\gamma^2}\frac{1}{\sqrt{2}}\begin{bmatrix}1\\1\end{bmatrix}^{\mathrm{T}}\begin{bmatrix}1\\3\end{bmatrix} + 0$$
$$= \frac{4}{2+\gamma^2}\begin{bmatrix}1\\1\end{bmatrix}.$$

The solution always exists as long as $\gamma^2 > 0$. It is a tapered-down form of the solution with $\gamma^2 = 0$ if all $\lambda_i \neq 0$. Therefore,

$$\mathbf{n} = \begin{bmatrix}1\\3\end{bmatrix} - \frac{4}{2+\gamma^2}\mathbf{E}\begin{bmatrix}1\\1\end{bmatrix} = \begin{bmatrix}1\\3\end{bmatrix} - \frac{4}{2+\gamma^2}\begin{bmatrix}2\\2\end{bmatrix},$$

so that, as $\gamma^2 \to \infty$, the solution $\tilde{\mathbf{x}}$ is minimized, becoming 0, and the residual is equal to $\mathbf{y}$.

2.5.3 Arbitrary systems

The singular vector expansion and singular value decomposition

It may be objected that this entire development is of little interest, because most problems, including those outlined in Chapter 1, produced $\mathbf{E}$ matrices that could not be guaranteed to have complete orthonormal sets of eigenvectors. Indeed, the problems considered produce matrices that are usually non-square, and for which the eigenvector problem is not even defined.

For arbitrary *square* matrices, the question of when a complete orthonormal set of eigenvectors exists is not difficult to answer, but becomes somewhat elaborate.[33] When a square matrix of dimension N is not symmetric, one must consider cases in which there are N distinct eigenvalues and where some are repeated, and the general approach requires the so-called Jordan form. But we will next find a way to avoid these intricacies, and yet deal with sets of simultaneous equations of arbitrary dimensions, not just square ones. The square, symmetric case nonetheless provides full analogues to all of the issues in the more general case, and the reader may find it helpful to refer back to this situation for insight.

Consider the possibility, suggested by the eigenvector method, of expanding the solution $\mathbf{x}$ in a set of orthonormal vectors. Equation (2.87) involves one vector, $\mathbf{x}$, of dimension N, and two vectors, $\mathbf{y}$, $\mathbf{n}$, of dimension M. We would like to use orthonormal basis vectors, but cannot expect, with two different vector dimensions involved, to use just one set: $\mathbf{x}$ can be expanded exactly in N, N-dimensional orthonormal vectors; and similarly, $\mathbf{y}$ and $\mathbf{n}$ can be exactly represented in M, M-dimensional orthonormal vectors. There are an infinite number of ways to select two such sets. But using the structure of $\mathbf{E}$, a particularly useful pair can be identified.

The simple development of the solutions in the square, symmetric case resulted from the theorem concerning the complete nature of the eigenvectors of such a matrix. So construct a new matrix,

$$\mathbf{B} = \begin{Bmatrix} \mathbf{0} & \mathbf{E}^{\mathrm{T}} \\ \mathbf{E} & \mathbf{0} \end{Bmatrix}, \tag{2.245}$$

which by definition is square (dimension $M + N$ by $M + N$) and symmetric. Thus, $\mathbf{B}$ satisfies the theorem just alluded to, and the eigenvalue problem,

$$\mathbf{B}\mathbf{q}_i = \lambda_i \mathbf{q}_i, \tag{2.246}$$

will give rise to $M + N$ orthonormal eigenvectors $\mathbf{q}_i$ (an orthonormal basis) whether or not the λ_i are distinct or non-zero. Writing out (2.246),

$$\begin{Bmatrix} \mathbf{0} & \mathbf{E}^{\mathrm{T}} \\ \mathbf{E} & \mathbf{0} \end{Bmatrix} \begin{bmatrix} q_{1i} \\ \cdot \\ q_{Ni} \\ q_{N+1,i} \\ \cdot \\ q_{N+M,i} \end{bmatrix} = \lambda_i \begin{bmatrix} q_{1i} \\ \cdot \\ q_{Ni} \\ q_{N+1,i} \\ \cdot \\ q_{N+M,i} \end{bmatrix}, \qquad i = 1, 2, \ldots, M + N, \tag{2.247}$$

where q_{pi} is the pth element of $\mathbf{q}_i$. Taking note of the zero matrices, (2.247) may be rewritten as

$$\mathbf{E}^{\mathrm{T}} \begin{bmatrix} q_{N+1,i} \\ \cdot \\ q_{N+M,i} \end{bmatrix} = \lambda_i \begin{bmatrix} q_{1i} \\ \cdot \\ q_{Ni} \end{bmatrix}, \tag{2.248}$$

$$\mathbf{E} \begin{bmatrix} q_{1i} \\ \cdot \\ q_{Ni} \end{bmatrix} = \lambda_i \begin{bmatrix} q_{N+1,i} \\ \cdot \\ q_{N+M,i} \end{bmatrix}, \qquad i = 1, 2, \ldots, M + N. \tag{2.249}$$

Define

$$\mathbf{u}_i = [q_{N+1,i} \quad \cdot \quad q_{N+M,i}]^{\mathrm{T}}, \quad \mathbf{v}_i = [q_{1,i} \quad \cdot \quad q_{N,i}]^{\mathrm{T}}, \quad \mathbf{q}_i = [\mathbf{v}_i^{\mathrm{T}} \quad \mathbf{u}_i^{\mathrm{T}}]^{\mathrm{T}}, \tag{2.250}$$

that is, defining the first N elements of $\mathbf{q}_i$ to be called $\mathbf{v}_i$ and the last M to be called $\mathbf{u}_i$, the two sets together being the "singular vectors." Then (2.248)–(2.249) are

$$\mathbf{E}\mathbf{v}_i = \lambda_i \mathbf{u}_i, \tag{2.251}$$

$$\mathbf{E}^{\mathrm{T}}\mathbf{u}_i = \lambda_i \mathbf{v}_i. \tag{2.252}$$

If (2.251) is left multiplied by $\mathbf{E}^{\mathrm{T}}$, and using (2.252), one has

$$\mathbf{E}^{\mathrm{T}}\mathbf{E}\mathbf{v}_i = \lambda_i^2 \mathbf{v}_i, \quad i = 1, 2, \ldots, N. \tag{2.253}$$

Similarly, left multiplying (2.252) by $\mathbf{E}$ and using (2.251) produces

$$\mathbf{E}\mathbf{E}^{\mathrm{T}}\mathbf{u}_i = \lambda_i^2 \mathbf{u}_i \quad i = 1, 2, \ldots, M. \tag{2.254}$$

These last two equations show that the $\mathbf{u}_i$, $\mathbf{v}_i$ each separately satisfy two independent eigenvector/eigenvalue problems of the square symmetric matrices $\mathbf{E}\mathbf{E}^{\mathrm{T}}$, $\mathbf{E}^{\mathrm{T}}\mathbf{E}$ and they can be separately given unit norm. The λ_i come in pairs as $\pm\lambda_i$ and the convention is made that only the non-negative ones are retained, as the $\mathbf{u}_i$, $\mathbf{v}_i$ corresponding to the singular values also differ at most by a minus sign, and hence are not independent of the ones retained.[34] If one of M, N is much smaller than the other, only the smaller eigenvalue/eigenvector problem needs to be solved for either of $\mathbf{u}_i$, $\mathbf{v}_i$; the other set is immediately calculated from (2.252) or (2.251). Evidently, in the limiting cases of either a single equation or a single unknown, the eigenvalue/eigenvector problem is purely scalar, no matter how large the other dimension.

In going from (2.248), (2.249) to (2.253), (2.254), the range of the index i has dropped from $M + N$ to M or N. The missing "extra" equations correspond to negative λ_i and carry no independent information.

Example *Consider the non-square, non-symmetric matrix*

$$\mathbf{E} = \left\{ \begin{matrix} 0 & 0 & 1 & -1 & 2 \\ 1 & 1 & 0 & 0 & 0 \\ 1 & -1 & 0 & 0 & 0 \\ 1 & 2 & 0 & 0 & 0 \end{matrix} \right\}.$$

Forming the larger matrix $\mathbf{B}$, *solve the eigenvector/eigenvalue problem that produces*

$$\mathbf{Q} = \left\{ \begin{matrix} -0.31623 & 0.63246 & -1.1796 \times 10^{-16} & -0.63246 & 0.31623 \\ -0.63246 & -0.31623 & -2.0817 \times 10^{-16} & 0.31623 & 0.63246 \\ 0.35857 & -0.22361 & 0.80178 & -0.22361 & 0.35857 \\ -0.11952 & -0.67082 & -0.26726 & -0.67082 & -0.11952 \\ 0.59761 & 0.00000 & -0.53452 & 0.00000 & 0.59761 \end{matrix} \right\},$$

$$
S = \left\{ \begin{array}{ccccc}
-2.6458 & 0 & 0 & 0 & 0 \\
0 & -1.4142 & 0 & 0 & 0 \\
0 & 0 & 0 & 0 & 0 \\
0 & 0 & 0 & 1.4142 & 0 \\
0 & 0 & 0 & 0 & 2.6458
\end{array} \right\},
$$

where Q *is the matrix whose columns are* q_i *and* S *is the diagonal matrix whose values are the corresponding eigenvalues. Note that one of the eigenvalues vanishes identically, and that the others occur in positive and negative pairs. The corresponding* q_i *differ only by sign changes in parts of the vectors, but they are all linearly independent. First define a* V *matrix from the first two rows of* Q,

$$
V = \left\{ \begin{array}{ccccc}
-0.31623 & 0.63246 & 0 & -0.63246 & 0.31623 \\
-0.63246 & -0.31623 & 0 & 0.31623 & 0.63246
\end{array} \right\}.
$$

Only two of the vectors are linearly independent (the zero-vector is not physically realizable). Similarly, the last three rows of Q *define a* U *matrix,*

$$
U = \left\{ \begin{array}{ccccc}
0.35857 & -0.22361 & 0.80178 & -0.22361 & 0.35857 \\
-0.11952 & -0.67082 & -0.26726 & -0.67082 & -0.11952 \\
0.59761 & 0.00000 & -0.53452 & 0.00000 & 0.59761
\end{array} \right\},
$$

in which only three columns are linearly independent. Retaining only the last two columns of V *and the last three of* U, *and column normalizing each to unity, produces the singular vectors.*

By convention, the λ_i are ordered in decreasing numerical value. Equations (2.251)–(2.252) provide a relationship between each u_i, v_i pair. But because $M \neq N$, generally, there will be more of one set than the other. The only way Eqs. (2.251)–(2.252) can be consistent is if $\lambda_i = 0$, $i > \min(M, N)$ (where $\min(M, N)$ is read as "the minimum of M and N"). Suppose $M < N$. Then (2.254) is solved for u_i, $i = 1, 2, \ldots, M$, and (2.252) is used to find the corresponding v_i. There are $N - M$ v_i not generated this way, but which can be found using the Gram–Schmidt method described on p. 22.

Let there be K non-zero λ_i; then

$$
E v_i \neq 0, \quad i = 1, 2, \ldots, K. \tag{2.255}
$$

These v_i are known as the "range vectors of E" or the "solution range vectors." For the remaining $N - K$ vectors v_i,

$$
E v_i = 0, \quad i = K + 1, \ldots N, \tag{2.256}
$$

which are known as the "nullspace vectors of **E**" or the "nullspace of the solution."
If $K < M$, there will be K of the $\mathbf{u}_i$ such that

$$\mathbf{E}^{\mathrm{T}}\mathbf{u}_i \neq 0, \quad i = 1, 2, \ldots, K, \tag{2.257}$$

which are the "range vectors of $\mathbf{E}^{\mathrm{T}}$," and $M - K$ of the $\mathbf{u}_i$ such that

$$\mathbf{E}^{\mathrm{T}}\mathbf{u}_i = 0, \quad i = K + 1, \ldots, M, \tag{2.258}$$

the "nullspace vectors of $\mathbf{E}^{\mathrm{T}}$" or the "data, or observation, nullspace vectors." The
"nullspace" of **E** is spanned by its nullspace vectors, and the "range" of **E** is spanned
by the range vectors, etc., in the sense, for example, that an arbitrary vector lying
in the range is perfectly described by a sum of the range vectors. We now have two
complete orthonormal sets in the two different spaces. Note that Eqs. (2.256) and
(2.258) imply that

$$\mathbf{E}\mathbf{v}_i = 0, \ \mathbf{u}_i^{\mathrm{T}}\mathbf{E} = 0, \quad i = K + 1, \ldots, N, \tag{2.259}$$

which expresses hard relationships among the columns and rows of **E**.

Because the $\mathbf{u}_i$, $\mathbf{v}_i$ are complete in their corresponding spaces, **x**, **y**, **n** can be
expanded without error:

$$\mathbf{x} = \sum_{i=1}^{N} \alpha_i \mathbf{v}_i, \quad \mathbf{y} = \sum_{j=1}^{M} \beta_i \mathbf{u}_i, \quad \mathbf{n} = \sum_{i=1}^{M} \gamma_i \mathbf{u}_i, \tag{2.260}$$

where **y** has been measured, so that we know $\beta_j = \mathbf{u}_j^{\mathrm{T}}\mathbf{y}$. To find **x**, we need α_i, and
to find **n**, we need the γ_i. Substitute (2.260) into (2.87), and using (2.251)–(2.252),

$$\sum_{i=1}^{N} \alpha_i \mathbf{E}\mathbf{v}_i + \sum_{i=1}^{M} \gamma_i \mathbf{u}_i = \sum_{i=1}^{K} \alpha_i \lambda_i \mathbf{u}_i + \sum_{i=1}^{M} \gamma_i \mathbf{u}_i \tag{2.261}$$

$$= \sum_{i=1}^{M} \left(\mathbf{u}_i^{\mathrm{T}}\mathbf{y}\right)\mathbf{u}_i.$$

Notice the differing upper limits on the summations. Because of the orthonormality
of the singular vectors, (2.261) can be solved as

$$\alpha_i \lambda_i + \gamma_i = \mathbf{u}_i^{\mathrm{T}}\mathbf{y}, \quad i = 1, 2, \ldots, M, \tag{2.262}$$

$$\alpha_i = \left(\mathbf{u}_i^{\mathrm{T}}\mathbf{y} - \gamma_i\right)/\lambda_i, \quad \lambda_i \neq 0, \ i = 1, 2, \ldots, K. \tag{2.263}$$

In these equations, if $\lambda_i \neq 0$, nothing prevents us from setting $\gamma_i = 0$, that is,

$$\mathbf{u}_i^{\mathrm{T}}\mathbf{n} = 0, \quad i = 1, 2, \ldots, K, \tag{2.264}$$

should we wish, which will have the effect of making the noise norm as small as
possible (there is arbitrariness in this choice, and later γ_i will be chosen differently).

Then (2.263) produces

$$\alpha_i = \frac{\mathbf{u}_i^{\mathrm{T}}\mathbf{y}}{\lambda_i}, \quad i = 1, 2, \ldots, K. \tag{2.265}$$

But, because $\lambda_i = 0$, $i > K$, the only solution to (2.262) for these values of i is $\gamma_i = \mathbf{u}_i^{\mathrm{T}}\mathbf{y}$, and α_i is indeterminate. These γ_i are non-zero, except in the event (unlikely with real data) that

$$\mathbf{u}_i^{\mathrm{T}}\mathbf{y} = 0, \quad i = K + 1, \ldots, N. \tag{2.266}$$

This last equation is a solvability condition – in direct analogy to (2.205).

The solution obtained in this manner now has the following form:

$$\tilde{\mathbf{x}} = \sum_{i=1}^{K} \frac{\mathbf{u}_i^{\mathrm{T}}\mathbf{y}}{\lambda_i}\mathbf{v}_i + \sum_{i=K+1}^{N} \alpha_i \mathbf{v}_i, \tag{2.267}$$

$$\tilde{\mathbf{y}} = \mathbf{E}\tilde{\mathbf{x}} = \sum_{i=1}^{K} \left(\mathbf{u}_i^{\mathrm{T}}\mathbf{y}\right)\mathbf{u}_i, \tag{2.268}$$

$$\tilde{\mathbf{n}} = \sum_{i=K+1}^{M} \left(\mathbf{u}_i^{\mathrm{T}}\mathbf{y}\right)\mathbf{u}_i. \tag{2.269}$$

The coefficients of the last $N - K$ of the $\mathbf{v}_i$ in Eq. (2.267), the solution nullspace vectors, are arbitrary, representing structures in the solution about which the equations provide no information. A nullspace is always present unless $K = N$. The solution residuals are directly proportional to the nullspace vectors of $\mathbf{E}^{\mathrm{T}}$ and will vanish only if $K = M$, or if the solvability conditions are met.

Just as in the simpler square symmetric case, no choice of the coefficients of the solution nullspace vectors can have any effect on the size of the residuals. If we choose once again to exercise Occam's razor, and regard the simplest solution as best, then setting the nullspace coefficients to zero gives

$$\tilde{\mathbf{x}} = \sum_{i=1}^{K} \frac{\mathbf{u}_i^{\mathrm{T}}\mathbf{y}}{\lambda_i}\mathbf{v}_i. \tag{2.270}$$

Along with (2.269), this is the "particular-SVD solution." It minimizes the residuals, and simultaneously produces the corresponding $\tilde{\mathbf{x}}$ with the smallest norm. If $\langle \mathbf{n} \rangle = 0$, the bias of (2.270) is evidently

$$\langle \tilde{\mathbf{x}} - \mathbf{x} \rangle = - \sum_{i=K+1}^{N} \alpha_i \mathbf{v}_i. \tag{2.271}$$

The solution uncertainty is

$$
\mathbf{P} = \sum_{i=1}^{K} \sum_{j=1}^{K} \mathbf{v}_i \frac{\mathbf{u}_i^{\mathrm{T}} \mathbf{R}_{nn} \mathbf{u}_j}{\lambda_i \lambda_j} \mathbf{v}_i^{\mathrm{T}} + \sum_{i=K+1}^{N} \sum_{j=K+1}^{N} \mathbf{v}_i \langle \alpha_i \alpha_j \rangle \mathbf{v}_j^{\mathrm{T}}. \tag{2.272}
$$

If the noise is white with variance σ_n^2 or, if a row scaling matrix $\mathbf{W}^{-\mathrm{T}/2}$ has been applied to make it so, then (2.272) becomes

$$
\mathbf{P} = \sum_{i=1}^{K} \frac{\sigma_n^2}{\lambda_i^2} \mathbf{v}_i \mathbf{v}_i^{\mathrm{T}} + \sum_{i=K+1}^{N} \langle \alpha_i^2 \rangle \mathbf{v}_i \mathbf{v}_i^{\mathrm{T}}, \tag{2.273}
$$

where it was also assumed that $\langle \alpha_i \alpha_j \rangle = \langle \alpha_i^2 \rangle \delta_{ij}$ in the nullspace. The influence of very small singular values on the uncertainty is plain: In the solution (2.267) or (2.270) there are error terms $\mathbf{u}_i^{\mathrm{T}} \mathbf{y}/\lambda_i$ that are greatly magnified by small or nearly vanishing singular values, introducing large terms proportional to σ_n^2/λ_i^2 into (2.273).

The structures dominating $\tilde{\mathbf{x}}$ are a competition between the magnitudes of $\mathbf{u}_i^{\mathrm{T}} \mathbf{y}$ and λ_i, given by the ratio, $\mathbf{u}_i^{\mathrm{T}} \mathbf{y}/\lambda_i$. Large λ_i can suppress comparatively large projections onto $\mathbf{u}_i$, and similarly, small, but non-zero λ_i may greatly amplify comparatively modest projections. In practice,[35] one is well-advised to study the behavior of both $\mathbf{u}_i^{\mathrm{T}} \mathbf{y}$, $\mathbf{u}_i^{\mathrm{T}} \mathbf{y}/\lambda_i$ as a function of i to understand the nature of the solution.

The decision to omit contributions to the residuals by the range vectors of $\mathbf{E}^{\mathrm{T}}$, as we did in Eqs. (2.264) and (2.269), needs to be examined. Should some other choice be made, the $\tilde{\mathbf{x}}$ norm would decrease, but the residual norm would increase. Determining the desirability of such a trade-off requires an understanding of the noise structure – in particular, (2.264) imposes rigid structures, and hence covariances, on the residuals.

2.5.4 The singular value decomposition

The singular vectors and values have been used to provide a convenient pair of orthonormal bases to solve an arbitrary set of simultaneous equations. The vectors and values have another use, however, in providing a decomposition of $\mathbf{E}$.

Define $\mathbf{\Lambda}$ as the $M \times N$ matrix whose diagonal elements are the λ_i, in order of descending values in the same order, $\mathbf{U}$ as the $M \times M$ matrix whose columns are the $\mathbf{u}_i$, $\mathbf{V}$ as the $N \times N$ matrix whose columns are the $\mathbf{v}_i$. As an example, suppose $M = 3$, $N = 4$; then

$$
\mathbf{\Lambda} = \begin{Bmatrix} \lambda_i & 0 & 0 & 0 \\ 0 & \lambda_2 & 0 & 0 \\ 0 & 0 & \lambda_3 & 0 \end{Bmatrix}.
$$

Alternatively, if $M = 4$, $N = 3$, then

$$\Lambda = \left\{ \begin{array}{ccc} \lambda_1 & 0 & 0 \\ 0 & \lambda_2 & 0 \\ 0 & 0 & \lambda_3 \\ 0 & 0 & 0 \end{array} \right\},$$

therefore extending the definition of a diagonal matrix to non-square ones.

Precisely as with matrix $\mathbf{G}$ considered above (see p. 75), the column orthonormality of $\mathbf{U}$, $\mathbf{V}$ implies that these matrices are orthogonal:

$$\mathbf{U}\mathbf{U}^{\mathrm{T}} = \mathbf{I}_M, \qquad \mathbf{U}^{\mathrm{T}}\mathbf{U} = \mathbf{I}_M, \tag{2.274}$$

$$\mathbf{V}\mathbf{V}^{\mathrm{T}} = \mathbf{I}_N, \qquad \mathbf{V}^{\mathrm{T}}\mathbf{V} = \mathbf{I}_N. \tag{2.275}$$

(It follows that $\mathbf{U}^{-1} = \mathbf{U}^{\mathrm{T}}$, etc.) As with $\mathbf{G}$ above, should one or more columns of $\mathbf{U}$, $\mathbf{V}$ be deleted, the matrices will become semi-orthogonal.

Equations (2.251)–(2.254) can be written compactly as:

$$\mathbf{E}\mathbf{V} = \mathbf{U}\Lambda, \qquad \mathbf{E}^{\mathrm{T}}\mathbf{U} = \mathbf{V}\Lambda^{\mathrm{T}}, \tag{2.276}$$

$$\mathbf{E}^{\mathrm{T}}\mathbf{E}\mathbf{V} = \mathbf{V}\Lambda^{\mathrm{T}}\Lambda, \qquad \mathbf{E}\mathbf{E}^{\mathrm{T}}\mathbf{U} = \mathbf{U}\Lambda\Lambda^{\mathrm{T}}. \tag{2.277}$$

Left multiply the first relation of (2.276) by $\mathbf{U}^{\mathrm{T}}$ and right multiply it by $\mathbf{V}^{\mathrm{T}}$, and, invoking Eq. (2.275),

$$\mathbf{U}^{\mathrm{T}}\mathbf{E}\mathbf{V} = \Lambda. \tag{2.278}$$

Therefore, $\mathbf{U}$, $\mathbf{V}$ diagonalize $\mathbf{E}$ (with "diagonal" having the extended meaning for a rectangular matrix as defined above).

Right multiplying the first relation of (2.276) by $\mathbf{V}^{\mathrm{T}}$ gives

$$\mathbf{E} = \mathbf{U}\Lambda\mathbf{V}^{\mathrm{T}}. \tag{2.279}$$

This last equation represents a product, called the "singular value decomposition" (SVD), of an arbitrary matrix, of two orthogonal matrices, $\mathbf{U}$, $\mathbf{V}$, and a usually non-square diagonal matrix, Λ.

There is one further step to take. Notice that for a rectangular Λ, as in the examples above, one or more rows or columns must be all zero, depending upon the shape of the matrix. In addition, if any of the $\lambda_i = 0$, $i < \min(M, N)$, the corresponding rows or columns of Λ will be all zeros. Let K be the number of non-vanishing singular values (the "rank" of $\mathbf{E}$). By inspection (multiplying it out), one finds that the last $N - K$ columns of $\mathbf{V}$ and the last $M - K$ columns of $\mathbf{U}$ are multiplied by zeros only. If these columns are dropped entirely from $\mathbf{U}$, $\mathbf{V}$ so that $\mathbf{U}$ becomes $M \times K$ and $\mathbf{V}$ becomes $N \times K$, and reducing Λ to a $K \times K$ square matrix, then the representation (2.279) remains exact, in the form

$$\mathbf{E} = \mathbf{U}_K \mathbf{\Lambda}_K \mathbf{V}_K^{\mathrm{T}} = \lambda_1 \mathbf{u}_1 \mathbf{v}_1^{\mathrm{T}} + \lambda_2 \mathbf{u}_2 \mathbf{v}_2^{\mathrm{T}} + \cdots + \lambda_K \mathbf{u}_K \mathbf{v}_K^{\mathrm{T}}, \tag{2.280}$$

with the subscript indicating the number of columns, where $\mathbf{U}_K$, $\mathbf{V}_K$ are then only semi-orthogonal, and $\mathbf{\Lambda}_K$ is now square. Equation (2.280) should be compared to (2.226).[36]

The SVD solution can be obtained by matrix manipulation, rather than vector by vector. Consider once again finding the solution to the simultaneous equations (2.87), but first write $\mathbf{E}$ in its reduced SVD,

$$\mathbf{U}_K \mathbf{\Lambda}_K \mathbf{V}_K^{\mathrm{T}} \mathbf{x} + \mathbf{n} = \mathbf{y}. \tag{2.281}$$

Left multiplying by $\mathbf{U}_K^{\mathrm{T}}$ and invoking the semi-orthogonality of $\mathbf{U}_K$ produces

$$\mathbf{\Lambda}_K \mathbf{V}_K^{\mathrm{T}} \mathbf{x} + \mathbf{U}_K^{\mathrm{T}} \mathbf{n} = \mathbf{U}_K^{\mathrm{T}} \mathbf{y}. \tag{2.282}$$

The inverse of $\mathbf{\Lambda}_K$ is easily computed, and

$$\mathbf{V}_K^{\mathrm{T}} \mathbf{x} + \mathbf{\Lambda}_K^{-1} \mathbf{U}_K^{\mathrm{T}} \mathbf{n} = \mathbf{\Lambda}_K^{-1} \mathbf{U}_K^{\mathrm{T}} \mathbf{y}. \tag{2.283}$$

But $\mathbf{V}_K^{\mathrm{T}} \mathbf{x}$ is the dot product of the first K of the $\mathbf{v}_i$ with the unknown $\mathbf{x}$. Equation (2.283) thus represents statements about the relationship between dot products of the unknown vector, $\mathbf{x}$, with a set of orthonormal vectors, and therefore must represent the expansion coefficients of the solution in those vectors. If we set

$$\mathbf{U}_K^{\mathrm{T}} \mathbf{n} = 0, \tag{2.284}$$

then

$$\mathbf{V}_K^{\mathrm{T}} \mathbf{x} = \mathbf{\Lambda}_K^{-1} \mathbf{U}_K^{\mathrm{T}} \mathbf{y}, \tag{2.285}$$

and hence

$$\tilde{\mathbf{x}} = \mathbf{V}_K \mathbf{\Lambda}_K^{-1} \mathbf{U}_K^{\mathrm{T}} \mathbf{y}. \tag{2.286}$$

Equation (2.286) is identical to the solution (2.270), and can be confirmed by writing it out explicitly. As with the square symmetric case, the contribution of any structure in $\mathbf{y}$ proportional to $\mathbf{u}_i$ depends upon the ratio of the projection $\mathbf{u}_i^{\mathrm{T}} \mathbf{y}$ to λ_i. Substituting (2.286) into (2.281),

$$\mathbf{U}_K \mathbf{\Lambda}_K \mathbf{V}_K^{\mathrm{T}} \mathbf{V}_K \mathbf{\Lambda}_K^{-1} \mathbf{U}_K^{\mathrm{T}} \mathbf{y} + \mathbf{n} = \mathbf{U}_K \mathbf{U}_K^{\mathrm{T}} \mathbf{y} + \mathbf{n} = \mathbf{y},$$

or

$$\tilde{\mathbf{n}} = \left(\mathbf{I} - \mathbf{U}_K \mathbf{U}_K^{\mathrm{T}} \right) \mathbf{y}. \tag{2.287}$$

Let the full $\mathbf{U}$ and $\mathbf{V}$ matrices be rewritten as

$$\mathbf{U} = \{\mathbf{U}_K \quad \mathbf{Q}_u\}, \ \mathbf{V} = \{\mathbf{V}_K \quad \mathbf{Q}_v\}, \tag{2.288}$$

where $\mathbf{Q}_u$, $\mathbf{Q}_v$ are the matrices whose columns are the corresponding nullspace vectors. Then

$$\mathbf{E}\tilde{\mathbf{x}} + \tilde{\mathbf{n}} = \mathbf{y}, \ \mathbf{E}\tilde{\mathbf{x}} = \tilde{\mathbf{y}}, \tag{2.289}$$

and

$$\tilde{\mathbf{y}} = \mathbf{U}_K \mathbf{U}_K^{\mathrm{T}} \mathbf{y}, \ \tilde{\mathbf{n}} = \mathbf{Q}_u \mathbf{Q}_u^{\mathrm{T}} \mathbf{y} = \sum_{j=K+1}^{N} \left(\mathbf{u}_j^{\mathrm{T}} \mathbf{y} \right) \mathbf{u}_j, \tag{2.290}$$

which is identical to (2.268). Note that

$$\mathbf{Q}_u \mathbf{Q}_u^{\mathrm{T}} = \left(\mathbf{I} - \mathbf{U}_K \mathbf{U}_K^{\mathrm{T}} \right), \ \mathbf{Q}_v \mathbf{Q}_v^{\mathrm{T}} = \left(\mathbf{I} - \mathbf{V}_K \mathbf{V}_K^{\mathrm{T}} \right), \tag{2.291}$$

which are idempotent ($\mathbf{V}_K \mathbf{V}_K^{\mathrm{T}}$ is matrix $\mathbf{H}$ of Eq. (2.97)). The two vector sets $\mathbf{Q}_u$, $\mathbf{Q}_v$ span the data and solution nullspaces respectively. The general solution is

$$\tilde{\mathbf{x}} = \mathbf{V}_K \mathbf{\Lambda}_K^{-1} \mathbf{U}_K \mathbf{y} + \mathbf{Q}_v \boldsymbol{\alpha}, \tag{2.292}$$

where $\boldsymbol{\alpha}$ is now restricted to being the vector of coefficients of the nullspace vectors. The solution uncertainty (2.272) is thus

$$\begin{aligned}
\mathbf{P} &= \mathbf{V}_K \mathbf{\Lambda}_K^{-1} \mathbf{U}_K^{\mathrm{T}} \langle \mathbf{n} \mathbf{n}^{\mathrm{T}} \rangle \mathbf{U}_K \mathbf{\Lambda}_K^{-1} \mathbf{V}_K^{\mathrm{T}} \\
&\quad + \mathbf{Q}_v \langle \boldsymbol{\alpha} \boldsymbol{\alpha}^{\mathrm{T}} \rangle \mathbf{Q}_G^{\mathrm{T}} = \mathbf{C}_{xx} + \mathbf{Q}_v \langle \boldsymbol{\alpha} \boldsymbol{\alpha}^{\mathrm{T}} \rangle \mathbf{Q}_v^{\mathrm{T}},
\end{aligned} \tag{2.293}$$

or

$$\mathbf{P} = \sigma_n^2 \mathbf{V}_K \mathbf{\Lambda}_K^{-2} \mathbf{V}_K^{\mathrm{T}} + \mathbf{Q}_v \langle \boldsymbol{\alpha} \boldsymbol{\alpha}^{\mathrm{T}} \rangle \mathbf{Q}_v^{\mathrm{T}}, \tag{2.294}$$

for white noise.

Least-squares solution of simultaneous solutions by SVD has several important advantages. Among other features, we can write down within one algebraic formulation the solution to systems of equations that can be under-, over-, or just-determined. Unlike the eigenvalue/eigenvector solution for an arbitrary square system, the singular values (eigenvalues) are always non-negative and real, and the singular vectors (eigenvectors) can always be made a complete orthonormal set. Furthermore, the relations (2.276) provide a specific, quantitative statement of the connection between a set of orthonormal structures in the data, and the corresponding presence of orthonormal structures in the solution. These relations provide a very powerful diagnostic method for understanding precisely why the solution takes on the form it does.

2.5.5 *Some simple examples: algebraic equations*

Example *The simplest underdetermined system is 1×2. Suppose $x_1 - 2x_2 = 3$, so that*

$$\mathbf{E} = \{1 \quad -2\}, \quad \mathbf{U} = \{1\}, \quad \mathbf{V} = \begin{Bmatrix} 0.447 & -0.894 \\ -0.894 & -0.447 \end{Bmatrix}, \quad \lambda_1 = 2.23,$$

where the second column of V is the nullspace of E. The general solution is $\tilde{\mathbf{x}} = [0.6, -1.2]^{\mathrm{T}} + \alpha_2 \mathbf{v}_2$. Because $K = 1$ is the only possible choice, this solution satisfies the equation exactly, and a data nullspace is not possible.

Example *The most elementary overdetermined problem is 2×1. Suppose that*

$$x_1 = 1,$$
$$x_1 = 3.$$

The appearance of two such equations is possible if there is noise in the observations, and they are properly written as

$$x_1 + n_1 = 1,$$
$$x_1 + n_2 = 3.$$

$\mathbf{E} = \{1, 1\}^{\mathrm{T}}, \mathbf{E}^{\mathrm{T}}\mathbf{E}$ *represents the eigenvalue problem of the smaller dimension, again 1×1, and*

$$\mathbf{U} = \begin{Bmatrix} 0.707 & -0.707 \\ 0.707 & 0.707 \end{Bmatrix}, \quad \mathbf{V} = \{1\}, \quad \lambda_1 = \sqrt{2},$$

where the second column of $\mathbf{U}$ lies in the data nullspace, there being no solution nullspace. The general solution is $\mathbf{x} = x_1 = 2$, which if substituted back into the original equations produces

$$\mathbf{E}\tilde{\mathbf{x}} = \begin{bmatrix} 2 & 2 \end{bmatrix}^{\mathrm{T}} = \tilde{\mathbf{y}}.$$

Hence there are residuals $\tilde{\mathbf{n}} = \tilde{\mathbf{y}} - \mathbf{y} = [1, -1]^{\mathrm{T}}$ that are necessarily proportional to $\mathbf{u}_2$ and thus orthogonal to $\tilde{\mathbf{y}}$. No other solution can produce a smaller l_2 norm residual than this one. The SVD provides a solution that compromises the contradiction between the two original equations.

Example *The possibility of $K < M$, $K < N$ simultaneously is also easily seen. Consider the system*

$$\begin{Bmatrix} 1 & -2 & 1 \\ 3 & 2 & 1 \\ 4 & 0 & 2 \end{Bmatrix} \mathbf{x} = \begin{bmatrix} 1 \\ -1 \\ 2 \end{bmatrix},$$

which appears superficially just-determined. But the singular values are $\lambda_1 = 5.67$, $\lambda_2 = 2.80$, $\lambda_3 = 0$. The vanishing of the third singular value means that the row and column vectors are not linearly independent sets – indeed, the third row vector is just the sum of the first two (but the third element of $\mathbf{y}$ is not the sum of the first two – making the equations inconsistent). Thus there are both solution and data nullspaces, which the reader might wish to find. With a vanishing singular value, $\mathbf{E}$ can be written exactly using only two columns of $\mathbf{U}$, $\mathbf{V}$ and the linear dependence of the equations is given explicitly as $\mathbf{u}_3^{\mathrm{T}}\mathbf{E} = 0$.

Example *Consider now the underdetermined system*

$$x_1 + x_2 - 2x_3 = 1,$$
$$x_1 + x_2 - 2x_3 = 2,$$

which has no conventional solution at all, being a contradiction, and is thus simultaneously underdetermined and incompatible. If one of the coefficients is modified by a very small quantity, $|\epsilon| > 0$, to produce

$$x_1 + x_2 - (2 + \epsilon)x_3 = 1,$$
$$x_1 + x_2 - 2x_3 = 2, \tag{2.295}$$

then not only is there a solution, there is an infinite number of them, which can be shown by computing the particular SVD solution and the nullspace. Thus the slightest perturbation in the coefficients has made the system jump from one having no solution to one having an infinite number – a disconcerting situation. The label for such a system is "ill-conditioned." How would we know the system is ill-conditioned? There are several indicators. First, the ratio of the two singular values is determined by ϵ. If we set $\epsilon = 10^{-10}$, the two singular values are $\lambda_1 = 3.46$, $\lambda_2 = 4.1 \times 10^{-11}$, an immediate warning that the two equations are nearly linearly dependent. (In a mathematical problem, the non-vanishing of the second singular value is enough to assure a solution. It is the inevitable slight errors in y that suggest sufficiently small singular values should be treated as though they were actually zero.)

Example *A similar problem exists with the system*

$$x_1 + x_2 - 2x_3 = 1,$$
$$x_1 + x_2 - 2x_3 = 1,$$

which has an infinite number of solutions. But the change to

$$x_1 + x_2 - 2x_3 = 1,$$
$$x_1 + x_2 - 2x_3 = 1 + \epsilon,$$

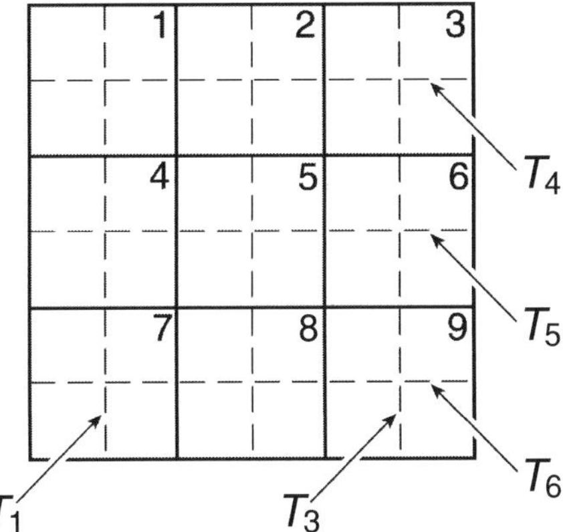

Figure 2.10 Tomographic problem with nine unknowns and only six integral constraints. Box numbers are in the upper right-corner of each, and the T_i are the measured integrals through various combinations of three boxes.

for arbitrarily small ϵ produces a system with no solutions in the conventional mathematical sense, although the SVD will handle the system in a sensible way, which the reader should confirm.

Problems like these are simple examples of the practical issues that arise once one recognizes that, unlike textbook problems, observational ones always contain inaccuracies; any discussion of how to handle data in the presence of mathematical relations must account for these inaccuracies as intrinsic – not as something to be regarded as an afterthought. But the SVD itself is sufficiently powerful that it always contains the information to warn of ill-conditioning, and by determination of K to cope with it – producing useful solutions.

Example *(The tomographic problem from Chapter 1.) A square box is made up of 3×3 unit dimension sub-boxes (Fig. 2.10). All rays are in the r_x or r_y directions. Therefore, the equations are*

$$\left\{\begin{array}{ccccccccc} 1 & 0 & 0 & 1 & 0 & 0 & 1 & 0 & 0 \\ 0 & 1 & 0 & 0 & 1 & 0 & 0 & 1 & 0 \\ 0 & 0 & 1 & 0 & 0 & 1 & 0 & 0 & 1 \\ 1 & 1 & 1 & 0 & 0 & 0 & 0 & 0 & 0 \\ 0 & 0 & 0 & 1 & 1 & 1 & 0 & 0 & 0 \\ 0 & 0 & 0 & 0 & 0 & 0 & 1 & 1 & 1 \end{array}\right\} \begin{bmatrix} x_1 \\ x_2 \\ . \\ . \\ . \\ x_9 \end{bmatrix} = \begin{bmatrix} 0 \\ 1 \\ 0 \\ 0 \\ 1 \\ 0 \end{bmatrix},$$

that is, $\mathbf{Ex} = \mathbf{y}$. There are six integrals (rays) across the nine boxes in which one seeks the corresponding value of x_i. $\mathbf{y}$ was calculated by assuming that the "true" *value is $x_5 = 1$, $x_i = 0$, $i \neq 5$. The SVD produces*

$$
\mathbf{U} = \left\{
\begin{array}{cccccc}
-0.408 & 0 & 0 & 0.816 & 0 & 0.408 \\
-0.408 & 0.703 & -0.0543 & -0.408 & -0.0549 & 0.408 \\
-0.408 & -0.703 & 0.0543 & -0.408 & 0.0549 & 0.408 \\
-0.408 & -0.0566 & 0.0858 & 0 & -0.81 & -0.408 \\
-0.408 & -0.0313 & -0.744 & 0 & 0.335 & -0.408 \\
-0.408 & 0.0879 & 0.658 & 0 & 0.475 & -0.408
\end{array}
\right\},
$$

$$
\mathbf{\Lambda} = \mathrm{diag}([2.45 \quad 1.73 \quad 1.73 \quad 1.73 \quad 1.73 \quad 0]),
$$

$$
\mathbf{V} = \left\{
\begin{array}{ccccccccc}
-0.333 & -0.0327 & 0.0495 & 0.471 & -0.468 & -0.38 & -0.224 & 0.353 & 0.353 \\
-0.333 & 0.373 & 0.0182 & -0.236 & -0.499 & 0.432 & 0.302 & -0.275 & 0.302 \\
-0.333 & -0.438 & 0.0808 & -0.236 & -0.436 & -0.0515 & -0.0781 & -0.0781 & -0.655 \\
-0.333 & -0.0181 & -0.43 & 0.471 & 0.193 & 0.519 & -0.361 & -0.15 & -0.15 \\
-0.333 & 0.388 & -0.461 & -0.236 & 0.162 & -0.59 & -0.0791 & -0.29 & -0.0791 \\
-0.333 & -0.424 & -0.398 & -0.236 & 0.225 & 0.0704 & 0.44 & 0.44 & 0.229 \\
-0.333 & 0.0507 & 0.38 & 0.471 & 0.274 & -0.139 & 0.585 & -0.204 & -0.204 \\
-0.333 & 0.457 & 0.349 & -0.236 & 0.243 & 0.158 & -0.223 & 0.566 & -0.223 \\
-0.333 & -0.355 & 0.411 & -0.236 & 0.306 & -0.0189 & -0.362 & -0.362 & 0.427
\end{array}
\right\}.
$$

The zeros appearing in $\mathbf{U}$, and in the last element of $\mathrm{diag}(\mathbf{\Lambda})$, are actually very small numbers ($O(10^{-16})$) or less). Rank $K = 5$ despite there being six equations – a consequence of redundancy in the integrals. Notice that there are four repeated λ_i, and the lack of expected simple symmetries in the corresponding $\mathbf{v}_i$ is a consequence of a random assignment in the eigenvectors.

Singular vector $\mathbf{u}_1$ just averages the right-hand side values, and the corresponding solution is completely uniform, proportional to $\mathbf{v}_1$. The average of $\mathbf{y}$ is often the most robust piece of information.

The "right" answer is $\mathbf{x} = [0, 0, 0, 0, 1, 0, 0, 0, 0]^{\mathrm{T}}$. The rank 5 answer by SVD is $\tilde{\mathbf{x}} = [-0.1111, \ 0.2222, -0.1111, 0.2222, 0.5556, 0.2222, -0.1111, 0.2222, -0.1111]^{\mathrm{T}}$, which exactly satisfies the same equations, with $\tilde{\mathbf{x}}^{\mathrm{T}}\tilde{\mathbf{x}} = 0.556 \ < \mathbf{x}^{\mathrm{T}}\mathbf{x}$. When mapped into two dimensions, $\tilde{\mathbf{x}}$ at rank 5 is

$$
r_y \uparrow \quad
\begin{bmatrix}
-0.11 & 0.22 & -0.11 \\
0.22 & 0.56 & 0.22 \\
-0.11 & 0.22 & -0.11
\end{bmatrix},
\qquad r_x \rightarrow
\tag{2.296}
$$

and is the minimum norm solution. The mapped $\mathbf{v}_6$, which belongs in the null-space, is

$$
r_y \uparrow \quad
\begin{bmatrix}
-0.38 & 0.43 & -0.05 \\
0.52 & -0.59 & 0.07 \\
-0.14 & 0.16 & -0.02
\end{bmatrix},
\qquad r_x \rightarrow
$$

and along with any remaining nullspace vectors produces a zero sum along any of the ray paths. $\mathbf{u}_6$ *is in the data nullspace.* $\mathbf{u}_6^{\mathrm{T}}\mathbf{E} = 0$ *shows that*

$$a\,(y_1 + y_2 + y_3) - a\,(y_4 + y_5 + y_6) = 0,$$

if there is to be a solution without a residual, or alternatively, that no solution would permit this sum to be non-zero. This requirement is physically sensible, as it says that the vertical and horizontal rays cover the same territory and must therefore produce the same sum travel times. It shows why the rank is 5, and not 6.

There is no noise in the problem as stated. The correct solution and the SVD solution differ by the nullspace vectors. One can easily confirm that $\tilde{\mathbf{x}}$ *is column 5 of* $\mathbf{V}_5\mathbf{V}_5^{\mathrm{T}}$.

Least-squares allows one to minimize (or maximize) anything one pleases. Suppose that for some reason we want the solution that minimizes the differences between the value in box 5 and its neighbors, perhaps as a way of finding a "smooth" solution. Let

$$\mathbf{W} = \begin{Bmatrix} -1 & 0 & 0 & 0 & 1 & 0 & 0 & 0 & 0 \\ 0 & -1 & 0 & 0 & 1 & 0 & 0 & 0 & 0 \\ 0 & 0 & -1 & 0 & 1 & 0 & 0 & 0 & 0 \\ 0 & 0 & 0 & -1 & 1 & 0 & 0 & 0 & 0 \\ 0 & 0 & 0 & 0 & 1 & -1 & 0 & 0 & 0 \\ 0 & 0 & 0 & 0 & 1 & 0 & -1 & 0 & 0 \\ 0 & 0 & 0 & 0 & 1 & 0 & 0 & -1 & 0 \\ 0 & 0 & 0 & 0 & 1 & 0 & 0 & 0 & -1 \\ 0 & 0 & 0 & 0 & 1 & 0 & 0 & 0 & 0 \end{Bmatrix}. \tag{2.297}$$

The last row is included to render $\mathbf{W}$ *a full-rank matrix. Then*

$$\mathbf{W}\mathbf{x} = [x_5 - x_1 \quad x_5 - x_2 \quad \ldots \quad x_5 - x_9 \quad x_5]^{\mathrm{T}}, \tag{2.298}$$

and we can minimize

$$J = \mathbf{x}^{\mathrm{T}}\mathbf{W}^{\mathrm{T}}\mathbf{W}\mathbf{x} \tag{2.299}$$

subject to $\mathbf{E}\mathbf{x} = \mathbf{y}$ *by finding the stationary value of*

$$J' = J - 2\boldsymbol{\mu}^{\mathrm{T}}\,(\mathbf{y} - \mathbf{E}\mathbf{x}). \tag{2.300}$$

The normal equations are then

$$\mathbf{W}^{\mathrm{T}}\mathbf{W}\mathbf{x} = \mathbf{E}^{\mathrm{T}}\boldsymbol{\mu}, \tag{2.301}$$

$$\mathbf{E}\mathbf{x} = \mathbf{y}, \tag{2.302}$$

and

$$\tilde{\mathbf{x}} = (\mathbf{W}^{\mathrm{T}}\mathbf{W})^{-1}\mathbf{E}^{\mathrm{T}}\boldsymbol{\mu}.$$

Then

$$\mathbf{E}(\mathbf{W}^{\mathrm{T}}\mathbf{W})^{-1}\mathbf{E}^{\mathrm{T}}\boldsymbol{\mu} = \mathbf{y}.$$

The rank of $\mathbf{E}(\mathbf{W}^{\mathrm{T}}\mathbf{W})^{-1}\mathbf{E}^{\mathrm{T}}$ *is* $K = 5 < M = 6$, *and so we need a generalized inverse,*

$$\tilde{\boldsymbol{\mu}} = (\mathbf{E}(\mathbf{W}^{\mathrm{T}}\mathbf{W})^{-1}\mathbf{E}^{\mathrm{T}})^{+}\mathbf{y} = \sum_{j=1}^{5} \mathbf{v}_i \frac{\mathbf{v}_i^{\mathrm{T}}\mathbf{y}}{\lambda_i}.$$

The nullspace of $\mathbf{E}(\mathbf{W}^{\mathrm{T}}\mathbf{W})^{-1}\mathbf{E}^{\mathrm{T}}$ *is the vector*

$$[-0.408 \quad -0.408 \quad -0.408 \quad 0.408 \quad 0.408 \quad 0.408]^{\mathrm{T}}, \tag{2.303}$$

which produces the solvability condition. Here, because $\mathbf{E}(\mathbf{W}^{\mathrm{T}}\mathbf{W})^{-1}\mathbf{E}^{\mathrm{T}}$ *is symmetric, the SVD reduces to the symmetric decomposition.*

Finally, the mapped $\tilde{\mathbf{x}}$ *is*

$$r_y \uparrow \begin{array}{c} r_x \rightarrow \\ \begin{bmatrix} -0.20 & 0.41 & -0.20 \\ 0.41 & 0.18 & 0.41 \\ -0.20 & 0.41 & -0.21 \end{bmatrix} \end{array},$$

and one cannot further decrease the sum-squared differences of the solution elements. One can confirm that this solution satisfies the equations. Evidently, it produces a minimum, not a maximum (it suffices to show that the eigenvalues of $\mathbf{W}^{\mathrm{T}}\mathbf{W}$ *are all non-negative). The addition of any of the nullspace vectors of* $\mathbf{E}$ *to* $\tilde{\mathbf{x}}$ *will necessarily increase the value of* J *and hence there is no bounded maximum. In real tomographic problems, the arc lengths making up matrix* $\mathbf{E}$ *are three-dimensional curves and depend upon the background index of refraction in the medium, which is usually itself determined from observations.[37] There are thus errors in* $\mathbf{E}$ *itself, rendering the problem one of non-linear estimation. Approaches to solving such problems are described in Chapter 3.*

Example *Consider the box reservoir problem with two sources ("end members") described in Eqs. (2.144)–(2.145), which was reduced to two equations in two unknowns by dividing through by one of the unknown fluxes,* J_0, *and solving for the ratios* J_1/J_0, J_2/J_0. *Suppose they are solved instead in their original form as two equations in three unknowns:*

$$J_1 + J_2 - J_0 = 0,$$
$$C_1 J_1 + C_2 J_2 - C_0 J_0 = 0.$$

To make it numerically definite, let $C_1 = 1$, $C_2 = 2$, $C_0 = 1.75$. The SVD produces

$$\mathbf{U} = \left\{ \begin{matrix} -0.51 & -0.86 \\ -0.86 & 0.51 \end{matrix} \right\}, \quad \mathbf{V} = \left\{ \begin{matrix} -0.41 & -0.89 & 0.20 \\ -0.67 & 0.44 & 0.59 \\ 0.61 & -0.11 & 0.78 \end{matrix} \right\}, \quad \Lambda = \mathrm{diag}\,([3.3, 0.39, 0])$$

(rounded). As the right-hand side of the governing equations vanishes, the coefficients of the range vectors, $\mathbf{v}_{1,2}$, must also vanish, and the only possible solution here is proportional to the nullspace vector, $\alpha_3 \mathbf{v}_3$, or $[J_1, J_2, J_0] = \alpha_3[0.20, 0.59, 0.78]^{\mathrm{T}}$, and α_3 is arbitrary. Alternatively, $J_1/J_0 = 0.25$, $J_2/J_0 = 0.75$.

Example *Consider the flow into a four-sided box with missing integration constant as described in Chapter 1. Total mass conservation and conservation of dye is denoted by C_i. Let the relative areas of each interface be $1, 2, 3, 1$ units respectively. Let the corresponding velocities on each side be $1, 1/2, -2/3, 0$ respectively, with the minus sign indicating a flow out. That mass is conserved is confirmed by*

$$1\,(1) + 2\left(\frac{1}{2}\right) + 3\left(\frac{-2}{3}\right) + 1\,(0) = 0.$$

Now suppose that the total velocity is not in fact known, but that an integration constant is missing on each interface, so that

$$1\left(\frac{1}{2} + b_1\right) + 2\,(1 + b_2) + 3\left(\frac{1}{3} + b_3\right) + 1\,(2 + b_4) = 0,$$

where the $b_i = [1/2, -1/2, -1, -2]$, but are here treated as unknown. Then the above equation becomes

$$b_1 + 2b_2 + 3b_3 + b_4 = -5.5,$$

or one equation in four unknowns. One linear combination of the unknown b_i can be determined. We would like more information. Suppose that a tracer of concentration $C_i = [2, 1, 3/2, 0]$ is measured at each side, and is believed conserved. The governing equation is

$$1\left(\frac{1}{2} + b_1\right)2 + 2\,(1 + b_2)\,1 + 3\left(\frac{1}{3} + b_3\right)\frac{3}{2} + 1\,(2 + b_4)\,0 = 0,$$

or

$$2b_1 + 2b_2 + 4.5b_3 + 0b_4 = -4.5,$$

giving a system of two equations in four unknowns:

$$\begin{Bmatrix} 1 & 2 & 3 & 1 \\ 2 & 2 & 4.5 & 0 \end{Bmatrix} \begin{bmatrix} b_1 \\ b_2 \\ b_3 \\ b_4 \end{bmatrix} = \begin{bmatrix} -5.5 \\ -4.5 \end{bmatrix}.$$

The SVD of the coefficient matrix, $\mathbf{E}$*, is*

$$\mathbf{E} = \begin{Bmatrix} -0.582 & -0.813 \\ 0.813 & 0.582 \end{Bmatrix} \begin{Bmatrix} 6.50 & 0 & 0 & 0 \\ 0 & 1.02 & 0 & 0 \end{Bmatrix}$$

$$\times \begin{Bmatrix} -0.801 & 0.179 & -0.454 & 0.347 \\ 0.009 & 0.832 & 0.429 & 0.340 \\ -0.116 & 0.479 & -0.243 & -0.835 \\ 0.581 & 0.215 & -0.742 & 0.259 \end{Bmatrix}.$$

The remainder of the solution is left to the reader.

2.5.6 Simple examples: differential and partial differential equations

Example *As an example of the use of this machinery with differential equations, consider*

$$\frac{d^2 x(r)}{dr^2} - k^2 x(r) = 0, \tag{2.304}$$

subject to initial and/or boundary conditions. Using one-sided, uniform discretization,

$$x((m+1)\,\Delta r) - (2 + k^2(\Delta r)^2)x(m\,\Delta r) + x((m-1)\,\Delta r) = 0, \tag{2.305}$$

at all interior points. Take the specific case, with two end conditions, $x(\Delta r) = 10$*,* $x(51\Delta r) = 1$*,* $\Delta r = 0.1$*. The numerical solution is depicted in Fig. 2.11 from the direct (conventional) solution to* $\mathbf{Ax} = \mathbf{y}$*. The first two rows of* $\mathbf{A}$ *were used to impose the boundary conditions on* $x(\Delta r)$*,* $x(51\Delta r)$*. The singular values of* $\mathbf{A}$ *are also plotted in Fig. 2.11. The range is over about two orders of magnitude, and there is no reason to suspect numerical difficulties. The first and last singular vectors* $\mathbf{u}_1$*,* $\mathbf{v}_1$*,* $\mathbf{u}_{51}$*,* $\mathbf{v}_{51}$ *are also plotted. One infers (by plotting additional such vectors), that the large singular values correspond to singular vectors showing a great deal of small-scale structure, and the smallest singular values correspond to the least structured (largest spatial scales) in both the solution and in the specific corresponding weighted averages of the equations. This result may be counterintuitive. But note that, in this problem, all elements of* $\mathbf{y}$ *vanish except the first two, which are being used to set the boundary conditions. We know from the analytical*

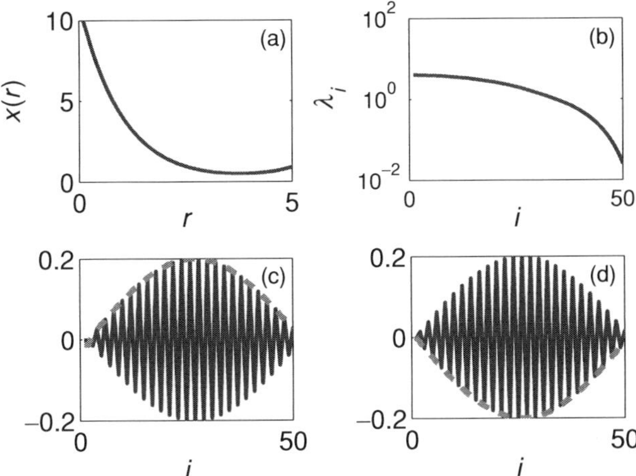

Figure 2.11 (a) $\bar{\mathbf{x}}$ from Eq. (2.304) by brute force from the simultaneous equations; (b) displays the corresponding singular values; all are finite (there is no nullspace); (c) shows $\mathbf{u}_1$ (solid curve), and $\mathbf{u}_{51}$ (dashed); (d) shows the corresponding $\mathbf{v}_1$, $\mathbf{v}_{51}$. The most robust information corresponds to the *absence* of small scales in the solution.

solution that the true solution is large-scale; most of the information contained in the differential equation (2.304) or its numerical counterpart (2.305) is an assertion that all small scales are absent; this information is the most robust and corresponds to the largest singular values. The remaining information, on the exact nature of the largest scales, which is contained in only two of the 51 equations – given by the boundary conditions, is extremely important, but is less robust than that concerning the absence of small scales. (Less "robust" is being used in the sense that small changes in the boundary conditions will lead to relatively large changes in the large-scale structures in the solution because of the division by relatively small λ_i.)

Example *Consider now the classical Neumann problem described in Chapter 1. The problem is to be solved on a 10×10 grid as in Eq. (1.17), $\mathbf{Ax} = \mathbf{b}$. The singular values of $\mathbf{A}$ are plotted in Fig. 2.12; the largest one is $\lambda_1 = 7.8$, and the smallest non-zero one is $\lambda_{99} = 0.08$. As expected, $\lambda_{100} = 0$. The singular vector $\mathbf{v}_{100}$ corresponding to the zero singular value is a constant; $\mathbf{u}_{100}$, also shown in Fig. 2.12, is not a constant, and has considerable structure – which provides the solvability condition for the Neumann problem, $\mathbf{u}_{100}^{\mathrm{T}}\mathbf{y} = 0$. The physical origin of the solvability condition is readily understood: Neumann boundary conditions prescribe boundary flux rates, and the sum of the interior source strengths plus the boundary flux rates must sum to zero, otherwise no steady state is possible. If the boundary conditions are homogeneous, then no flow takes place through the boundary, and the interior sources must sum to zero. In particular, the value of $\mathbf{u}_{100}$*

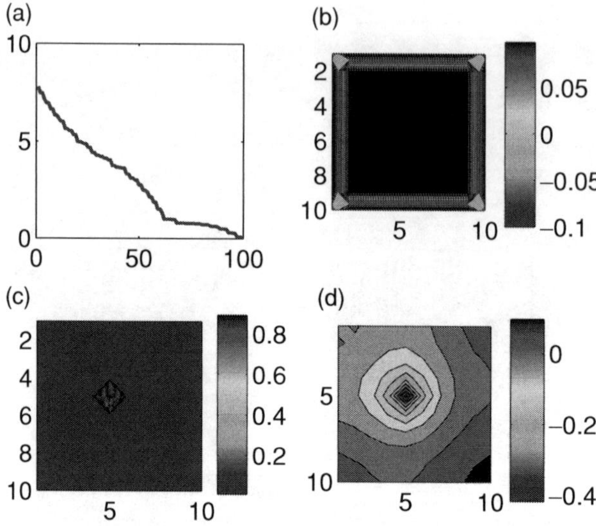

Figure 2.12 (a) Graph of the singular values of the coefficient matrix $\mathbf{A}$ of the numerical Neumann problem on a 10×10 grid. All λ_i are non-zero except the last one. (b) shows $\mathbf{u}_{100}$, the nullspace vector of $\mathbf{E}^{\mathrm{T}}$ defining the solvability or consistency condition for a solution through $\mathbf{u}_{100}^{\mathrm{T}}\mathbf{y} = 0$. Plotted as mapped onto the two-dimensional spatial grid (r_x, r_y) with $\Delta x = \Delta y = 1$. The interpretation is that the sum of the influx through the boundaries and from interior sources must vanish. Note that corner derivatives differ from other boundary derivatives by $1/\sqrt{2}$. The corresponding $\mathbf{v}_{100}$ is a constant, indeterminate with the information available, and not shown. (c) A source $\mathbf{b}$ (a numerical delta function) is present, not satisfying the solvability condition $\mathbf{u}_{100}^{\mathrm{T}}\mathbf{b} = 0$, because all boundary fluxes were set to vanish. (d) The particular SVD solution, $\tilde{\mathbf{x}}$, at rank $K = 99$. One confirms that $\mathbf{A}\tilde{\mathbf{x}} - \mathbf{b}$ is proportional to $\mathbf{u}_{100}$ as the source is otherwise inconsistent with no flux boundary conditions. With $\mathbf{b}$ a Kronecker delta function at one grid point, this solution is a numerical Green function for the Neumann problem and insulating boundary conditions. (See color figs.)

on the interior grid points is a constant. The Neumann problem is thus a forward one requiring coping with both a solution nullspace and a solvability condition.

2.5.7 Relation of least-squares to the SVD

What is the relationship of the SVD solution to the least-squares solutions? To some extent, the answer is already obvious from the orthonormality of the two sets of singular vectors: they *are* the least-squares solution, where it exists. When does the simple least-squares solution exist? Consider first the formally overdetermined problem, $M > N$. The solution (2.95) is meaningful if and only if the matrix inverse exists. Substituting the SVD for $\mathbf{E}$, one finds that

$$(\mathbf{E}^{\mathrm{T}}\mathbf{E})^{-1} = \left(\mathbf{V}_N \mathbf{\Lambda}_N^{\mathrm{T}} \mathbf{U}_N^{\mathrm{T}} \mathbf{U}_N \mathbf{\Lambda}_N \mathbf{V}_N^{\mathrm{T}}\right)^{-1} = \left(\mathbf{V}_N \mathbf{\Lambda}_N^2 \mathbf{V}_N^{\mathrm{T}}\right)^{-1}, \qquad (2.306)$$

where the semi-orthogonality of $\mathbf{U}_N$ has been used. Suppose that $K = N$, its maximum possible value; then $\mathbf{\Lambda}_N^2$ is $N \times N$ with *all non-zero diagonal elements* λ_i^2. The inverse in (2.306) may be found by inspection, using $\mathbf{V}_N^{\mathrm{T}} \mathbf{V}_N = \mathbf{I}_N$,

$$(\mathbf{E}^{\mathrm{T}}\mathbf{E})^{-1} = \mathbf{V}_N \mathbf{\Lambda}_N^{-2} \mathbf{V}_N^{\mathrm{T}}. \tag{2.307}$$

Then the solution (2.95) becomes

$$\tilde{\mathbf{x}} = \left(\mathbf{V}_N \mathbf{\Lambda}_N^{-2} \mathbf{V}_N^{\mathrm{T}}\right) \mathbf{V}_N \mathbf{\Lambda}_N \mathbf{U}_N^{\mathrm{T}} \mathbf{y} = \mathbf{V}_N \mathbf{\Lambda}_N^{-1} \mathbf{U}_N^{\mathrm{T}} \mathbf{y}, \tag{2.308}$$

which is identical to the SVD solution (2.286). If $K < N$, $\mathbf{\Lambda}_N^2$ has at least one zero on the diagonal, there is no matrix inverse, and the conventional least-squares solution is not defined. The condition for its existence is thus $K = N$, the so-called "full rank overdetermined" case. The condition $K < N$ is called "rank deficient." The dependence of the least-squares solution magnitude upon the possible presence of very small, but non-vanishing, singular values is obvious.

That the full-rank overdetermined case is unbiassed, as previously asserted (p. 47), can now be seen from

$$\langle \tilde{\mathbf{x}} - \mathbf{x} \rangle = \sum_{i=1}^{N} \frac{\left(\mathbf{u}_i^{\mathrm{T}} \langle \mathbf{y} \rangle\right)}{\lambda_i} \mathbf{v}_i - \mathbf{x} = \sum_{i=1}^{N} \frac{\mathbf{u}_i^{\mathrm{T}} \mathbf{y}_0}{\lambda_i} \mathbf{v}_i - \mathbf{x} = \mathbf{0},$$

with $\mathbf{y} = \mathbf{y}_0 + \mathbf{n}$, if $\langle \mathbf{n} \rangle = \mathbf{0}$, assuming that the correct $\mathbf{E}$ (model) is being used.

Now consider another problem, the conventional purely underdetermined least-squares one, whose solution is (2.167). When does that exist? Substituting the SVD,

$$\begin{aligned}
\tilde{\mathbf{x}} &= \mathbf{V}_M \mathbf{\Lambda}_M \mathbf{U}_M^{\mathrm{T}} \left(\mathbf{U}_M \mathbf{\Lambda}_M \mathbf{V}_M^{\mathrm{T}} \mathbf{V}_M \mathbf{\Lambda}_M^{\mathrm{T}} \mathbf{U}_M^{\mathrm{T}}\right)^{-1} \mathbf{y} \\
&= \mathbf{V}_M \mathbf{\Lambda}_M \mathbf{U}_M^{\mathrm{T}} \left(\mathbf{U}_M \mathbf{\Lambda}_M^2 \mathbf{U}_M^{\mathrm{T}}\right)^{-1} \mathbf{y}.
\end{aligned} \tag{2.309}$$

Again, the matrix inverse exists if and only if $\mathbf{\Lambda}_M^2$ has all non-zero diagonal elements, which occurs only when $K = M$. Under that specific condition by inspection,

$$\tilde{\mathbf{x}} = \mathbf{V}_M \mathbf{\Lambda}_M \mathbf{U}_M^{\mathrm{T}} \left(\mathbf{U}_M \mathbf{\Lambda}_M^{-2} \mathbf{U}_M^{\mathrm{T}}\right) \mathbf{y} = \mathbf{V}_M \mathbf{\Lambda}_M^{-1} \mathbf{U}_M^{\mathrm{T}} \mathbf{y}, \tag{2.310}$$

$$\tilde{\mathbf{n}} = 0, \tag{2.311}$$

which is once again the particular-SVD solution (2.286) – with the nullspace coefficients set to zero. This situation is usually referred to as the "full-rank underdetermined case." Again, the possible influence of small singular values is apparent and an arbitrary sum of nullspace vectors can be added to (2.310). The bias of (2.309) is given by the nullspace elements, and its uncertainty arises only from their contribution, because with $\tilde{\mathbf{n}} = \mathbf{0}$, the noise variance vanishes, and the particular-SVD solution covariance $\mathbf{C}_{xx}$ would be zero.

The particular-SVD solution thus coincides with the two simplest forms of least-squares solution, and generalizes both of them to the case where the matrix inverses do not exist. *All of the structure imposed by the SVD, in particular the restriction on the residuals in (2.264), is present in the least-squares solution.* If the system is not of full rank, then the simple least-squares solutions do not exist. *The SVD generalizes these results* by determining what it can: the elements of the solution lying in the range of $\mathbf{E}$, and an explicit structure for the resulting nullspace vectors.

The SVD provides a lot of flexibility. For example, it permits one to modify the simplest underdetermined solution (2.167) to remove its greatest shortcoming, the necessities that $\tilde{\mathbf{n}} = \mathbf{0}$ and that the residuals be orthogonal to all range vectors. One simply truncates the solution (2.270) at $K = K' < M$, thus assigning all vectors $\mathbf{v}_i$, $K' + i = 1, 2, \ldots, K$, to an "effective nullspace" (or substitutes K' for K everywhere). The residual is then

$$\tilde{\mathbf{n}} = \sum_{i=K'+1}^{M} \left(\mathbf{u}_i^{\mathrm{T}} \mathbf{y}\right) \mathbf{u}_i, \qquad (2.312)$$

with an uncertainty for $\tilde{\mathbf{x}}$ given by (2.293), but with the upper limit being K' rather than K. Such truncation has the effect of reducing the solution covariance contribution to the uncertainty, but increasing the contribution owing to the nullspace (and increasing the bias). In the presence of singular values small compared to σ_n, the resulting overall reduction in uncertainty may be very great – at the expense of a solution bias.

The general solution now consists of three parts,

$$\tilde{\mathbf{x}} = \sum_{i=1}^{K'} \frac{\mathbf{u}_i^{\mathrm{T}} \mathbf{y}}{\lambda_i} \mathbf{v}_i + \sum_{i=K'+1}^{K} \alpha_i \, \mathbf{v}_i + \sum_{i=K+1}^{N} \alpha_i \mathbf{v}_i, \qquad (2.313)$$

where the middle sum contains the terms appearing with singular values too small to be employed – for the given noise – and the third sum is the strict nullspace. Usually, one lumps the two nullspace sums together. The first sum, by itself, represents the particular-SVD solution in the presence of noise. Resolution and covariance matrices are modified by the substitution of K' for K.

This consideration is extremely important – it says that despite the mathematical condition $\lambda_i \neq 0$, some structures in the solution cannot be estimated with sufficient reliability to be useful. The "effective rank" is then not the same as the mathematical rank.

It was already noticed that the simplest form of least-squares does not provide a method to control the ratios of the solution and noise norms. Evidently, truncation of the SVD offers a simple way to do so – by reducing K'. It follows that the solution norm necessarily is reduced, and that the residuals must grow, along with the size of the solution nullspace. The issue of how to choose K', that is, "rank determination," in practice is an interesting one to which we will return (see p. 116).

2.5.8 Pseudo-inverses

Consider an arbitrary $M \times N$ matrix $\mathbf{E} = \mathbf{U}_K \mathbf{\Lambda}_K \mathbf{V}_K^{\mathrm{T}}$ and

$$\mathbf{E}\mathbf{x} + \mathbf{n} = \mathbf{y}.$$

Then if $\mathbf{E}$ is full-rank underdetermined, the minimum norm solution is

$$\tilde{\mathbf{x}} = \mathbf{E}^{\mathrm{T}}(\mathbf{E}\mathbf{E}^{\mathrm{T}})^{-1}\mathbf{y} = \mathbf{V}_K \mathbf{\Lambda}_K^{-1} \mathbf{U}_K^{\mathrm{T}} \mathbf{y}, \quad K = M,$$

and if it is full-rank overdetermined, the minimum noise solution is

$$\tilde{\mathbf{x}} = (\mathbf{E}^{\mathrm{T}}\mathbf{E})^{-1}\mathbf{E}^{\mathrm{T}}\mathbf{y} = \mathbf{V}_K \mathbf{\Lambda}_K^{-1} \mathbf{U}_K^{\mathrm{T}} \mathbf{y}, \quad K = N.$$

The first of these, the Moore–Penrose, or pseudo-inverse, $\mathbf{E}_1^+ = \mathbf{E}^{\mathrm{T}}(\mathbf{E}\mathbf{E}^{\mathrm{T}})^{-1}$ is sometimes also known as a "right-inverse," because $\mathbf{E}\mathbf{E}_1^+ = \mathbf{I}_M$. The second pseudo-inverse, $\mathbf{E}_2^+ = (\mathbf{E}^{\mathrm{T}}\mathbf{E})^{-1}\mathbf{E}^{\mathrm{T}}$, is a "left-inverse" as $\mathbf{E}_2^+\mathbf{E} = \mathbf{I}_N$. They can both be represented as $\mathbf{V}_K \mathbf{\Lambda}_K^{-1} \mathbf{U}_K^{\mathrm{T}}$, but with differing values of K. If $K < M, N$ neither of the pseudo-inverses exists, but $\mathbf{V}_K \mathbf{\Lambda}_K^{-1} \mathbf{U}_K^{\mathrm{T}} \mathbf{y}$ still provides the particular SVD solution. When $K = M = N$, one has a demonstration that the left and right inverses are identical; they are then written as $\mathbf{E}^{-1}$.

2.5.9 Row and column scaling

The effects on the least-squares solutions of the row and column scaling can now be understood. We discuss them in the context of noise covariances, but as always in least-squares, the weight matrices need no statistical interpretation, and can be chosen by the investigator to suit his or her convenience or taste.

Suppose we have two equations written as

$$\left\{\begin{matrix} 1 & 1 & 1 \\ 1 & 1.01 & 1 \end{matrix}\right\} \begin{bmatrix} x_1 \\ x_2 \\ x_3 \end{bmatrix} + \begin{bmatrix} n_1 \\ n_2 \end{bmatrix} = \begin{bmatrix} y_1 \\ y_2 \end{bmatrix},$$

where $\mathbf{R}_{nn} = \mathbf{I}_2$, $\mathbf{W} = \mathbf{I}_3$. The SVD of $\mathbf{E}$ is

$$\mathbf{U} = \left\{\begin{matrix} 0.7059 & -0.7083 \\ 0.7083 & 0.7059 \end{matrix}\right\}, \quad \mathbf{V} = \left\{\begin{matrix} 0.5764 & -0.4096 & 0.7071 \\ 0.5793 & 0.8151 & 0.0000 \\ 0.5764 & -0.4096 & -0.7071 \end{matrix}\right\},$$

$$\lambda_1 = 2.4536, \quad \lambda_2 = 0.0058.$$

The SVD solutions, choosing ranks $K' = 1, 2$ in succession, are very nearly (the numbers having been rounded)

$$\tilde{\mathbf{x}} \approx \left(\frac{y_1 + y_2}{2.45} \right) [0.58 \quad 0.58 \quad 0.58]^{\mathrm{T}}, \tag{2.314}$$

$$\tilde{\mathbf{x}} \approx \left(\frac{y_1 + y_2}{2.45} \right) [0.58 \quad 0.58 \quad 0.58]^{\mathrm{T}} + \left(\frac{y_1 - y_2}{0.0058} \right) [-0.41 \quad 0.82 \quad 0.41]^{\mathrm{T}},$$

respectively, so that the first term simply averages the two measurements, y_i, and the difference between them contributes – with great uncertainty – in the second term of the rank 2 solution owing to the very small singular value. The uncertainty is given by

$$(\mathbf{E}\mathbf{E}^{\mathrm{T}})^{-1} = \left\{ \begin{matrix} 1.51 \times 10^4 & -1.50 \times 10^4 \\ -1.50 \times 10^4 & 1.51 \times 10^4 \end{matrix} \right\}.$$

Now suppose that the covariance matrix of the noise is known to be

$$\mathbf{R}_{nn} = \left\{ \begin{matrix} 1 & 0.999999 \\ 0.999999 & 1 \end{matrix} \right\},$$

(an extreme case, chosen for illustrative purposes). Then, putting $\mathbf{W} = \mathbf{R}_{nn}$,

$$\mathbf{W}^{1/2} = \left\{ \begin{matrix} 1.0000 & 1.0000 \\ 0 & 0.0014 \end{matrix} \right\}, \qquad \mathbf{W}^{-\mathrm{T}/2} = \left\{ \begin{matrix} 1.0000 & 0 \\ -707.1063 & 707.1070 \end{matrix} \right\}.$$

The new system to be solved is

$$\left\{ \begin{matrix} 1.0000 \ 1.0000 \ 1.0000 \\ 0.0007 \ 7.0718 \ 0.0007 \end{matrix} \right\} \begin{bmatrix} x_1 \\ x_2 \\ x_3 \end{bmatrix} = \begin{bmatrix} y_1 \\ 707.1(-y_1 + y_2) \end{bmatrix}.$$

The SVD is

$$\mathbf{U} = \left\{ \begin{matrix} 0.1456 & 0.9893 \\ 0.9893 & -0.1456 \end{matrix} \right\}, \qquad \mathbf{V} = \left\{ \begin{matrix} 0.0205 & 0.7068 & 0.7071 \\ 0.9996 & -0.0290 & 0.0000 \\ 0.0205 & 0.7068 & -0.7071 \end{matrix} \right\},$$

$$\lambda_1 = 7.1450, \qquad \lambda_2 = 1.3996.$$

The second singular value is now much larger relative to the first one, and the two solutions are

$$\tilde{\mathbf{x}} \approx \frac{y_2 - y_1}{7.1} [0 \quad 1 \quad 0]^{\mathrm{T}}, \tag{2.315}$$

$$\tilde{\mathbf{x}} \approx \frac{y_2 - y_1}{7.1} [0 \quad 1 \quad 0]^{\mathrm{T}} + \frac{y_1 - 103 \, (y_2 - y_1)}{1.4} [0.71 \quad 0 \quad 0.71]^{\mathrm{T}},$$

and the rank 1 solution is given by the difference of the observations, in contrast to the unscaled solution. The result is quite sensible – the noise in the two equations is so nearly perfectly correlated that it can be removed by subtraction; the difference $y_2 - y_1$ is a nearly noise-free piece of information and accurately defines the appropriate structure in $\tilde{\mathbf{x}}$. In effect, the information provided in the row scaling with $\mathbf{R}$ permits the SVD to nearly eliminate the noise at rank 1 by an effective subtraction, whereas without that information, the noise is reduced in the solution (2.314) at rank 1 only by averaging.

At full rank, that is, $K = 2$, it can be confirmed that the solutions (2.314) and (2.315) are identical, as they must be. But the error covariances are quite different:

$$(\mathbf{E}'\mathbf{E}'^{\mathrm{T}})^{-1} = \begin{Bmatrix} 0.5001 & -0.707 \\ -0.707 & 0.5001 \end{Bmatrix}. \tag{2.316}$$

because the imposed covariance permits a large degree of noise suppression.

It was previously asserted (p. 67) that in a full-rank formally underdetermined system, row scaling is irrelevant to $\tilde{\mathbf{x}}$, $\tilde{\mathbf{n}}$, as may be seen as follows:

$$\begin{aligned}
\tilde{\mathbf{x}} &= \mathbf{E}'^{\mathrm{T}}(\mathbf{E}'\mathbf{E}'^{\mathrm{T}})^{-1}\mathbf{y}' \\
&= \mathbf{E}^{\mathrm{T}}\mathbf{W}^{-1/2}(\mathbf{W}^{-\mathrm{T}/2}\mathbf{E}\mathbf{E}^{\mathrm{T}}\mathbf{W}^{-1/2})^{-1}\mathbf{W}^{-\mathrm{T}/2}\mathbf{y} \\
&= \mathbf{E}^{\mathrm{T}}\mathbf{W}^{-1/2}\mathbf{W}^{1/2}(\mathbf{E}\mathbf{E}^{\mathrm{T}})^{-1}\mathbf{W}^{\mathrm{T}/2}\mathbf{W}^{-\mathrm{T}/2}\mathbf{y} \\
&= \mathbf{E}^{\mathrm{T}}(\mathbf{E}\mathbf{E}^{\mathrm{T}})^{-1}\mathbf{y},
\end{aligned} \tag{2.317}$$

but which is true only in the full rank situation.

There is a subtlety in row weighting. Suppose we have two equations of form

$$\begin{aligned}
10x_1 + 5x_2 + x_3 + n_1 &= 1, \\
100x_1 + 50x_2 + 10x_3 + n_2 &= 2,
\end{aligned} \tag{2.318}$$

after row scaling to make the expected noise variance in each the same. A rank 1 solution to these equations by SVD is $\tilde{\mathbf{x}} = [0.0165, 0.0083, 0.0017]^{\mathrm{T}}$, which produces residuals $\tilde{\mathbf{y}} - \mathbf{y} = [-0.79, 0.079]^{\mathrm{T}}$ – much smaller in the second equation than in the first one.

Consider that the second equation is ten times the first one – in effect saying that a measurement of ten times the values of $10x_1 + 5x_2 + x_3$ has the same noise in it as a measurement of one times this same linear combination. The second equation represents a much more accurate determination of this linear combination and the equation should be given much more weight in determining the unknowns – and both the SVD and ordinary least-squares does precisely that. To the extent that one finds this result undesirable (one should be careful about why it is so found), there is an easy remedy – divide the equations by their row norms $(\sum_j E_{ij}^2)^{1/2}$. But there will be a contradiction with any assertion that the noise in all equations was the same to begin with. Such row scaling is best regarded as non-statistical in nature.

An example of this situation is readily apparent in the box balances discussed in Chapter 1. Equations such as (1.32) could have row norms much larger than those (1.31) for the corresponding mass balance, simply because the tracer is measured by convention in its own units. If the tracer is, e.g., oceanic salt, values are, by convention, measured on the Practical Salinity Scale, and are near 35 (but are dimensionless). Because there is nothing fundamental about the choice of units, it seems unreasonable to infer that the requirement of tracer balance has an expected error 35 times smaller than for mass. One usually proceeds in the obvious way by dividing the tracer equations by their row norms as the first step. (This approach need have no underlying statistical validity, but is often done simply on the assumption that salt balance equations are unlikely to be 35 times more accurate than the mass ones.) The second step is to ask whether anything further can be said about the relative errors of mass and salt balance, which would introduce a second, purely statistical, row weight.

Column scaling

In the least-squares problem, we formally introduced a "column scaling" matrix $\mathbf{S}$. Column scaling operates on the SVD solution exactly as it does in the least-squares solution, to which it reduces in the two special cases already described. That is, we should apply the SVD to sets of equations only where any knowledge of the solution element size has been removed first. If the SVD has been computed for such a column (and row) scaled system, the solution is for the scaled unknown $\mathbf{x}'$, and the physical solution is

$$\tilde{\mathbf{x}} = \mathbf{S}^{\mathrm{T}/2}\tilde{\mathbf{x}}'. \tag{2.319}$$

But there are occasions, with underdetermined systems, where a non-statistical scaling may also be called for, the analogue to the situation considered above where a row scaling was introduced on the basis of possible non-statistical considerations.

Example *Suppose we have one equation in two unknowns:*

$$10x_1 + 1x_2 = 3. \tag{2.320}$$

The particular-SVD solution produces $\tilde{\mathbf{x}} = [0.2970, 0.0297]^{\mathrm{T}}$, *in which the magnitude of x_1 is much larger than that of x_2 and the result is readily understood. As we have seen, the SVD automatically finds the exact solution, subject to making the solution norm as small as possible. Because the coefficient of x_1 in (2.320) is ten times that of x_2, it is more efficient in minimizing the norm to give x_1 a larger value than x_2. Although we have demonstrated this dependence for a trivial example,* *similar behavior occurs for underdetermined systems in general. In many cases, this distribution of the elements of the solution vector* $\mathbf{x}$ *is desirable, the numerical value 10 appearing for good physical reasons. In other problems, the numerical values appearing in the coefficient matrix* $\mathbf{E}$ *are an "accident." In the box-balance example of Chapter 1, the distances defining the interfaces of the boxes are a consequence of the spatial distance between measurements. Unless one believed that velocities should be larger where the distances are greater or the fluid depth was greater, then the solutions may behave unphysically.*[38] *Indeed, in some situations the velocities are expected to be inverse to the fluid depth and such a prior statistical hypothesis is best imposed after one has removed the structural accidents from the system. (The tendency for the solutions to be proportional to the column norms is not rigid. In particular, the equations themselves may preclude the proportionality.) Take a positive definite, diagonal matrix* $\mathbf{S}$, *and rewrite (2.87) as*

$$\mathbf{E}\mathbf{S}^{T/2}\mathbf{S}^{-T/2}\mathbf{x} + \mathbf{n} = \mathbf{y}.$$

Then

$$\mathbf{E}'\mathbf{x}' + \mathbf{n} = \mathbf{y}, \ \ \mathbf{E}' = \mathbf{E}\mathbf{S}^{T/2}, \ \ \mathbf{x}' = \mathbf{S}^{-T/2}\mathbf{x}.$$

Solving

$$\tilde{\mathbf{x}}' = \mathbf{E}'^{T}(\mathbf{E}'\mathbf{E}'^{T})^{-1}\mathbf{y}, \ \ \tilde{\mathbf{x}} = \mathbf{S}^{T/2}\tilde{\mathbf{x}}'. \tag{2.321}$$

How should $\mathbf{S}$ *be chosen? Apply the recipe (2.321) for the one equation example of (2.320), with*

$$\mathbf{S} = \begin{Bmatrix} 1/a^2 & 0 \\ 0 & 1/b^2 \end{Bmatrix},$$

$$\mathbf{E}' = \begin{Bmatrix} 10/a & 1/b \end{Bmatrix}, \ \mathbf{E}'\mathbf{E}'^{T} = \frac{100}{a^2} + \frac{1}{b^2}, \tag{2.322}$$

$$(\mathbf{E}'\mathbf{E}'^{T})^{-1} = \frac{a^2 b^2}{100 b^2 + a^2}, \tag{2.323}$$

$$\tilde{\mathbf{x}}' = \begin{Bmatrix} 10/a \\ 1/b \end{Bmatrix} \frac{a^2 b^2}{100 b^2 + a^2} 3, \tag{2.324}$$

$$\tilde{\mathbf{x}} = \mathbf{S}^{T/2}\tilde{\mathbf{x}}' = \begin{Bmatrix} 10/a^2 \\ 1/b^2 \end{Bmatrix} \frac{a^2 b^2}{100 b^2 + a^2} 3. \tag{2.325}$$

The relative magnitudes of the elements of $\tilde{\mathbf{x}}$ *are proportional to* $10/a^2$, $1/b^2$. *To make the numerical values identical, choose* $a^2 = 10$, $b^2 = 1$, *that is, divide the elements of the first column of* $\mathbf{E}$ *by* $\sqrt{10}$ *and the second column by* $\sqrt{1}$. *The apparent rule (which is general) is to divide each column of* $\mathbf{E}$ *by the square root of its length. The square root of the length may be surprising, but arises because of the second multiplication by the elements of* $\mathbf{S}^{\mathrm{T}/2}$ *in (2.321). This form of column scaling should be regarded as "non-statistical," in that it is based upon inferences about the numerical magnitudes of the columns of* $\mathbf{E}$ *and does not employ information about the statistics of the solution. Indeed, its purpose is to prevent the imposition of structure on the solution for which no statistical basis has been anticipated. In general, the elements of* $\tilde{\mathbf{x}}$ *will not prove to be equal – because the equations themselves do not permit it.*

If the system is full-rank overdetermined, the column weights drop out, as claimed for least-squares above. To see this result, consider that, in the full-rank case,

$$
\begin{aligned}
\tilde{\mathbf{x}}' &= (\mathbf{E}'^{\mathrm{T}}\mathbf{E}')^{-1}\mathbf{E}'^{\mathrm{T}}\mathbf{y} \\
\tilde{\mathbf{x}} &= \mathbf{S}^{\mathrm{T}/2}(\mathbf{S}^{1/2}\mathbf{E}^{\mathrm{T}}\mathbf{E}\mathbf{S}^{\mathrm{T}/2})^{-1}\mathbf{S}^{1/2}\mathbf{E}^{\mathrm{T}}\mathbf{y} \\
&= \mathbf{S}^{\mathrm{T}/2}\mathbf{S}^{-\mathrm{T}/2}(\mathbf{E}^{\mathrm{T}}\mathbf{E})^{-1}\mathbf{S}^{-1/2}\mathbf{S}^{1/2}\mathbf{E}^{\mathrm{T}}\mathbf{y} = (\mathbf{E}^{\mathrm{T}}\mathbf{E})^{-1}\mathbf{E}^{\mathrm{T}}\mathbf{y}.
\end{aligned}
\tag{2.326}
$$

Usually row scaling is done prior to column scaling so that the row norms have a simple physical interpretation, but one can row normalize in the column normalized space.

2.5.10 Solution and observation resolution: data ranking

Typically, either or both of the set of vectors $\mathbf{v}_i$, $\mathbf{u}_i$ used to present $\mathbf{x}$, $\mathbf{y}$ will be deficient in the sense of the expansions in (2.187). It follows immediately from Eqs. (2.188) that the particular-SVD solution is

$$
\tilde{\mathbf{x}} = \mathbf{V}_K \mathbf{V}_K^{\mathrm{T}}\mathbf{x} = \mathbf{T}_v\mathbf{x},
\tag{2.327}
$$

and the data vector with which both it and the general solution are consistent is

$$
\tilde{\mathbf{y}} = \mathbf{U}_K \mathbf{U}_K^{\mathrm{T}}\mathbf{y} = \mathbf{T}_u\mathbf{y}.
\tag{2.328}
$$

It is convenient therefore, to define the solution and observation resolution matrices,

$$
\mathbf{T}_v = \mathbf{V}_K \mathbf{V}_K^{\mathrm{T}}, \qquad \mathbf{T}_u = \mathbf{U}_K \mathbf{U}_K^{\mathrm{T}}.
\tag{2.329}
$$

The interpretation of the solution resolution matrix is identical to that in the square-symmetric case (p. 77).

Interpretation of the data resolution matrix is slightly subtle. Suppose an element of $\mathbf{y}$ was fully resolved, that is, some column, j_0, of $\mathbf{U}_K \mathbf{U}_K^{\mathrm{T}}$ were all zeros except for diagonal element j_0, which is one. Then a change of unity in y_{j_0} would produce a change in $\tilde{\mathbf{x}}$ that would leave unchanged all other elements of $\tilde{\mathbf{y}}$. If element j_0 is *not* fully resolved, then a change of unity in observation y_{j_0} produces a solution that leads to changes in other elements of $\tilde{\mathbf{y}}$. Stated slightly differently, if y_i is not fully

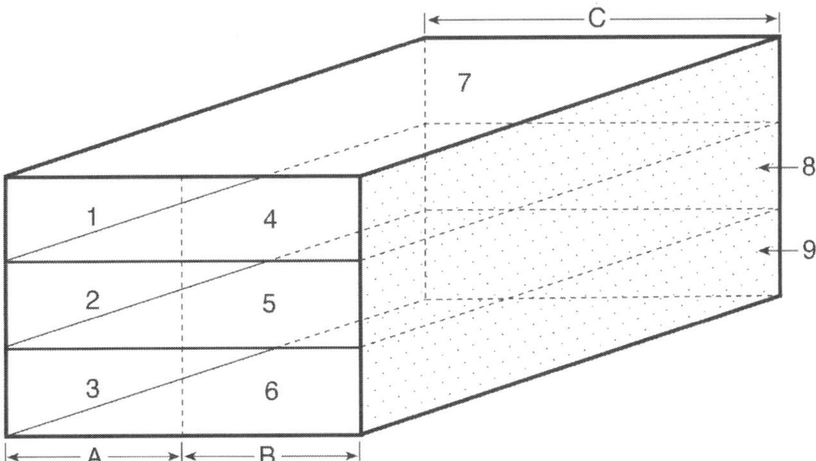

Figure 2.13 Box model with mass fluxes across the bounding interfaces $i =$ 1, ..., 9. Mass conservation equations are written for the total volume, for specifying the flux across the southern and northern boundaries separately, and for the mass balance in the three layers shown.

resolved, the system lacks adequate information to distinguish equation i from a linear dependence on one or more other equations.[39]

One can use these ideas to construct quantitative statements of which observations are the most important ("data ranking"). From (2.190), trace($\mathbf{T}_u$) = K and the relative contribution to the solution of any particular constraint is given by the corresponding diagonal element of $\mathbf{T}_u$.

Consider the example (2.318) without row weighting. At rank 1,

$$\mathbf{T}_u = \begin{Bmatrix} 0.0099 & 0.099 \\ 0.099 & 0.9901 \end{Bmatrix},$$

showing that the second equation has played a much more important role in the solution than the first one – despite the fact that we asserted the expected noise in both to be the same. The reason is that described above, the second equation in effect asserts that the measurement is 10 times more accurate than in the first equation – and the data resolution matrix informs us of that explicitly. The elements of $\mathbf{T}_u$ can be used to rank the data in order of importance to the final solution. All of the statements about the properties of resolution matrices made above apply to both $\mathbf{T}_u$, $\mathbf{T}_v$.

Example *A fluid flows into a box as depicted in Fig. 2.13, which is divided into three layers. The box is bounded on the left and right by solid walls. At its southern boundary, the inward (positive directed) mass flow has six unknown values, three each into the layers on the west (region A, q_i, $i = 1, 2, 3$), and three each into the* *layers on the east (region B, $q_i, i = 4, 5, 6$). At the northern boundary, there are only three unknown flow fields, for which a positive value denotes a flow outward (region C, $i = 7, 8, 9$). We write seven mass conservation equations. The first three represent mass balance in each of the three layers. The second three fix the mass transports across each of the sections A, B, C. The final equation asserts that the sum of the three mass transports across the sections must vanish (for overall mass balance). Note that the last equation is a linear combination of equations 4 to 6. Then the equations are*

$$\left\{ \begin{array}{ccccccccc} 1 & 0 & 0 & 1 & 0 & 0 & -1 & 0 & 0 \\ 0 & 1 & 0 & 0 & 1 & 0 & 0 & -1 & 0 \\ 0 & 0 & 1 & 0 & 0 & 1 & 0 & 0 & -1 \\ 1 & 1 & 1 & 0 & 0 & 0 & 0 & 0 & 0 \\ 0 & 0 & 0 & 1 & 1 & 1 & 0 & 0 & 0 \\ 0 & 0 & 0 & 0 & 0 & 0 & 1 & 1 & 1 \\ 1 & 1 & 1 & 1 & 1 & 1 & -1 & -1 & -1 \end{array} \right\} \mathbf{x} + \mathbf{n} = \mathbf{y}, \qquad (2.330)$$

which is 7×9. Inspection, or the singular values, demonstrate that the rank of the coefficient matrix, $\mathbf{E}$, is $K = 5$ because of the linear dependencies noted. Figures 2.14 and 2.15 display the singular vectors and the rank 5 resolution matrices for Eq. (2.330). Notice that the first pair of singular vectors, $\mathbf{u}_1, \mathbf{v}_1$ show that the mean mass flux (accounting for the sign convention making the northern flow positive outwards) is determined by a simple sum over all of the equations. Other linear combinations of the mass fluxes are determined from various sums and differences of the equations. It is left to the reader to determine whether they make physical sense. The resolution matrices show that none of the equations nor elements of the mass flux are fully resolved, and that all the equations contribute roughly equally.

If row and column scaling have been applied to the equations prior to application of the SVD, the covariance, uncertainty, and resolution expressions apply in those new, scaled spaces. The resolution in the original spaces is

$$\mathbf{T}_v = \mathbf{S}^{T/2} \mathbf{T}_{v'} \mathbf{S}^{-T/2}, \qquad (2.331)$$

$$\mathbf{T}_u = \mathbf{W}^{T/2} \mathbf{T}_{u'} \mathbf{W}^{-T/2}, \qquad (2.332)$$

so that

$$\tilde{\mathbf{x}} = \mathbf{T}_v \mathbf{x}, \qquad \tilde{\mathbf{y}} = \mathbf{T}_u \mathbf{y}, \qquad (2.333)$$

where $\mathbf{T}_{v'}, \mathbf{T}_{u'}$ are the expressions Eq. (2.329) in the scaled space. Some insight into the effects of weighting can be obtained by applying row and column scaling

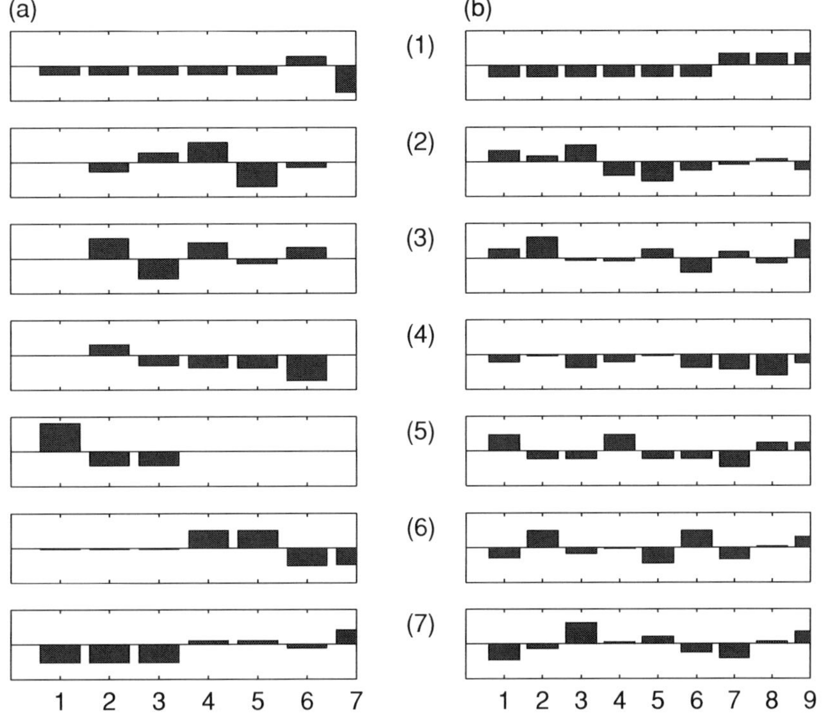

Figure 2.14 Singular vectors of the box flow coefficient matrix. Column (a) shows the $\mathbf{u}_i$, the column (b) the $\mathbf{v}_i$. The last two null space vectors $\mathbf{v}_8$, $\mathbf{v}_9$ are not shown, but $\mathbf{u}_6$, $\mathbf{u}_7$ do lie in the data nullspace, and $\mathbf{v}_6$, $\mathbf{v}_7$ are also in the solution nullspace. All plots have a full scale of ± 1.

to the simple physical example of Eq. (2.330). The uncertainty in the new space is $\mathbf{P} = \mathbf{S}^{1/2}\mathbf{P}'\mathbf{S}^{\mathrm{T}/2}$ where $\mathbf{P}'$ is the uncertainty in the scaled space.

We have seen an interpretation of three matrices obtained from the SVD: $\mathbf{V}_K\mathbf{V}_K^{\mathrm{T}}$, $\mathbf{U}_K\mathbf{U}_K^{\mathrm{T}}$, $\mathbf{V}_K\mathbf{\Lambda}_K^{-2}\mathbf{V}_K^{\mathrm{T}}$. The reader may well wonder, on the basis of the symmetries between solution and data spaces, whether there is an interpretation of the remaining matrix $\mathbf{U}_K\mathbf{\Lambda}_K^{-2}\mathbf{U}_K^{\mathrm{T}}$? To understand its use, recall the normal equations (2.163, 2.164) that emerged from the constrained objective function (2.149). They become, using the SVD for $\mathbf{E}$,

$$\mathbf{V}\mathbf{\Lambda}\mathbf{U}^{\mathrm{T}}\boldsymbol{\mu} = \mathbf{x}, \tag{2.334}$$

$$\mathbf{U}\mathbf{\Lambda}\mathbf{V}^{\mathrm{T}}\mathbf{x} = \mathbf{y}. \tag{2.335}$$

The pair of equations is always square, of dimension $M + N$. These equations show that $\mathbf{U}\mathbf{\Lambda}^2\mathbf{U}^{\mathrm{T}}\boldsymbol{\mu} = \mathbf{y}$. The particular SVD solution is

$$\boldsymbol{\mu} = \mathbf{U}_K\mathbf{\Lambda}_K^{-2}\mathbf{U}_K^{\mathrm{T}}\mathbf{y}, \tag{2.336}$$

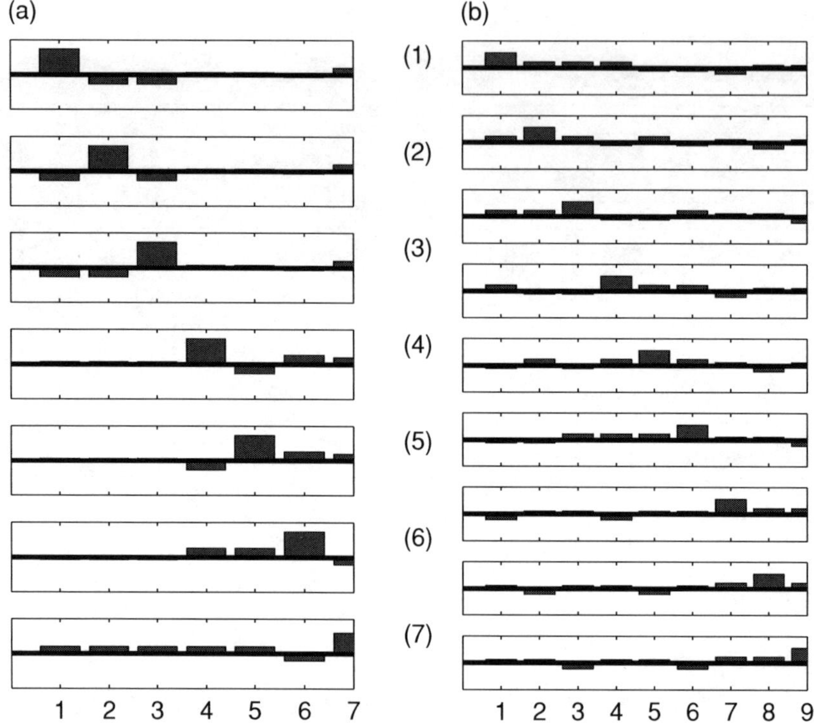

Figure 2.15 Rank $K = 5$ $\mathbf{T}_u$ (a) and $\mathbf{T}_v$ (b) for the box model coefficient matrix. Full scale is ± 1 in all plots.

involving the "missing" fourth matrix. Thus,

$$\frac{\partial J}{\partial \mathbf{y}} = 2\mathbf{U}_K \mathbf{\Lambda}_K^{-2} \mathbf{U}_K^{\mathrm{T}} \mathbf{y},$$

and, taking the second derivative,

$$\frac{\partial^2 J}{\partial \mathbf{y}^2} = 2\mathbf{U}_K \mathbf{\Lambda}_K^{-2} \mathbf{U}_K^{\mathrm{T}}, \tag{2.337}$$

which is the Hessian of J with respect to the data. If any of the λ_i become very small, the objective function will be extremely sensitive to small perturbations in $\mathbf{y}$ – producing an effective nullspace of the problem. Equation (2.337) supports the suggestion that perfect constraints can lead to difficulties.

2.5.11 Relation to tapered and weighted least-squares

In using least-squares, a shift was made from the simple objective functions (2.89) and (2.149) to the more complicated ones in (2.114) or (2.125). The change was made to permit a degree of control of the relative norms of $\tilde{\mathbf{x}}$, $\tilde{\mathbf{n}}$, and through the use of $\mathbf{W}$, $\mathbf{S}$ of the individual elements and the resulting uncertainties and covariances. Application of the weight matrices $\mathbf{W}$, $\mathbf{S}$ through their Cholesky decompositions to the equations prior to the use of the SVD is equally valid – thus providing the same amount of influence over the solution elements. The SVD provides its control over the solution norms, uncertainties and covariances through choice of the effective rank K'. This approach is different from the use of the extended objective functions (2.114), but the SVD is actually useful in understanding the effect of such functions.

Assume that any necessary $\mathbf{W}$, $\mathbf{S}$ have been applied. Then the full SVD, including zero singular values and corresponding singular vectors, is substituted into (2.116),

$$\tilde{\mathbf{x}} = (\gamma^2 \mathbf{I}_N + \mathbf{V}\mathbf{\Lambda}^{\mathrm{T}}\mathbf{\Lambda}\mathbf{V}^{\mathrm{T}})^{-1}\mathbf{V}\mathbf{\Lambda}^{\mathrm{T}}\mathbf{U}^{\mathrm{T}}\mathbf{y},$$

and

$$\tilde{\mathbf{x}} = \mathbf{V}(\mathbf{\Lambda}^{\mathrm{T}}\mathbf{\Lambda} + \gamma^2 \mathbf{I})^{-1}\mathbf{V}^{\mathrm{T}}\mathbf{V}\mathbf{\Lambda}^{\mathrm{T}}\mathbf{U}^{\mathrm{T}}\mathbf{y} \tag{2.338}$$

$$= \mathbf{V} \operatorname{diag}\left(\lambda_i^2 + \gamma^2\right)^{-1} \mathbf{\Lambda}^{\mathrm{T}}\mathbf{U}^{\mathrm{T}}\mathbf{y},$$

or

$$\tilde{\mathbf{x}} = \sum_{i=1}^{N} \frac{\lambda_i \left(\mathbf{u}_i^{\mathrm{T}}\mathbf{y}\right)}{\lambda_i^2 + \gamma^2}\mathbf{v}_i. \tag{2.339}$$

It is now apparent what the effect of "tapering" has done in least-squares. The word refers to the tapering down of the coefficients of the $\mathbf{v}_i$ by the presence of γ^2 from the values they would have in the "pure" SVD. In particular, the guarantee that matrices like $(\mathbf{E}^{\mathrm{T}}\mathbf{E} + \gamma^2 \mathbf{I})$ always have an inverse despite vanishing singular values, is seen to follow because the presence of $\gamma^2 > 0$ assures that the inverse of the sum always exists, irrespective of the rank of $\mathbf{E}$. The simple addition of a positive constant to the diagonal of a singular matrix is a well-known ad hoc method for giving it an approximate inverse. Such methods are a form of what is usually known as "regularization," and are procedures for suppressing nullspaces. Note that the coefficients of $\mathbf{v}_i$ vanish with λ_i and a solution nullspace still exists.

The residuals of the tapered least-squares solution can be written in various forms. Eqs. (2.117) are

$$\tilde{\mathbf{n}} = \gamma^2 \mathbf{U}(\gamma^2 \mathbf{I} + \mathbf{\Lambda}\mathbf{\Lambda}^{\mathrm{T}})^{-1}\mathbf{U}^{\mathrm{T}}\mathbf{y} \tag{2.340}$$

$$= \sum_{i=1}^{M} \frac{\left(\mathbf{u}_i^{\mathrm{T}}\mathbf{y}\right)\gamma^2}{\lambda_i^2 + \gamma^2}\mathbf{u}_i, \quad \gamma^2 > 0,$$

that is, the projection of the noise onto the range vectors $\mathbf{u}_i$ no longer vanishes. Some of the structure of the range of $\mathbf{E}^{\mathrm{T}}$ is being attributed to noise and it is no longer true that the residuals are subject to the rigid requirement (2.264) of having zero contribution from the range vectors. An increased noise norm is also deemed acceptable, as the price of keeping the solution norm small, by assuring that none of the coefficients in the sum (2.339) becomes overly large – values we can control by varying γ^2. The covariance of this solution about its mean (Eq. (2.118)) is readily rewritten as

$$
\begin{aligned}
\mathbf{C}_{xx} &= \sum_{i=1}^{N}\sum_{j=1}^{N} \frac{\lambda_i\lambda_j\mathbf{u}_i^{\mathrm{T}}\mathbf{R}_{nn}\mathbf{u}_j^{\mathrm{T}}}{(\lambda_i^2+\gamma^2)(\lambda_j^2+\gamma^2)}\mathbf{v}_i\mathbf{v}_j^{\mathrm{T}} \\
&= \sigma_n^2 \sum_{i=1}^{N} \frac{\lambda_i^2}{(\lambda_i^2+\gamma^2)^2}\mathbf{v}_i\mathbf{v}_i^{\mathrm{T}} \\
&= \sigma_n^2\mathbf{V}(\mathbf{\Lambda}^{\mathrm{T}}\mathbf{\Lambda}+\gamma^2\mathbf{I}_N)^{-1}\mathbf{\Lambda}^{\mathrm{T}}\mathbf{\Lambda}(\mathbf{\Lambda}^{\mathrm{T}}\mathbf{\Lambda}+\gamma^2\mathbf{I}_N)^{-1}\mathbf{V}^{\mathrm{T}},
\end{aligned}
\tag{2.341}
$$

where the second and third lines are again the special case of white noise. The role of γ^2 in controlling the solution variance, as well as the solution size, should be plain. The tapered least-squares solution is biassed – but the presence of the bias can greatly reduce the solution variance. Study of the solution as a function of γ^2 is known as "ridge regression." Elaborate techniques have been developed for determining the "right" value of γ^2.[40]

The uncertainty, $\mathbf{P}$, is readily found as

$$
\begin{aligned}
\mathbf{P} &= \gamma^2 \sum_{i=1}^{N}\frac{\mathbf{v}_i\mathbf{v}_i^{\mathrm{T}}}{(\lambda_i^2+\gamma^2)^2} + \sigma_n^2\sum_{i=1}^{N}\frac{\lambda_i^2\mathbf{v}_i\mathbf{v}_i^{\mathrm{T}}}{(\lambda_i^2+\gamma^2)^2} \\
&= \gamma^2\mathbf{V}(\mathbf{\Lambda}^{\mathrm{T}}\mathbf{\Lambda}+\gamma^2\mathbf{I})^{-2}\mathbf{V}^{\mathrm{T}} + \sigma_n^2\mathbf{V}(\mathbf{\Lambda}^{\mathrm{T}}\mathbf{\Lambda}+\gamma^2\mathbf{I})^{-1}\mathbf{\Lambda}^{\mathrm{T}}\mathbf{\Lambda}(\mathbf{\Lambda}^{\mathrm{T}}\mathbf{\Lambda}+\gamma^2\mathbf{I})^{-1}\mathbf{V}^{\mathrm{T}},
\end{aligned}
\tag{2.342}
$$

showing the variance reduction possible for finite γ^2 (reduction of the second term), and the bias error incurred in compensation in the first term.

The truncated SVD and the tapered SVD-tapered least-squares solutions produce the same qualitative effect – it is possible to increase the noise norm while decreasing the solution norm. Although the solutions differ somewhat, they both achieve the purpose stated above – to extend ordinary least-squares in such a way that one can control the relative noise and solution norms. The quantitative difference between them is readily stated – the truncated form makes a clear separation between range and nullspace in both solution and residual spaces: the basic SVD solution contains only range vectors and no nullspace vectors. The residual contains only nullspace vectors and no range vectors. The tapered form permits a merger of the two different sets of vectors: then both solution and residuals contain some contribution from both formal range and effective nullspaces ($0 < \gamma^2$).

We have already seen several times that preventing $\tilde{\mathbf{n}}$ from having any contribution from the range of $\mathbf{E}^{\mathrm{T}}$ introduces covariances into the residuals, with a consequent inability to produce values that are strictly white noise in character (although it is only a real issue as the number of degrees of freedom, $M - K$, goes toward zero). In the tapered form of least-squares, or the equivalent tapered SVD, contributions from the range vectors $\mathbf{u}_i$, $i = 1, 2, \ldots, K$, are permitted, and a potentially more realistic residual estimate is obtained. (There is usually no good reason why $\tilde{\mathbf{n}}$ is expected to be orthogonal to the range vectors.)

2.5.12 Resolution and variance of tapered solutions

The tapered least-squares solutions have an implicit nullspace, arising both from the terms corresponding to zero singular values, or from values small compared to γ^2. To obtain a measure of solution resolution when the $\mathbf{v}_i$ vectors have not been computed, consider a situation in which the true solution were $\mathbf{x}_{j_0} \equiv \delta_{j,j_0}$, that is, unity in the j_0 element and zero elsewhere. Then, in the absence of noise, the correct value of $\mathbf{y}$ would be

$$\mathbf{E}\mathbf{x}_{j_0} = \mathbf{y}_{j_0}, \tag{2.343}$$

defining $\mathbf{y}_{j_0}$. Suppose we actually knew (had measured) $\mathbf{y}_{j_0}$, what solution $\mathbf{x}_{j_0}$ would be obtained?

Assuming all covariance matrices have been applied and suppressing any primes, tapered least-squares (Eq. (2.120)) produces

$$\tilde{\mathbf{x}}_{j_0} = \mathbf{E}^{\mathrm{T}}(\mathbf{E}\mathbf{E}^{\mathrm{T}} + \gamma^2 \mathbf{I})^{-1}\mathbf{y}_{j_0} = \mathbf{E}^{\mathrm{T}}(\mathbf{E}\mathbf{E}^{\mathrm{T}} + \gamma^2 \mathbf{I})^{-1}\mathbf{E}\mathbf{x}_{j_0}, \tag{2.344}$$

which is row (or column) j_0 of

$$\mathbf{T}_v = \mathbf{E}^{\mathrm{T}}(\mathbf{E}\mathbf{E}^{\mathrm{T}} + \gamma^2 \mathbf{I})^{-1}\mathbf{E}. \tag{2.345}$$

Thus we can interpret any row or column of $\mathbf{T}_v$ as the solution for one in which a Kronecker delta was the underlying correct one. It is an easy matter, using the SVD of $\mathbf{E}$ and letting $\gamma^2 \to 0$, to show that (2.345) reduces to $\mathbf{V}\mathbf{V}^{\mathrm{T}}$, if $K = M$. These expressions apply in the row- and column-scaled space and are suitably modified to take account of any $\mathbf{W}, \mathbf{S}$ which may have been applied, as in Eqs. (2.331) and (2.332). An obvious variant of (2.345) follows from the alternative least-squares solution (2.127), with $\mathbf{W} = \gamma^2 \mathbf{I}$, $\mathbf{S} = \mathbf{I}$,

$$\mathbf{T}_v = (\mathbf{E}^{\mathrm{T}}\mathbf{E} + \gamma^2 \mathbf{I})^{-1}\mathbf{E}^{\mathrm{T}}\mathbf{E}. \tag{2.346}$$

Data resolution matrices are obtained similarly. Let $y_j = \delta_{jj_1}$. Equation (2.135) produces

$$\tilde{\mathbf{x}}_{j_1} = \mathbf{E}^{\mathrm{T}}(\mathbf{E}\mathbf{E}^{\mathrm{T}} + \gamma^2 \mathbf{I})^{-1}\mathbf{y}_{j_1}, \tag{2.347}$$

which if substituted into the original equations is

$$\mathbf{E}\tilde{\mathbf{x}}_{j_1} = \mathbf{E}\mathbf{E}^{\mathrm{T}}(\mathbf{E}\mathbf{E}^{\mathrm{T}} + \gamma^2\mathbf{I})^{-1}\mathbf{y}_{j_1}. \tag{2.348}$$

Thus,

$$\mathbf{T}_u = \mathbf{E}\mathbf{E}^{\mathrm{T}}(\mathbf{E}\mathbf{E}^{\mathrm{T}} + \gamma^2\mathbf{I})^{-1}. \tag{2.349}$$

The alternate form is

$$\mathbf{T}_u = \mathbf{E}(\mathbf{E}^{\mathrm{T}}\mathbf{E} + \gamma^2\mathbf{I})^{-1}\mathbf{E}^{\mathrm{T}}. \tag{2.350}$$

All of the resolution matrices reduce properly to either $\mathbf{U}\mathbf{U}^{\mathrm{T}}$ or $\mathbf{V}\mathbf{V}^{\mathrm{T}}$ as $\gamma^2 \to 0$ when the SVD for $\mathbf{E}$ is substituted, and $K = M$ or N as necessary.

2.6 Combined least-squares and adjoints

2.6.1 Exact constraints

Consider now a modest generalization of the constrained problem Eq. (2.87) in which the unknowns $\mathbf{x}$ are also meant to satisfy some constraints exactly, or nearly so, for example,

$$\mathbf{A}\mathbf{x} = \mathbf{b}. \tag{2.351}$$

In some contexts, (2.351) is referred to as the "model," a term also employed, confusingly, for the physics defining $\mathbf{E}$ and/or the statistics assumed to describe $\mathbf{x}, \mathbf{n}$. We will temporarily refer to Eq. (2.351) as "perfect constraints," as opposed to those involving $\mathbf{E}$, which generally always have a non-zero noise element.

An example of a model in these terms occurs in acoustic tomography (Chapter 1), where measurements exist of both density and velocity fields, and they are connected by dynamical relations; the errors in the relations are believed to be so much smaller than those in the data, that for practical purposes, the constraints (2.351) might as well be treated as though they are perfect.[41] But otherwise, the distinction between constraints (2.351) and the observations is an arbitrary one, and the introduction of an error term in the former, no matter how small, removes any particular reason to distinguish them: $\mathbf{A}$ may well be some subset of the rows of $\mathbf{E}$. What follows can in fact be obtained by imposing the zero noise limit for some of the rows of $\mathbf{E}$ in the solutions already described. Furthermore, whether the model should be satisfied exactly, or should contain a noise element too, is situation dependent. One should be wary of introducing exact equalities into estimation problems, because they carry the strong possibility of introducing small eigenvalues, or near singular relationships, into the solution, and which may dominate the results. Nonetheless, carrying one or more perfect constraints does produce some insight into how the system is behaving.

Several approaches are possible. Consider, for example, the objective function

$$J = (\mathbf{Ex} - \mathbf{y})^{\mathrm{T}}(\mathbf{Ex} - \mathbf{y}) + \gamma^2(\mathbf{Ax} - \mathbf{b})^{\mathrm{T}}(\mathbf{Ax} - \mathbf{b}), \qquad (2.352)$$

where $\mathbf{W}, \mathbf{S}$ have been previously applied if necessary, and γ^2 is retained as a trade-off parameter. This objective function corresponds to the requirement of a solution of the combined equation sets,

$$\left\{\begin{matrix}\mathbf{E}\\\mathbf{A}\end{matrix}\right\}\mathbf{x} + \begin{bmatrix}\mathbf{n}\\\mathbf{u}\end{bmatrix} = \begin{bmatrix}\mathbf{y}\\\mathbf{b}\end{bmatrix}, \qquad (2.353)$$

in which $\mathbf{u}$ is the model noise, and the weight given to the model is $\gamma^2\mathbf{I}$. For any finite γ^2, the perfect constraints are formally "soft" because they are being applied only as a minimized sum of squares. The solution follows immediately from (2.95) with

$$\mathbf{E} \longrightarrow \left\{\begin{matrix}\mathbf{E}\\\gamma\mathbf{A}\end{matrix}\right\}, \qquad \mathbf{y} \longrightarrow \left\{\begin{matrix}\mathbf{y}\\\gamma\mathbf{b}\end{matrix}\right\},$$

assuming the matrix inverse exists. As $\gamma^2 \to \infty$, the second set of equations is being imposed with arbitrarily great accuracy, and, barring numerical issues, becomes as close to exactly satisfied as one wants.

Alternatively, the model can be imposed as a hard constraint. All prior covariances and scalings having been applied, and Lagrange multipliers introduced, the problem is one with an objective function,

$$J = \mathbf{n}^{\mathrm{T}}\mathbf{n} - 2\boldsymbol{\mu}^{\mathrm{T}}(\mathbf{Ax} - \mathbf{b}) = (\mathbf{Ex} - \mathbf{y})^{\mathrm{T}}(\mathbf{Ex} - \mathbf{y}) - 2\boldsymbol{\mu}^{\mathrm{T}}(\mathbf{Ax} - \mathbf{b}), \qquad (2.354)$$

which is a variant of (2.149). But now, Eq. (2.351) is to be exactly satisfied, and the observations only approximately so.

Setting the derivatives of J with respect to $\mathbf{x}$, $\boldsymbol{\mu}$ to zero, gives the following normal equations:

$$\mathbf{A}^{\mathrm{T}}\boldsymbol{\mu} = \mathbf{E}^{\mathrm{T}}(\mathbf{Ex} - \mathbf{y}), \qquad (2.355)$$

$$\mathbf{Ax} = \mathbf{b}. \qquad (2.356)$$

Equation (2.355) represents the adjoint, or "dual" model, for the adjoint or dual solution $\boldsymbol{\mu}$, and the two equation sets are to be solved simultaneously for $\mathbf{x}$, $\boldsymbol{\mu}$. They are again $M + N$ equations in $M + N$ unknowns (M of the μ_i, N of the x_i), but need not be full-rank. The first set, sometimes referred to as the "adjoint model," determines $\boldsymbol{\mu}$ from the *difference between* $\mathbf{Ex}$ and $\mathbf{y}$. The last set is just the exact constraints.

We can most easily solve two extreme cases in Eqs. (2.355) and (2.356) – one in which $\mathbf{A}$ is square, $N \times N$, and of full-rank, and one in which $\mathbf{E}$ has this property.

In the first case,

$$\tilde{\mathbf{x}} = \mathbf{A}^{-1}\mathbf{b}, \tag{2.357}$$

and

$$\tilde{\boldsymbol{\mu}} = \mathbf{A}^{-T}(\mathbf{E}^{T}\mathbf{E}\mathbf{A}^{-1} - \mathbf{E}^{T})\mathbf{b}. \tag{2.358}$$

Here, the values of $\tilde{\mathbf{x}}$ are completely determined by the full-rank, perfect constraints and the minimization of the deviation from the observations is passive. The Lagrange multipliers or adjoint solution, however, are useful in providing the sensitivity information, $\partial J/\partial \mathbf{b} = 2\boldsymbol{\mu}$, as already discussed. The uncertainty of this solution is zero because of the full-rank perfect model assumption (2.356).

In the second case, from (2.355),

$$\tilde{\mathbf{x}} = (\mathbf{E}^{T}\mathbf{E})^{-1}[\mathbf{E}^{T}\mathbf{y} + \mathbf{A}^{T}\boldsymbol{\mu}] \equiv \tilde{\mathbf{x}}_{u} + (\mathbf{E}^{T}\mathbf{E})^{-1}\mathbf{A}^{T}\boldsymbol{\mu},$$

where $\tilde{\mathbf{x}}_{u} = (\mathbf{E}^{T}\mathbf{E})^{-1}\mathbf{E}^{T}\mathbf{y}$ is the ordinary, unconstrained least-squares solution. Substituting into (2.356) produces

$$\tilde{\boldsymbol{\mu}} = [\mathbf{A}(\mathbf{E}^{T}\mathbf{E})^{-1}\mathbf{A}^{T}]^{-1}(\mathbf{b} - \mathbf{A}\tilde{\mathbf{x}}_{u}), \tag{2.359}$$

and

$$\tilde{\mathbf{x}} = \tilde{\mathbf{x}}_{u} + (\mathbf{E}^{T}\mathbf{E})^{-1}\mathbf{A}^{T}[\mathbf{A}(\mathbf{E}^{T}\mathbf{E})^{-1}\mathbf{A}^{T}]^{-1}(\mathbf{b} - \mathbf{A}\tilde{\mathbf{x}}_{u}), \tag{2.360}$$

assuming $\mathbf{A}$ is full-rank underdetermined. The perfect constraints are underdetermined; their range is being fit perfectly, with its nullspace being employed to reduce the misfit to the data as far as possible. The uncertainty of this solution may be written as[42]

$$\mathbf{P} = D^{2}(\tilde{\mathbf{x}} - \mathbf{x}) \tag{2.361}$$
$$= \sigma^{2}\{(\mathbf{E}^{T}\mathbf{E})^{-1} - (\mathbf{E}^{T}\mathbf{E})^{-1}\mathbf{A}^{T}[\mathbf{A}(\mathbf{E}^{T}\mathbf{E})^{-1}\mathbf{A}^{T}]^{-1}\mathbf{A}(\mathbf{E}^{T}\mathbf{E})^{-1}\},$$

which represents a reduction in the uncertainty of the ordinary least-squares solution (first term on the right) by the information in the perfectly known constraints. The presence of $\mathbf{A}^{-1}$ in these solutions is a manifestation of the warning about the possible introduction of components dependent upon small eigenvalues of $\mathbf{A}$. If neither $\mathbf{E}^{T}\mathbf{E}$ nor $\mathbf{A}$ is of full-rank one can use, e.g., the SVD with the above solution; the combined $\mathbf{E}, \mathbf{A}$ may be rank deficient, or just-determined.

Example *Consider the least-squares problem of solving*

$$x_1 + n_1 = 1,$$
$$x_2 + n_2 = 1,$$
$$x_1 + x_2 + n_3 = 3,$$

with uniform, uncorrelated noise of variance 1 in each of the equations. The least-squares solution is then

$$\tilde{\mathbf{x}} = [1.3333 \quad 1.3333]^{\mathrm{T}},$$

with uncertainty

$$\mathbf{P} = \begin{Bmatrix} 0.6667 & -0.3333 \\ -0.333 & 0.6667 \end{Bmatrix}.$$

But suppose that it is known or desired that $x_1 - x_2 = 1$. Then (2.360) produces $\tilde{\mathbf{x}} = [1.8333 \quad 0.8333]^{\mathrm{T}}$, $\mu = 0.5$, $J' = 0.8333$, with uncertainty

$$\mathbf{P} = \begin{Bmatrix} 0.1667 & 0.1667 \\ 0.1667 & 0.1667 \end{Bmatrix}.$$

If the constraint is shifted to $x_1 - x_2 = 1.1$, the new solution is $\tilde{\mathbf{x}} = [1.8833 \quad 0.7833]^{\mathrm{T}}$ and the new objective function is $J' = 0.9383$, consistent with the sensitivity deduced from μ.

A more generally useful case occurs when the errors normally expected to be present in the supposedly exact constraints are explicitly acknowledged. If the exact constraints have errors either in the "forcing," $\mathbf{b}$, or in a mis-specification of $\mathbf{A}$, then we write

$$\mathbf{A}\mathbf{x} = \mathbf{b} + \mathbf{\Gamma}\mathbf{u}, \tag{2.362}$$

assuming that $\langle \mathbf{u} \rangle = 0$, $\langle \mathbf{u}\mathbf{u}^{\mathrm{T}} \rangle = \mathbf{Q}$. $\mathbf{\Gamma}$ is a known coefficient matrix included for generality. If, for example, the errors were thought to be the same in all equations, we could write $\mathbf{\Gamma} = [1, 1, \ldots 1]^{\mathrm{T}}$, and then $\mathbf{u}$ would be just a scalar. Let the dimension of $\mathbf{u}$ be $P \times 1$. Such representations are not unique and more will be said about them in Chapter 4. A hard constraint formulation can still be used, in which (2.362) is to be exactly satisfied, imposed through an objective function of form

$$J = (\mathbf{E}\mathbf{x} - \mathbf{y})^{\mathrm{T}}\mathbf{R}_{nn}^{-1}(\mathbf{E}\mathbf{x} - \mathbf{y}) + \mathbf{u}^{\mathrm{T}}\mathbf{Q}^{-1}\mathbf{u} - 2\mu^{\mathrm{T}}(\mathbf{A}\mathbf{x} - \mathbf{b} - \mathbf{\Gamma}\mathbf{u}). \tag{2.363}$$

Here, the noise error covariance matrix has been explicitly included. Finding the normal equations by setting the derivatives with respect to $(\mathbf{x}, \mathbf{u}, \mu)$ to zero produces

$$\mathbf{A}^{\mathrm{T}}\mu = \mathbf{E}^{\mathrm{T}}\mathbf{R}_{nn}^{-1}(\mathbf{E}\mathbf{x} - \mathbf{y}), \tag{2.364}$$

$$\mathbf{\Gamma}^{\mathrm{T}}\mu = \mathbf{Q}^{-1}\mathbf{u}, \tag{2.365}$$

$$\mathbf{A}\mathbf{x} + \mathbf{\Gamma}\mathbf{u} = \mathbf{b}. \tag{2.366}$$

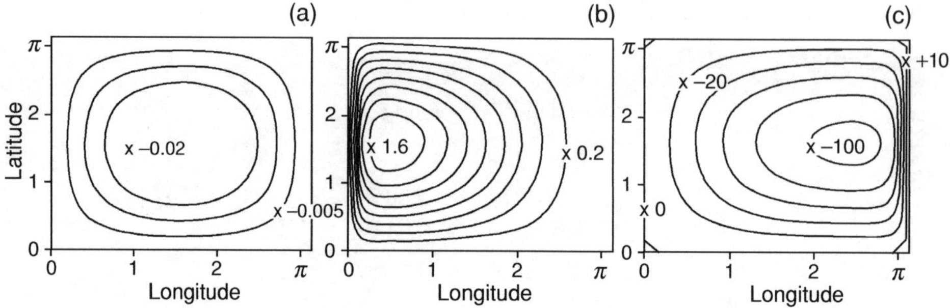

Figure 2.16 Numerical solution of the partial differential equation, Eq. (2.370). Panel (a) shows the imposed symmetric forcing – $\sin x \sin y$. (b) Displays the solution ϕ, and (c) shows the Lagrange multipliers, or adjoint solution, μ, whose structure is a near mirror image of ϕ. (Source: Schröter and Wunsch, 1986)

This system is $(2N + P)$ equations in $(2N + P)$ unknowns, where the first equation is again the adjoint system, and dependent upon $\mathbf{Ex} - \mathbf{y}$. Because $\mathbf{u}$ is a simple function of the Lagrange multipliers, the system is easily reduced to

$$\mathbf{A}^{\mathrm{T}}\mu = \mathbf{E}^{\mathrm{T}}\mathbf{R}_{nn}^{-1}(\mathbf{Ex} - \mathbf{y}), \tag{2.367}$$

$$\mathbf{Ax} + \mathbf{\Gamma Q \Gamma}^{\mathrm{T}}\mu = \mathbf{b}, \tag{2.368}$$

which is now $2N \times 2N$, the $\mathbf{u}$ having dropped out. If all matrices are full-rank, the solution is immediate; otherwise the SVD can be used.

To use a soft constraint methodology, write

$$J = (\mathbf{Ex} - \mathbf{y})^{\mathrm{T}}\mathbf{R}_{nn}^{-1}(\mathbf{Ex} - \mathbf{y}) + (\mathbf{Ax} - \mathbf{b} - \mathbf{\Gamma u})^{\mathrm{T}}\mathbf{Q}^{-1}(\mathbf{Ax} - \mathbf{b} - \mathbf{\Gamma u})^{\mathrm{T}}, \tag{2.369}$$

and find the normal equations. It is again readily confirmed that the solutions using (2.352) or (2.363) are identical, and the hard/soft distinction is seen again to be artificial. The soft constraint method can deal with perfect constraints, by letting $\|\mathbf{Q}^{-1}\| \to 0$ but stopping when numerical instability sets in. The resulting numerical algorithms fall under the general subject of "penalty" and "barrier" methods.[43] Objective functions like (2.363) and (2.369) will be used extensively in Chapter 4.

Example *Consider the partial differential equation*

$$\epsilon\nabla^2\phi + \frac{\partial\phi}{\partial x} = -\sin x \sin y. \tag{2.370}$$

A code was written to solve it by finite differences for the case $\epsilon = 0.05$ and $\phi = 0$ on the boundaries, $0 \le x \le \pi, 0 \le y \le \pi$, as depicted in Fig. 2.16. The discretized form of the model is then the perfect $N \times N$ constraint system

$$\mathbf{Ax} = \mathbf{b}, \quad \mathbf{x} = \{\phi_{ij}\}, \tag{2.371}$$

and **b** *is equivalently discretized* $-\sin x \sin y$. *The theory of partial differential equations shows that this system is full-rank and generally well-behaved. But let us pretend that this information is unknown to us, and seek the values of* **x** *that make the objective function,*

$$J = \mathbf{x}^{\mathrm{T}}\mathbf{x} - 2\boldsymbol{\mu}^{\mathrm{T}}(\mathbf{A}\mathbf{x} - \mathbf{b}), \qquad (2.372)$$

stationary with respect to **x,** $\boldsymbol{\mu}$, *that is the Eqs. (2.355) and (2.356) with* **E** $=$ **I,** **y** $=$ **0**. *Physically,* $\mathbf{x}^{\mathrm{T}}\mathbf{x}$ *is identified with the solution potential energy. The solution* $\boldsymbol{\mu}$, *corresponding to the solution of Fig. 2.16(b) is shown in Fig. 2.16(c). What is the interpretation? The Lagrange multipliers represent the sensitivity of the solution potential energy to perturbations in the forcing field. The sensitivity is greatest in the right-half of the domain, and indeed displays a boundary layer character. A physical interpretation of the Lagrange multipliers can be inferred, given the simple structure of the governing equation (2.370), and the Dirichlet boundary conditions. This equation is not self-adjoint; the adjoint partial differential equation is of form*

$$\epsilon\nabla^2 v - \frac{\partial v}{\partial x} = d, \qquad (2.373)$$

where d is a forcing term, subject to mixed boundary conditions, and whose discrete form is obtained by taking the transpose of the **A** *matrix of the discretization (see Appendix 2 to this chapter). The forward solution exhibits a boundary layer on the left-hand wall, while the adjoint solution has a corresponding behavior in the dual space on the right-hand wall. The structure of the* $\boldsymbol{\mu}$ *would evidently change if J were changed.*[44]

The original objective function J is very closely analogous to the Lagrangian (not to be confused with the Lagrange multiplier) in classical mechanics. In mechanics, the gradients of the Lagrangian commonly are virtual forces (forces required to enforce the constraints). The modified Lagrangian, J', is used in mechanics to impose various physical constraints, and the virtual force required to impose the constraints, for example, the demand that a particle follow a particular path, is the Lagrange multiplier.[45] In an economics/management context, the multipliers are usually called "shadow prices" as they are intimately related to the question of how much profit will change with a shift in the availability or cost of a product ingredient. The terminology "cost function" is a sensible substitute for what we call the "objective function."

More generally, there is a close connection between the stationarity requirements imposed upon various objective functions throughout this book, and the mathematics of classical mechanics. An elegant Hamiltonian formulation of the material is possible.

2.6.2 Relation to Green functions

Consider any linear set of simultaneous equations, involving an arbitrary matrix, $\mathbf{A}$,

$$\mathbf{A}\mathbf{x} = \mathbf{b}. \tag{2.374}$$

Write the adjoint equations for an arbitrary right-hand side,

$$\mathbf{A}^{\mathrm{T}}\mathbf{z} = \mathbf{r}. \tag{2.375}$$

Then the scalar relation

$$\mathbf{z}^{\mathrm{T}}\mathbf{A}\mathbf{x} - \mathbf{x}^{\mathrm{T}}\mathbf{A}^{\mathrm{T}}\mathbf{z} = 0 \tag{2.376}$$

(the "bilinear identity") implies that

$$\mathbf{z}^{\mathrm{T}}\mathbf{b} = \mathbf{x}^{\mathrm{T}}\mathbf{r}. \tag{2.377}$$

In the special case, $\mathbf{r} = \mathbf{0}$, we have

$$\mathbf{z}^{\mathrm{T}}\mathbf{b} = 0, \tag{2.378}$$

that is, $\mathbf{b}$, the right-hand side of the original equations (2.374), must be orthogonal to any solution of the homogeneous adjoint equations. (In SVD terms, this result is the solvability condition Eq. (2.266).) If $\mathbf{A}$ is of full rank, then there is no non-zero solution to the homogeneous adjoint equations.

Now assume that $\mathbf{A}$ is $N \times N$ of full rank. Add a single equation to (2.374) of the form

$$x_p = \alpha_p, \tag{2.379}$$

or

$$\mathbf{e}_p^{\mathrm{T}}\mathbf{x} = \alpha_p, \tag{2.380}$$

where $\mathbf{e}_p = \delta_{ip}$ and α_p is unknown. We also demand that Eq. (2.374) should remain exactly satisfied. The combined system of (2.374) and (2.380), written as

$$\mathbf{A}_1\mathbf{x} = \mathbf{b}_1, \tag{2.381}$$

is overdetermined. If it is to have a solution without any residual, it must still be orthogonal to any solution of the homogeneous adjoint equations,

$$\mathbf{A}_1^{\mathrm{T}}\mathbf{z} = \mathbf{0}. \tag{2.382}$$

There is only one such solution (because there is only one vector, $\mathbf{z} = \mathbf{u}_{N+1}$, in the null space of $\mathbf{A}_1^{\mathrm{T}}$). Write $\mathbf{u}_{N+1} = [\mathbf{g}_p, \gamma]^{\mathrm{T}}$, separating out the first N elements of

$\mathbf{u}_{N+1}$, calling them $\mathbf{g}_p$, and calling the one remaining element, γ. Thus Eq. (2.377) is

$$\mathbf{u}_{N+1}^{\mathrm{T}}\mathbf{b}_1 = \mathbf{g}_p^{\mathrm{T}}\mathbf{b} + \gamma\alpha_p = 0. \tag{2.383}$$

Choose $\gamma = -1$ (any other choice can be absorbed into $\mathbf{g}_p$). Then

$$\alpha_p = \mathbf{g}_p^{\mathrm{T}}\mathbf{b}. \tag{2.384}$$

If $\mathbf{g}_p$ were known, then α_p in (2.384) would be the only value consistent with the solutions to (2.374), and would be the correct value of x_p. But (2.382) is the same as

$$\mathbf{A}^{\mathrm{T}}\mathbf{g}_p = \mathbf{e}_p. \tag{2.385}$$

Because *all* elements x_p are needed, Eq. (2.385) is solved for all $p = 1, 2, \ldots, N$, that is,

$$\mathbf{A}^{\mathrm{T}}\mathbf{G} = \mathbf{I}_N, \tag{2.386}$$

which is N separate problems, each for the corresponding column of $\mathbf{G} = \{\mathbf{g}_1, \mathbf{g}_2, \ldots, \mathbf{g}_N\}$. Here, $\mathbf{G}$ is the Green function. With $\mathbf{G}$ known, we have immediately that

$$\mathbf{x} = \mathbf{G}^{\mathrm{T}}\mathbf{b}, \tag{2.387}$$

(from Eq. (2.384)). The Green function is an inverse to the adjoint equations; its significance here is that it generalizes in the continuous case to an operator inverse.[46]

2.7 Minimum variance estimation and simultaneous equations

The fundamental objective for least-squares is minimization of the noise norm (2.89), although we complicated the discussion somewhat by introducing trade-offs against $\|\tilde{\mathbf{x}}\|$, various weights in the norms, and even the restriction that $\tilde{\mathbf{x}}$ should satisfy certain equations exactly. Least-squares methods, whether used directly as in (2.95) or indirectly through the vector representations of the SVD, are fundamentally deterministic. Statistics were used only to understand the sensitivity of the solutions to noise, and to obtain measures of the expected deviation of the solution from some supposed truth.

But there is another, very different, approach to obtaining estimates of the solution to equation sets like (2.87), directed more clearly toward the physical goal: to find an estimate $\tilde{\mathbf{x}}$ that deviates as little as possible in the *mean-square* from the true solution. That is, we wish to minimize the statistical quantities $\langle(\tilde{x}_i - x_i)^2\rangle$ for all i. The next section is devoted to finding such an $\tilde{\mathbf{x}}$ (and the corresponding $\tilde{\mathbf{n}}$), through an excursion into statistical estimation theory. It is far from obvious that this $\tilde{\mathbf{x}}$ should bear any resemblance to one of the least-squares estimates; but as will be seen, under some circumstances the two are identical. Their possible identity is extremely useful, but has apparently led many investigators to seriously confuse the methodologies, and therefore the interpretation of the result.

2.7.1 The fundamental result

Suppose we are interested in making an estimate of a physical variable, $\mathbf{x}$, which might be a vector or a scalar, and is either constant or varying with space and time. To be definite, let $\mathbf{x}$ be a function of an independent variable $\mathbf{r}$, written discretely as $\mathbf{r}_j$ (it might be a vector of space coordinates, or a scalar time, or an accountant's label). Let us make some suppositions about what is usually called "prior information." In particular, suppose we have an estimate of the low-order statistics describing $\mathbf{x}$, that is, specifying its mean and second moments,

$$\langle \mathbf{x} \rangle = \mathbf{0}, \qquad \langle \mathbf{x}(\mathbf{r}_i)\mathbf{x}(\mathbf{r}_j)^{\mathrm{T}} \rangle = \mathbf{R}_{xx}(\mathbf{r}_i, \mathbf{r}_j). \tag{2.388}$$

To make this problem concrete, one might think of $\mathbf{x}$ as being the temperature anomaly (about the mean) at a fixed depth in a fluid (a scalar) and $\mathbf{r}_j$ a vector of horizontal positions; or conductivity in a well, where $\mathbf{r}_j$ would be the depth coordinate, and $\mathbf{x}$ is the vector of scalars at any location, $\mathbf{r}_p$, $x_p = x(\mathbf{r}_p)$. Alternatively, $\mathbf{x}$ might be the temperature at a fixed point, with r_j being the scalar of time. But if the field of interest is the velocity vector, then each element of $\mathbf{x}$ is itself a vector, and one can extend the notation in a straightforward fashion. To keep the notation a little cleaner, however, all elements of $\mathbf{x}$ are written as scalars.

Now suppose that we have some observations, y_i, as a function of the same coordinate $\mathbf{r}_i$, with a known, zero mean, and second moments

$$\mathbf{R}_{yy}(\mathbf{r}_i, \mathbf{r}_j) = \langle \mathbf{y}(\mathbf{r}_i)\mathbf{y}(\mathbf{r}_j)^{\mathrm{T}} \rangle, \quad \mathbf{R}_{xy}(\mathbf{r}_i, \mathbf{r}_j) = \langle \mathbf{x}(\mathbf{r}_i)\mathbf{y}(\mathbf{r}_j)^{\mathrm{T}} \rangle, \quad i, j = 1, 2, \ldots, M. \tag{2.389}$$

(The individual observation elements can also be vectors – for example, two or three components of velocity and a temperature at a point – but as with $\mathbf{x}$, the modifications required to treat this case are straightforward, and scalar observations are assumed.) Could the measurements be used to make an estimate of $\mathbf{x}$ at a point $\tilde{\mathbf{r}}_\alpha$ where no measurement is available? Or could many measurements be used to obtain a better estimate even at points where there exists a measurement? The idea is to exploit the concept that finite covariances carry predictive capabilities from known variables to unknown ones. A specific example would be to suppose the measurements are of temperature, $y(\mathbf{r}_j) = y_0(\mathbf{r}_j) + n(\mathbf{r}_j)$, where n is the noise and temperature estimates are sought at different locations, perhaps on a regular grid $\tilde{\mathbf{r}}_\alpha$,

$\alpha = 1, 2, \ldots, N$. This special problem is one of gridding or mapmaking (the tilde is placed on $\mathbf{r}_\alpha$ as a device to emphasize that this is a location where an estimate is sought; the numerical values of these places or labels are assumed known). Alternatively, and somewhat more interesting, perhaps the measurements are more indirect, with $y(r_i)$ representing a velocity field component at depth in a fluid and believed connected, through a differential equation, to the temperature field. We might want to estimate the temperature from measurements of the velocity.

Given the previous statistical discussion (p. 31), it is reasonable to ask for an estimate $\tilde{x}(\tilde{\mathbf{r}}_\alpha)$, whose dispersion about its true value, $x(\tilde{\mathbf{r}}_\alpha)$ is as small as possible, that is,

$$P\left(\tilde{\mathbf{r}}_\alpha, \tilde{\mathbf{r}}_\alpha\right) = \langle (\tilde{x}(\tilde{\mathbf{r}}_\alpha) - x(\tilde{\mathbf{r}}_\alpha))(\tilde{x}(\tilde{\mathbf{r}}_\beta) - x(\tilde{\mathbf{r}}_\beta)) \rangle |_{\tilde{\mathbf{r}}_\alpha = \tilde{\mathbf{r}}_\beta}$$

is to be minimized. If an estimate is needed at more than one point, $\tilde{\mathbf{r}}_\alpha$, the covariance of the errors in the different estimates would usually be required, too. Form a vector of values to be estimated, $\{\tilde{x}(\mathbf{r}_\alpha)\} \equiv \tilde{\mathbf{x}}$, and their uncertainty is

$$\begin{aligned}\mathbf{P}(\tilde{\mathbf{r}}_\alpha, \tilde{\mathbf{r}}_\beta) &= \langle (\tilde{x}(\tilde{\mathbf{r}}_\alpha) - x(\tilde{\mathbf{r}}_\alpha))(\tilde{x}(\tilde{\mathbf{r}}_\beta) - x(\tilde{\mathbf{r}}_\beta)) \rangle \\ &= \langle (\tilde{\mathbf{x}} - \mathbf{x})(\tilde{\mathbf{x}} - \mathbf{x})^{\mathrm{T}} \rangle, \quad \alpha, \beta = 1, 2, \ldots, N,\end{aligned} \tag{2.390}$$

where the *diagonal* elements, $\mathbf{P}(\tilde{\mathbf{r}}_\alpha, \tilde{\mathbf{r}}_\alpha)$, are to be *individually* minimized (not in the sum of squares). Thus a solution with *minimum variance about the correct value* is sought.

What should the relationship be between data and estimate? At least initially, a linear combination of data is a reasonable starting point,

$$\tilde{x}(\tilde{\mathbf{r}}_\alpha) = \sum_{j=1}^{M} B(\tilde{\mathbf{r}}_\alpha, \mathbf{r}_j) y(\mathbf{r}_j), \tag{2.391}$$

for all α, which makes the diagonal elements of $\mathbf{P}$ in (2.390) as small as possible. By letting $\mathbf{B}$ be an $N \times M$ matrix, all of the points can be handled at once:

$$\tilde{\mathbf{x}}(\tilde{\mathbf{r}}_\alpha) = \mathbf{B}(\tilde{\mathbf{r}}_\alpha, \mathbf{r}_j) \mathbf{y}(\mathbf{r}_j). \tag{2.392}$$

(This notation is redundant. Equation (2.392) is a shorthand for (2.391), in which the argument has been put into $\mathbf{B}$ explicitly as a reminder that there is a summation over all the data locations, $\mathbf{r}_j$, for all mapping locations, $\tilde{\mathbf{r}}_\alpha$, but it is automatically accounted for by the usual matrix multiplication convention. It suffices to write $\tilde{\mathbf{x}} = \mathbf{B}\mathbf{y}$.)

An important result, often called the "Gauss–Markov theorem," produces the values of $\mathbf{B}$ that will minimize the diagonal elements of $\mathbf{P}$.[47] Substituting (2.392) into (2.390) and expanding,

$$\mathbf{P}(\tilde{\mathbf{r}}_\alpha, \tilde{\mathbf{r}}_\beta) = \langle(\mathbf{B}(\tilde{\mathbf{r}}_\alpha, \mathbf{r}_j)\mathbf{y}(\mathbf{r}_j) - \mathbf{x}(\tilde{\mathbf{r}}_\alpha))(\mathbf{B}(\tilde{\mathbf{r}}_\beta, \mathbf{r}_l)\mathbf{y}(\mathbf{r}_l) - \mathbf{x}(\tilde{\mathbf{r}}_\beta))^{\mathrm{T}}\rangle$$

$$\equiv \langle(\mathbf{B}\mathbf{y} - \mathbf{x})(\mathbf{B}\mathbf{y} - \mathbf{x})^{\mathrm{T}}\rangle \tag{2.393}$$

$$= \mathbf{B}\langle\mathbf{y}\mathbf{y}^{\mathrm{T}}\rangle - \langle\mathbf{x}\mathbf{y}^{\mathrm{T}}\rangle\mathbf{B}^{\mathrm{T}} - \mathbf{B}\langle\mathbf{y}\mathbf{x}^{\mathrm{T}}\rangle + \langle\mathbf{x}\mathbf{x}^{\mathrm{T}}\rangle.$$

Using $\mathbf{R}_{xy} = \mathbf{R}_{yx}^{\mathrm{T}}$, Eq. (2.393) is

$$\mathbf{P} = \mathbf{B}\mathbf{R}_{yy}\mathbf{B}^{\mathrm{T}} - \mathbf{R}_{xy}\mathbf{B}^{\mathrm{T}} - \mathbf{B}\mathbf{R}_{xy}^{\mathrm{T}} + \mathbf{R}_{xx}. \tag{2.394}$$

Notice that because $\mathbf{R}_{xx}$ represents the moments of $\mathbf{x}$ evaluated at the estimation positions, it is a function of $\tilde{\mathbf{r}}_\alpha, \tilde{\mathbf{r}}_\beta$, whereas $\mathbf{R}_{xy}$ involves covariances of $\mathbf{y}$ at the data positions with $\mathbf{x}$ at the estimation positions, and is consequently a function $\mathbf{R}_{xy}(\tilde{\mathbf{r}}_\alpha, \mathbf{r}_j)$.

Now completing the square (Eq. (2.37)) (by adding and subtracting $\mathbf{R}_{xy}\mathbf{R}_{yy}^{-1}\mathbf{R}_{xy}^{\mathrm{T}}$), (2.394) becomes

$$\mathbf{P} = (\mathbf{B} - \mathbf{R}_{xy}\mathbf{R}_{yy}^{-1})\mathbf{R}_{yy}(\mathbf{B} - \mathbf{R}_{xy}\mathbf{R}_{yy}^{-1})^{\mathrm{T}} - \mathbf{R}_{xy}\mathbf{R}_{yy}^{-1}\mathbf{R}_{xy}^{\mathrm{T}} + \mathbf{R}_{xx}. \tag{2.395}$$

Setting $\tilde{\mathbf{r}}_\alpha = \tilde{\mathbf{r}}_\beta$ so that (2.395) is the variance of the estimate at point $\tilde{\mathbf{r}}_\alpha$ about its true value, and noting that all three terms in Eq. (2.395) are positive definite, minimization of any diagonal element of $\mathbf{P}$ is obtained by choosing $\mathbf{B}$ so that the first term vanishes, or

$$\mathbf{B}(\tilde{\mathbf{r}}_\alpha, \mathbf{r}_j) = \mathbf{R}_{xy}(\tilde{\mathbf{r}}_\alpha, \mathbf{r}_i)\,\mathbf{R}_{yy}(\mathbf{r}_i, \mathbf{r}_j)^{-1} = \mathbf{R}_{xy}\mathbf{R}_{yy}^{-1}. \tag{2.396}$$

(The diagonal elements of $(\mathbf{B} - \mathbf{R}_{xy}\mathbf{R}_{yy}^{-1})\mathbf{R}_{yy}(\mathbf{B} - \mathbf{R}_{xy}\mathbf{R}_{yy}^{-1})^{\mathrm{T}}$ need to be written out explicitly to see that Eq. (2.396) is necessary. Consider the 2×2 case: the first term of Eq. (2.395) is of the form

$$\begin{bmatrix} C_{11} & C_{12} \\ C_{21} & C_{22} \end{bmatrix} \begin{bmatrix} R_{11} & R_{12} \\ R_{21} & R_{22} \end{bmatrix} \begin{bmatrix} C_{11} & C_{12} \\ C_{21} & C_{22} \end{bmatrix}^{\mathrm{T}},$$

where $\mathbf{C} = \mathbf{B} - \mathbf{R}_{xy}\mathbf{R}_{yy}^{-1}$. Then, one has the diagonal of

$$\begin{Bmatrix} C_{11}^2 R_{11} + C_{12}C_{11}(R_{21} + R_{12}) + C_{12}^2 R_{22} & \cdot \\ \cdot & C_{21}^2 R_{11} + C_{21}C_{22}(R_{21} + R_{12}) + C_{22}^2 R_{22} \end{Bmatrix},$$

and these diagonals vanish (with $R_{11}, R_{22} > 0$, only if $C_{11} = C_{12} = C_{21} = C_{22} = 0$). Thus the minimum variance estimate is

$$\tilde{\mathbf{x}}(\tilde{\mathbf{r}}_\alpha) = \mathbf{R}_{xy}(\tilde{\mathbf{r}}_\alpha, \mathbf{r}_i)\,\mathbf{R}_{yy}^{-1}(\mathbf{r}_i, \mathbf{r}_j)\mathbf{y}(\mathbf{r}_j), \tag{2.397}$$

and the minimum of the diagonal elements of $\mathbf{P}$ is found by substituting back into (2.394), producing

$$\mathbf{P}(\tilde{\mathbf{r}}_\alpha, \tilde{\mathbf{r}}_\beta) = \mathbf{R}_{xx}(\tilde{\mathbf{r}}_\alpha, \tilde{\mathbf{r}}_\beta) - \mathbf{R}_{xy}(\tilde{\mathbf{r}}_\alpha, \mathbf{r}_j)\mathbf{R}_{yy}^{-1}(\mathbf{r}_j, \mathbf{r}_k)\mathbf{R}_{xy}^{\mathrm{T}}(\tilde{\mathbf{r}}_\beta, \mathbf{r}_k). \tag{2.398}$$

The bias of (2.398) is

$$\langle \tilde{\mathbf{x}} - \mathbf{x} \rangle = \mathbf{R}_{xy}\mathbf{R}_{yy}^{-1}\langle \mathbf{y} \rangle - \mathbf{x}. \qquad (2.399)$$

If $\langle \mathbf{y} \rangle = \mathbf{x} = 0$, the estimator is unbiassed, and called a "best linear unbiassed estimator," or "BLUE"; otherwise it is biassed. The whole development here began with the assumption that $\langle \mathbf{x} \rangle = \langle \mathbf{y} \rangle = 0$; what is usually done is to remove the *sample* mean from the observations $\mathbf{y}$, and to ignore the difference between the true and sample means. An example of using this machinery for mapping purposes will be seen in Chapter 3. Under some circumstances, this approximation is unacceptable, and the mapping error introduced by the use of the sample mean must be found. A general approach falls under the label of "kriging," which is also briefly discussed in Chapter 3.

2.7.2 Linear algebraic equations

The result shown in (2.396)–(2.398) is the abstract general case and is deceptively simple. Invocation of the physical problem of interpolating temperatures, etc., is not necessary: the only information actually used is that there are finite covariances between $\mathbf{x}$, $\mathbf{y}$, $\mathbf{n}$. Although mapping will be explicitly explored in Chapter 3, suppose instead that the observations are related to the unknown vector $\mathbf{x}$ as in our canonical problem, that is, through a set of linear equations, $\mathbf{Ex} + \mathbf{n} = \mathbf{y}$. The measurement covariance, $\mathbf{R}_{yy}$, can then be computed directly as

$$\mathbf{R}_{yy} = \langle (\mathbf{Ex} + \mathbf{n})(\mathbf{Ex} + \mathbf{n})^{\mathrm{T}} \rangle = \mathbf{E}\mathbf{R}_{xx}\mathbf{E}^{\mathrm{T}} + \mathbf{R}_{nn}. \qquad (2.400)$$

The unnecessary, but simplifying and often excellent, assumption was made that the cross-terms of form

$$\mathbf{R}_{xn} = \mathbf{R}_{nx}^{\mathrm{T}} = 0, \qquad (2.401)$$

so that

$$\mathbf{R}_{xy} = \langle \mathbf{x}(\mathbf{Ex} + \mathbf{n})^{\mathrm{T}} \rangle = \mathbf{R}_{xx}\mathbf{E}^{\mathrm{T}}, \qquad (2.402)$$

that is, there is no correlation between the measurement noise and the actual state vector (e.g., that the noise in a temperature measurement does not depend upon whether the true value is $10°$ or $25°$).

Under these circumstances, Eqs. (2.397) and (2.398) take on the following form:

$$\tilde{\mathbf{x}} = \mathbf{R}_{xx}\mathbf{E}^{\mathrm{T}}(\mathbf{E}\mathbf{R}_{xx}\mathbf{E}^{\mathrm{T}} + \mathbf{R}_{nn})^{-1}\mathbf{y}, \qquad (2.403)$$

$$\tilde{\mathbf{n}} = \mathbf{y} - \mathbf{E}\tilde{\mathbf{x}}, \qquad (2.404)$$

$$\mathbf{P} = \mathbf{R}_{xx} - \mathbf{R}_{xx}\mathbf{E}^{\mathrm{T}}(\mathbf{E}\mathbf{R}_{xx}\mathbf{E}^{\mathrm{T}} + \mathbf{R}_{nn})^{-1}\mathbf{E}\mathbf{R}_{xx}. \qquad (2.405)$$

These latter expressions are extremely important; they permit discussion of the solution to a set of linear algebraic equations in the presence of noise using information concerning the statistics of both the noise and the solution. Notice that they are *identical to the least-squares expression* (2.135) *if* $\mathbf{S} = \mathbf{R}_{xx}$, $\mathbf{W} = \mathbf{R}_{nn}$, except that there the uncertainty was estimated about the mean solution; here it is taken about the true one. As is generally true of all linear methods, the uncertainty, $\mathbf{P}$, is independent of the actual data, and can be computed in advance should one wish.

From the matrix inversion lemma, Eqs. (2.403)–(2.405) can be rewritten as

$$\tilde{\mathbf{x}} = \left(\mathbf{R}_{xx}^{-1} + \mathbf{E}^{\mathrm{T}}\mathbf{R}_{nn}^{-1}\mathbf{E}\right)^{-1}\mathbf{E}^{\mathrm{T}}\mathbf{R}_{nn}^{-1}\mathbf{y}, \tag{2.406}$$

$$\tilde{\mathbf{n}} = \mathbf{y} - \mathbf{E}\tilde{\mathbf{x}}, \tag{2.407}$$

$$\mathbf{P} = \left(\mathbf{R}_{xx}^{-1} + \mathbf{E}^{\mathrm{T}}\mathbf{R}_{nn}^{-1}\mathbf{E}\right)^{-1}. \tag{2.408}$$

Although these alternate forms are algebraically and numerically identical to Eqs. (2.403)–(2.405), the size of the matrices to be inverted changes from $M \times M$ matrices to $N \times N$, where $\mathbf{E}$ is $M \times N$ (but note that $\mathbf{R}_{nn}$ is $M \times M$; the efficacy of this alternate form may depend upon whether the *inverse* of $\mathbf{R}_{nn}$ is known). Depending upon the relative magnitudes of M, N, one form may be more preferable to the other. Finally, (2.408) has an important interpretation that we will discuss when we come to recursive methods. Recall, too, the options we had with the SVD of solving $M \times M$ or $N \times N$ problems. Note that in the limit of complete a priori ignorance of the solution, $\|\mathbf{R}_{xx}^{-1}\| \to 0$, Eqs. (2.406) and (2.408) reduce to

$$\tilde{\mathbf{x}} = \left(\mathbf{E}^{\mathrm{T}}\mathbf{R}_{nn}^{-1}\mathbf{E}\right)^{-1}\mathbf{E}^{\mathrm{T}}\mathbf{R}_{nn}^{-1}\mathbf{y},$$

$$\mathbf{P} = (\mathbf{E}^{\mathrm{T}}\mathbf{R}_{nn}^{-1}\mathbf{E})^{-1},$$

the conventional weighted least-squares solution, now with $\mathbf{P} = \mathbf{C}_{xx}$. More generally, the presence of finite $\mathbf{R}_{xx}^{-1}$ introduces a bias into the solution so that $\langle\tilde{\mathbf{x}}\rangle \neq \mathbf{x}$, which, however, produces a smaller solution variance than in the unbiased solution.

The solution shown in Eqs. (2.403)–(2.405) and (2.406)–(2.408) is an "estimator"; it was found by demanding a solution with the minimum dispersion about the true solution and it is found, surprisingly, identical to the tapered, weighted least-squares solution when $\mathbf{S} = \mathbf{R}_{xx}$, $\mathbf{W} = \mathbf{R}_{nn}$, the least-squares objective function weights are chosen. This correspondence of the two solutions often leads them to be seriously confused. It is essential to recognize that the logic of the derivations are quite distinct: we were free in the least-squares derivation to use weight matrices which were anything we wished – as long as appropriate inverses existed.

The correspondence of least-squares with what is usually known as minimum variance estimation can be understood by recognizing that the Gauss–Markov estimator was derived by minimizing a quadratic objective function. The least-squares estimate was obtained by minimizing a summation which is a sample *estimate* of the Gauss–Markov objective function when **S**, **W** are chosen properly.

2.7.3 Testing after the fact

As with any statistical estimator, an essential step after an apparent solution has been found is the testing that the behavior of $\tilde{\mathbf{x}}$, $\tilde{\mathbf{n}}$ is consistent with the assumed prior statistics reflected in $\mathbf{R}_{xx}$, $\mathbf{R}_{nn}$, and any assumptions about their means or other properties. Such a-posteriori checks are both necessary and very demanding. One sometimes hears it said that estimation using Gauss–Markov and related methods is "pulling solutions out of the air" because the prior covariance matrices $\mathbf{R}_{xx}$, $\mathbf{R}_{nn}$ often are only poorly known. But producing solutions that pass the test of consistency with the prior covariances can be very difficult. It is also true that the solutions tend to be somewhat insensitive to the details of the prior covariances and it is easy to become overly concerned with the detailed structure of $\mathbf{R}_{xx}$, $\mathbf{R}_{nn}$.

As stated previously, it is also rare to be faced with a situation in which one is truly ignorant of the covariances, true ignorance meaning that arbitrarily large or small numerical values of x_i, n_i would be acceptable. In the box inversions of Chapter 1 (to be revisited in Chapter 5), solution velocities of order 1000 cm/s might be regarded as absurd, and their absurdity is readily asserted by choosing $\mathbf{R}_{xx} = \mathrm{diag}(10\mathrm{cm/s})^2$, which reflects a mild belief that velocities are 0(10cm/s) with no known correlations with each other. Testing of statistical estimates against prior hypotheses is a highly developed field in applied statistics, and we leave it to the references already listed for their discussion. Should such tests be failed, one must reject the solutions $\tilde{\mathbf{x}}$, $\tilde{\mathbf{n}}$ and ask why they failed – as it usually implies an incorrect model, **E**, and the assumed statistics of solution and/or noise.

Example *The underdetermined system*

$$\left\{\begin{matrix} 1 & 1 & 1 & 1 \\ 1 & -1 & -1 & 1 \end{matrix}\right\} \mathbf{x} + \mathbf{n} = \begin{bmatrix} 1 \\ -1 \end{bmatrix},$$

with noise variance $\langle \mathbf{nn}^{\mathrm{T}} \rangle = 0.01\mathbf{I}$, *has a solution, if* $\mathbf{R}_{xx} = \mathbf{I}$, *of*

$$\tilde{\mathbf{x}} = \mathbf{E}^{\mathrm{T}}(\mathbf{EE}^{\mathrm{T}} + 0.01\mathbf{I})^{-1}\mathbf{y} = [0 \quad 0.4988 \quad 0.4988 \quad 0]^{\mathrm{T}},$$
$$\tilde{\mathbf{n}} = [0.0025, -0.0025]^{\mathrm{T}}.$$

If the solution was thought to be large scale and smooth, one might use the covariance

$$\mathbf{R}_{xx} = \left\{\begin{matrix} 1 & 0.999 & 0.998 & 0.997 \\ 0.999 & 1 & 0.999 & 0.998 \\ 0.998 & 0.999 & 1 & 0.999 \\ 0.997 & 0.998 & 0.999 & 1 \end{matrix}\right\},$$

which produces a solution

$$\tilde{\mathbf{x}} = [0.2402 \pm 0.028 \quad 0.2595 \pm 0.0264 \quad 0.2595 \pm 0.0264 \quad 0.2402 \pm 0.0283]^{\mathrm{T}},$$

$$\tilde{\mathbf{n}} = [0.0006 \quad -0.9615]^{\mathrm{T}},$$

having the desired large-scale property. (One might worry a bit about the structure of the residuals, but two equations are inadequate to draw any conclusions.)

2.7.4 Use of basis functions

A superficially different way of dealing with prior statistical information is often commonly used. Suppose that the indices of x_i refer to a spatial or temporal position, call it r_i, so that $x_i = x(r_i)$. Then it is often sensible to consider expanding the unknown $\mathbf{x}$ in a set of basis functions, F_j, for example, sines and cosines, Chebyschev polynomials, ordinary polynomials, etc. One might write

$$x(r_i) = \sum_{j=1}^{L} \alpha_j F_j(r_i),$$

or

$$\mathbf{x} = \mathbf{F}\boldsymbol{\alpha}, \quad \mathbf{F} = \left\{ \begin{array}{cccc} F_1(r_1) & F_2(r_1) & \cdots & F_L(r_1) \\ F_1(r_2) & F_2(r_2) & \cdots & F_L(r_2) \\ \cdot & \cdot & \cdot & \cdot \\ F_1(r_N) & F_2(r_N) & \cdots & F_L(r_N) \end{array} \right\}, \quad \boldsymbol{\alpha} = [\alpha_1 \cdots \alpha_L]^{\mathrm{T}},$$

which, when substituted into (2.87), produces

$$\mathbf{T}\boldsymbol{\alpha} + \mathbf{n} = \mathbf{y}, \quad \mathbf{T} = \mathbf{EF}. \tag{2.409}$$

If $L < M < N$, one can convert an underdetermined system into one which is formally overdetermined and, of course, the reverse is possible as well. It should be apparent, however, that the solution to (2.409) will have a covariance structure dictated in large part by that contained in the basis functions chosen, and thus there is no fundamental gain in employing basis functions, although they may be convenient, numerically or otherwise. If $\mathbf{P}_{\alpha\alpha}$ denotes the uncertainty of $\boldsymbol{\alpha}$, then

$$\mathbf{P} = \mathbf{F}\mathbf{P}_{\alpha\alpha}\mathbf{F}^{\mathrm{T}}, \tag{2.410}$$

is the uncertainty of $\tilde{\mathbf{x}}$. If there are special conditions applying to $\mathbf{x}$, such as boundary conditions at certain positions, r_B, a choice of basis function satisfying those conditions could be more convenient than appending them as additional equations.

Example *If, in the last example, one attempts a solution as a first-order polynomial,*

$$x_i = a + br_i, \quad r_1 = 0, \ r_2 = 1, \ r_3 = 2, \dots,$$

the system will become two equations in the two unknowns a, b:

$$\mathbf{EF}\begin{bmatrix} a \\ b \end{bmatrix} = \begin{Bmatrix} 4 & 6 \\ 0 & 0 \end{Bmatrix}\begin{bmatrix} a \\ b \end{bmatrix} + \mathbf{n} = \begin{bmatrix} 1 \\ -1 \end{bmatrix},$$

and if no prior information about the covariance of a, b is provided,

$$[\tilde{a}, \ \tilde{b}] = [0.0769, \ 0.1154],$$
$$\tilde{\mathbf{x}} = \begin{bmatrix} 0.0769 \pm 0.0077 & 0.1923 \pm 0.0192 & 0.3076 \pm 0.0308 & 0.4230 \pm 0.0423 \end{bmatrix}^{\mathrm{T}},$$
$$\tilde{\mathbf{n}} = [0.0002, \ -1.00]^{\mathrm{T}},$$

*which is also large scale and smooth, but different than that obtained above.
Although this latter solution has been obtained from a just-determined system,
it is not clearly "better." If a linear trend is expected in the solution, then the
polynomial expansion is certainly convenient – although such a structure can be
imposed through use of* $\mathbf{R}_{xx}$ *by specifying a growing variance with* r_i.

2.7.5 Determining a mean value

Let the measurements of the physical quantity continue to be denoted y_i and suppose
that each is made up of an unknown large-scale mean, m, plus a deviation from that
mean, θ_i. Then

$$m + \theta_i = y_i, \quad i = 1, 2, \dots, M, \tag{2.411}$$

or

$$\mathbf{D}m + \boldsymbol{\theta} = \mathbf{y}, \quad \mathbf{D} = [1 \quad 1 \quad 1 \quad \cdots \quad 1]^{\mathrm{T}}, \tag{2.412}$$

and we seek a best estimate, $\tilde{m}$, of m. In (2.411) or (2.412) the unknown $\mathbf{x}$ has
become the scalar m, and the deviation of the field from its mean is the noise, that
is, $\boldsymbol{\theta} \equiv \mathbf{n}$, whose true mean is zero. The problem is evidently a special case of the
use of basis functions, in which only one function – a zeroth-order polynomial, m,
is retained.

Set $\mathbf{R}_{nn} = \langle \boldsymbol{\theta}\boldsymbol{\theta}^{\mathrm{T}} \rangle$. If we were estimating a large-scale mean temperature in a fluid
flow filled with smaller-scale eddies, then $\mathbf{R}_{nn}$ is the sum of the covariance of the
eddy field plus that of observational errors and any other fields contributing to the
difference between y_i and the true mean m. To be general, suppose $\mathbf{R}_{xx} = \langle m^2 \rangle =$

m_0^2, and that, from (2.406),

$$\tilde{m} = \left\{ \frac{1}{m_0^2} + \mathbf{D}^{\mathrm{T}} \mathbf{R}_{nn}^{-1} \mathbf{D} \right\}^{-1} \mathbf{D}^{\mathrm{T}} \mathbf{R}_{nn}^{-1} \mathbf{y}$$
$$= \frac{1}{1/m_0^2 + \mathbf{D}^{\mathrm{T}} \mathbf{R}_{nn}^{-1} \mathbf{D}} \mathbf{D}^{\mathrm{T}} \mathbf{R}_{nn}^{-1} \mathbf{y}, \tag{2.413}$$

where $\mathbf{D}^{\mathrm{T}} \mathbf{R}_{nn}^{-1} \mathbf{D}$ is a scalar.[48] The expected uncertainty of this estimate is (2.408),

$$P = \left\{ \frac{1}{m_0^2} + \mathbf{D}^{\mathrm{T}} \mathbf{R}_{nn}^{-1} \mathbf{D} \right\}^{-1} = \frac{1}{1/m_0^2 + \mathbf{D}^{\mathrm{T}} \mathbf{R}_{nn}^{-1} \mathbf{D}}, \tag{2.414}$$

(also a scalar).

The estimates may appear somewhat unfamiliar; they reduce to more common expressions in certain limits. Let the θ_i be uncorrelated, with uniform variance σ^2; $\mathbf{R}_{nn}$ is then diagonal and (2.413) is

$$\tilde{m} = \frac{1}{\left(1/m_0^2 + M/\sigma^2\right)\sigma^2} \sum_{i=1}^{M} y_i = \frac{m_0^2}{\sigma^2 + M m_0^2} \sum_{i=1}^{M} y_i, \tag{2.415}$$

where the relations $\mathbf{D}^{\mathrm{T}}\mathbf{D} = M$, $\mathbf{D}^{\mathrm{T}}\mathbf{y} = \sum_{i=1}^{M} y_i$ were used. The expected value of the estimate is

$$\langle \tilde{m} \rangle = \frac{m_0^2}{\sigma^2 + M m_0^2} \sum_{i}^{M} \langle y_i \rangle = \frac{m_0^2}{\sigma^2 + M m_0^2} M m \neq m, \tag{2.416}$$

that is, it is biassed, as inferred above, unless $\langle y_i \rangle = 0$, implying $m = 0$. $\mathbf{P}$ becomes

$$P = \frac{1}{1/m_0^2 + M/\sigma^2} = \frac{\sigma^2 m_0^2}{\sigma^2 + M m_0^2}. \tag{2.417}$$

Under the further assumption that $m_0^2 \to \infty$,

$$\tilde{m} = \frac{1}{M} \sum_{i=1}^{M} y_i, \tag{2.418}$$

$$P = \sigma^2/M, \tag{2.419}$$

which are the ordinary average and its variance (the latter expression is the well-known "square root of M rule" for the standard deviation of an average; recall Eq. (2.42)); $\langle \tilde{m} \rangle$ in (2.418) is readily seen to be the true mean – this estimate has become unbiassed. However, the magnitude of (2.419) always exceeds that of (2.417) – acceptance of bias in the estimate (2.415) reduces the uncertainty of the result – a common trade-off.

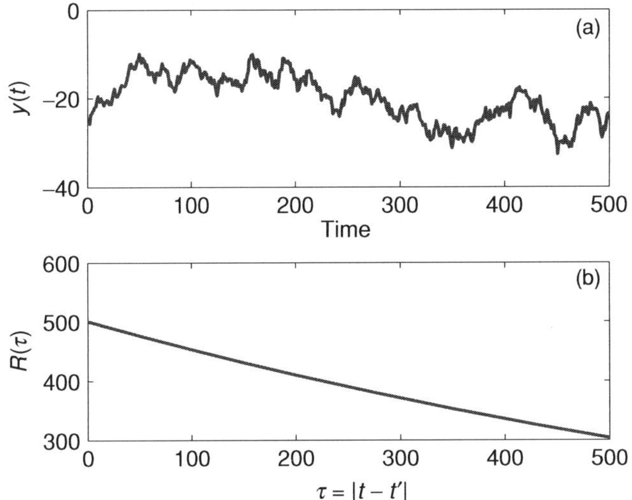

Figure 2.17 (a) Time series $y(t)$ whose mean is required. (b) The autocovariance $\langle y(t)y(t')\rangle$ as a function of $|t - t'|$ (in this special case, it does not depend upon t, t' separately.) The true mean of $y(t)$ is zero by construction.

Equations (2.413) and (2.414) are the more general estimation rule – accounting through $\mathbf{R}_{nn}$ for correlations in the observations and their irregular distribution. Because samples may not be independent, (2.419) may be extremely optimistic. Equation (2.414) gives one the appropriate expression for the variance when the data are correlated (that is, when there are fewer degrees of freedom than the number of sample points).

Example *The mean is needed for the $M = 500$ values of the measured time series, shown in Fig. 2.17. If one calculates the ordinary average, $\tilde{m} = -20.0$, the standard error, treating the measurements as uncorrelated, by Eq. (2.419) is ± 0.31. If, on the other hand, one uses the covariance function displayed in Fig. 2.17, and Eqs. (2.413) and (2.414) with $m_0^2 \to \infty$, one obtains $\tilde{m} = -23.7$, with a standard error of ± 20. The true mean of the time series is actually zero (it was generated that way), and one sees the dire effects of assuming uncorrelated measurement noise, when the correlation is actually very strong. Within two standard deviations (a so-called 95% confidence interval for the mean, if n is Gaussian), one finds, correctly, that the sample mean is indistinguishable from zero, whereas the mean assuming uncorrelated noise would appear to be very well determined and markedly different from zero.[49] (One might be tempted to apply a transformation to render the observations uncorrelated before averaging, and so treat the result as having M degrees-of-freedom. But recall, e.g., that for Gaussian variables (p. 39),* *the resulting numbers will have different variances, and one would be averaging apples and oranges.)*

The use of the prior estimate, m_0^2, is interesting. Letting m_0^2 go to infinity does not mean that an infinite mean is expected (Eq. (2.418) is finite). It is is merely a statement that there is no information whatsoever, before we start, as to the magnitude of the true average – it could be arbitrarily large (or small and of either sign) and if it came out that way, it would be acceptable. Such a situation is, of course, unlikely and even though we might choose not to use information concerning the probable size of the solution, we should remain aware that we could do so (the importance of the prior estimate diminishes as M grows – so that with an infinite amount of data it has no effect at all on the estimate). If a prior estimate of m itself is available, rather than just its mean square, the problem should be reformulated as one for the estimate of the perturbation about this value.

It is very important not to be tempted into making a first estimate of m_0^2 by using (2.418), substituting into (2.415), thinking to reduce the error variance. For the Gauss–Markov theorem to be valid, the prior information must be truly independent of the data being used.

2.8 Improving recursively

2.8.1 Least-squares

A common situation arises that one has a solution $\tilde{\mathbf{x}}$, $\tilde{\mathbf{n}}$, $\mathbf{P}$, and more information becomes available, often in the form of further noisy linear constraints. One way of using the new information is to combine the old and new equations into one larger system, and re-solve. This approach may well be the best one. Sometimes, however, perhaps because the earlier equations have been discarded, or for reasons of storage or both, one prefers to retain the information from the previous solution without having to re-solve the entire system. So-called recursive methods, in both least-squares and minimum variance estimation, provide the appropriate recipes.

Let the original equations be re-labeled so that we can distinguish them from those that come later, in the form

$$\mathbf{E}(1)\mathbf{x} + \mathbf{n}(1) = \mathbf{y}(1), \tag{2.420}$$

where the noise $\mathbf{n}(1)$ has zero mean and covariance matrix $\mathbf{R}_{nn}(1)$. Let the estimate of the solution to (2.420) from one of the estimators be written as $\tilde{\mathbf{x}}(1)$, with uncertainty $\mathbf{P}(1)$. To be specific, suppose (2.420) is full-rank overdetermined, and was solved using row-weighted least-squares, as

$$\tilde{\mathbf{x}}(1) = [\mathbf{E}(1)^{\mathrm{T}}\mathbf{R}_{nn}(1)^{-1}\mathbf{E}(1)]^{-1}\mathbf{E}(1)^{\mathrm{T}}\mathbf{R}_{nn}(1)^{-1}\mathbf{y}(1), \tag{2.421}$$

with corresponding $\mathbf{P}(1)$ (column weighting is redundant in the full-rank fully-determined case).

Some new observations, $\mathbf{y}(2)$, are obtained, with the error covariance of the new observations given by $\mathbf{R}_{nn}(2)$, so that the problem for the unknown $\mathbf{x}$ is

$$\left\{\begin{matrix}\mathbf{E}(1)\\\mathbf{E}(2)\end{matrix}\right\}\mathbf{x} + \begin{bmatrix}\mathbf{n}(1)\\\mathbf{n}(2)\end{bmatrix} = \begin{bmatrix}\mathbf{y}(1)\\\mathbf{y}(2)\end{bmatrix}, \tag{2.422}$$

where $\mathbf{x}$ is the same unknown. We assume $\langle\mathbf{n}(2)\rangle = \mathbf{0}$ and

$$\langle\mathbf{n}(1)\mathbf{n}(2)^{\mathrm{T}}\rangle = \mathbf{0}, \tag{2.423}$$

that is, *no correlation of the old and new measurement errors.* A solution to (2.422) should give a better estimate of $\mathbf{x}$ than (2.420) alone, because more observations are available. It is sensible to row weight the concatenated set to

$$\begin{bmatrix}\mathbf{R}_{nn}(1)^{-\mathrm{T}/2}\mathbf{E}(1)\\\mathbf{R}_{nn}(2)^{-\mathrm{T}/2}\mathbf{E}(2)\end{bmatrix}\mathbf{x} + \begin{bmatrix}\mathbf{R}_{nn}(1)^{-\mathrm{T}/2}\mathbf{n}(1)\\\mathbf{R}_{nn}(2)^{-\mathrm{T}/2}\mathbf{n}(2)\end{bmatrix} = \begin{bmatrix}\mathbf{R}_{nn}(1)^{-\mathrm{T}/2}\mathbf{y}(1)\\\mathbf{R}_{nn}(2)^{-\mathrm{T}/2}\mathbf{y}(2)\end{bmatrix}. \tag{2.424}$$

"Recursive weighted least-squares" seeks the solution to (2.424) without inverting the new, larger matrix, by taking advantage of the existing knowledge of $\tilde{\mathbf{x}}(1)$, $\mathbf{P}(1)$ – however they might actually have been obtained. The objective function corresponding to finding the minimum weighted error norm in (2.424) is

$$\begin{aligned}J = {}&(\mathbf{y}(1) - \mathbf{E}(1)\mathbf{x})^{\mathrm{T}}\mathbf{R}_{nn}(1)^{-1}(\mathbf{y}(1) - \mathbf{E}(1)\mathbf{x})\\&+ (\mathbf{y}(2) - \mathbf{E}(2)\mathbf{x})^{\mathrm{T}}\mathbf{R}_{nn}(2)^{-1}(\mathbf{y}(2) - \mathbf{E}(2)\mathbf{x}).\end{aligned} \tag{2.425}$$

Taking the derivatives with respect to $\mathbf{x}$, the normal equations produce a new solution,

$$\begin{aligned}\tilde{\mathbf{x}}(2) = {}&\{\mathbf{E}(1)^{\mathrm{T}}\mathbf{R}_{nn}(1)^{-1}\mathbf{E}(1) + \mathbf{E}(2)^{\mathrm{T}}\mathbf{R}_{nn}(2)^{-1}\mathbf{E}(2)\}^{-1}\\&\times \{\mathbf{E}(1)^{\mathrm{T}}\mathbf{R}_{nn}(1)^{-1}\mathbf{y}(1) + \mathbf{E}(2)^{\mathrm{T}}\mathbf{R}_{nn}(2)^{-1}\mathbf{y}(2)\}.\end{aligned} \tag{2.426}$$

This is the result from the brute-force re-solution. But one can manipulate (2.426) into[50] (see Appendix 3 to this chapter):

$$\begin{aligned}\tilde{\mathbf{x}}(2) &= \tilde{\mathbf{x}}(1) + \mathbf{P}(1)\mathbf{E}(2)^{\mathrm{T}}[\mathbf{E}(2)\mathbf{P}(1)\mathbf{E}(2)^{\mathrm{T}} + \mathbf{R}_{nn}(2)]^{-1}[\mathbf{y}(2) - \mathbf{E}(2)\tilde{\mathbf{x}}(1)]\\&= \tilde{\mathbf{x}}(1) + \mathbf{K}(2)[\mathbf{y}(2) - \mathbf{E}(2)\tilde{\mathbf{x}}(1)],\end{aligned} \tag{2.427}$$

$$\mathbf{P}(2) = \mathbf{P}(1) - \mathbf{K}(2)\mathbf{E}(2)\mathbf{P}(1), \tag{2.428}$$

$$\mathbf{K}(2) = \mathbf{P}(1)\mathbf{E}(2)^{\mathrm{T}}[\mathbf{E}(2)\mathbf{P}(1)\mathbf{E}(2)^{\mathrm{T}} + \mathbf{R}_{nn}(2)]^{-1}. \tag{2.429}$$

An alternate form for $\mathbf{P}(2)$, found from the matrix inversion lemma, is

$$\mathbf{P}(2) = [\mathbf{P}(1)^{-1} + \mathbf{E}(2)^{\mathrm{T}}\mathbf{R}_{nn}(2)^{-1}\mathbf{E}(2)]^{-1}. \tag{2.430}$$

A similar alternate for $\tilde{\mathbf{x}}(2)$, involving different dimensions of the matrices to be inverted, is also available from the matrix inversion lemma, but is generally less useful. (In some large problems, however, matrix inversion can prove less onerous than matrix multiplication.)

The solution (2.427) is just the least-squares solution to the full set, but rearranged after a bit of algebra. *The original data, $\mathbf{y}(1)$, and coefficient matrix, $\mathbf{E}(1)$, have disappeared, to be replaced by the first solution, $\tilde{\mathbf{x}}(1)$, and its uncertainty, $\mathbf{P}(1)$. That is to say, one need not retain the original data and $\mathbf{E}(1)$ for the new solution to be computed.* Furthermore, because the new solution depends only upon $\tilde{\mathbf{x}}(1)$, and $\mathbf{P}(1)$, the particular methodology originally employed for obtaining them is irrelevant: they might even have been obtained from an educated guess, or from some previous calculation of arbitrary complexity. If the initial set of equations (2.420) is actually underdetermined, and should it have been solved using the SVD, one must be careful that $\mathbf{P}(1)$ includes the estimated error owing to the missing nullspace. Otherwise, these elements would be assigned zero error variance, and the new data could never affect them. Similarly, the dimensionality and rank of $\mathbf{E}(2)$ is arbitrary, as long as the matrix inverse exists.

Example *Suppose we have a single measurement of a scalar, x, so that $x + n(1) = y(1)$, $\langle n(1) \rangle = 0$, $\langle n(1)^2 \rangle = R(1)$. Then an estimate of x is $\tilde{x}(1) = y(1)$, with uncertainty $P(1) = R(1)$. A second measurement then becomes available, $x + n(2) = y(2)$, $\langle n(2) \rangle = 0$, $\langle n(2)^2 \rangle = R(2)$. By Eq. (2.427), an improved solution is*

$$\tilde{x}(2) = y(1) + R(1)/(R(1) + R(2))(y(1) - y(2)),$$

with uncertainty by Eq. (2.430),

$$P(2) = 1/(1/R(1) + 1/R(2)) = R(1)R(2)/(R(1) + R(2)).$$

If $R(1) = R(2) = R$, we have $\tilde{x}(2) = (y(1) + y(2))/2$, $P(2) = R/2$. If there are M successive measurements all with the same error variance, R, one finds the last estimate is

$$\tilde{x}(M) = \tilde{x}(M-1) + R/(M-1)(R/(M-1) + R)^{-1} y(M)$$

$$= \tilde{x}(M-1) + \frac{1}{M}y(M)$$

$$= \frac{1}{M}(y(1) + y(2) + \cdots + y(M)),$$

with uncertainty

$$P(M) = \frac{1}{((M-1)/R + 1/R)} = \frac{R}{M},$$

the conventional average and its variance. Note that each new measurement is given a weight $1/M$ relative to the average, $\tilde{x}(M-1)$, already computed from the previous $M-1$ data points.

The structure of the improved solution (2.427) is also interesting and suggestive. It is made up of two terms: the previous estimate plus a term proportional to the difference between the new observations $\mathbf{y}(2)$, and *a prediction of what those observations should have been* were the first estimate the wholly correct one and the new observations perfect. It thus has the form of a "predictor-corrector." The difference between the prediction and the forecast can be called the "prediction error," but recall there is observational noise in $\mathbf{y}(2)$. The new estimate is a weighted average of this difference and the prior estimate, with the weighting depending upon the details of the uncertainty of prior estimate and new data. The behavior of the updated estimate is worth understanding in various limits. For example, suppose the initial uncertainty estimate is diagonal, $\mathbf{P}(1) = \Delta^2\mathbf{I}$. Then,

$$\mathbf{K}(2) = \mathbf{E}(2)^{\mathrm{T}}[\mathbf{E}(2)\mathbf{E}(2)^{\mathrm{T}} + \mathbf{R}_{nn}(2)/\Delta^2]^{-1}. \tag{2.431}$$

If the new observations are extremely accurate, the norm of $\mathbf{R}_{nn}(2)/\Delta^2$ is small, and if the second set of observations is full-rank underdetermined,

$$\mathbf{K}(2) \longrightarrow \mathbf{E}(2)^{\mathrm{T}}(\mathbf{E}(2)\mathbf{E}(2)^{\mathrm{T}})^{-1}$$

and

$$\begin{aligned}
\tilde{\mathbf{x}}(2) &= \tilde{\mathbf{x}}(1) + \mathbf{E}(2)^{\mathrm{T}}(\mathbf{E}(2)\mathbf{E}(2)^{\mathrm{T}})^{-1}[\mathbf{y}(2) - \mathbf{E}(2)\tilde{\mathbf{x}}(1)] \\
&= [\mathbf{I} - \mathbf{E}(2)^{\mathrm{T}}(\mathbf{E}(2)\mathbf{E}(2)^{\mathrm{T}})^{-1}\mathbf{E}(2)]\tilde{\mathbf{x}}(1) + \mathbf{E}(2)^{\mathrm{T}}(\mathbf{E}(2)\mathbf{E}(2)^{\mathrm{T}})^{-1}\mathbf{y}(2).
\end{aligned} \tag{2.432}$$

Now, $[\mathbf{I} - \mathbf{E}(2)^{\mathrm{T}}(\mathbf{E}(2)\mathbf{E}(2)^{\mathrm{T}})^{-1}\mathbf{E}(2)] = \mathbf{I}_N - \mathbf{V}\mathbf{V}^{\mathrm{T}} = \mathbf{Q}_v\mathbf{Q}_v^{\mathrm{T}}$, where $\mathbf{V}$ is the full-rank singular vector matrix for $\mathbf{E}(2)$, and it spans the nullspace of $\mathbf{E}(2)$ (see Eq. (2.291)). The update thus replaces, in the first estimate, all the structures given perfectly by the second set of observations, but retains those structures from the first estimate about which the new observations say nothing – a sensible result. (Compare Eq. (2.427) with Eq. (2.360).) At the opposite extreme, when the new observations are very noisy compared to the previous ones, $\|\mathbf{R}_{nn}/\Delta^2\| \to \infty$, $\|\mathbf{K}(2)\| \to 0$, and the first estimate is left unchanged.

The general case represents a weighted average of the previous estimate with elements found from the new data, with the weighting depending both upon the relative noise in each, and upon the structure of the observations relative to the structure of $\mathbf{x}$ as represented in $\mathbf{P}(1)$, $\mathbf{R}_{nn}(2)$, $\mathbf{E}(2)$. The matrix being inverted in (2.429) is the sum of the measurement error covariance, $\mathbf{R}_{nn}(2)$, and the error covariance of the "forecast" $\mathbf{E}(2)\tilde{\mathbf{x}}(1)$. To see this, let γ be the error component in $\tilde{\mathbf{x}}(1) = \mathbf{x}(1) + \gamma$, which by definition has covariance $\langle\gamma\gamma^{\mathrm{T}}\rangle = \mathbf{P}(1)$. Then the expected covariance of the error of prediction is $\langle \mathbf{E}(1)\boldsymbol{\gamma}\boldsymbol{\gamma}^{\mathrm{T}}\mathbf{E}(1)^{\mathrm{T}}\rangle = \mathbf{E}(1)\mathbf{P}(1)\mathbf{E}(1)^{\mathrm{T}}$, which appears in $\mathbf{K}(2)$. Because of the assumptions (2.423), and $\langle\boldsymbol{\gamma}(1)\mathbf{x}(1)^{\mathrm{T}}\rangle = \mathbf{0}$, it follows that

$$\langle \mathbf{y}(1)(\mathbf{y}(2) - \mathbf{E}(2)\tilde{\mathbf{x}}(1))\rangle = \mathbf{0}. \tag{2.433}$$

That is, the prediction error or "innovation," $\mathbf{y}(2) - \mathbf{E}(2)\tilde{\mathbf{x}}(1)$, is uncorrelated with the previous measurement.

The possibility of a recursion based on Eqs. (2.427) and (2.428) (or (2.430)) is obvious – all subscript 1 variables being replaced by subscript 2 variables, which in turn are replaced by subscript 3 variables, etc. The general form would be:

$$\tilde{\mathbf{x}}(m) = \tilde{\mathbf{x}}(m-1) + \mathbf{K}(m)[\mathbf{y}(m) - \mathbf{E}(m)\tilde{\mathbf{x}}(m-1)], \tag{2.434}$$

$$\mathbf{K}(m) = \mathbf{P}(m-1)\mathbf{E}(m)^{\mathrm{T}}[\mathbf{E}(m)\mathbf{P}(m-1)\mathbf{E}(m)^{\mathrm{T}} + \mathbf{R}_{nn}(m)]^{-1}, \tag{2.435}$$

$$\mathbf{P}(m) = \mathbf{P}(m-1) - \mathbf{K}(m)\mathbf{E}(m)\mathbf{P}(m-1), \quad m = 1, 2, \ldots, \tag{2.436}$$

where m conventionally starts with 0. An alternative form for Eq. (2.436) is, from (2.430),

$$\mathbf{P}(m) = [\mathbf{P}(m-1)^{-1} + \mathbf{E}(m)^{\mathrm{T}}\mathbf{R}_{nn}(m)^{-1}\mathbf{E}(m)]^{-1}. \tag{2.437}$$

The computational load of the recursive solution needs to be addressed. A least-squares solution does *not* require one to calculate the uncertainty $\mathbf{P}$ (although the utility of $\tilde{\mathbf{x}}$ without such an estimate is unclear). But to use the recursive form, one must have $\mathbf{P}(m-1)$, otherwise the update step, Eq. (2.434) cannot be used. In very large problems, such as appear in oceanography and meteorology (Chapter 6), the computation of the uncertainty, from (2.436) or (2.437), can become prohibitive. In such a situation, one might simply store all the data, and do one large calculation – if this is feasible. Normally, it will involve less pure computation than will the recursive solution which must repeatedly update $\mathbf{P}(m)$.

The comparatively simple interpretation of the recursive, weighted least-squares problem will be used in Chapter 4 to derive the Kalman filter and suboptimal filters in a very simple form. It also becomes the key to understanding "assimilation" schemes such as "nudging," "forcing to climatology," and "robust diagnostic" methods.

2.8.2 Minimum variance recursive estimates

The recursive least-squares result is identical to a recursive estimation procedure, if appropriate least-squares weight matrices were used. To see this result, suppose there exist two *independent* estimates of an unknown vector $\mathbf{x}$, denoted $\tilde{\mathbf{x}}_a$, $\tilde{\mathbf{x}}_b$ with estimated uncertainties $\mathbf{P}_a$, $\mathbf{P}_b$, respectively. They are either unbiassed, or have the same mean, $\langle \tilde{\mathbf{x}}_a \rangle = \langle \tilde{\mathbf{x}}_b \rangle = \mathbf{x}_B$. How should the two be combined to give a third estimate $\tilde{\mathbf{x}}^+$ with minimum error variance?

Try a linear combination,

$$\tilde{\mathbf{x}}^{\,!} = \mathbf{L}_a \tilde{\mathbf{x}}_a + \mathbf{L}_b \tilde{\mathbf{x}}_b. \tag{2.438}$$

If the new estimate is to be unbiassed, or is to retain the prior bias (that is, the same mean), it follows that,

$$\langle \tilde{\mathbf{x}}^+ \rangle = \mathbf{L}_a \langle \tilde{\mathbf{x}}_a \rangle + \mathbf{L}_b \langle \tilde{\mathbf{x}}_b \rangle, \tag{2.439}$$

or

$$\mathbf{x}_B = \mathbf{L}_a \mathbf{x}_B + \mathbf{L}_b \mathbf{x}_B, \tag{2.440}$$

or

$$\mathbf{L}_b = \mathbf{I} - \mathbf{L}_a. \tag{2.441}$$

Then the uncertainty is

$$\begin{aligned}
\mathbf{P}^+ &= \langle (\tilde{\mathbf{x}}^+ - \mathbf{x})(\tilde{\mathbf{x}}^+ - \mathbf{x})^{\mathrm{T}} \rangle = \langle (\mathbf{L}_a \tilde{\mathbf{x}}_a + (\mathbf{I} - \mathbf{L}_a)\tilde{\mathbf{x}}_b)(\mathbf{L}_a \tilde{\mathbf{x}}_a + (\mathbf{I} - \mathbf{L}_a)\tilde{\mathbf{x}}_b)^{\mathrm{T}} \rangle \\
&= \mathbf{L}_a \mathbf{P}_a \mathbf{L}_a^{\mathrm{T}} + (\mathbf{I} - \mathbf{L}_a)\mathbf{P}_b(\mathbf{I} - \mathbf{L}_a)^{\mathrm{T}},
\end{aligned} \tag{2.442}$$

where the independence assumption has been used to set $\langle (\tilde{\mathbf{x}}_a - \mathbf{x})(\tilde{\mathbf{x}}_b - \mathbf{x}) \rangle = \mathbf{0}$. $\mathbf{P}^+$ is positive definite; minimizing its diagonal elements with respect to $\mathbf{L}_a$ yields (after writing out the diagonal elements of the products)

$$\mathbf{L}_a = \mathbf{P}_b(\mathbf{P}_a + \mathbf{P}_b)^{-1}, \qquad \mathbf{L}_b = \mathbf{P}_a(\mathbf{P}_a + \mathbf{P}_b)^{-1}.$$

(Blithely differentiating and setting to zero produces the correct answer:

$$\frac{\partial(\operatorname{diag}\mathbf{P}^+)}{\partial \mathbf{L}_a} = \operatorname{diag}\left(\frac{\partial \mathbf{P}^+}{\partial \mathbf{L}_a}\right) = \operatorname{diag}\left[2\mathbf{P}_a \mathbf{L}_a - \mathbf{P}_b(\mathbf{I} - \mathbf{L}_a)\right] = 0,$$

or $\mathbf{L}_a = \mathbf{P}_b(\mathbf{P}_a + \mathbf{P}_b)^{-1}$.) The new combined estimate is

$$\tilde{\mathbf{x}}^+ = \mathbf{P}_b(\mathbf{P}_a + \mathbf{P}_b)^{-1}\tilde{\mathbf{x}}_a + \mathbf{P}_a(\mathbf{P}_a + \mathbf{P}_b)^{-1}\tilde{\mathbf{x}}_b. \tag{2.443}$$

This last expression can be rewritten by adding and subtracting $\tilde{\mathbf{x}}_a$ as

$$\begin{aligned}
\tilde{\mathbf{x}}^+ &= \tilde{\mathbf{x}}_a + \mathbf{P}_b(\mathbf{P}_a + \mathbf{P}_b)^{-1}\tilde{\mathbf{x}}_a \\
&\quad + \mathbf{P}_a(\mathbf{P}_a + \mathbf{P}_b)^{-1}\tilde{\mathbf{x}}_b - (\mathbf{P}_a + \mathbf{P}_b)(\mathbf{P}_a + \mathbf{P}_b)^{-1}\tilde{\mathbf{x}}_a \\
&= \tilde{\mathbf{x}}_a + \mathbf{P}_a(\mathbf{P}_a + \mathbf{P}_b)^{-1}(\tilde{\mathbf{x}}_b - \tilde{\mathbf{x}}_a).
\end{aligned} \tag{2.444}$$

Notice, in particular, the re-appearance of a predictor-corrector form relative to $\tilde{\mathbf{x}}_a$.

The uncertainty of the estimate (2.444) is easily evaluated as

$$\mathbf{P}^+ = \mathbf{P}_a - \mathbf{P}_a(\mathbf{P}_a + \mathbf{P}_b)^{-1}\mathbf{P}_a. \tag{2.445}$$

or, by straightforward application of the matrix inversion lemma,

$$\mathbf{P}^+ = \left(\mathbf{P}_a^{-1} + \mathbf{P}_b^{-1}\right)^{-1}. \tag{2.446}$$

The uncertainty is again independent of the observations. Equations (2.444)–(2.446) are the general rules for combining two estimates with uncorrelated errors.

Now suppose that $\tilde{\mathbf{x}}_a$ and its uncertainty are known, but that instead of $\tilde{\mathbf{x}}_b$ there are measurements,

$$\mathbf{E}(2)\mathbf{x} + \mathbf{n}(2) = \mathbf{y}(2), \tag{2.447}$$

with $\langle \mathbf{n}(2) \rangle = 0$, $\langle \mathbf{n}(2)\mathbf{n}(2)^{\mathrm{T}} \rangle = \mathbf{R}_{nn}(2)$. From this second set of observations, we *estimate* the solution, using the minimum variance estimator (2.406, 2.408) with no use of the solution variance; that is, let $\|\mathbf{R}_{xx}^{-1}\| \to 0$. The reason for suppressing $\mathbf{R}_{xx}$, which logically could come from $\mathbf{P}_a$, is to maintain the independence of the previous and the new estimates. Then

$$\tilde{\mathbf{x}}_b = [\mathbf{E}(2)^{\mathrm{T}}\mathbf{R}_{nn}(2)^{-1}\mathbf{E}(2)]^{-1}\mathbf{E}(2)^{\mathrm{T}}\mathbf{R}_{nn}(2)^{-1}\mathbf{y}(2), \tag{2.448}$$

$$\mathbf{P}_b = [\mathbf{E}(2)^{\mathrm{T}}\mathbf{R}_{nn}(2)^{-1}\mathbf{E}(2)]^{-1}. \tag{2.449}$$

Substituting (2.448), (2.449) into (2.444), (2.445), and using the matrix inversion lemma, (see Appendix 3 to this chapter) gives

$$\tilde{\mathbf{x}}^+ = \tilde{\mathbf{x}}_a + \mathbf{P}_a\mathbf{E}(2)^{\mathrm{T}}[\mathbf{E}(2)\mathbf{P}_a\mathbf{E}(2)^{\mathrm{T}} + \mathbf{R}_{nn}(2)]^{-1}(\mathbf{y}(2) - \mathbf{E}(2)\tilde{\mathbf{x}}_a), \tag{2.450}$$

$$\mathbf{P}^+ = [\mathbf{P}_a^{-1} + \mathbf{E}(2)^{\mathrm{T}}\mathbf{R}_{nn}(2)^{-1}\mathbf{E}(2)]^{-1}, \tag{2.451}$$

which is the same as (2.434), (2.437) and thus *a recursive minimum variance estimate coincides with a corresponding weighted least-squares recursion.* The new covariance may also be confirmed to be that in either of Eqs. (2.436) or (2.437). Notice that if $\tilde{\mathbf{x}}_a$ was itself estimated from an earlier set of observations, then those data have disappeared from the problem, with all the information derived from them contained in $\tilde{\mathbf{x}}_a$ and $\mathbf{P}_a$. Thus, again, earlier data can be wholly discarded after use. It does not matter where $\tilde{\mathbf{x}}_a$ originated, whether from over- or underdetermined equations or a pure guess – as long as $\mathbf{P}_a$ is realistic. Similarly, expression (2.450) remains valid whatever the dimensionality or rank of $\mathbf{E}(2)$ as long as the inverse matrix exists. The general implementation of this sequence for a continuing data stream corresponds to Eqs. (2.434)–(2.437).

2.9 Summary

This chapter has not exhausted the possibilities for inverse methods, and the techniques will be extended in several directions in the next chapters. Given the lengthy nature of the discussion so far, however, some summary of what has been accomplished may be helpful.

The focus is on making inferences about parameters or fields, $\mathbf{x}$, $\mathbf{n}$ satisfying linear relationships of the form

$$\mathbf{Ex} + \mathbf{n} = \mathbf{y}.$$

Such equations arise as we have seen, from both "forward" and "inverse" problems, but the techniques for estimating $\mathbf{x}$, $\mathbf{n}$ and their uncertainty are useful whatever the physical origin of the equations. Two methods for estimating $\mathbf{x}$, $\mathbf{n}$ have been the focus of the chapter: least-squares (including the singular value decomposition) and the Gauss–Markov or minimum variance technique. Least-squares, in any of its many guises, is a very powerful method – but its power and ease of use have (judging from the published literature) led many investigators into serious confusion about what they are doing. This confusion is compounded by the misunderstandings about the difference between an inverse problem and an inverse method.

An attempt is made therefore, to emphasize the two distinct roles of least-squares: as a method of *approximation*, and as a method of *estimation*. It is only in the second formulation that it can be regarded as an inverse method. A working definition of an inverse method is a technique able to estimate unknown parameters or fields of a model, while producing an estimate of the uncertainties of the results. Solution plus uncertainty are the fundamental requirements. There are many desirable additional features of inverse methods. Among them are: (1) separation of nullspace uncertainties from observational noise uncertainties; (2) the ability to rank the data in its importance to the solution; (3) the ability to use prior statistical knowledge; (4) understanding of solution structures in terms of data structure; (5) the ability to trade resolution against variance. (The list is not exhaustive. For example, we will briefly examine in Chapter 3 the use of inequality information.) As with all estimation methods, one also trades computational load against the need for information. (The SVD, for example, is a powerful form of least-squares, but requires more computation than do other forms.) The Gauss–Markov approach has the strength of forcing explicit use of prior statistical information and is directed at the central goal of obtaining $\mathbf{x}$, $\mathbf{n}$ with the smallest mean-square error, and for this reason might well be regarded as the default methodology for linear inverse problems. It has the added advantage that we know we can obtain precisely the same result with appropriate versions of least-squares, including the SVD, permitting the use of least-squares algorithms, but at the risk of losing sight of what we are actually attempting. A limitation is that the underlying probability densities of solution and noise have to be unimodal (so that a minimum variance estimate makes sense). If unimodality fails, one must look to other methods.

The heavy emphasis here on noise and uncertainty may appear to be a tedious business. But readers of the scientific literature will come to recognize how qualitative much of the discussion is – the investigator telling a story about what he or she thinks is going on with no estimate of uncertainties, and no attempt to resolve quantitatively differences with previous competing estimates. In a quantitative subject, such vagueness is ultimately intolerable.

A number of different procedures for producing estimates of the solution to a set of noisy simultaneous equations of arbitrary dimension have been described here. The reader may wonder which of the variants makes the most sense to use in practice. Because, in the presence of noise one is dealing with a statistical estimation problem, there is no single "best" answer, and one must be guided by model context and goals. A few general remarks might be helpful.

In any problem where data are to be used to make inferences about physical parameters, one typically needs some approximate idea of just how large the solution is likely to be and how large the residuals probably are. In this nearly agnostic case, where almost nothing else is known and the problem is very large, the weighted, tapered least-squares solution is a good first choice – it is easily and efficiently computed and coincides with the Gauss–Markov and tapered SVD solutions, if the weight matrices are the appropriate covariances. Sparse matrix methods exist for its solution should that be necessary.[51] Coincidence with the Gauss–Markov solution means one can reinterpret it as a minimum-variance or maximum-likelihood solution (see Appendix 1 to this chapter) should one wish.

It is a comparatively easy matter to vary the trade-off parameter, γ^2, to explore the consequences of any errors in specifying the noise and solution variances. Once a value for γ^2 is chosen, the tapered SVD can then be computed to understand the relationships between solution and data structures, their resolution and their variance. For problems of small to moderate size (the meaning of "moderate" is constantly shifting, but it is difficult to examine and interpret matrices of more than about 500×500), the SVD, whether in the truncated or tapered forms is probably the method of choice – because it provides the fullest information about data and its relationship to the solution. Its only disadvantages are that one can easily be overwhelmed by the available information, particularly if a range of solutions must be examined. The SVD has a flexibility beyond even what we have discussed – one could, for example, change the degree of tapering in each of the terms of (2.338) and (2.339) should there be reason to repartition the variance between solution and noise, or some terms could be dropped out of the truncated form at will – should the investigator know enough to justify it.

To the extent that either or both of $\mathbf{x}$, $\mathbf{n}$ have expected structures expressible through covariance matrices, these structures can be removed from the problem through the various weight matrix and/or the Cholesky decomposition. The resulting problem is then one in completely unstructured (equivalent to white noise) elements $\mathbf{x}$, $\mathbf{n}$. In the resulting scaled and rotated systems, one can use the simplest of all objective functions. Covariance, resolution, etc., in the original spaces of $\mathbf{x}$, $\mathbf{n}$ is readily recovered by appropriately applying the weight matrices to the results of the scaled/rotated space.

Both ordinary weighted least-squares and the SVD applied to row- and column-weighted equations are best thought of as approximation, rather than estimation, methods. In particular, the truncated SVD does not produce a minimum variance estimate the way the tapered version can. The tapered SVD (along with the Gauss–Markov estimate, or the tapered least-squares solutions) produce the minimum variance property by tolerating a bias in the solution. Whether the bias is more desirable than a larger uncertainty is a decision the user must make. But the reader is warned against the belief that there is any single best method.

Appendix 1. Maximum likelihood

The estimation procedures used in this book are based primarily upon the idea of minimizing the variance of the estimate about the true value. Alternatives exist. For example, given a set of observations with known joint probability density, one can use a principle of "maximum likelihood." This very general and powerful principle attempts to find those estimated parameters that render the actual observations the most likely to have occurred. By way of motivation, consider the simple case of uncorrelated jointly normal stationary time series, x_i, where

$$\langle x_i \rangle = m, \quad \langle (x_i - m)(x_j - m) \rangle = \sigma^2 \delta_{ij}.$$

The corresponding joint probability density for $\mathbf{x} = [x_1, x_2, \ldots, x_N]^{\mathrm{T}}$ can be written as

$$p_{\mathbf{x}}(\mathbf{X}) = \frac{1}{(2\pi)^{N/2} \sigma^N} \tag{2.452}$$
$$\times \exp\left\{ -\frac{1}{2\sigma^2} [(X_1 - m)^2 + (X_2 - m)^2 + \cdots + (X_N - m)^2] \right\}.$$

Substitution of the observed values, $X_1 = x_1$, $X_2 = x_2$, $\ldots$, into Eq. (2.452) permits evaluation of the probability that these particular values occurred. Denote the corresponding probability density as L. One can demand those values of m, σ rendering the value of L as large as possible. L will be a maximum if $\log(L)$ is as large as possible: that is, we seek to maximize

$$\log (L) = -\frac{1}{2\sigma^2}[(x_1 - m)^2 + (x_2 - m)^2 + \cdots + (x_N - m)^2]$$
$$+ N \log (\sigma) + \frac{N}{2} \log(2\pi),$$

with respect to m, σ. Setting the corresponding partial derivatives to zero and solving produces

$$\tilde{m} = \frac{1}{N} \sum_{i=1}^{M} x_i, \quad \tilde{\sigma}^2 = \frac{1}{N} \sum_{i=1}^{N} (x_i - \tilde{m})^2.$$

That is, the usual sample mean, and biassed sample variance maximize the probability of the observed data actually occurring. A similar calculation is readily carried out using correlated normal, or any random variables with a different probability density.

Likelihood estimation, and its close cousin, Bayesian methods, are general powerful estimation methods that can be used as an alternative to almost everything covered in this book.[52] Some will prefer that route, but the methods used here are adequate for a wide range of problems.

Appendix 2. Differential operators and Green functions

Adjoints appear prominently in the theory of differential operators and are usually discussed independently of any optimization problem. Many of the concepts are those used in defining Green functions.

Suppose we want to solve an ordinary differential equation,

$$\frac{du (\xi)}{d\xi} + \frac{d^2 u (\xi)}{d\xi^2} = \rho(\xi), \tag{2.453}$$

subject to boundary conditions on $u (\xi)$ at $\xi = 0, L$. To proceed, seek first a solution to

$$\alpha \frac{\partial v(\xi, \xi_0)}{\partial \xi} + \frac{\partial^2 v(\xi, \xi_0)}{\partial \xi^2} = \delta(\xi_0 - \xi), \tag{2.454}$$

where α is arbitrary for the time being. Multiply (2.453) by v, and (2.454) by u, and subtract:

$$v(\xi, \xi_0)\frac{du(\xi)}{d\xi} + v(\xi, \xi_0)\frac{d^2 u(\xi)}{d\xi^2} - u(\xi)\alpha \frac{\partial v(\xi, \xi_0)}{\partial \xi} - u (\xi) \frac{\partial^2 v(\xi, \xi_0)}{\partial \xi^2}$$
$$= v(\xi, \xi_0)\rho (\xi) - u(\xi)v(\xi, \xi_0). \tag{2.455}$$

Integrate this last equation over the domain,

$$\int_0^L \left\{ v(\xi, \xi_0)\frac{\mathrm{d}u\,(\xi)}{\mathrm{d}\xi} + v(\xi, \xi_0)\frac{\mathrm{d}^2 u\,(\xi)}{\mathrm{d}\xi^2} - u(\xi)\alpha\frac{\partial v(\xi, \xi_0)}{\partial \xi} - u\,(\xi)\frac{\partial^2 v(\xi, \xi_0)}{\partial \xi^2} \right\} \mathrm{d}\xi \tag{2.456}$$

$$= \int_0^L \{v(\xi, \xi_0)\rho(\xi) - u\,(\xi)\,\delta(\xi_0 - \xi)\}\mathrm{d}\xi, \tag{2.457}$$

or

$$\int_0^L \frac{\mathrm{d}}{\mathrm{d}\xi}\left\{ v\frac{\mathrm{d}u}{\mathrm{d}\xi} - \alpha u\frac{\mathrm{d}v}{\mathrm{d}\xi} \right\} \mathrm{d}\xi + \int_0^L \left\{ u\frac{\partial^2 v(\xi, \xi_0)}{\partial \xi^2} - u\frac{\partial^2 v(\xi, \xi_0)}{\partial \xi^2} \right\} \mathrm{d}\xi$$

$$= \int_0^L v(\xi, \xi_0)\rho\,(\xi)\,\mathrm{d}\xi - u(\xi_0). \tag{2.458}$$

If we choose $\alpha = -1$, then the first term on the left-hand side is integrable, as

$$\int_0^L \frac{\mathrm{d}}{\mathrm{d}\xi}\,\{uv\}\,\mathrm{d}\xi = uv|_0^L, \tag{2.459}$$

as is the second term on the left,

$$\int_0^L \frac{\mathrm{d}}{\mathrm{d}\xi}\left\{ u\frac{\partial v}{\partial \xi} - v\frac{\partial u}{\partial \xi} \right\} \partial\xi = \left[u\frac{\partial v}{\partial \xi} - v\frac{\partial u}{\partial \xi} \right]_0^L, \tag{2.460}$$

and thus,

$$u(\xi_0) = \int_0^L v(\xi, \xi_0)\rho\,(\xi)\,\mathrm{d}\xi + uv|_0^L + \left[u\frac{\mathrm{d}v}{\mathrm{d}\xi} - v\frac{\mathrm{d}u}{\mathrm{d}\xi} \right]_0^L. \tag{2.461}$$

Because the boundary conditions on v were not specified, we are free to choose them such that $v = 0$ on $\xi = 0, L$, e.g., the boundary terms reduce simply to $[u\mathrm{d}v/\mathrm{d}\xi]_0^L$, which is then known.

Here, v is the adjoint solution to Eq. (2.454), with $\alpha = -1$, defining the adjoint equation to (2.453); it was found by requiring that the terms on the left-hand side of Eq. (2.458) should be exactly integrable. v is also the problem Green function (although the Green function is sometimes defined so as to satisfy the forward operator, rather than the adjoint one). Textbooks show that for a general differential operator, $\mathcal{L}$, the requirement that v should render the analogous terms integrable is that

$$u^{\mathrm{T}}\mathcal{L}v = v^{\mathrm{T}}\mathcal{L}^{\mathrm{T}}u, \tag{2.462}$$

where here the superscript T denotes the adjoint. Equation (2.462) defines the adjoint operator (compare to (2.376)).

Appendix 3. Recursive least-squares and Gauss–Markov solutions

The recursive least-squares solution Eq. (2.427) is appealingly simple. Unfortunately, obtaining it from the concatenated least-squares form (2.426),

$$\tilde{\mathbf{x}}(2) = \{\mathbf{E}(1)^{\mathrm{T}}\mathbf{R}_{nn}(1)^{-1}\mathbf{E}(1) + \mathbf{E}(2)^{\mathrm{T}}\mathbf{R}_{nn}(2)^{-1}\mathbf{E}(2)\}^{-1}$$
$$\times \{\mathbf{E}(1)^{\mathrm{T}}\mathbf{R}_{nn}(1)^{-1}\mathbf{y}(1) + \mathbf{E}(2)^{\mathrm{T}}\mathbf{R}_{nn}(2)^{-1}\mathbf{y}\}$$

is not easy. First note that

$$\tilde{\mathbf{x}}(1) = [\mathbf{E}(1)^{\mathrm{T}}\mathbf{R}_{nn}(1)^{-1}\mathbf{E}(1)]^{-1}\mathbf{E}(1)^{\mathrm{T}}\mathbf{R}_{nn}(1)^{-1}\mathbf{y}(1) \qquad (2.463)$$
$$= \mathbf{P}(1)\mathbf{E}(1)^{\mathrm{T}}\mathbf{R}_{nn}(1)^{-1}\mathbf{y}(1),$$

where

$$\mathbf{P}(1) = [\mathbf{E}(1)^{\mathrm{T}}\mathbf{R}_{nn}(1)^{-1}\mathbf{E}(1)]^{-1},$$

are the solution and uncertainty of the overdetermined system from the first set of observations alone. Then

$$\tilde{\mathbf{x}}(2) = \{\mathbf{P}(1)^{-1} + \mathbf{E}(2)^{\mathrm{T}}\mathbf{R}_{nn}(2)^{-1}\mathbf{E}(2)\}^{-1}$$
$$\times \{\mathbf{E}(1)^{\mathrm{T}}\mathbf{R}_{nn}(1)^{-1}\mathbf{y}(1) + \mathbf{E}(2)^{\mathrm{T}}\mathbf{R}_{nn}(2)^{-1}\mathbf{y}(2)\}.$$

Apply the matrix inversion lemma, in the form Eq. (2.35), to the first bracket (using $\mathbf{C} \to \mathbf{P}(1)^{-1}, \mathbf{B} \to \mathbf{E}(2), \mathbf{A} \to \mathbf{R}_{nn}(2)$),

$$\tilde{\mathbf{x}}(2) = \{\mathbf{P}(1) - \mathbf{P}(1)\mathbf{E}(2)^{\mathrm{T}}[\mathbf{E}(2)\mathbf{P}(1)\mathbf{E}(2)^{\mathrm{T}} + \mathbf{R}_{nn}(2)]^{-1}\mathbf{E}(2)\mathbf{P}(1)\}$$
$$\times \{\mathbf{E}(1)^{\mathrm{T}}\mathbf{R}_{nn}(1)^{-1}\mathbf{y}(1)\} + \{\mathbf{P}(1) - \mathbf{P}(1)\mathbf{E}(2)^{\mathrm{T}}[\mathbf{E}(2)\mathbf{P}(1)\mathbf{E}(2)^{\mathrm{T}}$$
$$+ \mathbf{R}_{nn}(2)]^{-1}\mathbf{E}(2)\mathbf{P}(1)\}\{\mathbf{E}(2)^{\mathrm{T}}\mathbf{R}_{nn}(2)^{-1}\mathbf{y}(2)\}$$
$$= \tilde{\mathbf{x}}(1) - \mathbf{P}(1)\mathbf{E}(2)^{\mathrm{T}}[\mathbf{E}(2)\mathbf{P}(1)\mathbf{E}(2)^{\mathrm{T}} + \mathbf{R}_{nn}(2)]^{-1}\mathbf{E}(2)\tilde{\mathbf{x}}(1)$$
$$+ \mathbf{P}(1)\mathbf{E}(2)^{\mathrm{T}}\{\mathbf{I} - [\mathbf{E}(2)\mathbf{P}(1)\mathbf{E}(2)^{\mathrm{T}} + \mathbf{R}_{nn}(2)]^{-1}\mathbf{E}(2)\mathbf{P}(1)\mathbf{E}(2)^{\mathrm{T}}\}$$
$$\times \mathbf{R}_{nn}(2)^{-1}\mathbf{y}(2),$$

using (2.463) and factoring $\mathbf{E}(2)^{\mathrm{T}}$ in the last line. Using the identity

$$[\mathbf{E}(2)\mathbf{P}(1)\mathbf{E}(2)^{\mathrm{T}} + \mathbf{R}_{nn}(2)]^{-1}[\mathbf{E}(2)\mathbf{P}(1)\mathbf{E}(2)^{\mathrm{T}} + \mathbf{R}_{nn}(2)] = \mathbf{I},$$

and substituting for $\mathbf{I}$ in the previous expression, factoring, and collecting terms, gives

$$\tilde{\mathbf{x}}(2) = \tilde{\mathbf{x}}(1) + \mathbf{P}(1)\mathbf{E}(2)^{\mathrm{T}}[\mathbf{E}(2)\mathbf{P}(1)\mathbf{E}(2)^{\mathrm{T}} + \mathbf{R}_{nn}(2)]^{-1}[\mathbf{y}(2) - \mathbf{E}(2)\tilde{\mathbf{x}}(1)],$$
$$(2.464)$$

which is the desired expression. The new uncertainty is given by (2.428) or (2.430).

Manipulation of the recursive Gauss–Markov solution (2.443) or (2.444) is similar, involving repeated use of the matrix inversion lemma. Consider Eq. (2.443) with $\mathbf{x}_b$ from Eq. (2.448),

$$\tilde{\mathbf{x}}^+ = \left(\mathbf{E}(2)^{\mathrm{T}}\mathbf{R}_{nn}^{-1}\mathbf{E}(2)\right)^{-1}\left[\mathbf{P}_a + \mathbf{E}(2)^{\mathrm{T}}\mathbf{R}_{nn}^{-1}\mathbf{E}(2)\right]^{-1}\tilde{\mathbf{x}}_a$$

$$+ \mathbf{P}_a\left[\mathbf{P}_a + \mathbf{E}(2)^{\mathrm{T}}\mathbf{R}_{nn}^{-1}\mathbf{E}(2)\right]^{-1}\left(\mathbf{E}(2)^{\mathrm{T}}\mathbf{R}_{nn}\mathbf{E}(2)\right)^{-1}\mathbf{E}(2)^{\mathrm{T}}\mathbf{R}_{nn}^{-1}\mathbf{y}(2).$$

Using Eq. (2.36) on the first term (with $\mathbf{A} \to (\mathbf{E}(2)^{\mathrm{T}}\mathbf{R}_{nn}^{-1}\mathbf{E}(2))^{-1}$, $\mathbf{B} \to \mathbf{I}$, $\mathbf{C} \to \mathbf{P}_a$), and on the second term with $\mathbf{C} \to (\mathbf{E}(2)\mathbf{R}_{nn}^{-1}\mathbf{E}(2))$, $\mathbf{A} \to \mathbf{P}_a$, $\mathbf{B} \to \mathbf{I}$, this last expression becomes

$$\tilde{\mathbf{x}}^+ = \left[\mathbf{P}_a^{-1} + \mathbf{E}(2)^{\mathrm{T}}\mathbf{R}_{nn}^{-1}\mathbf{E}(2)\right]^{-1}\left[\mathbf{P}_a^{-1}\tilde{\mathbf{x}}_a + \mathbf{E}(2)^{\mathrm{T}}\mathbf{R}_{nn}^{-1}\mathbf{y}(2)\right],$$

yet another alternate form. By further application of the matrix inversion lemma,[53] this last expression can be manipulated into Eq. (2.450), which is necessarily the same as (2.464).

These expressions have been derived assuming that matrices such as $\mathbf{E}(2)^{\mathrm{T}}\mathbf{R}_{nn}^{-1}\mathbf{E}(2)$ are non-singular (full-rank overdetermined). If they are singular, they can be inverted using a generalized inverse, but taking care that $\mathbf{P}(1)$ includes the nullspace contribution (e.g., from Eq. (2.272)).

Notes

1 Noble and Daniel (1977), Strang (1988).
2 Lawson and Hanson (1995).
3 "Positive definite" will be defined below. Here it suffices to mean that $\mathbf{c}^{\mathrm{T}}\mathbf{W}\mathbf{c}$ should never be negative, for any $\mathbf{c}$.
4 Golub and Van Loan (1996).
5 Haykin (2002).
6 Lawson and Hanson (1995), Golub and van Loan (1996), Press *et al.* (1996), etc.
7 Determinants are used only rarely in this book. Their definition and properties are left to the references, as they are usually encountered in high school mathematics.
8 Rogers (1980) is an entire volume of matrix derivative identities, and many other useful properties are discussed by Magnus and Neudecker (1988).
9 Magnus and Neudecker (1988, p. 183).
10 Liebelt (1967, Sections 1–19).
11 The history of this not-very-obvious identity is discussed by Haykin (2002).
12 A good statistics text such as Cramér (1946), or one on regression such as Seber and Lee (2003), should be consulted.
13 Feller (1957) and Jeffreys (1961) represent differing philosophies. Jaynes (2003) forcefully and colorfully argues the case for so-called Bayesian inference (following Jeffreys), and it seems likely that this approach to statistical inference will ultimately become the default method; Gauch (2003) has a particularly clear account of Bayesian methods. For most of the methods in this book, however, we use little more than the first moments of probability distributions, and hence can ignore the underlying philosophical debate.
14 It follows from the Cauchy–Schwarz inequality: Consider $\langle(ax' + y')^2\rangle = a\langle x'^2\rangle + \langle y'^2\rangle + 2a\langle x'y'\rangle \geq 0$ for any constant a. Choose $a = -\langle x'y'\rangle/\langle x'^2\rangle$, and one has $-\langle x'y'\rangle^2/\langle x'^2\rangle + \langle y'^2\rangle \geq 0$, or $1 \geq \langle x'y'\rangle^2/(\langle x'^2\rangle\langle y'^2\rangle)$. Taking the square root of both sides, the required result follows.
15 Draper and Smith (1998), Seber and Lee (2003).

16 Numerical schemes for finding $\mathbf{C}_{\xi\xi}^{1/2}$ are described by Lawson and Hanson (1995) and Golub and Van Loan (1996)

17 Cramér (1946) discusses what happens when the determinant of $\mathbf{C}_{\xi\xi}$ vanishes, that is, if $\mathbf{C}_{\xi\xi}$ is singular.

18 Bracewell (2000).

19 Cramér (1946).

20 In problems involving time, one needs to be clear that "stationary" is not the same idea as "steady."

21 If the means and variances are independent of i, j and the first cross-moment is dependent only upon $|i - j|$, the process x is said to be stationary in the "wide-sense." If all higher moments also depend only on $|i - j|$, the process is said to be stationary in the "strict-sense," or, more simply, just stationary. A Gaussian process has the unusual property that wide-sense stationarity implies strict-sense stationarity.

22 The terminology "least-squares" is reserved in this book, conventionally, for the minimization of discrete sums such as Eq. (2.89). This usage contrasts with that of Bennett (2002) who applies it to continuous integrals, such as $\int_a^b (u(q) - r(q))^2 \, dq$, leading to the calculus of variations and Euler–Lagrange equations.

23 Box *et al.* (1994), Draper and Smith (1998), or Seber and Lee (2003), are all good starting points.

24 Draper and Smith (1998, Chapter 3) and the references given there.

25 Gill *et al.* (1986).

26 Wunsch and Minster (1982).

27 Morse and Feshbach (1953, p. 238), Strang (1988).

28 See Sewell (1987) for an interesting discussion.

29 But the matrix transpose is not what the older literature calls the "adjoint matrix," which is quite different. In the more recent literature the latter has been termed the "adjugate" matrix to avoid confusion.

30 In the meteorological terminology of Sasaki (1970) and others, exact relationships are called "strong" constraints, and those imposed in the mean-square are "weak" ones.

31 Claerbout (2001) displays more examples, and Lanczos (1961) gives a very general discussion of operators and their adjoints, Green functions, and their adjoints. See also the appendix to this chapter.

32 Wiggins (1972).

33 Brogan (1991) has a succinct discussion.

34 Lanczos (1961, pp. 117–18), sorts out the sign dependencies.

35 Lawson and Hanson (1995).

36 The singular value decomposition for arbitrary non-square matrices is apparently due to the physicist-turned-oceanographer Carl Eckart (Eckart and Young, 1939; see the discussion in Klema and Laub, 1980; Stewart, 1993; or Haykin, 2002). A particularly lucid account is given by Lanczos (1961) who, however, fails to give the decomposition a name. Other references are Noble and Daniel (1977), Strang (1988), and many recent books on applied linear algebra. The crucial role it plays in inverse methods appears to have been first noticed by Wiggins (1972).

37 Munk *et al.* (1995).

38 In physical oceanography, the distance would be that traveled by a ship between stops for measurement, and the water depth is clearly determined by the local topography.

39 Menke (1989).

40 Hansen (1992), or Lawson and Hanson (1995). Hansen's (1992) discussion is particularly interesting because he exploits the "generalized SVD," which is used to simultaneously diagonalize two matrices.

41 Munk and Wunsch (1982).

42 Seber and Lee (2003).

43 Luenberger (2003).

44 In oceanographic terms, the exact constraints describe the Stommel Gulf Stream solution. The eastward intensification of the adjoint solution corresponds to the change in sign of β in the adjoint model. See Schröter and Wunsch (1986) for details and an elaboration to a non-linear situation.

45 Lanczos (1960) has a good discussion.

46 See Lanczos (1961, Section 3.19).

47 The derivation follows Liebelt (1967).

48 Cf. Bretherton *et al.* (1976).

49 The time series was generated as $y_t = 0.999y_{t-1} + \theta_t$, $\langle \theta_t \rangle = 0$, $\langle \theta_t^2 \rangle = 1$, a so-called autoregressive process of order 1 (AR(1)). The covariance $\langle y_i y_j \rangle$ can be determined analytically; see Priestley (1982, p. 119). Many geophysical processes obey similar rules.

50 Stengel (1986); Brogan (1991).

51 Paige and Saunders (1982).

52 See especially, van Trees (2001).

53 Liebelt (1967, p. 164).

3

Extensions of methods

In this chapter we extend and apply some of the methods developed in Chapter 2. The problems discussed there raise a number of issues concerning models and data that are difficult to address with the mathematical machinery already available. Among them are the introduction of left and right eigenvectors, model nonlinearity, the potential use of inequality constraints, and sampling adequacy.

3.1 The general eigenvector/eigenvalue problem

To understand some recent work on so-called pseudospectra and some surprising recent results on fluid instability, it helps to review the more general eigenvector/ eigenvalue problem for arbitrary, *square,* matrices. Consider

$$\mathbf{E}\mathbf{g}_i = \lambda_i \mathbf{g}_i, \quad i = 1, 2, \ldots, N. \tag{3.1}$$

If there are no repeated eigenvalues, λ_i, then it is possible to show that there are always N independent $\mathbf{g}_i$, which are a basis, but which are *not usually orthogonal.* Because in most problems dealt with in this book, small perturbations can be made to the elements of $\mathbf{E}$ without creating any physical damage, it suffices here to assume that such perturbations can always ensure that there are N distinct λ_i. (Failure of the hypothesis leads to the Jordan form, which requires a somewhat tedious discussion.) A matrix $\mathbf{E}$ is "normal" if its eigenvectors form an orthonormal basis. Otherwise it is "non-normal." (Any matrix of forms $\mathbf{A}\mathbf{A}^{\mathrm{T}}$, $\mathbf{A}^{\mathrm{T}}\mathbf{A}$, is necessarily normal.)

Denote $\mathbf{G} = \{g_i\}$, $\mathbf{\Lambda} = \mathrm{diag}(\lambda_i)$. It follows immediately that $\mathbf{E}$ can be diagonalized:

$$\mathbf{G}^{-1}\mathbf{E}\mathbf{G} = \mathbf{\Lambda}, \tag{3.2}$$

but for a non-normal matrix, $\mathbf{G}^{-1} \neq \mathbf{G}^{\mathrm{T}}$. The general decomposition is

$$\mathbf{E} = \mathbf{G}\mathbf{\Lambda}\mathbf{G}^{-1}.$$

Matrix $\mathbf{E}^{\mathrm{T}}$ has a different set of spanning eigenvectors, but the same eigenvalues:

$$\mathbf{E}^{\mathrm{T}}\mathbf{f}_j = \lambda_j \mathbf{f}_j, \quad j = 1, 2, \ldots, N, \tag{3.3}$$

which can be written

$$\mathbf{f}_j^{\mathrm{T}}\mathbf{E} = \lambda_j \mathbf{f}_j^{\mathrm{T}}. \tag{3.4}$$

The $\mathbf{g}_i$ are hence known as the "right eigenvectors," and the $\mathbf{f}_i$ as the "left eigenvectors." Multiplying (3.1) on the left by $\mathbf{f}_j^{\mathrm{T}}$ and (3.4) on the right by $\mathbf{g}_i$ and subtracting shows

$$0 = (\lambda_i - \lambda_j)\mathbf{f}_j^{\mathrm{T}}\mathbf{g}_i, \tag{3.5}$$

or

$$\mathbf{f}_j^{\mathrm{T}}\mathbf{g}_i = 0, \quad i \neq j. \tag{3.6}$$

That is to say, the left and right eigenvectors are orthogonal for different eigenvalues, but $\mathbf{f}_j^{\mathrm{T}}\mathbf{g}_j \neq 0$. (In general the eigenvectors and eigenvalues are complex even for purely real $\mathbf{E}$. Note that some software automatically conjugates a transposed complex vector or matrix, and the derivation of (3.6) shows that it applies to the *non-conjugated* variables.)

Consider now a "model,"

$$\mathbf{A}\mathbf{x} = \mathbf{b}. \tag{3.7}$$

The norm of $\mathbf{b}$ is supposed bounded, $\|\mathbf{b}\| \leq b$, and the norm of $\mathbf{x}$ will be

$$\|\mathbf{x}\| = \|\mathbf{A}^{-1}\mathbf{b}\|. \tag{3.8}$$

What is the relationship of $\|\mathbf{x}\|$ to $\|\mathbf{b}\|$?

Let $\mathbf{g}_i$ be the right eigenvectors of $\mathbf{A}$. Write

$$\mathbf{b} = \sum_{i=1}^{N} \beta_i \mathbf{g}_i, \tag{3.9}$$

$$\mathbf{x} = \sum_{i=1}^{N} \alpha_i \mathbf{g}_i. \tag{3.10}$$

If the $\mathbf{g}_i$ were orthogonal, $|\beta_i| \leq \|\mathbf{b}\|$. But as they are not orthonormal, the β_i will need to be found through a system of simultaneous equations (2.3) (recall the discussion in Chapter 2 of the expansion of an arbitrary vector in non-orthogonal vectors) and no simple bound on the β_i is then possible; some may be very large. Substituting into (3.7),

$$\sum_{i=1}^{N} \alpha_i \lambda_i \mathbf{g}_i = \sum_{i=1}^{N} \beta_i \mathbf{g}_i. \tag{3.11}$$

A term-by-term solution is evidently no longer possible because of the lack of orthogonality. But multiplying on the left by $\mathbf{f}_j^T$, and invoking (3.6), produces

$$\alpha_j \lambda_j \mathbf{f}_j^T \mathbf{g}_j = \beta_j \mathbf{f}_j^T \mathbf{g}_j, \tag{3.12}$$

or

$$\alpha_j = \beta_j / \lambda_j, \quad \lambda_j \neq 0. \tag{3.13}$$

Even if the λ_j are all of the same order, the possibility that some of the β_j are very large implies that eigenstructures in the solution may be much larger than $\|\mathbf{b}\|$. This possibility becomes very interesting when we turn to time-dependent systems. At the moment, note that partial differential equations that are self-adjoint produce discretizations that have coefficient matrices $\mathbf{A}$, such that $\mathbf{A}^T = \mathbf{A}$. Thus self-adjoint systems have normal matrices, and the eigencomponents of the solution are all immediately bounded by $\|\mathbf{b}\| / \lambda_i$. Non-self-adjoint systems produce non-normal coefficient matrices and so can therefore unexpectedly generate very large eigenvector contributions.

Example *The equations*

$$\left\{ \begin{array}{cccc} -0.32685 & 0.34133 & 0.69969 & 0.56619 \\ -4.0590 & 0.80114 & 3.0219 & 1.3683 \\ -3.3601 & 0.36789 & 2.6619 & 1.2135 \\ -5.8710 & 1.0981 & 3.9281 & 1.3676 \end{array} \right\} \mathbf{x} = \left[\begin{array}{c} 0.29441 \\ -1.3362 \\ 0.71432 \\ 1.6236 \end{array} \right],$$

are readily solved by ordinary Gaussian elimination (or matrix inversion). If one attempts to use an eigenvector expansion, it is found from Eq. (3.1) (up to rounding errors) that $\lambda_i = [2.3171, 2.2171, -0.1888, 0.1583]^T$, and the right eigenvectors, $\mathbf{g}_i$, corresponding to the eigenvalues are

$$\mathbf{G} = \left\{ \begin{array}{cccc} 0.32272 & 0.33385 & 0.20263 & 0.36466 \\ 0.58335 & 0.58478 & 0.27357 & -0.23032 \\ 0.46086 & 0.46057 & 0.53322 & 0.75938 \\ 0.58581 & 0.57832 & -0.77446 & -0.48715 \end{array} \right\},$$

and $\mathbf{g}_{1,2}$ are nearly parallel. If one expands the right-hand side, $\mathbf{y}$, of the above equations in the $\mathbf{g}_i$, the coefficients, $\beta = [25.2230, -24.7147, -3.3401, 2.9680]^T$, and the first two are markedly greater than any of the elements in $\mathbf{y}$ or in the $\mathbf{g}_i$. Note that, in general, such arbitrary matrices will have complex eigenvalues and eigenvectors (in conjugate pairs).

3.2 Sampling

In Chapter 2, on p. 129, we discussed the problem of making a uniformly gridded map from irregularly spaced observations. But not just any set of observations proves adequate to the purpose. The most fundamental problem generally arises under the topic of "sampling" and "sampling error." This subject is a large and interesting one in its own right,[1] and we can only outline the basic ideas.

The simplest and most fundamental idea derives from consideration of a one-dimensional continuous function, $f(q)$, where q is an arbitrary independent variable, usually either time or space, and $f(q)$ is supposed to be sampled uniformly at intervals, Δq, an infinite number of times to produce the infinite set of sample values $\{f(n\Delta q)\}$, $-\infty \le n \le \infty$, n integer. The sampling theorem, or sometimes the "Shannon–Whittaker sampling theorem"[2] is a statement of the conditions under which $f(q)$ should be reconstructible from the sample values. Let the Fourier transform of $f(q)$ be defined as

$$\hat{f}(r) = \int_{-\infty}^{\infty} f(q)e^{2i\pi rq}\,dq, \tag{3.14}$$

and assumed to exist. The sampling theorem asserts that a necessary and sufficient condition to perfectly reconstruct $f(q)$ from its samples is that

$$|\hat{f}(r)| = 0, \quad |r| \ge 1/(2\Delta q). \tag{3.15}$$

It produces the Shannon–Whittaker formula for the reconstruction

$$f(q) = \sum_{n=-\infty}^{\infty} f(n\Delta q)\frac{\sin[(2\pi/2\Delta q)(q - n\Delta q)]}{(2\pi/2\Delta q)(q - n\Delta q)}. \tag{3.16}$$

Mathematically, the Shannon–Whittaker result is surprising – because it provides a condition under which a function at an uncountable infinity of points – the continuous line – can be perfectly reconstructed from information known only at a countable infinity, $n\Delta q$, of them. For present purposes, an intuitive interpretation is all we seek and this is perhaps best done by considering a special case in which the conditions of the theorem are violated.

Figure 3.1 displays an ordinary sinusoid whose Fourier transform can be represented as

$$\hat{f}(r) = \tfrac{1}{2}(\delta(r - r_0) - \delta(r + r_0)), \tag{3.17}$$

which is sampled as depicted, and in violation of the sampling theorem (δ is the Dirac delta-function). It is quite clear that there is at least one more perfect sinusoid, the one depicted with the dashed line, which is completely consistent with all the sample points and which cannot be distinguished from it using the measurements

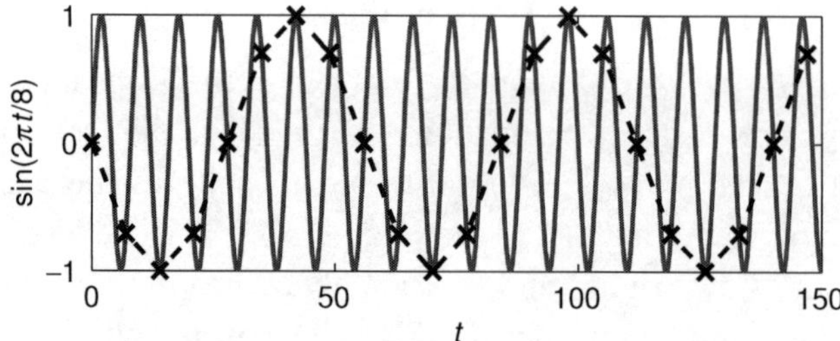

Figure 3.1 Effects of undersampling a periodic function: solid curve is $y(t) = \sin(2\pi t/8)$ sampled at time intervals of $\Delta t = 0.1$. The dashed curve is the same function, but sampled at intervals $\Delta t = 7$. With this undersampling, the curve of frequency $s = 1/8$ time units is aliased into one that appears to have a frequency $s_a = 1/8 - 1/7 = 1/56 < 1/14$. That is, the aliased curve appears to have a period of 56 time units.

alone. A little thought shows that the apparent frequency of this new sinusoid is

$$r_a = r_0 \pm \frac{n}{\Delta q}, \qquad (3.18)$$

such that

$$|r_a| \leq \frac{1}{2\Delta q}. \qquad (3.19)$$

The samples cannot distinguish the true high frequency sinusoid from this low frequency one, and the high frequency can be said to masquerade or "alias" as the lower frequency one.[3] The Fourier transform of a sampled function is easily seen to be periodic with period $1/\Delta q$ in the transform domain, that is, in the r space.[4] Because of this periodicity, there is no point in computing its values for frequencies outside $|r| \leq 1/2\Delta q$ (we make the convention that this "baseband," i.e., the fundamental interval for computation, is symmetric about $r = 0$, over a distance $1/2\Delta q$; see Fig. 3.2). Frequencies of absolute value larger than $1/2\Delta q$, the so-called Nyquist frequency, cannot be distinguished from those in the baseband, and alias into it. Figure 3.2 shows a densely sampled, non-periodic function and its Fourier transform compared to that obtained from the undersampled version overlain. Undersampling is a very unforgiving practice.

The consequences of aliasing range from the negligible to the disastrous. A simple example is that of the principal lunar tide, usually labeled M_2, with a period of 12.42 hours, $r = 1.932$ cycles/day. An observer measures the height of sea level at a fixed time, say 10 a.m. each day so that $\Delta q = 1$ day. Applying the formula (3.18), the apparent frequency of the tide will be 0.0676 cycles/day for a period of about 14.8 days ($n = 2$). To the extent that the observer understands what is going on, he or she will not conclude that the principal lunar tide has a period of

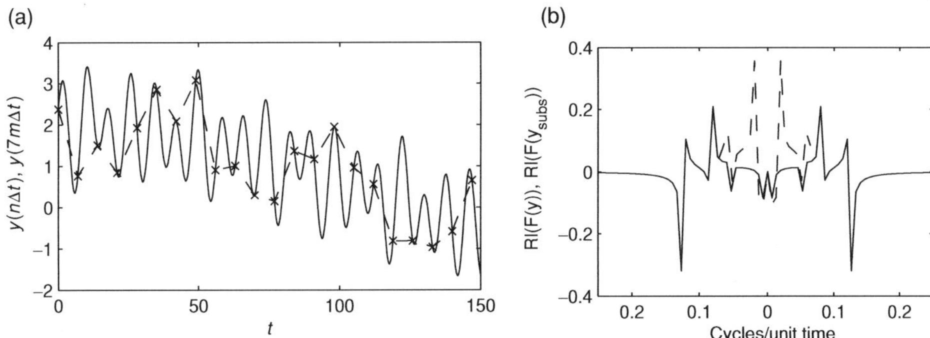

Figure 3.2 (a) A non-periodic function sampled at intervals $\Delta t = 0.1$, and the same function sampled at intervals $\Delta t = 7$ time units. (b) The real part of the Fourier components of the two functions shown in (a). The subsampled function has a Fourier transform confined to $|s| \leq 1/(2.7)$ (dashed) while that of the original, more densely sampled, function (solid) extends to $|s| \leq 1/0.1 = 10$, most of which is not displayed. The subsampled function has a very different Fourier transform from that of the original densely sampled one. Both transforms are periodic in frequency, s, with period equal to the width of their corresponding basebands. (This periodicity is suppressed in the plot.) Note in particular how erroneous an estimate of the temporal derivative of the undersampled function would be in comparison to that of the highly sampled one.

14.8 days, but will realize that the true period can be computed through (3.18) from the apparent one. But without that understanding, some bizarre theory might be produced.[5]

The reader should object that the Shannon–Whittaker theorem applies only to an infinite number of perfect samples and that one never has either perfect samples or an infinite number of them. In particular, it is true that if the duration of the data in the q domain is finite, then it is impossible for the Fourier transform to vanish over any finite interval, much less the infinite interval above the Nyquist frequency.[6] Nonetheless, the rule of thumb that results from (3.16) has been found to be quite a good one. The deviations from the assumptions of the theorem are usually dealt with by asserting that sampling should be done so that

$$\Delta q \ll 1/2r_0. \tag{3.20}$$

Many extensions and variations of the sampling theorem exist – taking account of the finite time duration , the use of "burst-sampling" and known function derivatives, etc.[7] Most of these variations are sensitive to noise. There are also extensions to multiple dimensions,[8] which are required for mapmaking purposes. Because failure to acknowledge the possibility that a signal is undersampled is so dire, one concludes that consideration of sampling is critical to any discussion of field data.

3.2.1 One-dimensional interpolation

Let there be two observations $[y_1, y_2]^T = [x_1 + n_1, x_2 + n_2]^T$ located at positions $[r_1, r_2]^T$, where n_i are the observation noise. We require an estimate of $x(\tilde{r})$, where $r_1 < \tilde{r} < r_2$. The formula (3.16) is unusable – there are only two noisy observations, not an infinite number of perfect ones. We could try using linear interpolation:

$$\tilde{x}(\tilde{r}) = \frac{|r_2 - \tilde{r}|}{|r_2 - r_1|} \, y(r_1) + \frac{|r_1 - \tilde{r}|}{|r_2 - r_1|} \, y(r_2). \tag{3.21}$$

If there are N data points, r_i, $1 = 1, 2, \ldots, N$, then another possibility is Aitken–Lagrange interpolation:[9]

$$\tilde{x}(\tilde{r}) = \sum_{j=1}^{N} l_j(\tilde{r}) y_j, \tag{3.22}$$

$$l_j(\tilde{r}) = \frac{(\tilde{r} - r_1) \cdots (\tilde{r} - r_M)}{(r_j - r_1) \cdots (r_j - r_{j-1})(r_j - r_{j+1}) \cdots (r_j - r_M)}. \tag{3.23}$$

Equations (3.21)–(3.23) are only two of many possible interpolation formulas. When would one be better than the other? How good are the estimates? To answer these questions, let us take a different tack, and employ the Gauss–Markov theorem, assuming we know something about the necessary covariances.

Suppose either $\langle x \rangle = \langle n \rangle = 0$ or that a known value has been removed from both (this just keeps our notation a bit simpler). Then

$$\mathbf{R}_{xy}(\tilde{r}, r_j) \equiv \langle x(\tilde{r}) y(r_j) \rangle = \langle x(\tilde{r})(x(r_j) + n(r_j)) \rangle = \mathbf{R}_{xx}(\tilde{r}, r_j), \tag{3.24}$$

$$\mathbf{R}_{yy}(r_i, r_j) \equiv \langle (x(r_i) + n(r_i))(x(r_j) + n(r_j)) \rangle \tag{3.25}$$

$$= \mathbf{R}_{xx}(r_i, r_j) + \mathbf{R}_{nn}(r_i, r_j), \tag{3.26}$$

where it has been assumed that $\langle x(r)n(q) \rangle = 0$.

From (2.396), the best linear interpolator is

$$\tilde{\mathbf{x}} = \mathbf{B}\mathbf{y}, \qquad \mathbf{B}(\tilde{r}, \mathbf{r}) = \sum_{j=1}^{M} \mathbf{R}_{xx}(\tilde{r}, r_j) \{ \mathbf{R}_{xx} + \mathbf{R}_{nn} \}_{ji}^{-1}, \tag{3.27}$$

($\{ \mathbf{R}_{xx} + \mathbf{R}_{nn} \}_{ji}^{-1}$ means the ji element of the inverse matrix) and the minimum possible error that results is

$$\mathbf{P}(\tilde{r}_\alpha, \tilde{r}_\beta) = \mathbf{R}_{xx}(\tilde{r}_\alpha, \tilde{r}_\beta) - \sum_{j}^{M} \sum_{i}^{M} \mathbf{R}_{xx}(\tilde{r}_\alpha, r_j) \{ \mathbf{R}_{xx} + \mathbf{R}_{nn} \}_{ji}^{-1} \mathbf{R}_{xx}(r_i, \tilde{r}_\beta), \tag{3.28}$$

and $\tilde{\mathbf{n}} = \mathbf{y} - \tilde{\mathbf{x}}$.

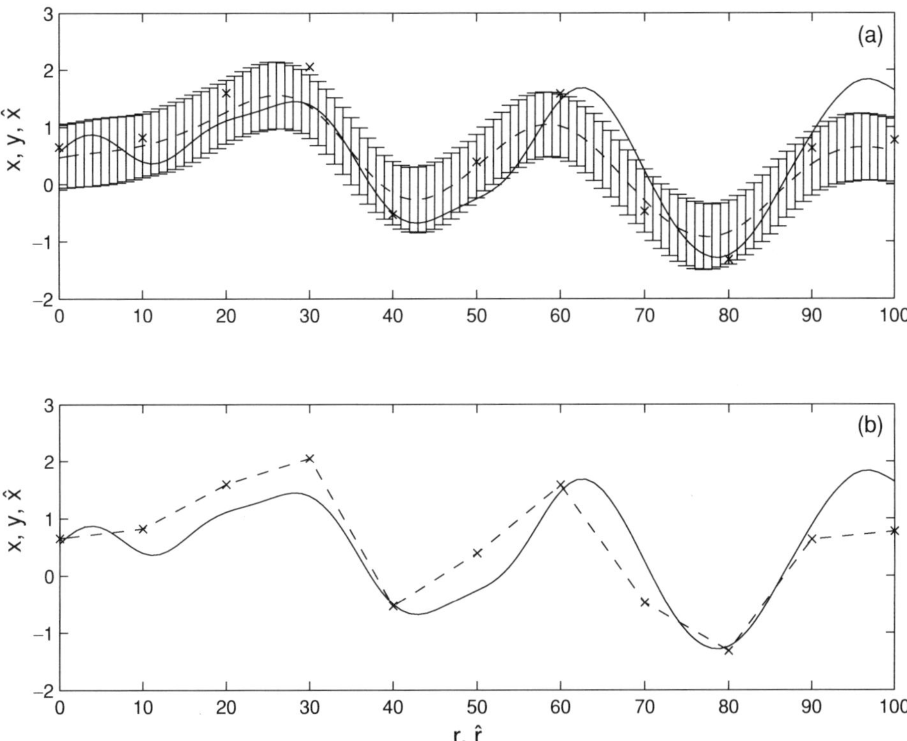

Figure 3.3 In both panels, the solid curve is the "true" curve, $x(r)$, from which noisy samples (denoted "x") have been obtained. $x(r)$ was generated to have a true covariance $S = \exp(-r^2/100)$, and the "data" values, $y(r_i) = x(r_i) + n_i$, where $\langle n_i \rangle = 0$, $\langle n_i n_j \rangle = (1/4)\delta_{ij}$, were generated from a Gaussian probability density. In (b), linear interpolation is used to generate the estimated values of $x(r)$ (dashed line). The estimates are identical to the observations at $r = r_i$. In (a), objective mapping was used to make the estimates (dashed line). Note that $\tilde{x}(r_i) \neq y(r_i)$, and that an error bar is available – as plotted. The true values are generally within one standard deviation of the estimated value (but about 35% of the estimated values would be expected to lie outside the error bars), and the estimated value is within two standard deviations of the correct one everywhere. The errors in the estimates, $\tilde{x}(r_i) - x(r_i)$, are clearly spatially correlated, and can be inferred from Eq. (3.28) (not shown). The values of $x(r)$ were generated to have the inferred covariance S, by forming the matrix, $\mathbf{S} = \text{toeplitz}\,(S(r_i, r_j))$, and obtaining its symmetric factorization, $\mathbf{S} = \mathbf{U}\Lambda\mathbf{U}^{\mathrm{T}}$, $\mathbf{x}(r) = \mathbf{U}\Lambda\alpha$, where the elements of α are pseudo-random numbers.

Results for both linear interpolation and objective mapping are shown in Fig. 3.3. Notice that, like other interpolation methods, the optimal one is a linear combination of the data. If any other set of weights $\mathbf{B}$ is chosen, then the interpolation is not as good as it could be in the mean-square error sense; the error of any such scheme can be obtained by substituting it into (2.394) and evaluating the result (the true covariances still need to be known).

Looking back now at the two familiar formulas in (3.21) and (3.22), it is clear what is happening: they represent a choice of $\mathbf{B}$. Unless the covariance is such as to produce one of the two sets of weights as the optimum choice, neither Aitken–Lagrange nor linear (nor any other common choice, like a spline) is the best one could do. Alternatively, if any of (3.21)–(3.23) were thought to be the best one, they are equivalent to specifying the solution and noise covariances.

If interpolation is done for two points, $\tilde{r}_\alpha$, $\tilde{r}_\beta$, the error of the two estimates will usually be correlated, and represented by $\mathbf{P}(\tilde{r}_\alpha, \tilde{r}_\beta)$. Knowledge of the correlations between the errors in different interpolated points is often essential – for example, if one wishes to interpolate to uniformly spaced grid points so as to make estimates of derivatives of x. Such derivatives might be numerically meaningless if the mapping errors are small scale (relative to the grid spacing) and of large amplitude. But if the mapping errors are large scale compared to the grid, the derivatives may tend to remove the error and produce better estimates than for x itself.

Both linear and Aitken–Lagrange weights will produce estimates that are exactly equal to the observed values if $\tilde{r}_\alpha = r_p$, that is, on the data points. Such a result is characteristic of "true interpolation." If no noise is present, then the observed value is the correct one to use at a data point. In contrast, the Gauss–Markov estimate will differ from the data values at the data points, because the estimator attempts to reduce the noise in the data by averaging over all observations, not just the one. The Gauss–Markov estimate is thus not a true interpolator; it is instead a "smoother." One can recover true interpolation if $\|\mathbf{R}_{nn}\| \to 0$, although the matrix being inverted in (3.27) and (3.28) can become singular. The weights $\mathbf{B}$ can be fairly complicated if there is any structure in either of $\mathbf{R}_{xx}$, $\mathbf{R}_{nn}$. The estimator takes explicit account of the expected spatial structure of both $\mathbf{x}$, $\mathbf{n}$ to weight the data in such a way as to most effectively "kill" the noise relative to the signal. One is guaranteed that no other linear filter can do better.

If $\|\mathbf{R}_{nn}\| \gg \|\mathbf{R}_{xx}\|$, $\tilde{\mathbf{x}} \to \mathbf{0}$, manifesting the bias in the estimator; this bias was deliberately introduced so as to minimize the uncertainty (minimum variance about the true value). Thus, estimated values of zero-mean processes tend toward zero, particularly far from the data points. For this reason, it is common to use expressions such as (2.413) to first remove the mean, prior to mapping the residual, and re-adding the estimated mean at the end. The interpolated values of the residuals are nearly unbiased, because their true mean is nearly zero. Rigorous estimates of $\mathbf{P}$ for this approach require some care, as the mapped residuals contain variances owing to the uncertainty of the estimated mean,[10] but the corrections are commonly ignored.

As we have seen, the addition of small positive numbers to the diagonal of a matrix usually renders it non-singular. In the formally noise-free case, $\mathbf{R}_{nn} \to \mathbf{0}$, and one has the prospect that $\mathbf{R}_{xx}$ by itself may be singular. To understand the meaning of this situation, consider the general case, involving both matrices. Then the symmetric form of the SVD of the sum of the two matrices is

$$\mathbf{R}_{xx} + \mathbf{R}_{nn} = \mathbf{U}\Lambda\mathbf{U}^{\mathrm{T}}. \tag{3.29}$$

If the sum covariance is positive definite, Λ will be square with $K = M$ and the inverse will exist. If the sum is not positive definite, but is only semi-definite, one or more of the singular values will vanish. The meaning is that there are *possible* structures in the data that have been assigned to neither the noise field nor the solution field. This situation is realistic only if one is truly confident that $\mathbf{y}$ does not contain such structures. In that case, the solution

$$\tilde{\mathbf{x}} = \mathbf{R}_{xx}(\mathbf{R}_{xx} + \mathbf{R}_{nn})^{-1}\mathbf{y} = \mathbf{R}_{xx}(\mathbf{U}\Lambda^{-1}\mathbf{U}^{\mathrm{T}})\mathbf{y} \tag{3.30}$$

will have components of the form $0/0$, the denominator corresponding to the zero singular values and the numerator to the absent, impossible, structures of $\mathbf{y}$. One can arrange that the ratio of these terms should be set to zero (e.g., by using the SVD). But such a delicate balance is not necessary. If one simply adds a small white noise covariance to $\mathbf{R}_{xx} + \mathbf{R}_{nn} \to \mathbf{R}_{xx} + \mathbf{R}_{nn} + \epsilon^2\mathbf{I}$, or $\mathbf{R}_{xx} \to \mathbf{R}_{xx} + \epsilon^2\mathbf{I}$, one is assured, by the discussion of tapering, that the result is no longer singular – all structures in the field are being assigned either to the noise or the solution (or in part to both).

Anyone using a Gauss–Markov estimator to make maps must check that the result is consistent with the prior estimates of $\mathbf{R}_{xx}$, $\mathbf{R}_{nn}$. Such checks include determining whether the differences between the mapped values at the data points and the observed values have numerical values consistent with the assumed noise variance; a further check involves the sample autocovariance of $\tilde{\mathbf{n}}$ and its test against $\mathbf{R}_{nn}$ (see books on regression for such tests). The mapped field should also have a variance and covariance consistent with the prior estimate $\mathbf{R}_{xx}$. If these tests are not passed, the entire result should be rejected.

3.2.2 Higher-dimensional mapping

We can now immediately write down the optimal interpolation formulas for an arbitrary distribution of data in two or more dimensions. Let the positions where data are measured be the set $\mathbf{r}_j$ with measured value $\mathbf{y}(\mathbf{r}_j)$, containing noise $\mathbf{n}$. It is assumed that aliasing errors are unimportant. The mean value of the field is first estimated and subtracted from the measurements and we proceed as though the true mean were zero.[11]

As in the case where the positions are scalars, one minimizes the expected mean-square difference between the estimated and the true field $\mathbf{x}(\tilde{\mathbf{r}}_\alpha)$. The result is (3.27)

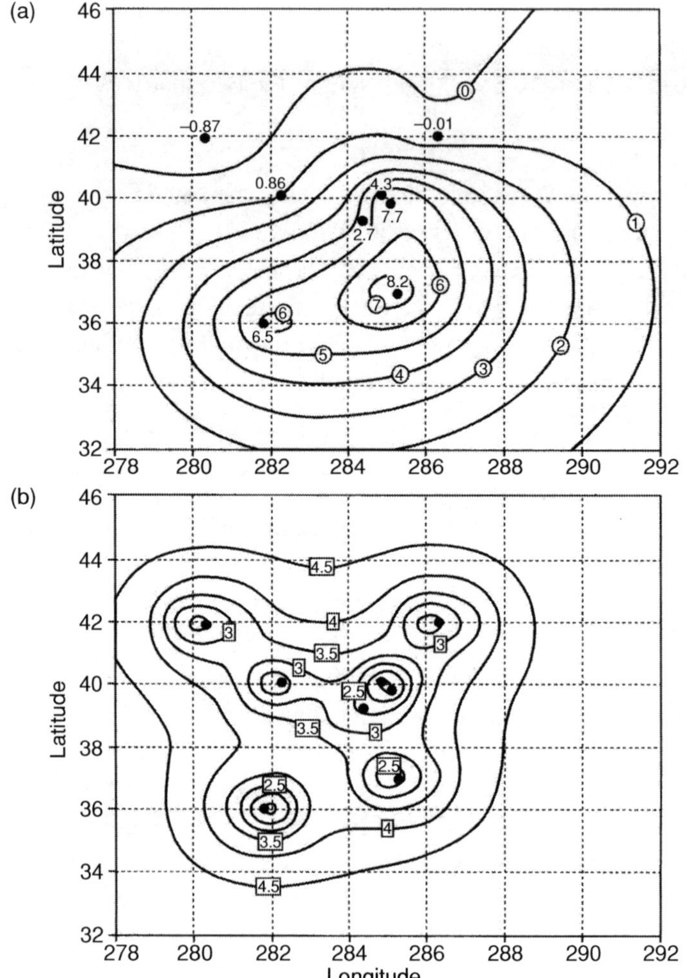

Figure 3.4 (a) Observations, shown as solid dots, from which a uniformly gridded map is desired. Contours were constructed using a fixed covariance and the Gauss–Markov estimate Eq. (3.27). Noise was assumed white with a variance of 1. (b) Expected standard error of the mapped field in (a). Values tend, far from the observations points, to a variance of 25, which was the specified field variance, and hence the largest expected error is $\sqrt{25}$. Note the minima centered on the data points.

and (3.28), except that now everything is a function of the vector positions. If the field being mapped is also a vector (e.g., two components of velocity) with known covariances between the two components, then the elements of $\mathbf{B}$ become matrices. The observations could also be vectors at each point.

An example of a two-dimensional map is shown in Fig. 3.4. The "data points," $y(\mathbf{r}_i)$, are the dots, while estimates of $\tilde{x}(\mathbf{r}_i)$ on the uniform grid were wanted.

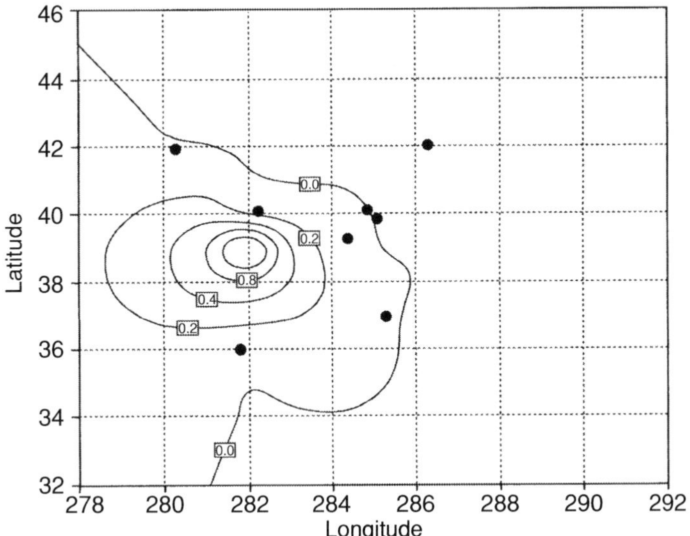

Figure 3.5 One of the rows of **P** corresponding to the grid point in Fig. 3.4 at 39°N, 282°E, displaying the expected correlations that occur in the errors of the mapped field. These errors would be important, e.g., in any use that differentiated the mapped field. For plotting purposes, the variance was normalized to 1.

The a priori noise was set to $\langle \mathbf{n} \rangle = \mathbf{0}$, $\mathbf{R}_{nn} = \langle n_i n_j \rangle = \sigma_n^2 \delta_{ij}$, $\sigma_n^2 = 1$, and the true field covariance was $\langle \mathbf{x} \rangle = \mathbf{0}$, $\mathbf{R}_{xx} = \langle \mathbf{x}(\mathbf{r}_i)\mathbf{x}(\mathbf{r}_j) \rangle = P_0 \exp -|\mathbf{r}_i - \mathbf{r}_j|^2/L_2$, $P_0 = 25$, $L_2 = 100$. Figure 3.4 also shows the estimated values and Figs. 3.4 and 3.5 show the error variance estimate of the mapped values. Notice that, far from the data points, the estimated values are 0: the mapped field goes asymptotically to the estimated true mean, with the error variance rising to the full value of 25, which cannot be exceeded. That is to say, when we are mapping far from any data point, the only real information available is provided by the prior statistics – that the mean is 0, and the variance about that mean is 25. So the expected uncertainty of the mapped field, in the absence of data, cannot exceed the prior estimate of how far from the mean the true value is likely to be. The best estimate is then the mean itself.

A complex error structure of the mapped field exists – even in the vicinity of the data points. Should a model be "driven" by this mapped field, one would need to make some provision in the model accounting for the spatial dependence in the expected errors of this forcing.

In practice, most published objective mapping (often called "OI" for "objective interpolation," although as we have seen, it is not true interpolation) has been based upon simple analytical statements of the covariances $\mathbf{R}_{xx}$, $\mathbf{R}_{nn}$ as used in the example: that is, they are commonly assumed to be spatially stationary and isotropic (depending upon $|\mathbf{r}_i - \mathbf{r}_j|$ and not upon the two positions separately nor upon their orientation). The use of analytic forms removes the necessity for finding, storing, and computing with the potentially very large $M \times M$ data covariance matrices in which hypothetically every data or grid point has a different covariance with every other data or grid point. But the analytical convenience often distorts the solutions, as many fluid flows and other fields are neither spatially stationary nor isotropic.[12]

3.2.3 Mapping derivatives

A common problem in setting up fluid models is the need to specify the fields of quantities such as temperature, density, etc., on a regular model grid. The derivatives of these fields must be specified for use in advection–diffusion equations,

$$\frac{\partial C}{\partial t} + \mathbf{v} \cdot \nabla C = K \nabla^2 C, \tag{3.31}$$

where C is any scalar field of interest. Suppose the spatial derivative is calculated as a one-sided difference,

$$\frac{\partial C(\tilde{r}_1)}{\partial r} \sim \frac{C(\tilde{r}_1) - C(\tilde{r}_2)}{\tilde{r}_1 - \tilde{r}_2}. \tag{3.32}$$

Then it is attractive to subtract the two estimates made from Eq. (3.27) , producing

$$\frac{\partial C(\tilde{r}_1)}{\partial r} \sim \frac{1}{\Delta r}(\mathbf{R}_{xx}(\tilde{r}_1, \, r_j) - \mathbf{R}_{xx}(\tilde{r}_2, \, r_j))(\mathbf{R}_{xx} + \mathbf{R}_{nn})^{-1}\mathbf{y}. \tag{3.33}$$

Alternatively, an estimate of $\partial C / \partial r$ could be made directly from (2.392), using $\mathbf{x} = C(r_1) - C(r_2)$. $\mathbf{R}_{yy} = \mathbf{R}_{xx} + \mathbf{R}_{nn}$, which describes the data, does not change. $\mathbf{R}_{xy}$ does change:

$$\mathbf{R}_{xy} = \langle (C(\tilde{r}_1) - C(\tilde{r}_2))(C(r_j) + n(r_j)) \rangle = \mathbf{R}_{xx}(\tilde{r}_1, \, r_j) - \mathbf{R}_{xx}(\tilde{r}_2, \, r_j), \tag{3.34}$$

which when substituted into (2.396) produces (3.33). *Thus, the optimal map of the finite difference field is simply the difference of the mapped values.* More generally, the optimally mapped value of any linear combination of the values is that linear combination of the maps.[13]

3.3 Inequality constraints: non-negative least-squares

In many estimation problems, it is useful to be able to impose inequality constraints upon the solutions. Problems involving tracer concentrations, for example, usually demand that they remain positive; empirical eddy diffusion coefficients are sometimes regarded as acceptable only when non-negative; in some fluid flow problems we may wish to impose directions, but not magnitudes, upon velocity fields.

Such needs lead to consideration of the forms

$$\mathbf{Ex} + \mathbf{n} = \mathbf{y}, \qquad (3.35)$$

$$\mathbf{Gx} \geq \mathbf{h}, \qquad (3.36)$$

where the use of a greater-than inequality to represent the general case is purely arbitrary; multiplication by minus 1 readily reverses it. $\mathbf{G}$ is of dimension $M_2 \times N$.

Several cases need to be distinguished. (A) Suppose $\mathbf{E}$ is full rank and fully determined; then the SVD solution to (3.35) by itself is $\tilde{\mathbf{x}}, \tilde{\mathbf{n}}$, and there is no solution nullspace. Substitution of the solution into (3.36) shows that the inequalities are either satisfied or that some are violated. In the first instance, the problem is solved, and the inequalities bring no new information. In the second case, the solution must be modified and, necessarily, $\|\tilde{\mathbf{n}}\|$ will increase, given the noise-minimizing nature of the SVD solution. It is also possible that the inequalities are contradictory, in which case there is no solution.

(B) Suppose that $\mathbf{E}$ is formally underdetermined – so that a solution nullspace exists. If the particular SVD solution violates one or more of the inequalities and requires modification, two subcases can be distinguished: (1) Addition of one or more nullspace vectors permits the inequalities to be satisfied. Then the solution residual norm will be unaffected, but $\|\tilde{\mathbf{x}}\|$ will increase. (2) The nullspace vectors by themselves are unable to satisfy the inequality constraints, and one or more range vectors are required to do so. Then both $\|\tilde{\mathbf{x}}\|, \|\tilde{\mathbf{n}}\|$ will increase.

Case (A) is the conventional one.[14] The so-called Kuhn–Tucker–Karush theorem is a requirement for a solution $\tilde{\mathbf{x}}$ to exist. Its gist is as follows: Let $M \geq N$ and $\mathbf{E}$ be full rank; there are no $\mathbf{v}_i$ in the solution nullspace. If there is a solution, there must exist a vector, $\mathbf{q}$, of dimension M_2 such that

$$\mathbf{E}^{\mathrm{T}}(\mathbf{E}\tilde{\mathbf{x}} - \mathbf{y}) = \mathbf{G}^{\mathrm{T}}\mathbf{q}, \qquad (3.37)$$

$$\mathbf{Gx} - \mathbf{h} = \mathbf{r}, \qquad (3.38)$$

where the M_2 elements of $\mathbf{q}$ are divided into two groups. For group 1, of dimension m_1,

$$r_i = 0, \quad q_i \geq 0, \qquad (3.39)$$

and for group 2, of dimension $m_2 = M_2 - m_1$,

$$r_i > 0, \quad q_i = 0. \qquad (3.40)$$

To understand this theorem, recall that in the solution to the ordinary overdetermined least-squares problem, the left-hand side of (3.37) vanishes identically (2.91

and 2.265), being the projection of the residuals onto the range vectors, $\mathbf{u}_i$, of $\mathbf{E}^T$. If this solution violates one or more of the inequality constraints, structures that produce increased residuals must be introduced into the solution.

Because there are no nullspace $\mathbf{v}_i$, the rows of $\mathbf{G}$ may each be expressed exactly by an expansion in the range vectors. In the second group of indices, the corresponding inequality constraints are already satisfied by the ordinary least-squares solution, and no modification of the structure proportional to $\mathbf{v}_i$ is required. In the first group of indices, the inequality constraints are marginally satisfied, at equality, only by permitting violation of the demand (2.91) that the residuals should be orthogonal to the range vectors of $\mathbf{E}$. If the ordinary least-squares solution violates the inequality, the minimum modification required to it pushes the solution to the edge of the acceptable bound, but at the price of increasing the residuals proportional to the corresponding $\mathbf{u}_i$. The algorithm consists of finding the two sets of indices and then the smallest coefficients of the $\mathbf{v}_i$ corresponding to the group 1 indices required to just satisfy any initially violated inequality constraints. A canonical special case, to which more general problems can be reduced, is based upon the solution to $\mathbf{G} = \mathbf{I}$, $\mathbf{h} = \mathbf{0}$ – called "non-negative least-squares."[15] The requirement, $\mathbf{x} \geq 0$, is essential in many problems involving tracer concentrations, which are necessarily positive.

The algorithm can be extended to the underdetermined/rank-deficient case in which the addition, to the original basic SVD solution, of appropriate amounts of the nullspace of $\mathbf{v}_i$ is capable of satisfying any violated inequality constraints.[16] One simply chooses the smallest mean-square solution coefficients necessary to push the solution to the edge of the acceptable inequalities, producing the smallest norm. The residuals of the original problem do not increase – because only nullspace vectors are being used. $\mathbf{G}$ must have a special structure for this to be possible.

The algorithm can be further generalized[17] by considering the general case of rank-deficiency/underdeterminism where the nullspace vectors by themselves are inadequate to produce a solution satisfying the inequalities. In effect, any inequalities "left over" are satisfied by invoking the smallest perturbations necessary to the coefficients of the range vectors $\mathbf{v}_i$.

3.4 Linear programming

In a number of important geophysical fluid problems, the objective functions are linear rather than quadratic functions. Scalar property fluxes such as heat, C_i, are carried by a fluid flow at rates $\sum C_i x_i$, which are linear functions of $\mathbf{x}$. If one sought the extreme fluxes of C, it would require finding the extremal values of the corresponding linear function. Least-squares does not produce useful answers in such problems because linear objective functions achieve their minima or maxima only at plus or minus infinity – unless the elements of $\mathbf{x}$ are bounded. The methods of *linear programming* are directed at finding extremal properties of linear objective functions subject to bounding constraints. Such problems can be written as

$$\text{minimize}: \quad J = \mathbf{c}^{\mathrm{T}}\mathbf{x},$$

$$\mathbf{E}_1\mathbf{x} = \mathbf{y}_1, \tag{3.41}$$

$$\mathbf{E}_2\mathbf{x} \geq \mathbf{y}_2, \tag{3.42}$$

$$\mathbf{E}_3\mathbf{x} \leq \mathbf{y}_3, \tag{3.43}$$

$$\mathbf{a} \leq \mathbf{x} \leq \mathbf{b}, \tag{3.44}$$

that is, as a collection of equality and inequality constraints of both greater than or less than form, plus bounds on the individual elements of $\mathbf{x}$. In distinction to the least-squares and minimum variance equations that have been discussed so far, these constraints are hard ones; they cannot be violated even slightly in an acceptable solution.

Linear programming problems are normally reduced to what is referred to as a *canonical form*, although different authors use different definitions of what it is. But all such problems are reducible to

$$\text{minimize}: \quad J = \mathbf{c}^{\mathrm{T}}\mathbf{x}, \tag{3.45}$$

$$\mathbf{E}\mathbf{x} \leq \mathbf{y} \tag{3.46}$$

$$\mathbf{x} \geq \mathbf{0}. \tag{3.47}$$

The use of a minimum rather than a maximum is readily reversed by introducing a minus sign, and the inequality is similarly readily reversed. The last relationship, requiring purely positive elements in $\mathbf{x}$, is obtained without difficulty by translation.

Linear programming problems are widespread in many fields including, especially, financial and industrial management where they are used to maximize profits, or minimize costs, in, say, a manufacturing process. Necessarily then, the amount of a product of each type is positive, and the inequalities reflect such things as the need to consume no more than the available amounts of raw materials. In some cases, J is then literally a "cost" function. General methodologies were first developed during World War II in what became known as "operations research" ("operational research" in the UK),[18] although special cases were known much earlier. Since then, because of the economic stake in practical use of linear programming, immense effort has been devoted both to textbook discussion and efficient, easy-to-use software.[19] Given this accessible literature and software, the methodologies of solution are not described here, and only a few general points are made.

The original solution algorithm invented by G. Dantzig is usually known as the "simplex method" (a simplex is a convex geometric shape). It is a highly efficient search method conducted along the bounding constraints of the problem. In general, it is possible to show that the outcome of a linear programming problem falls into several distinct categories: (1) The system is "infeasible," meaning that it is contradictory and there is no solution; (2) the system is unbounded, meaning that the minimum lies at negative infinity; (3) there is a unique minimizing solution; and (4) there is a unique finite minimum, but it is achieved by an infinite number of solutions $\mathbf{x}$.

The last situation is equivalent to observing that if there are two minimizing solutions, there must be an infinite number of them because then any linear combination of the two solutions is also a solution. Alternatively, if one makes up a matrix from the coefficients of $\mathbf{x}$ in Eqs. (3.45)–(3.47), one can determine if it has a nullspace. If one or more such vectors exists, it is also orthogonal to the objective function, and it can be assigned an arbitrary amplitude without changing J. One distinguishes between *feasible solutions*, meaning those that satisfy the inequality and equality constraints but which are not minimizing, and *optimal solutions*, which are both feasible and minimize the objective function.

An interesting and useful feature of a linear programming problem is that Eqs. (3.45)–(3.47) have a "dual":

$$\text{maximize:} \quad J_2 = \mathbf{y}^{\mathrm{T}}\boldsymbol{\mu}, \tag{3.48}$$

$$\mathbf{E}^{\mathrm{T}}\boldsymbol{\mu} \geq \mathbf{c}, \tag{3.49}$$

$$\boldsymbol{\mu} \geq \mathbf{0}. \tag{3.50}$$

It is possible to show that the minimum of J must equal the maximum of J_2. The reader may want to compare the structure of the original (the "primal") and dual equations with those relating the Lagrange multipliers to $\mathbf{x}$ discussed in Chapter 2. In the present case, the important relationship is

$$\frac{\partial J}{\partial y_i} = \mu_i. \tag{3.51}$$

That is, in a linear program, the dual solution provides the sensitivity of the objective function to perturbations in the constraint parameters $\mathbf{y}$. Duality theory pervades optimization problems, and the relationship to Lagrange multipliers is no accident.[20] Some simplex algorithms, called the "dual simplex," take advantage of the different dimensions of the primal and dual problems to accelerate solution. In recent years much attention has focussed upon a new, non-simplex method of solution[21] known as the "Karmarkar" or "interior point" method.

Linear programming is also valuable for solving estimation or approximation problems in which norms other than the 2-norms, which have been the focus of this book, are used. For example, suppose that one sought the solution to the constraints $\mathbf{Ex} + \mathbf{n} = \mathbf{y}$, $M > N$, but subject not to the conventional minimum of $J = \sum_i n_i^2$, but that of $J = \sum_i |n_i|$ (a 1-norm). Such norms are less sensitive to outliers than are the 2-norms and are said to be "robust." The maximum likelihood idea connects 2-norms to Gaussian statistics, and similarly, 1-norms are related to maximum likelihood with exponential statistics.[22] Reduction of such problems to linear programming is carried out by setting $n_i = n_i^+ - n_i^-$, $n_i^+ \geq 0$, $n_i^- \geq 0$, and the objective function is

$$\min: \ J = \sum_i \left(n_i^+ + n_i^-\right). \tag{3.52}$$

Other norms, the most important[23] of which is the so-called infinity norm, which minimizes the maximum element of an objective function ("mini-max" optimization), are also reducible to linear programming.

3.5 Empirical orthogonal functions

Consider an arbitrary $M \times N$ matrix $\mathbf{M}$. Suppose the matrix was representable, accurately, as the product of two vectors,

$$\mathbf{M} \approx \mathbf{ab}^{\mathrm{T}},$$

where $\mathbf{a}$ was $M \times 1$, and $\mathbf{b}$ was $N \times 1$. Approximation is intended in the sense that

$$\|\mathbf{M} - \mathbf{ab}^{\mathrm{T}}\| < \varepsilon,$$

for some acceptably small ε. Then one could conclude that the MN elements of $\mathbf{A}$ contain only $M + N$ pieces of information contained in $\mathbf{a}$, $\mathbf{b}$. Such an inference has many uses, including the ability to recreate the matrix accurately from only $M + N$ numbers, to physical interpretations of the meaning of $\mathbf{a}$, $\mathbf{b}$. More generally, if one pair of vectors is inadequate, some small number might suffice:

$$\mathbf{M} \approx \mathbf{a}_1 \mathbf{b}_1^{\mathrm{T}} + \mathbf{a}_2 \mathbf{b}_2^{\mathrm{T}} + \cdots + \mathbf{a}_K \mathbf{b}_K^{\mathrm{T}}. \tag{3.53}$$

A general mathematical approach to finding such a representation is through the SVD in a form sometimes known as the "Eckart–Young–Mirsky theorem."[24] This theorem states that the most efficient representation of a matrix in the form

$$\mathbf{M} \approx \sum_i^K \lambda_i \mathbf{u}_i \mathbf{v}_i^{\mathrm{T}}, \tag{3.54}$$

where the $\mathbf{u}_i$, $\mathbf{v}_i$ are orthonormal, is achieved by choosing the vectors to be the singular vectors, with λ_i providing the amplitude information (recall Eq. (2.227)).

The connection to regression analysis is readily made by noticing that the sets of singular vectors are the eigenvectors of the two matrices $\mathbf{MM}^{\mathrm{T}}$, $\mathbf{M}^{\mathrm{T}}\mathbf{M}$

(Eqs. (2.253) and (2.254)). If each row of $\mathbf{M}$ is regarded as a set of observations at a fixed coordinate, then $\mathbf{M}\mathbf{M}^\mathrm{T}$ is just proportional to the sample second-moment matrix of all the observations, and its eigenvectors, $\mathbf{u}_i$, are the EOFs. Alternatively, if each column is regarded as the observation set for a fixed coordinate, then $\mathbf{M}^\mathrm{T}\mathbf{M}$ is the corresponding sample second-moment matrix, and the $\mathbf{v}_i$ are the EOFs.

A large literature provides various statistical rules for use of EOFs. For example, the rank determination in the SVD becomes a test of the statistical significance of the contribution of singular vectors to the structure of $\mathbf{M}$.[25] In the wider context, however, one is dealing with the problem of efficient relationships amongst variables known or suspected to carry mutual correlations. Because of its widespread use, this subject is plagued by multiple discovery and thus multiple jargon. In different contexts and details (e.g., how the matrix is weighted), the problem is known as that of "principal components,"[26] "empirical orthogonal functions" (EOFs), the "Karhunen–Loève" expansion (in mathematics and electrical engineering),[27] "proper orthogonal decomposition,"[28] etc. Examples of the use of EOFs will be provided in Chapter 6.

3.6 Kriging and other variants of Gauss–Markov estimation

A variant of the Gauss–Markov mapping estimators, often known as "kriging" (named for David Krige, a mining geologist), addresses the problem of a spatially varying mean field, and is a generalization of the ordinary Gauss–Markov estimator.[29]

Consider the discussion on p. 132 of the fitting of a set of functions $f_i(\mathbf{r})$ to an observed field $y(\mathbf{r}_j)$. That is, we put

$$y(\mathbf{r}_j) = \mathbf{F}\alpha + q(\mathbf{r}_j), \tag{3.55}$$

where $\mathbf{F}(\mathbf{r}) = \{f_i(\mathbf{r})\}$ is a set of basis functions, and one seeks the expansion coefficients, α, and q such that the data, y, are *interpolated* (meaning reproduced exactly) at the observation points, although there is nothing to prevent further breaking up q into signal and noise components. If there is only one basis function – for example a constant – one is doing kriging, which is the determination of the mean prior to objective mapping of q, as discussed above. If several basis functions are being used, one has "universal kriging." The main issue concerns the production of an adequate statement of the expected error, given that the q are computed from a preliminary regression to determine the α.[30] The method is often used in situations where large-scale trends are expected in the data, and where one wishes to estimate and remove them before analyzing and mapping the q.

Because the covariances employed in objective mapping are simple to use and interpret only when the field is spatially stationary, much of the discussion of kriging uses instead what is called the "variogram," defined as $V = \langle (y(\mathbf{r}_i) - y(\mathbf{r}_j))(y(\mathbf{r}_i) - y(\mathbf{r}_j)) \rangle$, which is related to the covariance, and which is often encountered in turbulence theory as the "structure function." Kriging is popular in geology and hydrology, and deserves wider use.

3.7 Non-linear problems

The least-squares solutions examined thus far treat the coefficient matrix $\mathbf{E}$ as given. But in many of the cases encountered in practice, the elements of $\mathbf{E}$ are computed from data and are imperfectly specified. It is well known in the regression literature that treating $\mathbf{E}$ as known, even if $\mathbf{n}$ is increased beyond the errors contained in $\mathbf{y}$, can lead to significant bias errors in the least-squares and related solutions, particularly if $\mathbf{E}$ is nearly singular.[31] The problem is known as that of "errors in regressors or errors in variables" (EIV); it manifests itself in the classical simple least-squares problem (p. 43), where a straight line is fitted to data of the form $y_i = a + bt_i$, but where the measurement positions, t_i, are partly uncertain rather than perfect.

In general terms, when $\mathbf{E}$ has errors, the model statement becomes

$$(\tilde{\mathbf{E}} + \Delta\tilde{\mathbf{E}})\tilde{\mathbf{x}} = \tilde{\mathbf{y}} + \Delta\tilde{\mathbf{y}}, \tag{3.56}$$

where one seeks estimates, $\tilde{\mathbf{x}}$, $\Delta\tilde{\mathbf{E}}$, $\Delta\tilde{\mathbf{y}}$, where the old $\mathbf{n}$ is now broken into two parts: $\Delta\tilde{\mathbf{E}}\tilde{\mathbf{x}}$ and $\Delta\tilde{\mathbf{y}}$. If such estimates can be made, the result can be used to rewrite (3.56) as

$$\tilde{\mathbf{E}}\tilde{\mathbf{x}} = \tilde{\mathbf{y}}, \tag{3.57}$$

where the relation is to be exact. That is, one seeks to modify the elements of $\mathbf{E}$ such that the observational noise in it is reduced to zero.

3.7.1 Total least-squares

For some problems of this form, the method of total least-squares (TLS) is a powerful and interesting method. It is worth examining briefly to understand why it is not always immediately useful, and to motivate a different approach.[32]

The SVD plays a crucial role in TLS. Consider that in Eq. (2.17) the vector $\mathbf{y}$ was written as a sum of the column vectors of $\mathbf{E}$; to the extent that the column space does not fully describe $\mathbf{y}$, a residual must be left by the solution $\tilde{\mathbf{x}}$, and ordinary least-squares can be regarded as producing a solution in which a new estimate, $\tilde{\mathbf{y}} = \mathbf{E}\tilde{\mathbf{x}}$, of $\mathbf{y}$ is made; $\mathbf{y}$ is changed, but the elements of $\mathbf{E}$ are untouched. But suppose it were possible to introduce small changes in both the column vectors of $\mathbf{E}$, as well as in $\mathbf{y}$, such that the column vectors of the modified $\mathbf{E} + \Delta\mathbf{E}$ produced a spanning vector space for $\mathbf{y} + \Delta \mathbf{y}$, where both $\|\Delta \mathbf{y}\|$, $\|\Delta \mathbf{E}\|$ were "small," then the problem as stated would be solved.

The simplest problem to analyze is the full-rank, formally overdetermined one. Let $M \geq N = K$. Then, if we form the $M \times (N + 1)$ augmented matrix

$$\mathbf{E}_a = \{\mathbf{E} \quad \mathbf{y}\},$$

the solution sought is such that

$$\{\tilde{\mathbf{E}} \quad \tilde{\mathbf{y}}\} \begin{bmatrix} \tilde{\mathbf{x}} \\ -1 \end{bmatrix} = \mathbf{0} \tag{3.58}$$

(exactly). If this solution is to exist, $[\tilde{\mathbf{x}} - 1]^{\mathrm{T}}$ must lie in the nullspace of $\{\tilde{\mathbf{E}} \quad \tilde{\mathbf{y}}\}$. A solution is thus ensured by forming the SVD of $\{\mathbf{E} \quad \mathbf{y}\}$, setting $\lambda_{N+1} = 0$, and forming $\{\tilde{\mathbf{E}} \quad \tilde{\mathbf{y}}\}$ out of the remaining singular vectors and values. Then $[\tilde{\mathbf{x}} \quad - 1]^{\mathrm{T}}$ is the nullspace of the modified augmented matrix, and must therefore be proportional to the nullspace vector $\mathbf{v}_{N+1}$. Also,

$$\{\Delta \tilde{\mathbf{E}} \quad \Delta \tilde{\mathbf{y}}\} = -\mathbf{u}_{N+1} \lambda_{N+1} \mathbf{v}_{N+1}^{\mathrm{T}}. \tag{3.59}$$

Various complications can be considered, for example, if the last element of $\mathbf{v}_{N+1} = 0$; this and other special cases are discussed in the reference (see note 24). Cases of non-uniqueness are treated by selecting the solution of minimum norm. A simple generalization applies to the underdetermined case: if the rank of the augmented matrix is p, one reduces the rank by one to $p - 1$.

The TLS solution just summarized applies only to the case in which the errors in the elements of $\mathbf{E}$ and $\mathbf{y}$ are uncorrelated and of equal variance and in which there are no required structures – for example, where certain elements of $\mathbf{E}$ must always vanish. More generally, changes in some elements of $\mathbf{E}$ require, for reasons of physics, specific corresponding changes in other elements of $\mathbf{E}$ and in $\mathbf{y}$, and vice versa. The fundamental difficulty is that the model (3.56) presents a non-linear estimation problem with correlated variables, and its solution requires modification of the linear procedures developed so far.

3.7.2 *Method of total inversion*

The simplest form of TLS does not readily permit the use of correlations and prior variances in the parameters appearing in the coefficient matrix and does not provide any way of maintaining the zero structure there. Methods exist that permit accounting for prior knowledge of covariances.[33] Consider a set of non-linear constraints in a vector of unknowns $\mathbf{x}$,

$$\mathbf{g}(\mathbf{x}) + \mathbf{u} = \mathbf{q}. \tag{3.60}$$

This set of equations is the generalization of the linear models hitherto used; $\mathbf{u}$ again represents any expected error in the specification of the model. An example of a scalar non-linear model is

$$8x_1^2 + x_2^2 + u = 4.$$

In general, there will be some expectations about the behavior of $\mathbf{u}$. Without loss of generality, take its expected value to be zero, and its covariance is $\mathbf{Q} = \langle \mathbf{u}\mathbf{u}^{\mathrm{T}} \rangle$. There is nothing to prevent us from combining $\mathbf{x}, \mathbf{u}$ into one single set of unknowns $\boldsymbol{\xi}$, and indeed if the model has some unknown parameters, $\boldsymbol{\xi}$ might as well include those as well. So (3.60) can be written as

$$\mathcal{L}(\boldsymbol{\xi}) = \mathbf{0}. \tag{3.61}$$

In addition, it is supposed that a reasonable initial estimate $\tilde{\boldsymbol{\xi}}(0)$ is available, with uncertainty $\mathbf{P}(0) \equiv \langle (\boldsymbol{\xi} - \tilde{\boldsymbol{\xi}}(0))(\boldsymbol{\xi} - \tilde{\boldsymbol{\xi}}(0))^{\mathrm{T}} \rangle$ (or the covariances of the $\mathbf{u}, \mathbf{x}$ could be specified separately if their uncertainties are not correlated). An objective function whose minimum is sought is written,

$$J = \mathcal{L}(\boldsymbol{\xi})^{\mathrm{T}}\mathbf{Q}^{-1}\mathcal{L}(\boldsymbol{\xi}) + (\boldsymbol{\xi} - \tilde{\boldsymbol{\xi}}(0))^{\mathrm{T}}\mathbf{P}(0)^{-1}(\boldsymbol{\xi} - \tilde{\boldsymbol{\xi}}(0)). \tag{3.62}$$

The presence of the weight matrices $\mathbf{Q}, \mathbf{P}(0)$ permits control of the elements most likely to change, specification of elements that should not change at all (e.g., by introducing zeros into $\mathbf{P}(0)$), as well as the stipulation of covariances. It can be regarded as a generalization of the process of minimizing objective functions, which led to least-squares in previous chapters and is sometimes known as the "method of total inversion."[34]

Consider an example for the two simultaneous equations

$$2x_1 + x_2 + n_1 = 1, \tag{3.63}$$

$$0 + 3x_2 + n_2 = 2, \tag{3.64}$$

where all the numerical values except the zero are now regarded as in error to some degree. One way to proceed is to write the coefficients of $\mathbf{E}$ in the specific perturbation form (3.56). For example, write $E_{11} = 2 + \Delta E_{11}$, and define the unknowns $\boldsymbol{\xi}$ in terms of the ΔE_{ij}. For illustration retain the full non-linear form by setting

$$\xi_1 = E_{11}, \quad \xi_2 = E_{12}, \quad \xi_3 = E_{21}, \quad \xi_4 = E_{22}, \quad \xi_5 = x_1, \quad \xi_6 = x_2,$$
$$u_1 = n_1, \quad u_2 = n_2.$$

The equations are then

$$\xi_1\xi_5 + \xi_2\xi_6 + u_1 - 1 = 0, \tag{3.65}$$

$$\xi_3\xi_5 + \xi_4\xi_6 + u_2 - 2 = 0. \tag{3.66}$$

The y_i are being treated as formally fixed, but u_1, u_2 represent their possible errors (the division into knowns and unknowns is not unique). Let there be an initial estimate,

$$\xi_1 = 2 \pm 1, \quad \xi_2 = 2 \pm 2, \quad \xi_3 = 0 \pm 0,$$
$$\xi_4 = 3.5 \pm 1, \quad \xi_5 = x_1 = 0 \pm 2, \quad \xi_6 = 0 \pm 2,$$

with no imposed correlations so that $\mathbf{P}(0) = \text{diag}\,([1, 4, 0, 1, 4, 4])$; the zero represents the requirement that E_{21} remain unchanged. Let $\mathbf{Q} = \text{diag}\,([2, 2])$. Then a useful objective function is

$$
\begin{aligned}
J = {} & (\xi_1\xi_5 + \xi_2\xi_6 - 1)^2/2 \\
& + (\xi_3\xi_5 + \xi_4\xi_6 - 2)^2/2 + (\xi_1 - 2)^2 + (\xi_2 - 2)^2/4 \\
& + 10^6\xi_3^2 + (\xi_4 - 3.5)^2 + \xi_5^2/4 + \xi_6^2/4.
\end{aligned}
\tag{3.67}
$$

The 10^6 in front of the term in ξ_3^2 is a numerical approximation to the infinite value implied by a zero uncertainty in this term (an arbitrarily large value can cause numerical instability, characteristic of penalty and barrier methods).[35]

Such objective functions define surfaces in spaces of the dimension of $\boldsymbol{\xi}$. Most procedures require the investigator to make a first guess at the solution, $\tilde{\boldsymbol{\xi}}(0)$, and attempt to minimize J by going downhill from the guess. Various deterministic search algorithms have been developed and are variants of steepest descent, conjugate gradient, Newton and quasi-Newton methods. The difficulties are numerous. Some methods require computation or provision of the gradients of J with respect to $\boldsymbol{\xi}$, and the computational cost may become very great. The surfaces on which one is seeking to go downhill may become extremely tortuous, or very slowly changing. The search path can fall into local holes that are not the true minima. Non-linear optimization is something of an art. Nonetheless, existing techniques are very useful. The minimum of J corresponds to finding the solution of the non-linear normal equations that would result from setting the partial derivatives to zero.

Let the true minimum be at $\boldsymbol{\xi}^*$. Assuming that the search procedure has succeeded, the objective function is locally

$$J = \text{constant} + (\boldsymbol{\xi} - \boldsymbol{\xi}^*)^{\mathrm{T}}\mathcal{H}(\boldsymbol{\xi} - \boldsymbol{\xi}^*) + \Delta J, \tag{3.68}$$

where $\mathcal{H}$ is the Hessian and ΔJ is a correction – assumed to be small. In the linear least-squares problem (2.89), the Hessian is evidently $\mathbf{E}^{\mathrm{T}}\mathbf{E}$, the second derivative of the objective function with respect to $\mathbf{x}$. The supposition is then that near the true optimum, the objective function is locally quadratic with a small correction. To the extent that this supposition is true, the result can be analyzed in terms of the behavior of $\mathcal{H}$ as though it represented a locally defined version of $\mathbf{E}^{\mathrm{T}}\mathbf{E}$. In particular, if $\mathcal{H}$ has a nullspace, or small eigenvalues, one can expect to see all the issues arising that we dealt with in Chapter 2, including ill-conditioning and solution variances that may become large in some elements. The machinery used in Chapter 2 (row and column scaling, nullspace suppression, etc.) thus becomes immediately relevant here and can be used to help conduct the search and to understand the solution.

Example *It remains to find the minimum of J in (3.67).[36] Most investigators are best-advised to tackle such problems by using one of the many general purpose numerical routines written by experts.[37] Here, a quasi-Newton method was employed to produce*

$$E_{11} = 2.0001, \quad E_{12} = 1.987, \quad E_{21} = 0.0,$$
$$E_{22} = 3.5237, \quad x_1 = -0.0461, \quad x_2 = 0.556,$$

and the minimum of $J = 0.0802$. The inverse Hessian at the minimum is

$$\mathcal{H}^{-1} = \left\{ \begin{array}{cccccc} 0.4990 & 0.0082 & -0.0000 & -0.0014 & 0.0061 & 0.0005 \\ 0.0082 & 1.9237 & 0.0000 & 0.0017 & -0.4611 & -0.0075 \\ -0.0000 & 0.0000 & 0.0000 & -0.0000 & -0.0000 & 0.0000 \\ -0.0014 & 0.0017 & -0.0000 & 0.4923 & 0.0623 & -0.0739 \\ 0.0061 & -0.4611 & -0.0000 & 0.0623 & 0.3582 & -0.0379 \\ 0.0005 & -0.0075 & 0.0000 & -0.0739 & -0.0379 & 0.0490 \end{array} \right\}.$$

The eigenvalues and eigenvectors of $\mathcal{H}$ are

$$\lambda_i = \left[2.075 \times 10^6 \quad 30.4899 \quad 4.5086 \quad 2.0033 \quad 1.9252 \quad 0.4859 \right],$$

$$V = \left\{ \begin{array}{cccccc} 0.0000 & -0.0032 & 0.0288 & 0.9993 & 0.0213 & 0.0041 \\ -0.0000 & 0.0381 & -0.2504 & 0.0020 & 0.0683 & 0.9650 \\ -1.0000 & 0.0000 & 0.0000 & -0.0000 & -0.0000 & 0.0000 \\ 0.0000 & 0.1382 & 0.2459 & -0.0271 & 0.9590 & -0.0095 \\ -0.0000 & 0.1416 & -0.9295 & 0.0237 & 0.2160 & -0.2621 \\ 0.0000 & 0.9795 & 0.1095 & 0.0035 & -0.1691 & 0.0017 \end{array} \right\}.$$

The large jump from the first eigenvalue to the others is a reflection of the conditioning problem introduced by having one element, ξ_3, with almost zero uncertainty. It is left to the reader to use this information about $\mathcal{H}$ to compute the uncertainty of the solution in the neighborhood of the optimal values – this would be the new uncertainty, $\mathbf{P}(1)$. A local resolution analysis follows from that of the SVD, employing knowledge of the $\mathbf{V}$. The particular system is too small for a proper statistical test of the result against the prior covariances, but the possibility should be clear. If $\mathbf{P}(0)$, etc., are simply regarded as non-statistical weights, we are free to experiment with different values until a pleasing solution is found.

3.7.3 Variant non-linear methods, including combinatorial ones

As with the linear least-squares problems discussed in Chapter 2, many possibilities exist for objective functions that are non-linear in either data constraint terms or the model, and there are many variations on methods for searching for objective function minima.

A very interesting and useful set of methods has been developed comparatively recently, called "combinatorial optimization." Combinatorial methods do not promise that the true minimum is found – merely that it is highly probable – because they search the space of solutions in clever ways that make it unlikely that one is very far from the true optimal solution. Two such methods, simulated annealing and genetic algorithms, have recently attracted considerable attention.[38] Simulated annealing searches randomly for solutions that reduce the objective function from a present best value. Its clever addition to purely random guessing is a willingness to accept the occasional uphill solution – one that raises the value of the objective function – as a way of avoiding being trapped in purely local minima. The probability of accepting an uphill value and the size of the trial random perturbations depend upon a parameter, a temperature defined in analogy to the real temperature of a slowly cooling (annealing) solid.

Genetic algorithms, as their name would suggest, are based upon searches generated in analogy to genetic drift in biological organisms.[39] The recent literature is large and sophisticated, and this approach is not pursued here.

Notes

1 Freeman (1965), Jerri (1977), Butzer and Stens (1992), or Bracewell (2000).
2 In the Russian literature, Kotel'nikov's theorem.
3 Aliasing is familiar as the stroboscope effect. Recall the appearance of the spokes of a wagon wheel in the movies. The spokes can appear to stand still, or move slowly forward or backward, depending upon the camera shutter speed relative to the true rate at which the spokes revolve. (The terminology is apparently due to John Tukey.)
4 Hamming (1973) and Bracewell (2000) have particularly clear discussions.
5 There is a story, perhaps apocryphal, that a group of investigators was measuring the mass flux of the Gulf Stream at a fixed time each day. They were preparing to publish the exciting discovery that there was a strong 14-day periodicity to the flow, before someone pointed out that they were aliasing the tidal currents of period 12.42 hours.
6 It follows from the so-called Paley–Wiener criterion, and is usually stated in the form that "timelimited signals cannot be bandlimited."
7 Landau and Pollack (1962), Freeman (1965), Jerri (1977).
8 Petersen and Middleton (1962). An application, with discussion of the noise sensitivity, may be found in Wunsch (1989).
9 Davis and Polonsky (1965).
10 See Ripley (2004, Section 5.2).
11 Bretherton *et al.* (1976).
12 See Fukumori *et al.* (1991).
13 Luenberger (1969).
14 See Strang (1988) or Lawson and Hanson (1995); the standard full treatment is Fiacco and McCormick (1968).

15 Lawson and Hanson (1974).
16 Fu (1981).
17 Tziperman and Hecht (1987).
18 Dantzig (1963).
19 For example, Bradley *et al.* (1977), Luenberger (2003), and many others.
20 See Cacuci (1981), Hall and Cacuci (1984), Strang (1986), Rockafellar (1993), Luenberger (1997).
21 One of the few mathematical algorithms ever to be written up on the front page of the *New York Times* (November 19, 1984, story by J. Gleick) – a reflection of the huge economic importance of linear programs in industry.
22 Arthanari and Dodge (1993).
23 Wagner (1975), Arthanari and Dodge (1993).
24 Van Huffel and Vandewalle (1991).
25 The use of EOFs, with various normalizations, scalings, and in various row/column physical spaces, is widespread – for example, Wallace (1972), Wallace and Dickinson (1972), and many others.
26 Jolliffe (2002), Preisendorfer (1988), Jackson (2003).
27 Davenport and Root (1958), Wahba (1990).
28 Berkooz *et al.*(1993).
29 Armstrong (1989), David (1988), Ripley (2004).
30 Ripley (2004).
31 For example, Seber and Lee (2003).
32 Golub and van Loan (1996), Van Huffel and Vandewalle (1991).
33 Tarantola and Valette (1982), Tarantola (1987).
34 Tarantola and Valette (1982) labeled the use of similar objective functions and the determination of the minimum as the *method of total inversion*, although they considered only the case of perfect model constraints.
35 Luenberger (1984).
36 Tarantola and Valette (1982) suggested using a linearized search method, iterating from the initial estimate, which must be reasonably close to the correct answer. The method can be quite effective (e.g., Wunsch and Minster, 1982; Mercier *et al.*, 1993). In a wider context, however, their method is readily recognizable as a special case of the many known methods for minimizing a general objective function.
37 Numerical Algorithms Group (2005), Press *et al.* (1996).
38 For simulated annealing, the literature starts with Pincus (1968) and Kirkpatrick *et al.* (1983), and general discussions can be found in van Laarhoven and Aarts (1987), Ripley (2004), and Press *et al.* (1996). A simple oceanographic application to experiment design was discussed by Barth and Wunsch (1990).
39 Goldberg (1989), Holland (1992), Denning (1992).

4

The time-dependent inverse problem: state estimation

4.1 Background

The discussion so far has treated models and data that most naturally represent a static world. Much data, however, describe systems that are changing in time to some degree. Many familiar differential equations represent phenomena that are intrinsically time-dependent; such as the wave equation,

$$\frac{1}{c^2}\frac{\partial^2 x\,(t)}{\partial t^2} - \frac{\partial^2 x(t)}{\partial r^2} = 0. \tag{4.1}$$

One may well wonder if the methods described in Chapter 2 have any use with data thought to be described by (4.1). An approach to answering the question is to recognize that t is simply another coordinate, and can be regarded, e.g., as the counterpart of one of the space coordinates encountered in the previous discussion of two-dimensional partial differential equations. From this point of view, time-dependent systems are nothing but versions of the systems already developed. (The statement is even more obvious for the simpler equation,

$$\frac{\mathrm{d}^2 x\,(t)}{\mathrm{d}t^2} = q(t), \tag{4.2}$$

where the coordinate that is labeled t is arbitrary, and need not be time.)

On the other hand, time often has a somewhat different flavor to it than does a spatial coordinate because it has an associated direction. The most obvious example occurs when one has data up to and including some particular time t, and one asks for a *forecast* of some elements of the system at a future time $t' > t$. Even this role of time is not unique: one could imagine a completely equivalent spatial forecast problem, in which, e.g., one required extrapolation of the map of an ore body beyond some area in which measurements exist. In state estimation, time does not introduce truly novel problems. The main issue is really a computational one: problems in two or more spatial dimensions, when time-dependent, typically generate system

178

dimensions that are too large for conventionally available computer systems. To deal with the computational load, state estimation algorithms are sought that are computationally more efficient than what can be achieved with the methods used so far. Consider, as an example,

$$\frac{\partial C}{\partial t} = \kappa \nabla^2 C, \tag{4.3}$$

a generalization of the Laplace equation (a diffusion equation). Using a one-sided time difference, and the discrete form of the two-dimensional Laplacian in Eq. (1.13), one has

$$\frac{C_{ij}((n+1)\Delta t) - C_{ij}(n\Delta t)}{\Delta t} = \kappa\{C_{i+1,j}(n\Delta t) - 2C_{i,j}(n\Delta t) + C_{i-1,j}(n\Delta t)$$
$$+ C_{i,j+1}(n\Delta t) - 2C_{i,j}(n\Delta t) + C_{i,j-1}(n\Delta t)\}.$$
$$\tag{4.4}$$

If there are N^2 elements defining C_{ij} at each time $n\Delta t$, then the number of elements over the entire time span of T time steps would be TN^2, which grows rapidly as the number of time steps increases. Typically the relevant observation numbers also grow rapidly through time. On the other hand, the operation

$$\mathbf{x} = \mathrm{vec}(C_{ij}(n\Delta t)), \tag{4.5}$$

renders Eq. (4.4) in the familiar form

$$\mathbf{A}_1\mathbf{x} = \mathbf{0}, \tag{4.6}$$

and with some boundary conditions, some initial conditions and/or observations, and a big enough computer, one could use without change any of the methods of Chapter 2. As $T,\ N$ grow, however, even the largest available computer becomes inadequate. Methods are sought that can take advantage of special structures built into time evolving equations to reduce the computational load. (Note, however, that $\mathbf{A}_1$ is very sparse.)

This chapter is in no sense exhaustive; many entire books are devoted to the material and its extensions. The intention is to lay out the fundamental ideas, which are algorithmic rearrangements of methods already described in Chapters 2 and 3, with the hope that they will permit the reader to penetrate the wider literature. Several very useful textbooks are available for readers who are not deterred by discussions in contexts differing from their own applications.[1] Most of the methods now being used in fields involving large-scale fluid dynamics, such as oceanography and meteorology, have been known for years under the general headings of control theory and control engineering. The experience in these latter areas is very helpful; the main issues in applications to fluid problems concern the size of the models and data sets encountered: they are typically many orders of magnitude larger than anything contemplated by engineers. In meteorology, specialized techniques used for forecasting are commonly called "data assimilation."[2] The reader may find it helpful to keep in mind, through the details that follow, that almost all methods in actual use are, beneath the mathematical disguises, nothing but versions of least-squares fitting of models to data, but reorganized so as to increase the efficiency of solution, or to minimize storage requirements, or to accommodate continuing data streams.

Several notation systems are in wide use. The one chosen here is taken directly from the control theory literature; it is simple and adequate.[3]

4.2 Basic ideas and notation

4.2.1 Models

In the context of this chapter, "models" is used to mean statements about the connections between the system variables in some place at some time, and those in all other places and times. Maxwell's equations are a model of the behavior of time-dependent electromagnetic disturbances. These equations can be used to connect the magnetic and electric fields everywhere in space and time. Other physical systems are described by the Schrödinger, elastic, or fluid-dynamical equations. Static situations are special limiting cases, e.g., for an electrostatic field in a container with known boundary conditions.

A useful concept is that of the system "state." By that is meant the internal information at a single moment in time required to forecast the system one small time step into the future. The time evolution of a system described by the tracer diffusion equation (4.3), inside a closed container can be calculated with arbitrary accuracy at time $t + \Delta t$, if one knows $C(\mathbf{r}, t)$ and the boundary conditions $C_B(t)$, as $\Delta t \to 0$. $C(\mathbf{r}, t)$ is the state variable (the "internal" information), with the boundary conditions being regarded as separate externally provided variables (but the distinction is to some degree arbitrary, as we will see). In practice, because such quantities as initial and boundary conditions, container shape, etc., are obtained from measurements, and are thus always imperfectly known, the problems are conceptually identical to those already considered.

Consider any model, whether time dependent or steady, but rendered in discrete form. The "state vector" $\mathbf{x}(t)$ (where t is discrete) is defined as those elements of the model employed to describe fully the physical state of the system at any time and all places as required by the model in use. For the discrete Laplace/Poisson equation in Chapter 1, $\mathbf{x} = \mathrm{vec}(C_{ij})$ is the state vector. In a fluid model, the state vector might consist of three components of velocity, pressure, and temperature at each of millions of grid points, and it would be a function of time, $\mathbf{x}(t)$, as well. (One might want to regard the complete description,

$$\mathbf{x}_B = [\mathbf{x}(1\Delta t)^{\mathrm{T}}, \mathbf{x}(2\Delta t)^{\mathrm{T}}, \dots, \mathbf{x}(T\Delta t)^{\mathrm{T}}]^{\mathrm{T}}, \tag{4.7}$$

as the state vector, but by convention, it refers to the subvectors, $\mathbf{x}(t = n\Delta t)$, each of which, given the boundary conditions, is sufficient to compute any future one.)

Consider a partial differential equation,

$$\frac{\partial}{\partial t}(\nabla_h^2 p) + \beta \frac{\partial p}{\partial \eta} = 0, \tag{4.8}$$

subject to boundary conditions. ∇_h is the two-dimensional gradient operator. For the moment, t is a continuous variable. Suppose it is solved by an expansion,

$$p(\xi, \eta, t) = \sum_{j=1}^{N/2} [a_j(t)\cos(\mathbf{k}_j \cdot \mathbf{r}) + b_j(t)\sin(\mathbf{k}_j \cdot \mathbf{r})]. \tag{4.9}$$

$[\mathbf{k}_j = (k_\xi, k_\eta), \mathbf{r} = (\xi, \eta)]$, then $\mathbf{a}(t) = [a_1(t), b_1(t), \cdots a_j(t), b_j(t), \dots]^{\mathrm{T}}$. The $\mathbf{k}_j$ are chosen to be periodic in the domain. The a_i, b_i are a partial-discretization, reducing the time-dependence to a finite set of coefficients. Substitute into Eq. (4.8),

$$\sum \{-|\mathbf{k}_j|^2 (\dot{a}_j \cos(\mathbf{k}_j \cdot \mathbf{r}) + \dot{b}_j \sin(\mathbf{k}_j \cdot \mathbf{r}))$$
$$+ \beta k_{1j}[-a_j \sin(\mathbf{k}_j \cdot \mathbf{r}) + b_j \cos(\mathbf{k}_j \cdot \mathbf{r})]\} = 0.$$

The dot indicates a time-derivative, and k_{1j} is the η component of $\mathbf{k}_j$. Multiply this last equation through first by $\cos(\mathbf{k}_j \cdot \mathbf{r})$ and integrate over the domain:

$$-|\mathbf{k}_j|^2 \dot{a}_j + \beta k_{1j} b_j = 0.$$

Multiply by $\sin(\mathbf{k}_j \cdot \mathbf{r})$ and integrate again to give

$$|\mathbf{k}_j|^2 \dot{b}_j + \beta k_{1j} a_j = 0,$$

or

$$\frac{d}{dt}\begin{bmatrix} a_j \\ b_j \end{bmatrix} = \left\{ \begin{matrix} 0 & \beta k_{1j}/|\mathbf{k}_j|^2 \\ -\beta k_{1j}/|\mathbf{k}_j|^2 & 0 \end{matrix} \right\} \begin{bmatrix} a_j \\ b_j \end{bmatrix}.$$

Each pair of a_j, b_j satisfies a system of ordinary differential equations in time, and each can be further discretized, so that

$$\begin{bmatrix} a_j(n\Delta t) \\ b_j(n\Delta t) \end{bmatrix} = \left\{ \begin{matrix} 1 & \Delta t\beta k_{1j}/|\mathbf{k}_j|^2 \\ -\Delta t\beta k_{1j}/|\mathbf{k}_j|^2 & 1 \end{matrix} \right\} \begin{bmatrix} a_j((n-1)\Delta t) \\ b_j((n-1)\Delta t) \end{bmatrix}.$$

The state vector is then the collection

$$\mathbf{x}(n\Delta t) = [a_1(n\Delta t), b_1(n\Delta t), a_2(n\Delta t), b_2(n\Delta t), \dots]^{\mathrm{T}},$$

at time $t = n\Delta t$. Any adequate discretization can provide the state vector; it is not unique, and careful choice can greatly simplify calculations.

In the most general terms, we can write any discrete model as a set of functional relations:

$$\mathcal{L}[\mathbf{x}(0), \ldots, \mathbf{x}(t - \Delta t), \mathbf{x}(t), \mathbf{x}(t + \Delta t), \ldots, x(t_f), \ldots,$$
$$\mathbf{B}(t)\mathbf{q}(t), \mathbf{B}(t + \Delta t)\mathbf{q}(t + \Delta t), \ldots, t] = 0, \tag{4.10}$$

where $\mathbf{B}(t)\mathbf{q}(t)$ represents a general, canonical, form for boundary and initial conditions/sources/sinks. A time-dependent model is a set of rules for computing the state vector at time $t = n\Delta t$, from knowledge of its values at time $t - \Delta t$ and the externally imposed forces and boundary conditions. We almost always choose the time units so that $\Delta t = 1$, and t becomes an integer (the context will normally make clear whether t is continuous or discrete). The static system equation,

$$\mathbf{A}\mathbf{x} = \mathbf{b}, \tag{4.11}$$

is a special case. In practice, the collection of relationships (4.10) always can be rewritten as a time-stepping rule – for example,

$$\mathbf{x}(t) = \mathbf{L}(\mathbf{x}(t - 1), \mathbf{B}(t - 1)\mathbf{q}(t - 1), t - 1), \quad \Delta t = 1, \tag{4.12}$$

or, if the model is linear,

$$\mathbf{x}(t) = \mathbf{A}(t - 1)\mathbf{x}(t - 1) + \mathbf{B}(t - 1)\mathbf{q}(t - 1). \tag{4.13}$$

If the model is time invariant, $\mathbf{A}(t) = \mathbf{A}$, and $\mathbf{B}(t) = \mathbf{B}$. $\mathbf{A}(t)$ is called the "state transition matrix." It is generally true that any linear discretized model can be put into this canonical form, although it may take some work. By the same historical conventions described in Chapter 1, solution of systems like (4.12), subject to appropriate initial and boundary conditions, constitutes the forward, or direct, problem. Note that, in general, $\mathbf{x}(0) = \mathbf{x}_0$ has subcomponents that formally precede $t = 0$.

Example *The straight-line model, discussed in Chapter 1 satisfies the rule*

$$\frac{\mathrm{d}^2 \xi}{\mathrm{d}t^2} = 0, \tag{4.14}$$

which can be discretized as

$$\xi(t + \Delta t) - 2\xi(t) + \xi(t - \Delta t) = 0, \tag{4.15}$$

Define

$$x_1(t) = \xi(t), \quad x_2(t) = \xi(t - \Delta t),$$

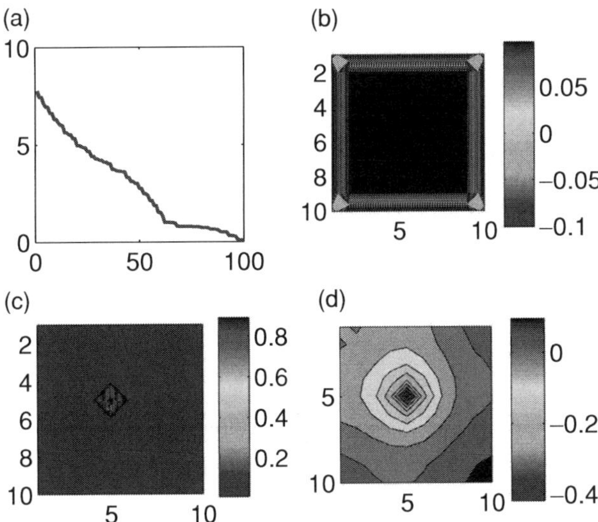

Figure 2.12 (a) Graph of the singular values of the coefficient matrix $\mathbf{A}$ of the numerical Neumann problem on a 10×10 grid. All λ_i are non-zero except the last one. (b) shows $\mathbf{u}_{100}$, the nullspace vector of $\mathbf{E}^{\mathrm{T}}$ defining the solvability or consistency condition for a solution through $\mathbf{u}_{100}^{\mathrm{T}}\mathbf{y} = 0$. Plotted as mapped onto the two-dimensional spatial grid (r_x, r_y) with $\Delta x = \Delta y = 1$. The interpretation is that the sum of the influx through the boundaries and from interior sources must vanish. Note that corner derivatives differ from other boundary derivatives by $1/\sqrt{2}$. The corresponding $\mathbf{v}_{100}$ is a constant, indeterminate with the information available, and not shown. (c) A source $\mathbf{b}$ (a numerical delta function) is present, not satisfying the solvability condition $\mathbf{u}_{100}^{\mathrm{T}}\mathbf{b} = 0$, because all boundary fluxes were set to vanish. (d) The particular SVD solution, $\tilde{\mathbf{x}}$, at rank $K = 99$. One confirms that $\mathbf{A}\tilde{\mathbf{x}} - \mathbf{b}$ is proportional to $\mathbf{u}_{100}$ as the source is otherwise inconsistent with no flux boundary conditions. With $\mathbf{b}$ a Kronecker delta function at one grid point, this solution is a numerical Green function for the Neumann problem and insulating boundary conditions.

This plate section is also available for download in colour from
www.cambridge.org/9781107406063

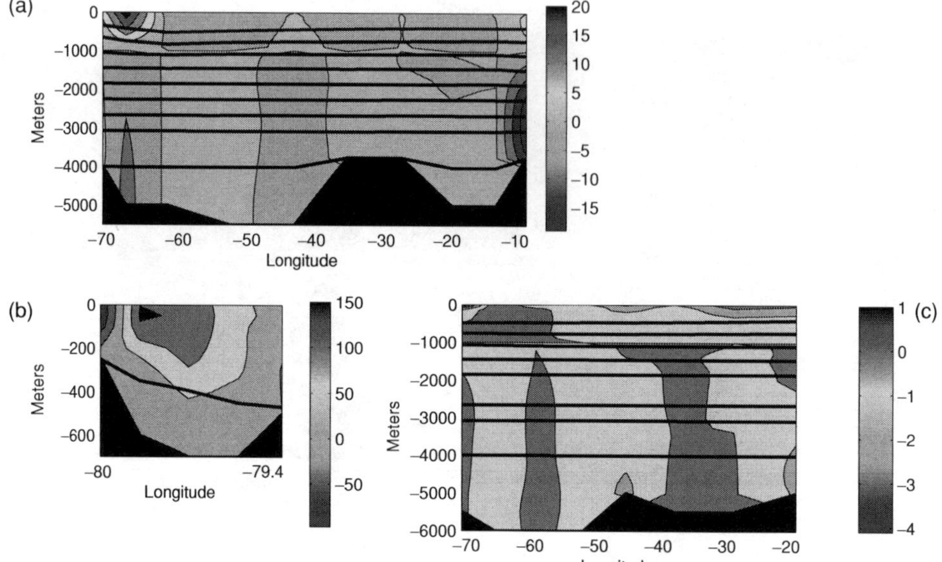

Figure 6.2 Geostrophic velocity (colors, in cm/s) relative to a 1000-decibar reference level, or the bottom, whichever is shallower, and isopycnal surfaces (thick lines) used to define ten layers in the constraints. Part (a) is for 36° N, (b) for the Florida Straits, and (c) for 24° N east of the Florida Straits. Levels-of-no-motion corresponding to the initial calculation are visible at 1000 m in the deep sections, but lie at the bottom in the Florida Straits. Note the greatly varying longitude, depth, and velocity scales.

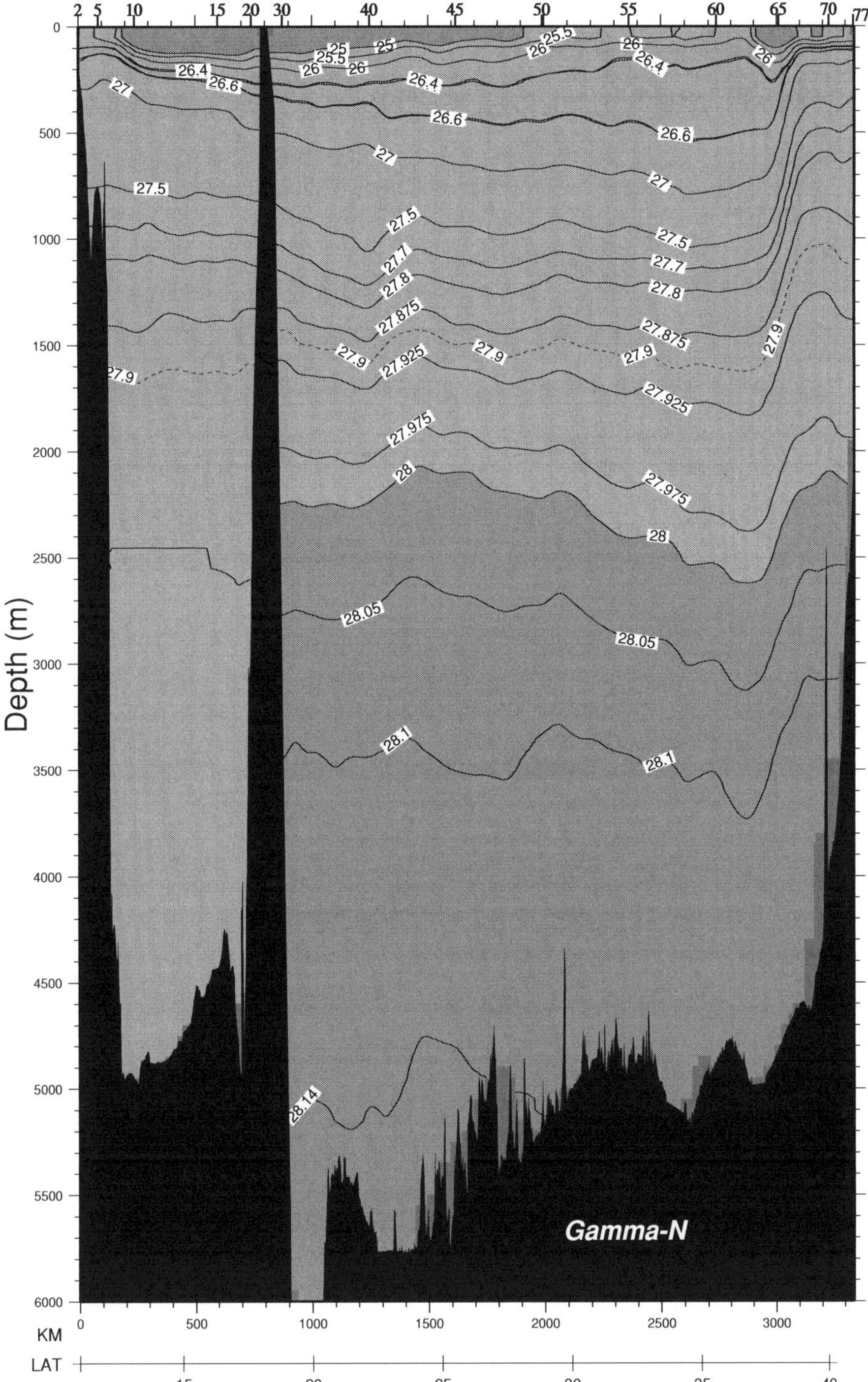

Figure 6.17 Density field constructed from the measurements at the positions in Fig. 6.16. Technically, these contours are of so-called neutral density, but for present purposes they are indistinguishable from the potential density.

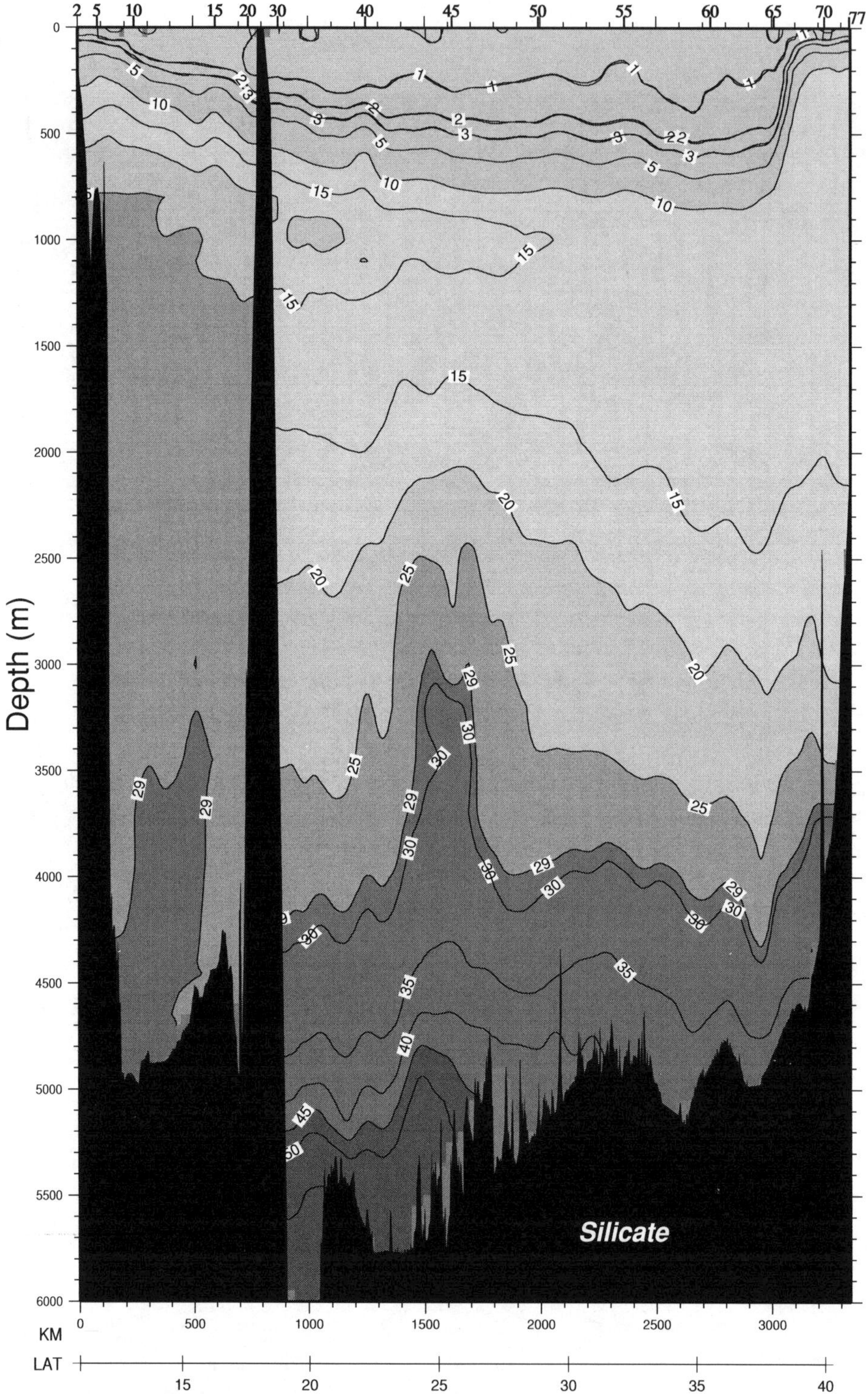

Figure 6.18 Same as Fig. 6.17, except showing the silicate concentration – a passive tracer – in μ moles/kg.

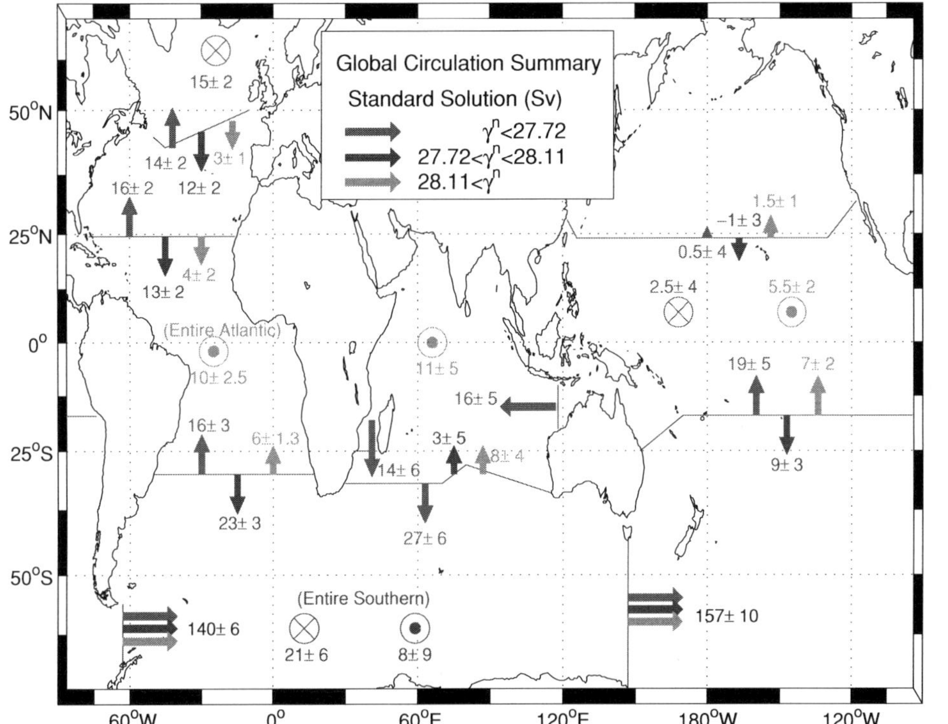

Figure 6.24 Mass flux in the Ganachaud (2003a) solution. Red, blue, and green arrows depict the vertically and horizontally averaged mass flux between the neutral surfaces noted. (Source: Ganachaud, 2003a)

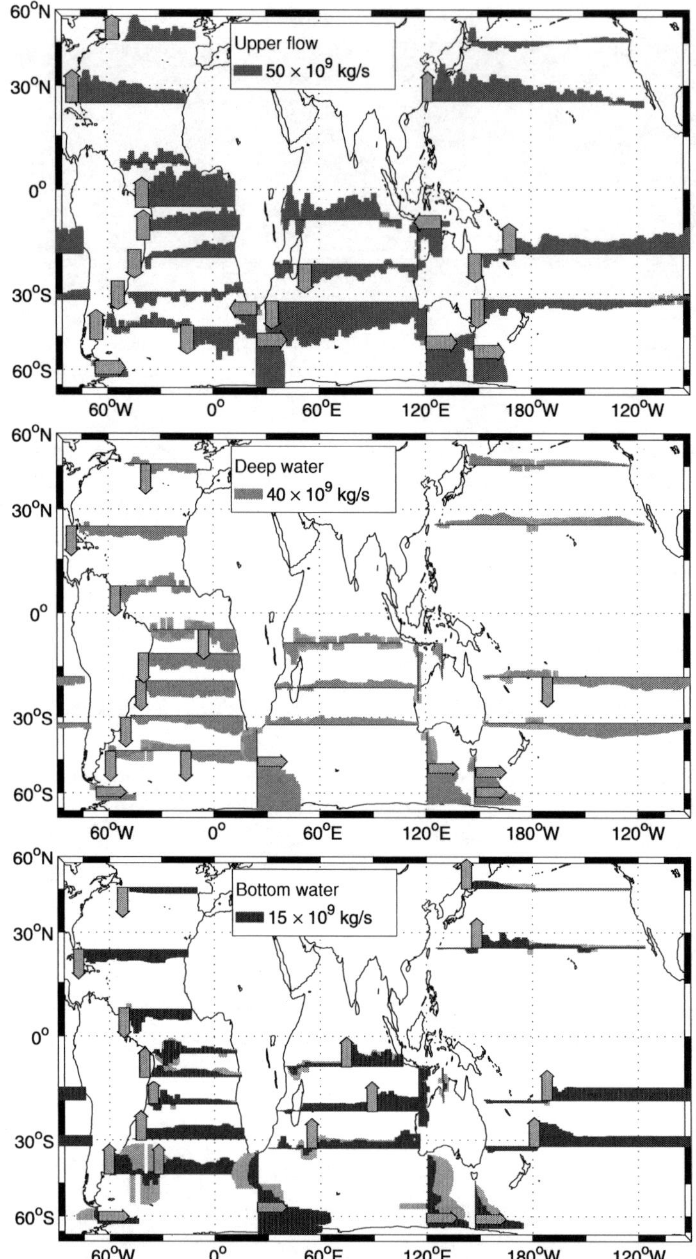

Figure 6.25 Mass transports integrated from west to east and north to south for the solution displayed in Fig. 6.24. Light shading shows the estimated standard errors. Arrows denote the major currents of the system. (Source: Ganachaud, 2003a)

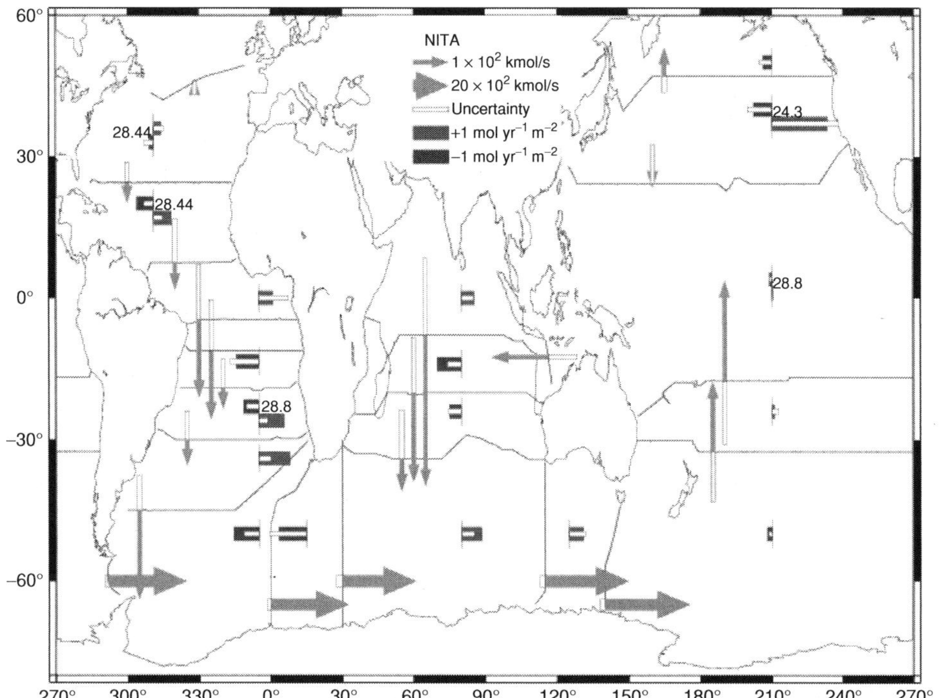

Figure 6.26 Integrated nitrate flux corresponding to the mass flux in Figs. 6.24 and 6.25. Although there is some resemblance to the mass fluxes, significant differences in the net movement of nitrate occur – owing to the spatially varying concentration of nitrate.

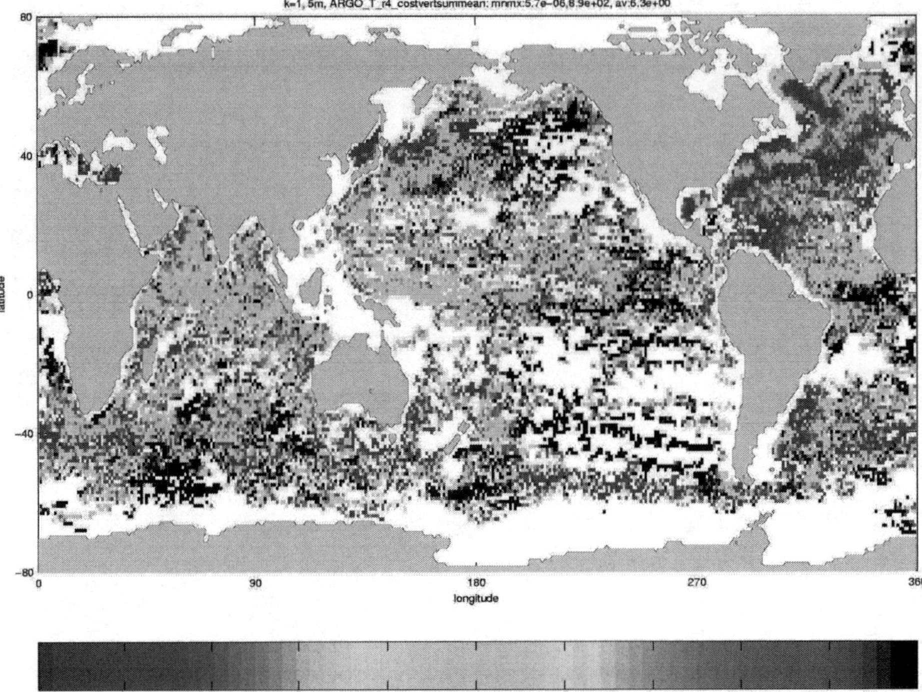

Figure 7.2 Misfit to so-called ARGO float temperature profiles. These instruments produce a vertical profile of temperature in the ocean above about 2000 m at pre-determined intervals (of order 10 days). The misfits shown here are weighted by the estimated errors in both model and data, and then averaged over the entire water column.

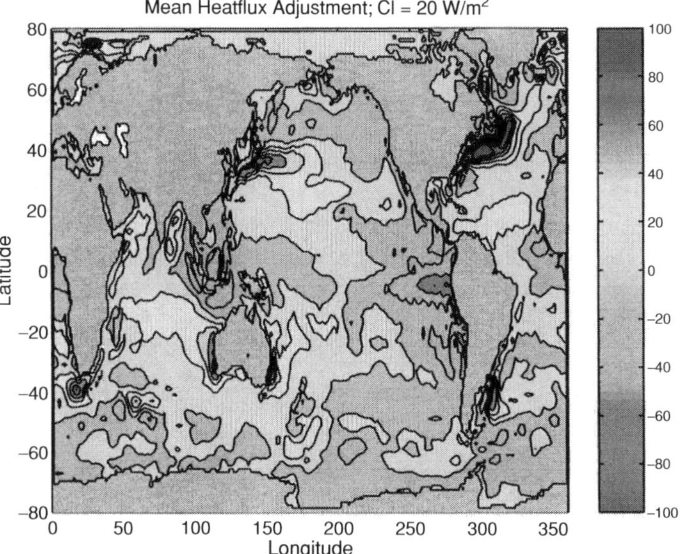

Figure 7.3 Mean changes in (a) net surface heat exchange (W/m^2), determined from six years of meteorological estimates and an ocean model. (Source: Stammer *et al.*, 2002)

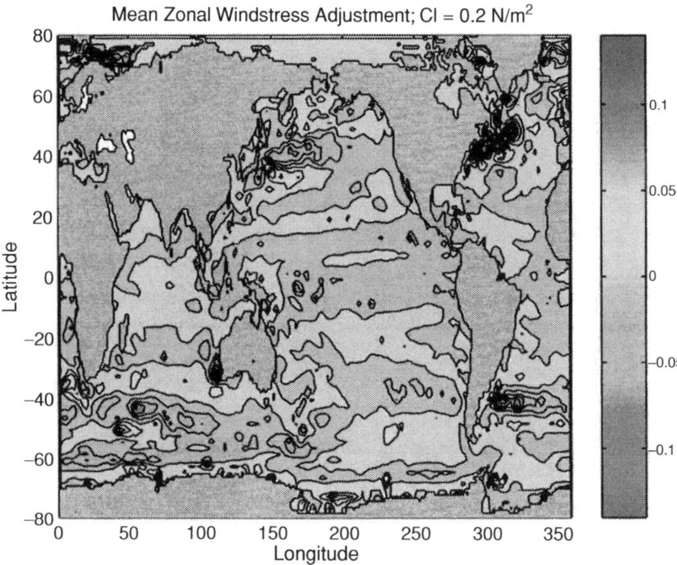

Figure 7.4 Adjusted wind stress (part of the control vector) in N/m^2. (Source: Stammer *et al.*, 2002)

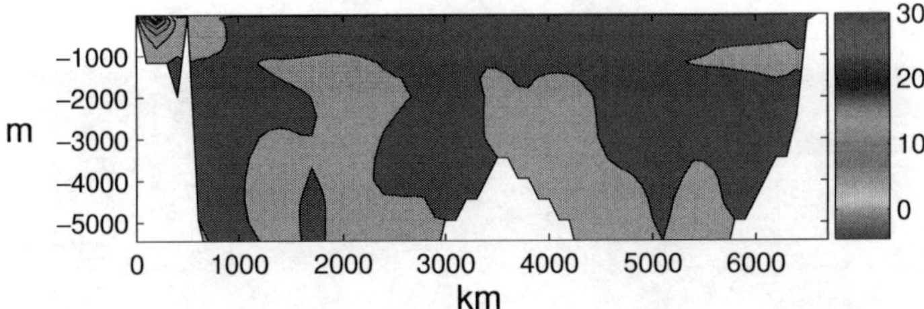

Figure 7.6 Twelve-year mean velocity across the North Atlantic Ocean (cm/s) at 26° N from a 13-year optimization of a 1 degree horizontal resolution (23 vertical layers) general circulation model and a large global data set. Red region is moving southward, remaining regions are all moving northward. Note the absence (recall Chapter 6) of any obvious level-of-no-motion, although there is a region of weak mean meridional flow near 1100 m depth. The northward-flowing Gulf Stream is visible on the extreme western side.

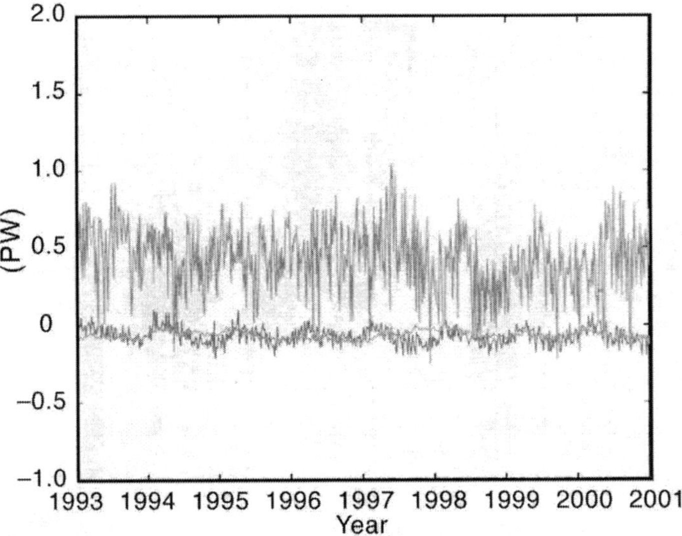

Figure 7.7 Time dependence in components of the meridional flux of heat across 21° S, in this case from eight years of analysis. The different colors correspond to different spatial integrals of the heat budget. (Source: Stammer *et al.*, 2003, Fig. 14)

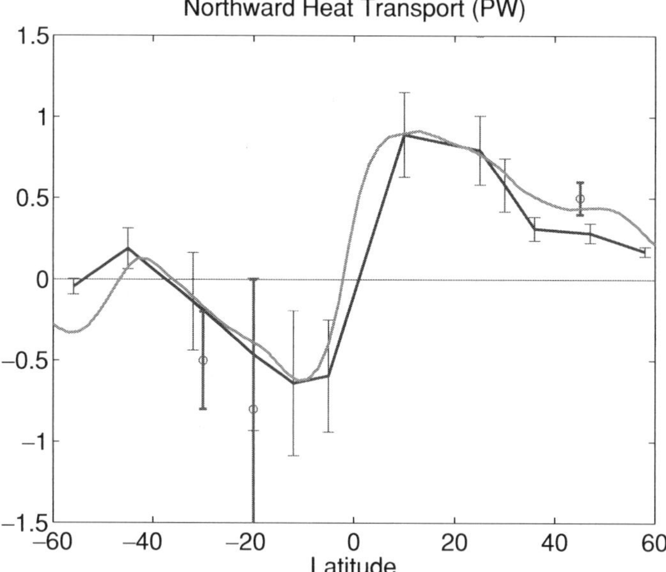

Figure 7.8 The blue curve shows the time and zonally integrated global meridional flux of heat (enthalpy) in the constrained model. The green curve shows the same field, estimated instead by integrating the fluxes through the surface. The two curves differ because heat entering from the surface can be stored rather than necessarily simply transported. Red bars are from the calculation of Ganachaud and Wunsch (2000) described in Chapter 6. Blue bars are the temporal standard deviation of the model heat flux and are an incomplete rendering of the uncertainty of the model result. (Source: Stammer *et al.*, 2003)

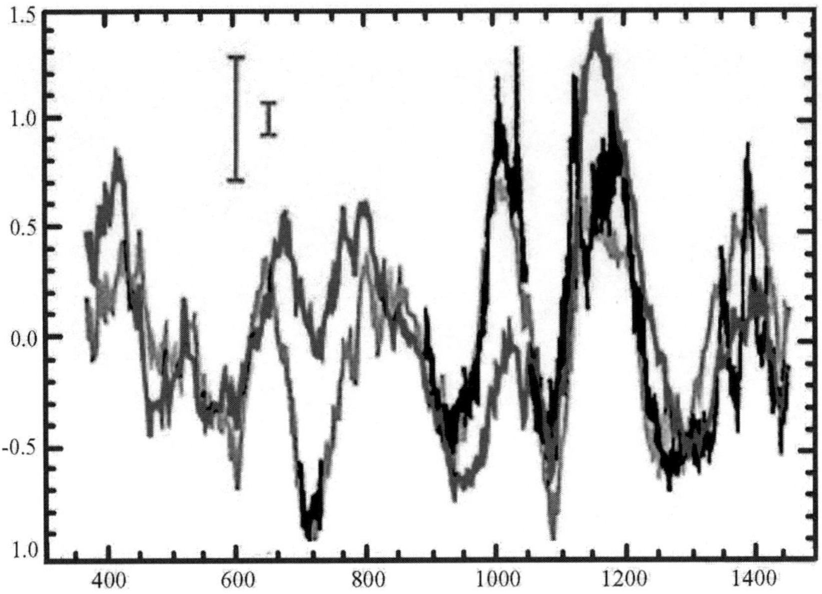

Figure 7.10 Comparison of temperature at 200 m depth at 8° N, 180° E of data (black), unconstrained model (red) and model constrained using altimetric data (blue). See Fukumori *et al.* (1999) for details and further comparisons. An approximate filter/smoother combination was used. (Source: Fukumori *et al.*, 1999)

and $t \to n\Delta t$. *One has*

$$\mathbf{x}(t) = \mathbf{A}\mathbf{x}(t - \Delta t),$$

where

$$\mathbf{A} = \begin{Bmatrix} 2 & -1 \\ 1 & 0 \end{Bmatrix}, \tag{4.16}$$

which is of the standard form (4.13), with $\mathbf{B} = \mathbf{0}$. *Let* $\mathbf{x}(0) = [1, 0]^{\mathrm{T}}$. *Then* $\mathbf{A}\mathbf{x}(0) = \mathbf{x}(1) = [2, 1]^{\mathrm{T}}$, $\mathbf{x}(2) = [3, 2]^{\mathrm{T}}$, *etc., and the slope and intercept are both* 1. $\mathbf{x}(0) = [\xi(0), \xi(-1)]^{\mathrm{T}}$ *involves an element preceding* $t = 0$.

Example *The mass–spring oscillator satisfies the differential equation*

$$m\frac{\mathrm{d}^2\xi(t)}{\mathrm{d}t^2} + r\frac{\mathrm{d}\xi(t)}{\mathrm{d}t} + k\xi(t) = q(t),$$

where r *is a damping constant. A one-sided time discretization produces*

$$m(\xi(t + \Delta t) - 2\xi(t) + \xi(t - \Delta t)) + r\Delta t(\xi(t) - \xi(t - \Delta t)) + k(\Delta t)^2\xi(t)$$
$$= q(t)(\Delta t)^2,$$

or

$$\xi(t) = \left(2 - \frac{r\Delta t}{m} - \frac{k(\Delta t)^2}{m}\right)\xi(t - \Delta t)$$
$$+ \left(\frac{r\Delta t}{m} - 1\right)\xi(t - 2\Delta t) + (\Delta t)^2\frac{q(t - \Delta t)}{m}, \tag{4.17}$$

which is

$$\begin{bmatrix} \xi(t) \\ \xi(t - \Delta t) \end{bmatrix} = \begin{Bmatrix} 2 - \frac{r}{m}\Delta t - \frac{k}{m}(\Delta t)^2 & \frac{r\Delta t}{m} - 1 \\ 1 & 0 \end{Bmatrix} \begin{bmatrix} \xi(t - \Delta t) \\ \xi(t - 2\Delta t) \end{bmatrix}$$
$$+ \begin{bmatrix} (\Delta t)^2\frac{q(t - \Delta t)}{m} \\ 0 \end{bmatrix}, \tag{4.18}$$

and is the canonical form with $\mathbf{A}$ *independent of time, where*

$$\mathbf{x}(t) = [\xi(t) \quad \xi(t - \Delta t)]^{\mathrm{T}}, \qquad \mathbf{B}(t)\mathbf{q}(t) = [(\Delta t)^2\, q(t)/m \quad 0]^{\mathrm{T}}.$$

Example *A difference equation important in time-series analysis*[4] *is*

$$\xi(t) + a_1\xi(t - 1) + a_2\xi(t - 2) + \cdots + a_N\xi(t - N) = \eta(t), \tag{4.19}$$

where $\eta(t)$ is a zero-mean, white-noise process (Eq. (4.19) is an example of an autoregressive process (AR)). To put this into the canonical form, write[5]

$$x_1(t) = \xi(t - N),$$
$$x_2(t) = \xi(t - N + 1),$$
$$\vdots$$
$$x_N(t) = \xi(t - 1),$$
$$x_N(t + 1) = -a_1 x_N(t) - a_2 x_{N-1}(t) \cdots - a_N x_1(t) + \eta(t).$$

It follows that $x_1(t + 1) = x_2(t)$, etc., or

$$\mathbf{x}(t) = \left\{ \begin{array}{cccccc} 0 & 1 & 0 & \cdots & 0 & 0 \\ 0 & 0 & 1 & \cdots & 0 & 0 \\ \cdot & \cdot & \cdot & \cdots & \cdot & \cdot \\ -a_N & -a_{N-1} & -a_{N-2} & \cdots & -a_2 & -a_1 \end{array} \right\} \mathbf{x}(t - 1) + \begin{bmatrix} 0 \\ 0 \\ \cdot \\ 1 \end{bmatrix} \eta(t - 1). \quad (4.20)$$

$\mathbf{A}$ is known as a "companion" matrix. Here, $\mathbf{B}(t) = [0 \quad 0 \quad \cdot \quad 1]^{\mathrm{T}}$, $\mathbf{q}(t) = \eta(t)$.

Given that most time-dependent models can be written as in (4.12) or (4.13), the forward-model solution involves marching forward from known initial conditions at $t = 0$, subject to specified boundary values. So, for example, the linear model (4.13), with given initial conditions $\mathbf{x}(0) = \mathbf{x}_0$, involves the sequence

$$\mathbf{x}(1) = \mathbf{A}(0)\,\mathbf{x}_0 + \mathbf{B}(0)\,\mathbf{q}(0),$$
$$\mathbf{x}(2) = \mathbf{A}(1)\,\mathbf{x}(1) + \mathbf{B}(1)\,\mathbf{q}(1),$$
$$= \mathbf{A}(1)\,\mathbf{A}(0)\,\mathbf{x}_0 + \mathbf{A}(1)\,\mathbf{B}(0)\,\mathbf{q}(0) + \mathbf{B}(1)\,\mathbf{q}(1),$$
$$\vdots$$
$$\mathbf{x}(t_f) = \mathbf{A}(t_f - 1)\,\mathbf{x}(t_f - 1) + \mathbf{B}(t_f - 1)\,\mathbf{q}(t_f - 1)$$
$$= \mathbf{A}(t_f - 1)\,\mathbf{A}(t_f - 2) \cdots \mathbf{A}(0)\,\mathbf{x}_0 + \cdots .$$

Most of the basic ideas can be understood in the notationally simplest case of time-independent $\mathbf{A}$, $\mathbf{B}$, and that is usually the situation we will address with little loss of generality, so that $\mathbf{A}(t)\,\mathbf{A}(t - 1) = \mathbf{A}^2$, etc. Figure 4.1 depicts the time history for the mass–spring oscillator, with the parameter choice $\Delta t = 1$, $k = 0.1$, $m = 1$, $r = 0$, so that

$$\mathbf{A} = \left\{ \begin{array}{cc} 1.9 & -1 \\ 1 & 0 \end{array} \right\}, \qquad \mathbf{B}\mathbf{q}(t) = \begin{bmatrix} 1 \\ 0 \end{bmatrix} u(t),$$

where $\langle u(t)^2 \rangle = 1$, a random variable. The initial conditions were $\mathbf{x}(0) = [\xi(0)\ \xi(-1)]^{\mathrm{T}}$.

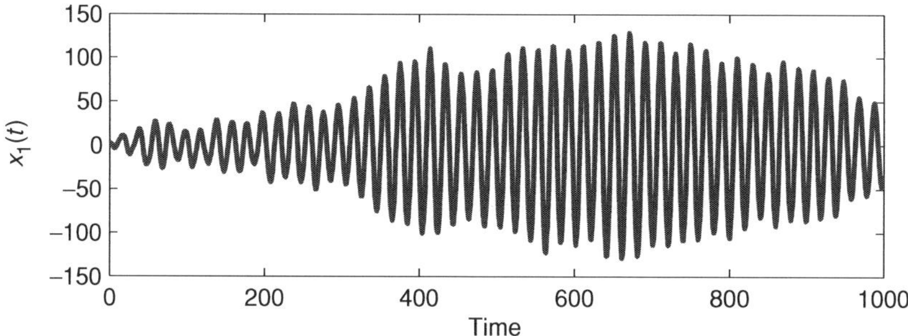

Figure 4.1 Time history of $x_1(t)$ for the linear oscillator with $\Delta t = 1, k = 0.1, m = 1, r = 0$ driven by a random sequence of zero mean and unit variance. Note the buildup in amplitude from the accumulating uncorrelated forcing increments.

It is important to recognize that this time-stepping procedure cannot be used if some of the elements of the initial conditions, $\mathbf{x}(0)$, are replaced, e.g., with elements of $\mathbf{x}(t_f)$, or more generally with elements of $\mathbf{x}(t)$ for arbitrary t. That is, the amount of information may be the same, and fully adequate, but not useful in straightforward time-stepping. Many of the algorithms developed here are directed at these less-conventional cases.

$\mathbf{A}$ is necessarily square. It is also often true that $\mathbf{A}^{-1}$ exists, and, if not, a generalized inverse can be used. If $\mathbf{A}^{-1}$ can be computed, one can contemplate the possibility (important later) of running a model backward in time, for example as

$$\mathbf{x}(t-1) = \mathbf{A}^{-1}\mathbf{x}(t) - \mathbf{A}^{-1}\mathbf{B}(t-1)\mathbf{q}(t-1).$$

Such a computation may be inaccurate if carried on for long times, but the same may well be true of the forward model.

Some attention must be paid to the structure of $\mathbf{B}(t)\mathbf{q}(t)$. The partitioning into these elements is not unique and can be done to suit one's convenience. The dimension of $\mathbf{B}$ is that of the size of the state vector by the dimension of $\mathbf{q}$, which typically would reflect the number of independent degrees of freedom in the forcing/boundary conditions. ("Forcing" is hereafter used to include boundary conditions, sources and sinks, and anything normally prescribed externally to the model.) Consider the model grid points displayed in Fig. 4.2. Suppose that the boundary grid points are numbered 1–5, 6, 10, 46–50, and all others are interior. If there are no interior forces, and all boundary values have a time history $q(t)$, then we could take

$$\mathbf{B} = [1 \quad 1 \quad 1 \quad 1 \quad 1 \quad 1 \quad 0 \quad 0 \quad 0 \cdots 1 \quad 1]^{\mathrm{T}}, \tag{4.21}$$

where the ones occur at the boundary points, and the zeros at the interior ones.

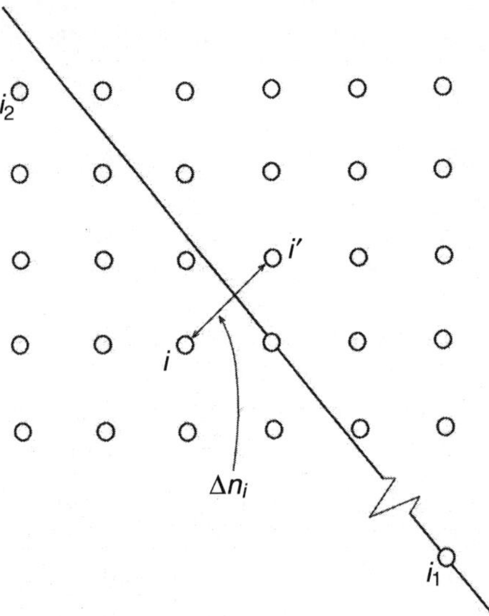

Figure 4.2 (a) Simple numerical grid for use of discrete form of model; $\times$ denote boundary grid points, and o are interior ones. Numbering is sequential down the columns, as shown. (b) Tomographic integral is assumed given between i_1, i_2, and the model values at the grid points would be used to calculate its predicted value.

Suppose, instead, that boundary grid point 2 has values $q_1(t)$, interior point 7 has a forcing history $q_2(t)$, and all others are unforced, then

$$\mathbf{B}\mathbf{q}(t) = \begin{Bmatrix} 0 & 1 & 0 & 0 & 0 & 0 & 0 & \cdot & 0 \\ 0 & 0 & 0 & 0 & 0 & 0 & 1 & \cdot & 0 \end{Bmatrix}^{\mathrm{T}} \begin{bmatrix} q_1(t) \\ q_2(t) \end{bmatrix}. \tag{4.22}$$

A time-dependent $\mathbf{B}$ would correspond to time-evolving positions at which forces were prescribed – a somewhat unusual situation. It would be useful, for example, if one were driving a fluid model with a heat flux or stress in the presence of a prescribed moving ice cover. One could also impose initial conditions using a time-dependent $\mathbf{B}(t)$, which would vanish after $t = 0$.

As with steady models, care is needed in understanding the propagation of errors in time and space. If we have some knowledge of the initial oceanic state, $\tilde{\mathbf{x}}(0)$, and are doing an experiment at a later time t, the prior information – the estimated initial conditions – carries information in addition to what is currently being measured. We seek to combine the two sets of information. How does information propagate forward in time? Formally, the rule (4.12) tells us exactly what to do. But because there are always errors in $\tilde{\mathbf{x}}(0)$, we need to be careful about assuming that a model computation of $\tilde{\mathbf{x}}(t)$ is useful. Depending upon the details of the model, the behavior of the errors through time can be distinguished: (1) The model has decaying components. If the amplitudes of these components are partially erroneous, then for large enough t, these elements will have diminished, perhaps to the point where they are negligible. (2) The model has neutral components. At time t, the erroneous elements have amplitudes the same as they were at $t = 0$. (3) The model has unstable components; at time t any erroneous parts may have grown to swamp everything else.

Realistic models, particularly fluid ones, can contain all three types of behavior simultaneously. It thus becomes necessary to determine which of the elements of the forecast $\tilde{\mathbf{x}}(t)$ can be used to help estimate the system state by combination with new data, and which elements should be suppressed as partially or completely erroneous. Simply assuming that all components are equally accurate can be a disastrous recipe.

Before proceeding, we reiterate the point that time need not be accorded a privileged position. Form the inclusive state vector, $\mathbf{x}_B$, defined in Eq. (4.7). Then, as in Eq. (4.6), models of the form (4.13) can be written in the "whole-domain" form,

$$\mathbf{A}_B \mathbf{x}_B = \mathbf{d}_B$$

$$
\mathbf{A}_B = \left\{
\begin{array}{cccccccc}
-\mathbf{A} & \mathbf{I} & \mathbf{0} & \cdot & \cdot & \mathbf{0} & \mathbf{0} \\
\mathbf{0} & -\mathbf{A} & \mathbf{I} & \mathbf{0} & \cdot & \mathbf{0} & \mathbf{0} \\
\cdot & \cdot & \cdot & \cdot & \cdot & -\mathbf{A} & \mathbf{I}
\end{array}
\right\}, \quad
\mathbf{d}_B = \begin{bmatrix} \mathbf{Bq}(0) \\ \mathbf{Bq}(1) \\ \vdots \end{bmatrix}, \tag{4.23}
$$

plus initial conditions, which is no different, except for its possibly enormous size, from that of a static system and can be handled by any of the methods of earlier chapters if the computational capacity is sufficient. If time-stepping is impossible because the initial condition is replaced by knowledge of $\mathbf{x}(t')$, $t' \neq 0$, the whole-domain form may be very attractive. Note the block-banded, sparse, nature of $\mathbf{A}_B$.

4.2.2 How to find the matrix $\mathbf{A}(t)$

Most modern large-scale time-evolving models, even if completely linear, are written in computer code, typically in languages such as Fortran90 or C/C++. The state transition matrix is not normally explicitly constructed; instead, the individual elements of $x_i(t)$ are time-stepped to produce the $x_j(t+1)$, $\Delta t = 1$, usually using various vectorizations. $\mathbf{A}(t)$ is often neither required nor constructed, as all one cares about is the result of its operation on $\mathbf{x}(t)$, as generated from the model code. If one requires an explicit $\mathbf{A}(t)$ but has only the forward code, several methods can be used. For simplicity let $\mathbf{Bq}(t) = \mathbf{0}$ (the more general approach is obvious).

(1) Solve Eq. (4.13) N times, starting at time $t = 0$, subject to $\mathbf{x}^{(i)}(0) =$ column i of $\mathbf{I}$ – that is, the model is stepped forward for N different initial conditions corresponding to the N different problems of unit initial condition at a single grid or boundary point, with zero initial conditions everywhere else. Let each column of $\mathbf{G}(t, 0)$ correspond to the appropriate value of $\mathbf{x}(t)$ – that is,

$$\mathbf{G}(0, 0) = \mathbf{I},$$
$$\mathbf{G}(1, 0) = \mathbf{A}(0)\mathbf{G}(0, 0),$$
$$\mathbf{G}(2, 0) = \mathbf{A}(1)\mathbf{G}(1, 0) = \mathbf{A}(1)\mathbf{A}(0),$$
$$\vdots$$
$$\mathbf{G}(t, 0) = \mathbf{A}(t - 1)\mathbf{A}(t - 2) \cdots \mathbf{A}(0).$$

We refer to $\mathbf{G}(t, 0)$ as a *unit solution*; it is closely related to the Green function discussed in Chapter 2. The solution for arbitrary initial conditions is then

$$\mathbf{x}(t) = \mathbf{G}(t, 0)\mathbf{x}(0), \tag{4.24}$$

and the modification for $\mathbf{Bq} \neq 0$ is straightforward. $\mathbf{A}(t)$ can be readily reconstructed from $\mathbf{G}(t, 0)$, most simply if $\mathbf{A}$ is time-independent and if one time-step is numerically accurate enough to represent $\mathbf{G}$. Otherwise, multiple time steps can be used until a sufficiently large change in $\mathbf{G}$ is produced.

Several other methods exist to obtain $\mathbf{A}$ from an existing computer model, but consider now only the case of a steady model, with no time-dependence in the governing matrices $(\mathbf{A}, \mathbf{B})$. We continue to simplify by setting $\mathbf{B} = \mathbf{0}$.

(2) Define N independent initial condition vectors $\mathbf{x}_0^{(i)}$, $i = 1, 2, \ldots, N$, and form a matrix,

$$\mathbf{X}_0 = \left\{ \mathbf{x}_0^{(i)} \right\}.$$

Time-step the model once, equivalent to

$$\mathbf{X}_1 = \mathbf{A}\mathbf{X}_0,$$

and invert $\mathbf{X}_0$:

$$\mathbf{A} = \mathbf{X}_1\mathbf{X}_0^{-1}. \tag{4.25}$$

The inverse will exist by the assumption of independence (a basis) in the initial condition vectors. One must again run the model N times in this approach. If $\mathbf{x}_0^{(i)} = \delta_{ij}$, $\mathbf{X}_1 = \mathbf{G}$.

Again, the changes from $\mathbf{X}_0$ to $\mathbf{X}_1$ may be too small for adequate numerical accuracy, and one might use multiple time-steps, computing, for time-independent $\mathbf{A}$,

$$\mathbf{X}_n = \mathbf{A}^n \mathbf{X}_0,$$

which would determine $\mathbf{A}^n$, and $\mathbf{A}$ itself can be found by one of the matrix root algorithms,[6] or by redefining unit time to $n\Delta t$. One might also define $\mathbf{A}^n$ as the average value of $\mathbf{A}(n)\mathbf{A}(n-1)\cdots\mathbf{A}(0)$.

(3) Suppose the statistics of the solutions are known, e.g.,

$$\mathbf{R}(0) = \langle \mathbf{x}(t)\mathbf{x}(t)^{\mathrm{T}} \rangle, \quad \mathbf{R}(1) = \langle \mathbf{x}(t+1)\mathbf{x}(t)^{\mathrm{T}} \rangle,$$

perhaps because the model has been run many times from different initial conditions – making it possible to estimate these from stored output. Then we note that

$$\langle \mathbf{x}(t+1)\mathbf{x}(t)^{\mathrm{T}} \rangle = \mathbf{A}\langle \mathbf{x}(t)\mathbf{x}(t)^{\mathrm{T}} \rangle,$$

or

$$\mathbf{R}(1) = \mathbf{A}\mathbf{R}(0),$$

and

$$\mathbf{A} = \mathbf{R}(1)\mathbf{R}(0)^{-1}. \tag{4.26}$$

That is to say, knowledge of these covariances is equivalent to knowledge of the model itself (and vice versa).[7] Multiple time steps can again be used if necessary to infer that $\mathbf{A}^n = \mathbf{R}(n)\mathbf{R}(0)^{-1}$. By writing $\langle \mathbf{x}(t+1)\mathbf{x}(t)^{\mathrm{T}} \rangle = \mathbf{R}(1)$, etc., stationarity is implied. More generally, one may have $\langle \mathbf{x}(t+1)\mathbf{x}(t)^{\mathrm{T}} \rangle = \mathbf{R}(t, 1)$. $\mathbf{R}(0)$ must be non-singular.

Note that determination of $\mathbf{B}$ can be done analogously – using a spanning set of $\mathbf{q}^{(i)}$ as initial conditions, setting $\mathbf{x}(0) = 0$.

(4) Automatic/algorithmic differentiation (AD) tools exist[8] that can take computer code (e.g., Fortran, C, Matlab®) for the forward model, and produce by analysis of the code, equivalent computer code (e.g., Fortran) for construction of $\mathbf{A}$. Some codes preferentially produce $\mathbf{A}^{\mathrm{T}}$, but transposition then can be employed. An example is provided in the Appendix to this chapter.

If the model is fully time-dependent, then $\mathbf{A}(t)$ has to be deduced at each time-step, as above. For some purposes, one might seek temporal averages, defining $\bar{\mathbf{A}}$ as

$$\bar{\mathbf{A}}^n = \mathbf{A}(n-1)\mathbf{A}(n-2)\cdots\mathbf{A}(1)\mathbf{A}(0).$$

Reintroduction of $\mathbf{B}$ is easily accommodated.

4.2.3 Observations and data

Here, observations are introduced into the modeling discussion so that they stand on an equal footing with the set of model equations (4.12) or (4.13). Observations will be represented as a set of linear simultaneous equations at time $t = n\Delta t$,

$$\mathbf{E}(t)\mathbf{x}(t) + \mathbf{n}(t) = \mathbf{y}(t), \tag{4.27}$$

which is a straightforward generalization of the previous static systems where t did not appear explicitly; here, $\mathbf{E}$ is sometimes called the "design" or "observation" matrix. The notation used in Chapter 2 to discuss recursive estimation was chosen deliberately to be the same as used here.

The requirement that the observations be linear combinations of the state-vector elements can be relaxed if necessary, but most common observations are of that form. An obvious exception would be the situation in which the state vector included fluid velocity components, $u(t)$, $v(t)$, but an instrument measuring speed, $\sqrt{(u(t)^2 + v(t)^2)}$, would produce a non-linear relation between $y_i(t)$ and the state vector. Such systems are usually handled by some form of linearization.[9]

To be specific, the noise $\mathbf{n}(t)$ is supposed to have zero mean and known second-moment matrix

$$\langle \mathbf{n}(t)\rangle = 0, \qquad \langle \mathbf{n}(t)\mathbf{n}(t)^{\mathrm{T}}\rangle = \mathbf{R}(t). \tag{4.28}$$

But

$$\langle \mathbf{n}(t)\mathbf{n}(t')^{\mathrm{T}}\rangle = \mathbf{0}, \quad t \neq t'. \tag{4.29}$$

That is, the observational noise should not be correlated from one measurement time to another; there is a considerable literature on how to proceed when this crucial assumption fails (called the "colored-noise" problem[10]). Unless specifically stated otherwise, we will assume that Eq. (4.29) is valid.

The matrix $\mathbf{E}(t)$ can accommodate almost any form of linear measurement. If, at some time, there are no measurements, then $\mathbf{E}(t)$ vanishes, along with $\mathbf{R}(t)$. If a single element $x_i(t)$ is measured, then $\mathbf{E}(t)$ is a row vector that is zero everywhere except in column i, where it is 1. It is particularly important to recognize that many measurements are weighted averages of the state-vector elements. Some measurements – for example, tomographic ones[11] as described in Chapter 1 – are explicitly spatial averages (integrals) obtained by measuring some property along a ray traveling between two points (see Fig. 4.2). Any such data representing spatially filtered versions of the state vector can be written as

$$y(t) = \sum \alpha_j x_j(t), \tag{4.30}$$

where the α_j are the averaging weights.

Point observations often occur at positions not coincident with model grid positions (although many models, e.g., spectral ones, do not use grids). Then (4.27) is an interpolation rule, possibly either very simple or conceivably a full-objective mapping calculation, of the value of the state vector at the measurement point. Often the number of model grid points vastly exceeds the number of the data grid points; thus, it is convenient that the formulation (4.27) requires interpolation from the dense model grid to the sparse data positions (see Fig. 4.2). (In the unusual situation where the data density is greater than the model grid density, one can restructure the problem so the interpolation goes the other way.) More complex filtered measurements exist. In particular, one may have measurements of a state vector only in specific wavenumber bands; but such "band-passed" observations are automatically in the form (4.27).

As with the model (4.23), the observations of the combined state vector can be concatenated into a single observational set,

$$\mathbf{E}_B \mathbf{x}_B + \mathbf{n}_B = \mathbf{y}_B, \tag{4.31}$$

where

$$\mathbf{E}_B = \begin{Bmatrix} \mathbf{I} & 0 & 0 & \cdot & 0 \\ 0 & \mathbf{E}(1) & 0 & \cdot & 0 \\ 0 & 0 & \mathbf{E}(2) & \cdot & \cdot \\ \cdot & \cdot & \cdot & 0 & \mathbf{E}(t_f) \end{Bmatrix}, \quad \mathbf{n}_B = \begin{bmatrix} \mathbf{n}(0) \\ \mathbf{n}(1) \\ \vdots \\ \mathbf{n}(t_f) \end{bmatrix}, \quad \mathbf{y}_B = \begin{bmatrix} \tilde{\mathbf{x}}(0) \\ \mathbf{y}(1) \\ \vdots \\ \mathbf{y}(t_f) \end{bmatrix}.$$

Here the initial conditions have been combined with the observations. $\mathbf{E}_B$ is block-banded and sparse. If the size is no problem, the combined and concatenated model and observations could be dealt with using any of the methods of Chapter 2. The rest of this chapter can be thought of as an attempt to produce from the model/data combination the same type of estimates as were found useful in Chapter 2, but exploiting the special structure of matrices $\mathbf{A}_B$ and $\mathbf{E}_B$ so as to avoid having to store them all at once in the computer.

As one example of how the combined model and observation equations can be used together, consider the situation in which only the initial conditions $\mathbf{x}(0)$ are unknown. The unit solution formulation of p. 188 leads to a particularly simple reduced form. One has immediately,

$$\mathbf{y}(t) = \mathbf{E}(t)\mathbf{G}(t, 0)\mathbf{x}(0) + \mathbf{n}(t), \quad t = 1, 2, \ldots, t_f, \tag{4.32}$$

which is readily solved in whole-domain form for $\mathbf{x}(0)$. If only a subset of the $\mathbf{x}(0)$ are thought to be non-zero, then the columns of $\mathbf{G}$ need to be computed only for those elements.[12]

4.3 Estimation

4.3.1 Model and data consistency

In many scientific fields, the central issue is to develop understanding from the combination of a skillful model with measurements. The model is intended to encompass all one's theoretical knowledge about how the system behaves, and the data are the complete observational knowledge of the same system. If done properly, and in the absence of contradictions between theory and experiment, inferences from the model/data combination should be no worse, and may well be very much better than those made from either alone. It is the latter possibility that motivates the development of state estimation procedures. "Best-estimates" made by combining models with observations are often used to forecast a system (e.g., to land an airplane), but this is by no means the major application.

Such model/data problems are ones of statistical inference with a host of specific subtypes. Some powerful techniques are available, but like any powerful tools (a chain saw, for example), they can be dangerous to the user! In general, one is confronted with a two-stage problem. Stage 1 involves developing a suitable model that is likely to be consistent with the data. "Consistent" means that, within the estimated data errors, the model is likely to be able to describe the features of interest. Obtaining the data errors is itself a serious modeling problem. Stage 2 produces the actual estimate with its error estimates.

One can go very badly wrong at stage 1, before any computation takes place. If elastic wave propagation is modeled using the equations of fluid dynamics, estimation methods will commonly produce some kind of "answer," but one that would be nonsensical. Model failure can, of course, be much more subtle, in which some omitted, supposed secondary element (e.g., a time-dependence) proves to be critical to a description of the data. Good technique alerts users to the presence of such failures, along with clues as to what should be changed in the model. But these issues do not, however, apply only to the model. The assertion that a particular data set carries a signal of a particular kind can prove to be false in a large number of ways. A temperature signal thought to represent the seasonal cycle might prove, on careful examination, to be dominated by higher or lower frequency structures, and thus its use with an excellent model of annual variation might prove disastrous. Whether this situation is to be regarded as a data or as a model issue is evidently somewhat arbitrary.

Thus stage 1 of any estimation problem has to involve understanding of whether the data and the model are physically and statistically consistent. If they are not, one should stop and reconsider. Often where they are believed to be generally consistent up to certain quantitative adjustments, one can combine the two stages. A model may have adjustable parameters (turbulent mixing coefficients, boundary condition errors, etc.) that could bring the model and data into consistency, in which case the estimation procedure becomes, in part, an attempt to find those parameters in addition to the state. Alternatively, the data error covariance, $\mathbf{R}(t)$, may be regarded as incompletely known, and one might seek, as part of the state estimation procedure, to improve one's estimate of it. (Problems like this one fall under the subject of "adaptive filtering.")

Assuming for now that the model and data are likely to prove consistent, one can address what might be thought of as a form of interpolation: given a set of observations in space and time as described by Eq. (4.27), use the dynamics as described by the model (4.12) or (4.13) to estimate various state-vector elements at various times of interest. Yet another, less familiar, problem recognizes that some of the forcing terms $\mathbf{B}(t-1)\mathbf{q}(t-1)$ are partially or wholly unknown (e.g., the wind stress boundary conditions over the ocean are imperfectly known), and one might seek to estimate them from whatever ocean observations are available and from the known model dynamics.

The forcing terms – representing boundary conditions as well as interior sources/sinks and forces – almost always need to be divided into two parts: the known and the unknown. The latter will often be perturbations about the known values. Thus, rewrite (4.13) in the modified form

$$\mathbf{x}(t) = \mathbf{A}(t-1)\mathbf{x}(t-1) + \mathbf{B}(t-1)\mathbf{q}(t-1) + \mathbf{\Gamma}(t-1)\mathbf{u}(t-1), \quad \Delta t = 1,$$

$$(4.33)$$

where now $\mathbf{B}(t)\mathbf{q}(t)$ represent the known forcing terms and $\mathbf{\Gamma}(t)\mathbf{u}(t)$ the unknown ones, which we will generally refer to as the "controls," or "control terms." $\mathbf{\Gamma}(t)$ is known and plays the same role for $\mathbf{u}(t)$ as does $\mathbf{B}(t)$ for $\mathbf{q}(t)$. Usually $\mathbf{B}(t)$, $\mathbf{\Gamma}(t)$ will be treated as time independent, but this simplification is not necessary. Almost always, we can make some estimate of the size of the control terms, as, for example,

$$\langle \mathbf{u}(t) \rangle = \mathbf{0}, \qquad \langle \mathbf{u}(t)\mathbf{u}(t)^{\mathrm{T}} \rangle = \mathbf{Q}(t). \tag{4.34}$$

The controls have a second, somewhat different, role: they can also represent the model error. All models are inaccurate to a degree – approximations are always made to the equations describing any particular physical situation. One can expect that the person who constructed the model has some idea of the size and structure of the physics or chemistry, or biology, etc. that have been omitted or distorted in the model construction. In this context, $\mathbf{Q}(t)$ represents the covariance of the model error, and the control terms represent the missing physics. The assumption $\langle \mathbf{u}(t) \rangle = \mathbf{0}$ must be critically examined in this case, and, in the event of failure, some modification of the model must be made or the control variance artificially modified to include what is a model bias error. But the most serious problem is that models are rarely produced with *any* quantitative description of their accuracy beyond one or two examples of comparison with known solutions. One is left to determine $\mathbf{Q}(t)$ by guesswork. Getting beyond such guesses is again a problem of adaptive estimation.

Collecting the standard equations of model and data:

$$\mathbf{x}(t) = \mathbf{A}(t-1)\mathbf{x}(t-1) + \mathbf{Bq}(t-1) + \mathbf{\Gamma u}(t-1), \quad t = 1, 2, \ldots, t_f, \quad (4.35)$$

$$\mathbf{E}(t)\mathbf{x}(t) + \mathbf{n}(t) = \mathbf{y}(t), \quad t = 1, 2, \ldots, t_f, \quad \Delta t = 1, \quad (4.36)$$

$$\mathbf{n}(t) = \mathbf{0}, \quad \langle \mathbf{n}(t)\mathbf{n}(t)^{T} \rangle = \mathbf{R}(t), \quad \langle \mathbf{n}(t)\mathbf{n}(t')^{T} \rangle = \mathbf{0}, \, t \neq t', \quad (4.37)$$

$$\langle \mathbf{u}(t) \rangle = \mathbf{0}, \quad \langle \mathbf{u}(t)\mathbf{u}(t)^{T} \rangle = \mathbf{Q}(t), \quad (4.38)$$

$$\tilde{\mathbf{x}}(0) = \mathbf{x}_0, \quad \langle (\tilde{\mathbf{x}}(0) - \mathbf{x}(0))(\tilde{\mathbf{x}}(0) - \mathbf{x}(0))^{T} \rangle = \mathbf{P}(0), \quad (4.39)$$

where t_f defines the endpoint of the interval of interest. The last equation, (4.39), treats the initial conditions of the model as a special case – the uncertain initialization problem, where $\mathbf{x}(0)$ is the true initial condition and $\tilde{\mathbf{x}}(0) = \mathbf{x}_0$ is the value actually used but with uncertainty $\mathbf{P}(0)$. Alternatively, one could write

$$\mathbf{E}(0)\mathbf{x}(0) + \mathbf{n}(0) = \mathbf{x}_0, \quad \mathbf{E}(0) = \mathbf{I}, \quad \langle \mathbf{n}(0)\mathbf{n}(0)^{T} \rangle = \mathbf{P}(0), \quad (4.40)$$

and include the initial conditions as a special case of the observations – recognizing explicitly that initial conditions are often obtained that way. (Compare Eq. (4.31).)

This general form permits one to grapple with reality. In the spirit of ordinary least-squares and its intimate cousin, minimum-error variance estimation, consider the general problem of finding state vectors and controls, $\mathbf{u}(t)$, that minimize an objective function,

$$J = [\mathbf{x}(0) - \mathbf{x}_0]^{T}\mathbf{P}(0)^{-1}[\mathbf{x}(0) - \mathbf{x}_0]$$
$$+ \sum_{t=1}^{t_f}[\mathbf{E}(t)\mathbf{x}(t) - \mathbf{y}(t)]^{T}\mathbf{R}(t)^{-1}[\mathbf{E}(t)\mathbf{x}(t) - \mathbf{y}(t)] \quad (4.41)$$
$$+ \sum_{t=0}^{t_f-1}\mathbf{u}(t)^{T}\mathbf{Q}(t)^{-1}\mathbf{u}(t),$$

subject to the model, Eqs. (4.35), (4.38) and (4.39). As written here, this choice of an objective function is somewhat arbitrary but perhaps reasonable as the direct analogue to those used in Chapter 2. It seeks a state vector $\mathbf{x}(t)$, $t = 0, 1, \ldots, t_f$, and a set of controls, $\mathbf{u}(t)$, $t = 0, 1, \ldots, t_f - 1$, that satisfy the model and that agree with the observations to an extent determined by the weight matrices $\mathbf{R}(t)$ and $\mathbf{Q}(t)$, respectively. From the previous discussions of least-squares and minimum-error variance estimation, the minimum-square requirement, Eq. (4.41), will produce a solution identical to that derived from minimum variance estimation by the specific choice of the weight matrices as the corresponding prior uncertainties, $\mathbf{R}(t)$, $\mathbf{Q}(t)$, $\mathbf{P}(0)$. In a Gaussian system, it also proves to be the maximum likelihood estimate. The introduction of the controls, $\mathbf{u}(t)$, into the objective function represents an acknowledgment that arbitrarily large controls (forces) would not usually be an acceptable solution; they should be consistent with $\mathbf{Q}(t)$.

Note on notation   As in Chapter 2, any values of $\mathbf{x}(t)$, $\mathbf{u}(t)$ minimizing J will be written $\tilde{\mathbf{x}}(t)$, $\tilde{\mathbf{u}}(t)$ and these symbols sometimes will be substituted into Eq. (4.41) if it helps their clarity.

Much of the rest of this chapter is directed at solving the problem of finding the minimum of J subject to the solution satisfying the model. Notice that J involves the state vector, the controls, and the observations over the entire time period under consideration, $t = 0, 1, \ldots, t_f$. This type of objective function is the one usually of most interest to scientists attempting to understand their system – in which data are stored and employed over a finite time. In some other applications, most notably forecasting, which is taken up immediately below, one has only the past measurements available; this situation proves to be a special case of the more general one.

Although we will not keep repeating the warning each time an objective function such as Eq. (4.41) is encountered, the reader is reminded of a general message from Chapter 2: *The assumption that the model and observations are consistent and that the minimum of the objective function produces a meaningful and useful estimate must always be tested after the fact.* That is, at the minimum of J, $\tilde{\mathbf{u}}(t)$ must prove consistent with $\mathbf{Q}(t)$, and $\tilde{\mathbf{x}}(t)$ must produce residuals consistent with $\mathbf{R}(t)$. Failure of these and other posterior tests should lead to rejection of the model. As always, one can thus reject a model (which includes $\mathbf{Q}(t)$, $\mathbf{R}(t)$) on the basis of a failed consistency with observations. But a model is never "correct" or "valid," merely "consistent." (See Note 8, Chapter 1.)

4.3.2 The Kalman filter

We begin with a special case. Suppose that by some means, at time $t = 0$, $\Delta t = 1$, we have an unbiassed estimate, $\tilde{\mathbf{x}}(0)$, of the state vector with uncertainty $\mathbf{P}(0)$. At time $t = 1$, observations from Eq. (4.36) are available. How would the information available best be used to estimate $\mathbf{x}(1)$?

The model permits a forecast of what $\mathbf{x}(1)$ should be, were $\tilde{\mathbf{x}}(0)$ known perfectly,

$$\tilde{\mathbf{x}}(1, -) = \mathbf{A}\tilde{\mathbf{x}}(0) + \mathbf{B}\mathbf{q}(0), \tag{4.42}$$

where the unknown control terms have been replaced by the best estimate we can make of them – their mean, which is zero, and $\mathbf{A}$ has been assumed to be time independent. A minus sign has been introduced into the argument of $\tilde{\mathbf{x}}(1, -)$ to show that *no data at $t = 1$ have yet been used to make the estimate* at $t = 1$, in a notation we will generally use. How good is this forecast?

Suppose the erroneous components of $\tilde{\mathbf{x}}(0)$ are

$$\gamma(0) = \tilde{\mathbf{x}}(0) - \mathbf{x}(0), \tag{4.43}$$

then the erroneous components of the forecast are

$$\gamma(1) \equiv \tilde{\mathbf{x}}(1, -) - \mathbf{x}(1) = \mathbf{A}\tilde{\mathbf{x}}(0) + \mathbf{B}\mathbf{q}(0) - (\mathbf{A}\mathbf{x}(0) + \mathbf{B}\mathbf{q}(0) + \mathbf{\Gamma}\mathbf{u}(0))$$
$$= \mathbf{A}\gamma(0) - \mathbf{\Gamma}\mathbf{u}(0), \tag{4.44}$$

that is, composed of two distinct elements: the propagated erroneous portion of $\tilde{\mathbf{x}}(0)$, and the unknown control term. Their second moments are

$$\begin{aligned}
\langle \gamma(1)\gamma(1)^{\mathrm{T}} \rangle &= \langle (\mathbf{A}\gamma(0) - \mathbf{\Gamma}\mathbf{u}(0))(\mathbf{A}\gamma(0) - \mathbf{\Gamma}u(0))^{\mathrm{T}} \rangle \\
&= \mathbf{A}\langle \gamma(0)\mathbf{P}(0)\gamma(0)^{\mathrm{T}} \rangle \mathbf{A}^{\mathrm{T}} + \mathbf{\Gamma}\langle \mathbf{u}(0)\mathbf{u}(0)^{\mathrm{T}} \rangle \mathbf{\Gamma}^{\mathrm{T}} \\
&= \mathbf{A}\mathbf{P}(0)\mathbf{A}^{\mathrm{T}} + \mathbf{\Gamma}\mathbf{Q}(0)\mathbf{\Gamma}^{\mathrm{T}} \\
&\equiv \mathbf{P}(1, -),
\end{aligned} \tag{4.45}$$

by the definitions of $\mathbf{P}(0)$, $\mathbf{Q}(0)$ and the assumption that the unknown controls are not correlated with the error in the state estimate at $t = 0$. We now have an estimate of $\mathbf{x}(1)$ with uncertainty $\mathbf{P}(1, -)$ and a set of observations,

$$\mathbf{E}(1)\mathbf{x}(1) + \mathbf{n}(1) = \mathbf{y}(1). \tag{4.46}$$

To combine the two sets of information, use the recursive least-squares solution from Eqs. (2.434)–(2.436). By assumption, the uncertainty in $\mathbf{y}(1)$ is uncorrelated with that in $\tilde{\mathbf{x}}(1, -)$. Making the appropriate substitutions into those equations,

$$\begin{aligned}
\tilde{\mathbf{x}}(1) &= \tilde{\mathbf{x}}(1, -) + \mathbf{K}(1)\left[\mathbf{y}(1) - \mathbf{E}(1)\tilde{\mathbf{x}}(1, -)\right], \\
\mathbf{K}(1) &= \mathbf{P}(1, -)\mathbf{E}(1)^{\mathrm{T}}\left[\mathbf{E}(1)\mathbf{P}(1, -)\mathbf{E}(1)^{\mathrm{T}} + \mathbf{R}(1)\right]^{-1},
\end{aligned} \tag{4.47}$$

with new uncertainty

$$\mathbf{P}(1) = \mathbf{P}(1, -) - \mathbf{K}(1)\mathbf{E}(1)\mathbf{P}(1, -). \tag{4.48}$$

(Compare to the discussion on p. 139.) Equation (4.47) is best interpreted as being the average of the model estimate with the estimate obtained from the data alone, but disguised by rearrangement.

Thus there are four steps:

1. Make a forecast using the model (4.35) with the unknown control terms $\mathbf{\Gamma}\mathbf{u}$ set to zero.
2. Calculate the uncertainty of this forecast, Eq. (4.45), which is made up of the sum of the errors owing to initial conditions and to missing controls.
3. Do a weighted average (4.47) of the forecast with the observations, the weighting being chosen to reflect the relative uncertainties.
4. Compute the uncertainty of the final weighted average, Eq. (4.48).

Such a computation is called a "Kalman filter";[13] it is conventionally given a more formal derivation. $\mathbf{K}$ is called the "Kalman gain." At the stage where the forecast (4.42) has already been made, the problem was reduced to finding the minimum of the objective function,

$$J = [\tilde{\mathbf{x}}(1, -) - \tilde{\mathbf{x}}(1)]^{\mathrm{T}} \mathbf{P}(1, -)^{-1} [\tilde{\mathbf{x}}(1, -) - \tilde{\mathbf{x}}(1)]$$
$$+ [\mathbf{y}(1) - \mathbf{E}(1)\tilde{\mathbf{x}}(1)]^{\mathrm{T}} \mathbf{R}(1)^{-1} [\mathbf{y}(1) - \mathbf{E}(1)\tilde{\mathbf{x}}(1)], \tag{4.49}$$

which is a variation of the objective function used to define the recursive least-squares algorithm (Eq. 2.425). In this final stage, the explicit model has disappeared, being present only implicitly through the uncertainty $\mathbf{P}(1, -)$. After the averaging step, all of the information about the observations has been used too, and is included in $\tilde{\mathbf{x}}(t)$, $\mathbf{P}(t)$ and the data can be discarded. For clarity, tildes have been placed over all appearances of $\mathbf{x}(t)$.

A complete recursion can now be defined through Eqs. (4.42)–(4.48), replacing all the $t = 0$ variables with $t = 1$ variables, the $t = 1$ variables becoming $t = 2$ variables, etc. In terms of arbitrary t, the recursion is:

$$\tilde{\mathbf{x}}(t, -) = \mathbf{A}(t - 1)\tilde{\mathbf{x}}(t - 1) + \mathbf{B}(t - 1)\mathbf{q}(t - 1), \tag{4.50}$$
$$\mathbf{P}(t, -) = \mathbf{A}(t - 1)\mathbf{P}(t - 1)\mathbf{A}(t - 1)^{\mathrm{T}} + \mathbf{\Gamma}\mathbf{Q}(t - 1)\mathbf{\Gamma}^{\mathrm{T}}, \tag{4.51}$$
$$\tilde{\mathbf{x}}(t) = \tilde{\mathbf{x}}(t, -) + \mathbf{K}(t)[\mathbf{y}(t) - \mathbf{E}(t)\tilde{\mathbf{x}}(t, -)], \tag{4.52}$$
$$\mathbf{K}(t) = \mathbf{P}(t, -)\mathbf{E}(t)^{\mathrm{T}}[\mathbf{E}(t)\mathbf{P}(t, -)\mathbf{E}(t)^{\mathrm{T}} + \mathbf{R}(t)]^{-1}, \tag{4.53}$$
$$\mathbf{P}(t) = \mathbf{P}(t, -) - \mathbf{K}(t)\mathbf{E}(t)\mathbf{P}(t, -), \quad t = 1, 2, \ldots, t_f. \tag{4.54}$$

These equations are those for the complete Kalman filter. Note that some authors prefer to write equations for $\tilde{\mathbf{x}}(t + 1, -)$ in terms of $\tilde{\mathbf{x}}(t)$, etc. Equation (2.36) permits the rewriting of Eq. (4.54) as

$$\mathbf{P}(t) = [\mathbf{P}(t, -)^{-1} + \mathbf{E}(t)^{\mathrm{T}}\mathbf{R}(t)^{-1}\mathbf{E}(t)]^{-1}, \tag{4.55}$$

and an alternate form for the gain is[14]

$$\mathbf{K}(t) = \mathbf{P}(t)\mathbf{E}(t)^{\mathrm{T}}\mathbf{R}(t)^{-1}, \tag{4.56}$$

These rewritten forms are often important for computational efficiency and accuracy. Note that in the special case where the observations are employed one-at-a-time, $\mathbf{E}(t)$ is a simple row vector, $\mathbf{E}(t)\mathbf{P}(t, -)\mathbf{E}(t)^{\mathrm{T}} + \mathbf{R}(t)$ is a scalar, and no matrix inversion is required in Eqs. (4.50)–(4.54). The computation would then be dominated by matrix multiplications. Such a strategy demands that the noise be uncorrelated from one observation to another, or removed by "pre-whitening," which, however, itself often involves a matrix inversion. Various re-arrangements are worth examining in large problems.[15]

Notice that the model is being satisfied exactly; in the terminology introduced in Chapter 2, it is a hard constraint. But as was true with the static models, the hard constraint description is misleading, as the presence of the terms in $\mathbf{u}(t)$ means that model errors are permitted. Notice too, that $\mathbf{u}(t)$ has not been estimated.

Example *Consider again the mass–spring oscillator described earlier, with time history in Fig. 4.1. It was supposed that the initial conditions were erroneously provided as $\tilde{\mathbf{x}}(0) = [10, 10]^{\mathrm{T}}$, $\mathbf{P}(0) = \mathrm{diag}([100, 100])$, but that the forcing was completely unknown. Observations of $x_1(t)$ were provided at every time step with a noise variance $R = 50$. The Kalman filter was computed by (4.50)–(4.54) and used to estimate the position at each time step. The result for part of the time history is in Fig. 4.3(a), showing the true value and the estimated value of component $x_1(t)$. The time history of the uncertainty of $x_1(t)$, $\sqrt{P_{11}(t)}$, is also depicted and rapidly reaches an asymptote. Overall, the filter manages to track the position of the oscillator everywhere within two standard deviations.*

If observations are not available at some time step, t, the best estimate reduces to that from the model forecast alone, $\mathbf{K}(t) = \mathbf{0}$, $\mathbf{P}(t) = \mathbf{P}(t, -)$ and one simply proceeds. Typically in such situations, the error variances will grow from the accumulation of the unknown $\mathbf{u}(t)$, at least, until such times as an observation does become available. If $\mathbf{u}(t)$ is purely random, the system will undergo a form of random walk.[16]

Example *Consider again the problem of fitting a straight line to data, as discussed in Chapter 2, but now in the context of a Kalman filter, using the canonical form derived from (4.50)–(4.54). "Data" were generated from the state transition matrix of Eq. (4.16) and an unforced model, as depicted in Fig. 4.4. The observation equation is*

$$y(t) = x_1(t) + n(t),$$

that is, $\mathbf{E}(t) = \{1 \quad 0\}$, $R(t) = 50$, *but observations were available only every 25th time step. There were no unknown control disturbances – that is $Q(t) = 0$, but the initial state estimate was set erroneously as $\tilde{\mathbf{x}}(0) = [30, 10]^{\mathrm{T}}$, with an uncertainty $\mathbf{P}(0) = \mathrm{diag}([900, 900])$. The result of the computation for the fit is shown in Fig. 4.5 for 100 time steps. Note that with the Kalman filter the estimate diverges rapidly from the true value (although well within the estimated error) and is brought discontinuously toward the true value when the first observations become available.*

If the state vector is redefined to consist of the two model parameters a, b, then $\mathbf{x} = [a \quad b]^{\mathrm{T}}$ and $\mathbf{A} = \mathbf{I}$. Now the observation matrix is $\mathbf{E} = [1 \quad t]$ – that is, time-dependent. The state vector has changed from a time-varying one to a constant. The incorrect estimates $\tilde{\mathbf{x}}(0) = [10 \quad 10]^{\mathrm{T}}$ were used, with

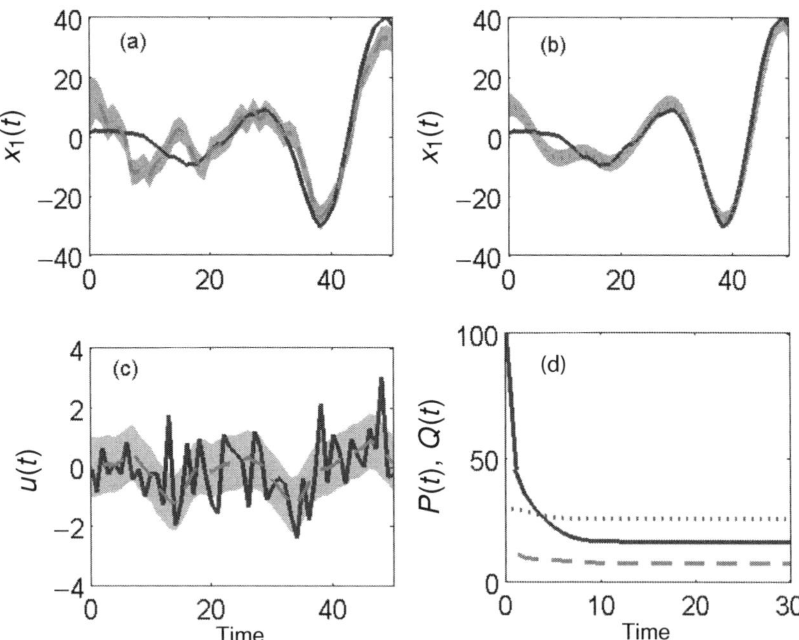

Figure 4.3 (a) Forward run (solid line) of a forced mass–spring oscillator with $r = 0$, $k = 0.1$ and initial condition $\mathbf{x}(0) = [1, 0]^{\mathrm{T}}$. The dashed line is a Kalman filter estimate, $\tilde{x}_1(t)$ started with $\tilde{\mathbf{x}}(0) = [10, 10]$, $\mathbf{P}(0) = \mathrm{diag}([100, 100])$. "Observations" were provided for $x_1(t)$ at every time step, but corrupted with white noise of variance $R = 50$. The shaded band is the one-standard deviation error bar for $\tilde{x}_1(t)$ computed from $\sqrt{P_{11}(t)}$ in the Kalman filter. Rapid convergence toward the true value occurs despite the high noise level. (b) The dotted line now shows $\tilde{x}_1(t, +)$ from the RTS smoothing algorithm. The solid line is again the "truth." Although only the first 50 points are shown, the Kalman filter was run out to $t = 300$, where the smoother was started. The band is the one standard deviation of the smoothed estimate from $\sqrt{P_{11}(t, +)}$ and is smaller than $\sqrt{P_{11}(t)}$. The smoothed estimate is closer to the true value almost everywhere. As with the filter, the smoothed estimate is consistent with the true values within two standard deviations. (c) Estimated $\tilde{u}(t)$ (dashed) and its standard error from the smoother. The solid line is the "true" value (which is itself white noise). That $\tilde{u}(t)$ lacks the detailed structure of the true $u(t)$ is a consequence of the inability of the mass–spring oscillator to respond instantaneously to a white noise forcing. Rather, it responds to an integrated value. (d) The solid line is $P_{11}(t)$, dashed is $P_{11}(t, +)$, and dotted curve is $30Q(t, +)$ with the scale factor used to make it visible. (Squares of values shown as bands in the other panels.) Note the rapid tendency towards a steady state. Values are largest at $t = 0$ as data are only available in the future, not the past. Q is multiplied by a large factor to make it visible.

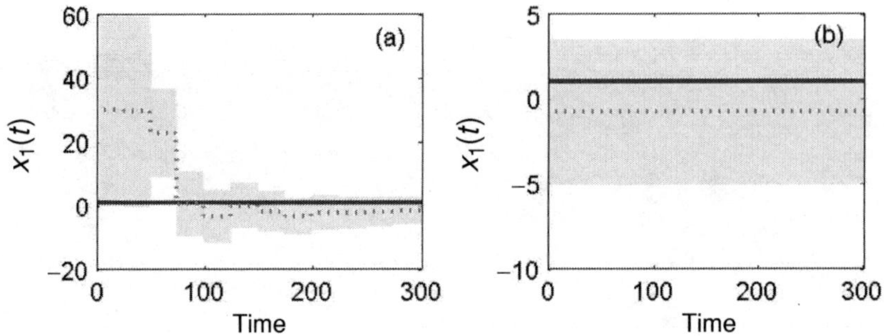

Figure 4.4 (a) $x_1(t)$ (solid) and $\tilde{x}_1(t) = \tilde{a}$ (dotted) from the straight-line model and the Kalman filter estimate when the state vector was defined to be the intercept and slope, and $\mathbf{E}(t) = [1, t]$. (b) Smoothed estimate, $\tilde{x}_1(t, +)$, (dotted) and its uncertainty corresponding to (a).

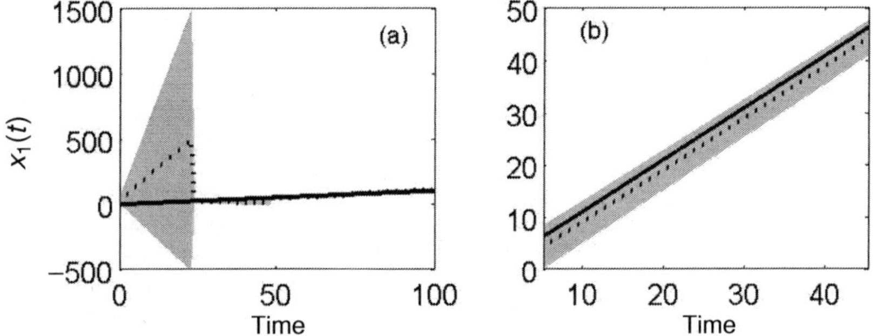

Figure 4.5 A straight line computed from the statespace model with $\mathbf{A}$ from Eq. (4.16) with no forcing. (a) The solid line shows the true values. Noisy observations were provided every 25th point, and the initial condition was set erroneously to $\tilde{\mathbf{x}}(0) = [30, 10]^{\mathrm{T}}$ with $\mathbf{P}(0) = \mathrm{diag}([900, 900])$. The estimation error grows rapidly away from the incorrect initial conditions until the first observations are obtained. Estimate is shown as the dashed line. The gray band is the one standard deviation error bar. (b) Result of applying the RTS smoother to the data and model in (a).

$\mathbf{P}(0) = \mathrm{diag}([10, 10])$ *(the correct values are* $a = 1$, $b = 2$*) and with the time histories of the estimates depicted in Fig. 4.3. At the end of 100 time steps, we have* $\tilde{a} = 1.85 \pm 2.0$, $\tilde{b} = 2.0 \pm 0.03$, *both of which are consistent with the correct values. For reasons the reader might wish to think about, the uncertainty of the intercept is much greater than for the slope.*

Example *For the mass–spring oscillator in Fig. 4.3, it was supposed that the same noisy observations were available, but only at every 25th time step. In general, the presence of the model error, or control uncertainty, accumulates over the 25 time steps as the model is run forward without observations. The expected error of such a system is shown for 100 time steps in Fig. 4.6(d). Notice (1) the growing*

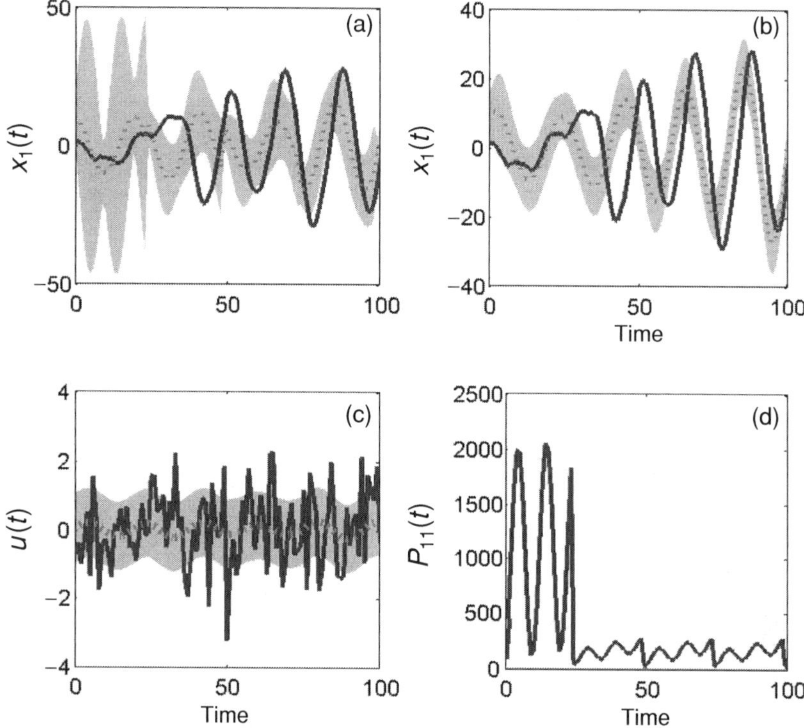

Figure 4.6 For the same model as in Fig. 4.3, except that noisy observations were available only every 25th point ($R = 50$). (a) Shows the correct trajectory of $x_1(t)$ for 100 time steps, the dotted line shows the filter estimate and the shaded band is the standard error of the estimate. (b) Displays the correct value of $x_1(t)$ compared to the (dotted) RTS smoother value with its standard error. (c) Is the estimated control (dotted) with its standard error, and the true value applied to mass–spring oscillator. (d) Shows the behavior of $P_{11}(t)$ for the Kalman filter with very large values (oscillating with twice the frequency of the oscillator) and which become markedly reduced as soon as the first observations become available at the 25th point.

envelope as uncertainty accumulates faster than the observations can reduce it; (2) the periodic nature of the error within the growing envelope; and (3) that the envelope appears to be asymptoting to a fixed upper bound for large t. The true and estimated time histories for a portion of the time history are shown in Fig. 4.6(d). As expected, with fewer available observations, the misfit of the estimated and true values is larger than with data at every time step. At every 25th point, the error norm drops as observations become available, but with the estimated value of $x_1(t)$ undergoing a jump when the observation is available.

If the observation is that of the velocity $x_1(t) - x_2(t) = \xi(t) - \xi(t-1)$, then $E = \{1 \quad -1\}$. A portion of the time history of the Kalman filtered estimate with a velocity observation available only at every 25th point may be seen in Fig. 4.7.

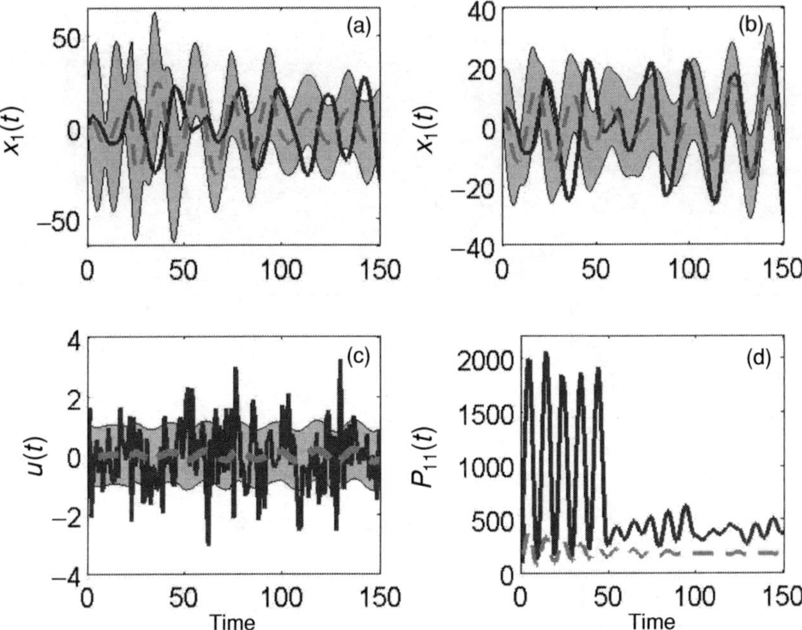

Figure 4.7 (a) $x_1(t)$ and Kalman filter estimate (dashed) when the noisy observations ($R = 50$) are of the velocity ($\mathbf{E} = [1, -1]$) every 25th point for the mass–spring oscillator. (b) RTS smoother estimate (dashed) and its uncertainty corresponding to (a). (c) The estimated control (dotted) and the correct value (solid). As seen previously, the high frequency variability in $u(t)$ is not detected by the moving mass, but only an integrated version. (d) $P_{11}(t)$ (solid) corresponding to the standard error in (a), and $P_{11}(t, +)$ (dashed) corresponding to that in (b).

Velocity observations are evidently useful for estimating position, owing to the connection between velocity and position provided by the model and is a simple example of how observations of almost anything can be used to improve a state estimate.

A number of more general reformulations of the equations into algebraically equivalent forms are particularly important. In one form, one works not with the covariances, $\mathbf{P}(t, -)$, ..., but with their inverses, the so-called information matrices, $\mathbf{P}(t, -)^{-1}$, etc. (See Eq. (4.55).) This "information filter" form may be more efficient if, for example, the information matrices are banded and sparse while the covariance matrices are not. Or, if the initial conditions are infinitely uncertain, $\mathbf{P}(0)^{-1}$ can be set to the zero matrix. In another formulation, one uses the square roots (Cholesky decomposition) of the covariance matrices rather than the matrices themselves. This "square root filter" is important, as there is a tendency for the computation of the updated values of $\mathbf{P}$ in Eq. (4.54) to become non-positive-definite owing to round-off errors, and the square root formulation guarantees a positive definite result.[17]

The Kalman filter does *not* produce the minimum of the objective function Eq. (4.41) because the data from times later than t are not being used to make estimates of the earlier values of the state vector or of $\mathbf{u}(t)$. At each step, the filter is instead minimizing an objective function of the form in Eq. (4.50). To obtain the needed minimum, we have to consider what is called the "smoothing problem," to which we will turn in a moment. Note too, that the time history of $\mathbf{x}(t)$ does not satisfy a known equation at the time observations are introduced. When no observation is available, the time evolution obeys the model equation with zero control term; the averaging step of the filter, however, leads to a change between t and $t - 1$ that compensates for the accumulated error. The evolution equation is no longer satisfied in this interval.

The Kalman filter is, nonetheless, extremely important in practice for many problems. In particular, if one must literally make a forecast (e.g., such filters are used to help land airplanes or, in a primitive way, to forecast the weather), then the future data are simply unavailable, and the state estimate made at time t, using data up to and including time t, is the best one can do.[18]

For estimation, the Kalman filter is only a first step – owing to its failure to use data from the formal future. It also raises questions about computational feasibility. As with all recursive estimators, the uncertainties $\mathbf{P}(t, -)$, $\mathbf{P}(t)$ must be available so as to form the weighted averages. If the state vector contains N elements, then the model (4.50) requires multiplying an N-dimensional vector by an $N \times N$ matrix at each time step. The covariance update (4.51) requires updating each of N columns of $\mathbf{P}(t)$ in the same way, and then doing it again (i.e., in practice, one forms $\mathbf{A}(t)\mathbf{P}(t)$, transposes it, and forms $\mathbf{A}(t)(\mathbf{A}(t)\mathbf{P}(t))^{\mathrm{T}}$, equivalent to running the model $2N$ times at each time step). In many applications, particularly in geophysical fluids, this covariance update step dominates the calculation, renders it impractical, and leads to some of the approximate methods taken up presently.

The Kalman filter was derived heuristically as a simple generalization of the ideas used in Chapter 2. Unsurprisingly, the static inverse results are readily recovered from the filter in various limits. As one example, consider the nearly noise-free case in which both process and observation noise are very small, i.e. $\|\mathbf{Q}\|$, $\|\mathbf{R}\| \to 0$. Then if $\mathbf{P}(t, -)$ is nearly diagonal, $\mathbf{P}(t, -) \sim \delta^2 \mathbf{I}$, and

$$\mathbf{K}(t) \longrightarrow \mathbf{E}^{\mathrm{T}}(\mathbf{E}\mathbf{E}^{\mathrm{T}})^{-1},$$

assuming existence of the inverse, and

$$
\begin{aligned}
\tilde{\mathbf{x}}(t) &\sim \mathbf{A}\tilde{\mathbf{x}}(t - 1) + \mathbf{B}\mathbf{q}(t - 1) \\
&\quad + \mathbf{E}^{\mathrm{T}}(\mathbf{E}\mathbf{E}^{\mathrm{T}})^{-1}\{\mathbf{y}(t) - \mathbf{E}[\mathbf{A}\tilde{\mathbf{x}}(t - 1) + \mathbf{B}\mathbf{q}(t - 1)]\} \\
&= \mathbf{E}^{\mathrm{T}}(\mathbf{E}\mathbf{E}^{\mathrm{T}})^{-1}\mathbf{y}(t) \\
&\quad + [\mathbf{I} - \mathbf{E}^{\mathrm{T}}(\mathbf{E}\mathbf{E}^{\mathrm{T}})^{-1}\mathbf{E}][\mathbf{A}\tilde{\mathbf{x}}(t - 1) + \mathbf{B}\mathbf{q}(t - 1)].
\end{aligned}
\tag{4.57}
$$

$\mathbf{E}^{\mathrm{T}}(\mathbf{E}\mathbf{E}^{\mathrm{T}})^{-1}\mathbf{y}(t)$ is just the expression in Eq. (2.95) for the direct estimate of $\mathbf{x}(t)$ from a set of underdetermined full-rank, noise-free observations. It is the static estimate we would use at time t if no dynamics were available. The columns of $\mathbf{I} - \mathbf{E}^{\mathrm{T}}(\mathbf{E}\mathbf{E}^{\mathrm{T}})^{-1}\mathbf{E}$ are the nullspace of $\mathbf{E}$ (recall the definition of $\mathbf{H}$ in Eq. (2.97)) and (4.57) thus employs only those elements of the forecast lying in the nullspace of the observations – a sensible result given that the observations here produce perfect estimates of components of $\mathbf{x}(t+1)$ in the range of $\mathbf{E}$. Thus, in this particular limit, the Kalman filter computes from the noise-free observations those elements of $\mathbf{x}(t+1)$ that it can, and for those which it cannot, it forecasts them from the dynamics. The reader ought to examine other limiting cases – retaining process and/or observational noise – including the behavior of the error covariance propagation.

Example *It is interesting to apply some of these expressions to the simple problem of finding the mean of a set of observations, considered before on p. 133. The model is of an unchanging scalar mean,*

$$x(t) = x(t-1),$$

observed in the presence of noise,

$$y(t) = x(t) + n(t),$$

where $\langle n(t) \rangle = 0$, $\langle n(t)^2 \rangle = R$, so $E = 1$, $A = 1$, $Q = 0$, $t = 0, 1, \ldots, m-1$. In contrast to the situation on p. 133, the machinery used here requires that the noise be uncorrelated: $\langle n(t)n(t') \rangle = 0$, $t \neq t'$, although as already mentioned, methods exist to overcome this restriction. Suppose that the initial estimate of the mean is 0 – that is, $\tilde{x}(0) = 0$, with uncertainty $P(0)$. Equation (4.51) is $P(t, -) = P(t-1)$, and the Kalman filter uncertainty, in the form (4.55), is

$$\frac{1}{P(t)} = \frac{1}{P(t-1)} + \frac{1}{R},$$

a difference equation, with known initial condition, whose solution by inspection is

$$\frac{1}{P(t)} = \frac{t}{R} + \frac{1}{P(0)}.$$

Using (4.52) with $E = 1$, and successively stepping forward, produces[19]

$$\tilde{x}(m-1) = \frac{R}{R + mP(0)} \left\{ \frac{P(0)}{R} \sum_{j=0}^{m-1} y(j) \right\}, \tag{4.58}$$

whose limit as $t \to \infty$ is

$$\tilde{x}(m-1) \longrightarrow \frac{1}{m} \sum_{j=0}^{m-1} y(j),$$

the simple average, with uncertainty $P(t) \to 0$, as $t \to \infty$. If there is no useful estimate available of $P(0)$, rewrite Eq. (4.58) as

$$\tilde{x}(m-1) = \frac{R}{R/P(0)+m} \left\{ \frac{1}{R} \sum_{j=0}^{m-1} y(j) \right\}, \qquad (4.59)$$

and take the agnostic limit, $1/P(0) \to 0$, or

$$\tilde{x}(m-1) = \frac{1}{m} \left\{ \sum_{j=0}^{m-1} y(j) \right\}, \qquad (4.60)$$

which is wholly conventional. (Compare these results to those on p. 133. The problem and result here are necessarily identical to those, except that now $x(t)$ is identified explicitly as a state vector rather than as a constant. Kalman filters with static models reduce to ordinary least-squares solutions.)

4.3.3 The smoothing problem

The Kalman filter permits one to make an optimal forecast from a linear model, subject to the accuracy of the various assumptions being made. Between observation times, the state estimate evolves smoothly according to the model dynamics. But when observations become available, the averaging can draw the combined state estimate abruptly towards the observations, and in the interval between the last unobserved state and the new one, model evolution is not followed. To obtain a state trajectory that is both consistent with model evolution and the data at all times, the state estimate jumps at the observation times need to be removed, and the problem solved as originally stated. Minimization of J in Eq. (4.41) subject to the model is still the goal. Begin the discussion again with a one-step process,[20] for the problem Eqs. (4.35)–(4.39), but where there are only two times involved, $t = 0,\ 1$. There is an initial estimate $\tilde{\mathbf{x}}(0)$, $\tilde{\mathbf{u}}(0) \equiv 0$ with uncertainties $\mathbf{P}(0)$, $\mathbf{Q}(0)$ for the initial state and control vectors respectively, a set of measurements at time-step 1, and the model. The objective function is

$$\begin{aligned}
J = &[\tilde{\mathbf{x}}(0,+) - \tilde{\mathbf{x}}(0)]^{\mathrm{T}} \mathbf{P}(0)^{-1} [\tilde{\mathbf{x}}(0,+) - \tilde{\mathbf{x}}(0)] \\
&+ [\tilde{\mathbf{u}}(0,+) - \tilde{\mathbf{u}}(0)]^{\mathrm{T}} \mathbf{Q}(0)^{-1} [\tilde{\mathbf{u}}(0,+) - \tilde{\mathbf{u}}(0)] \\
&+ [\mathbf{y}(1) - \mathbf{E}(1)\tilde{\mathbf{x}}(1)]^{\mathrm{T}} \mathbf{R}(1)^{-1} [\mathbf{y}(1) - \mathbf{E}(1)\tilde{\mathbf{x}}(1)],
\end{aligned} \qquad (4.61)$$

subject to the model

$$\tilde{\mathbf{x}}(1) = \mathbf{A}(0)\,\tilde{\mathbf{x}}(0, +) + \mathbf{B}(0)\mathbf{q}(0) + \mathbf{\Gamma}\tilde{\mathbf{u}}(0, +), \tag{4.62}$$

with the weight matrices again chosen as the inverses of the prior covariances. Tildes have now been placed on all estimated quantities. A minimizing solution to this objective function would produce a new estimate of $\mathbf{x}(0)$, denoted $\tilde{\mathbf{x}}(0, +)$, with error covariance $\mathbf{P}(0, +)$; the $+$ denotes use of *future* observations, $\mathbf{y}(1)$, in the estimate. On the other hand, we would still denote the estimate at $t = 1$ as $\tilde{\mathbf{x}}(1)$, coinciding with the Kalman filter estimate, because only data prior to and at the same time would have been used. The estimate $\tilde{\mathbf{x}}(1)$ must be given by Eq. (4.47), but it remains to improve $\tilde{\mathbf{u}}(0)$, $\tilde{\mathbf{x}}(0)$, while simultaneously eliminating the problem of the estimated state vector jump at the filter averaging (observation) time.

The basic issue can be understood by observing that the initial estimates $\tilde{\mathbf{u}}(0) = \mathbf{0}$, $\tilde{\mathbf{x}}(0)$ lead to a model forecast that disagrees with the final best estimate $\tilde{\mathbf{x}}(1)$. If either of $\tilde{\mathbf{u}}(0)$, or $\tilde{\mathbf{x}}(0)$ were known perfectly, the forecast discrepancy could be ascribed to the other one, permitting ready computation of the new required value. In practice, both are somewhat uncertain, and the modification must be partitioned between them; one would not be surprised to find that the partitioning proves to be proportional to their initial uncertainties.

To find the stationary point (we will not trouble to prove it a minimum rather than a maximum), set the differential of J with respect to $\tilde{\mathbf{x}}(0, +)$, $\tilde{\mathbf{x}}(1)$, $\tilde{\mathbf{u}}(0, +)$ to zero (recall Eq. 2.91),

$$\begin{aligned}
\frac{\mathrm{d}J}{2} ={}& \mathrm{d}\tilde{\mathbf{x}}(0, +)^{\mathrm{T}}\mathbf{P}(0)^{-1}[\tilde{\mathbf{x}}(0, +) - \tilde{\mathbf{x}}(0)] \\
&+ \mathrm{d}\tilde{\mathbf{u}}(0, +)^{\mathrm{T}}\mathbf{Q}(0)^{-1}[\tilde{\mathbf{u}}(0, +) - \tilde{\mathbf{u}}(0)] \\
&- \mathrm{d}\tilde{\mathbf{x}}(1)^{\mathrm{T}}\mathbf{E}(1)^{\mathrm{T}}\mathbf{R}(1)^{-1}[\mathbf{y}(1) - \mathbf{E}(1)\tilde{\mathbf{x}}(1)] = 0.
\end{aligned} \tag{4.63}$$

The coefficients of the differentials cannot be set to zero separately because they are connected via the model (4.62), which provides the relationship

$$\mathrm{d}\tilde{\mathbf{x}}(1) = \mathbf{A}(0)\,\mathrm{d}\tilde{\mathbf{x}}(0, +) + \mathbf{\Gamma}(0)\,\mathrm{d}\tilde{\mathbf{u}}(0, +). \tag{4.64}$$

Eliminating $\mathrm{d}\tilde{\mathbf{x}}(1)$,

$$\begin{aligned}
\frac{\mathrm{d}J}{2} ={}& \mathrm{d}\tilde{\mathbf{x}}(0, +)^{\mathrm{T}}\{\mathbf{P}(0)^{-1}\,[\tilde{\mathbf{x}}(0, +) - \tilde{\mathbf{x}}(0)] \\
&\qquad - \mathbf{A}(0)^{\mathrm{T}}\mathbf{E}(1)^{\mathrm{T}}\mathbf{R}(1)^{-1}\,[\mathbf{y}(1) - \mathbf{E}(1)\tilde{\mathbf{x}}(1)]\} \\
&+ \mathrm{d}\tilde{\mathbf{u}}(0, +)^{\mathrm{T}}\{\mathbf{Q}(0)^{-1}\,[\tilde{\mathbf{u}}(0, +) - \tilde{\mathbf{u}}(0)] \\
&\qquad + \mathbf{\Gamma}^{\mathrm{T}}(0)\mathbf{E}(1)^{\mathrm{T}}\mathbf{R}(1)^{-1}\,[\mathbf{y}(1) - \mathbf{E}(1)\tilde{\mathbf{x}}(1)]\}.
\end{aligned} \tag{4.65}$$

dJ vanishes only if the coefficients of $\mathrm{d}\tilde{\mathbf{x}}(0, +)$, $\mathrm{d}\tilde{\mathbf{u}}(0, +)$ separately vanish, yielding

$$\tilde{\mathbf{x}}(0, +) = \tilde{\mathbf{x}}(0) + \mathbf{P}(0)\mathbf{A}(0)^{\mathrm{T}}\mathbf{E}(1)^{\mathrm{T}}\mathbf{R}(1)^{-1}\left[\mathbf{y}(1) - \mathbf{E}(1)\tilde{\mathbf{x}}(1)\right], \tag{4.66}$$

$$\tilde{\mathbf{u}}(0, +) = \tilde{\mathbf{u}}(0) + \mathbf{Q}(0)\boldsymbol{\Gamma}(0)^{\mathrm{T}}\mathbf{E}(1)^{\mathrm{T}}\mathbf{R}(1)^{-1}\left[\mathbf{y}(1) - \mathbf{E}(1)\tilde{\mathbf{x}}(1)\right], \tag{4.67}$$

and

$$\begin{aligned}
\tilde{\mathbf{x}}(1) = \tilde{\mathbf{x}}(1, -) &+ \mathbf{P}(1, -)\mathbf{E}(1)^{\mathrm{T}}[\mathbf{E}(1)\mathbf{P}(1, -)\mathbf{E}(1)^{\mathrm{T}} + \mathbf{R}(1)]^{-1} \\
&\times [\mathbf{y}(1) - \mathbf{E}(1)\tilde{\mathbf{x}}(1, -)],
\end{aligned} \tag{4.68}$$

using the previous definitions of $\tilde{\mathbf{x}}(1, -)$, $\mathbf{P}(1, -)$. As anticipated, Eq. (4.68) is recognizable as the Kalman filter estimate. At this point we are essentially done: an estimate has been produced not only of $\mathbf{x}(1)$, but an improvement has been made in the prior estimate of $\mathbf{x}(0)$ using the future measurements, and the control term has been estimated. Notice that the corrections to $\tilde{\mathbf{u}}(0)$, $\tilde{\mathbf{x}}(0)$ are proportional to $\mathbf{Q}(0)$, $\mathbf{P}(0)$, respectively, as anticipated. The uncertainties of these latter quantities are still needed.

First rewrite the estimates (4.66) and (4.67) as

$$\begin{aligned}
\tilde{\mathbf{x}}(0, +) = \tilde{\mathbf{x}}(0) + \mathbf{L}(1)\left[\tilde{\mathbf{x}}(1) - \tilde{\mathbf{x}}(1, -)\right], \quad &\mathbf{L}(1) = \mathbf{P}(0)\mathbf{A}(0)^{\mathrm{T}}\mathbf{P}(1, -)^{-1}, \\
\tilde{\mathbf{u}}(0, +) = \tilde{\mathbf{u}}(0) + \mathbf{M}(1)\left[\tilde{\mathbf{x}}(1) - \tilde{\mathbf{x}}(1, -)\right], \quad &\mathbf{M}(1) = \mathbf{Q}(0)\boldsymbol{\Gamma}(0)^{\mathrm{T}}\mathbf{P}(1, -)^{-1},
\end{aligned} \tag{4.69}$$

which can be done by extended, but uninteresting, algebraic manipulation. The importance of these latter two expressions is that both $\tilde{\mathbf{x}}(0, +)$, $\tilde{\mathbf{u}}(0, +)$ are expressed in terms of their prior estimates in a weighted average with the difference between the prediction of the state at $t = 1$, $\tilde{\mathbf{x}}(1, -)$ and what was actually estimated there following the data use, $\tilde{\mathbf{x}}(1)$. (But the data do not appear explicitly in (4.69).) It is also possible to show that

$$\begin{aligned}
\mathbf{P}(0, +) &= \mathbf{P}(0) + \mathbf{L}(1)[\mathbf{P}(1) - \mathbf{P}(1, -)]\mathbf{L}(1)^{\mathrm{T}}, \\
\mathbf{Q}(0, +) &= \mathbf{Q}(0) + \mathbf{M}(1)[\mathbf{P}(1) - \mathbf{P}(1, -)]\mathbf{M}(1)^{\mathrm{T}}.
\end{aligned} \tag{4.70}$$

Based upon this one-step derivation, a complete recursion for any time interval can be inferred. Suppose that the Kalman filter has been run all the way to a terminal time, t_f. The result is $\tilde{\mathbf{x}}(t_f)$ and its variance $\mathbf{P}(t_f)$. With no future data available, $\tilde{\mathbf{x}}(t_f)$ cannot be further improved. At time $t_f - 1$, we have an estimate $\tilde{\mathbf{x}}(t_f - 1)$ with uncertainty $\mathbf{P}(t_f - 1)$, which could be improved by knowledge of the future observations at t_f. But this situation is precisely the one addressed by the objective function (4.61), replacing $t = 1 \to t_f$, and $t = 0 \to t_f - 1$. Now having improved the estimate at $t_f - 1$ and calling it $\tilde{\mathbf{x}}(t_f - 1, +)$ with uncertainty $\mathbf{P}(t_f - 1, +)$, this new estimate is used to improve the prior estimate $\tilde{\mathbf{x}}(t_f - 2)$, and step all the way back to $t = 0$. The complete recursion is

$$
\tilde{\mathbf{x}}(t, +) = \tilde{\mathbf{x}}(t) + \mathbf{L}(t + 1)\left[\tilde{\mathbf{x}}(t + 1, +) - \tilde{\mathbf{x}}(t + 1, -)\right],
$$
$$
\mathbf{L}(t + 1) = \mathbf{P}(t)\mathbf{A}(t)^{\mathrm{T}} \mathbf{P}(t + 1, -)^{-1}, \tag{4.71}
$$
$$
\tilde{\mathbf{u}}(t, +) = \tilde{\mathbf{u}}(t) + \mathbf{M}(t + 1)\left[\tilde{\mathbf{x}}(t + 1, +) - \tilde{\mathbf{x}}(t + 1, -)\right],
$$
$$
\mathbf{M}(t + 1) = \mathbf{Q}(t)\boldsymbol{\Gamma}(t)^{\mathrm{T}}\mathbf{P}(t + 1, -)^{-1}, \tag{4.72}
$$
$$
\mathbf{P}(t, +) = \mathbf{P}(t) + \mathbf{L}(t + 1)[\mathbf{P}(t + 1, +) - \mathbf{P}(t + 1, -)]\mathbf{L}(t + 1)^{\mathrm{T}}, \tag{4.73}
$$
$$
\mathbf{Q}(t, +) = \mathbf{Q}(t) + \mathbf{M}(t + 1)[\mathbf{P}(t + 1, +) - \mathbf{P}(t + 1, -)]\mathbf{M}(t + 1)^{\mathrm{T}}, \tag{4.74}
$$

with $\tilde{\mathbf{x}}(t_f, +) \equiv \tilde{\mathbf{x}}(t_f)$, $\mathbf{P}(t_f, +) \equiv \mathbf{P}(t_f)$, for $t = 0, 1, \ldots, t_f - 1$.

This recipe, which uses the Kalman filter on a first forward sweep to the end of the available data, and which then successively improves the prior estimates by sweeping backwards, is called the "RTS algorithm" or smoother.[21] The particular form has the advantage that the data are not involved in the backward sweep, because all of the available information has been used in the filter calculation. It does have the potential disadvantage of requiring the storage at each time step of $\mathbf{P}(t)$. ($\mathbf{P}(t, -)$ is readily recomputed, without $\mathbf{y}(t)$, from (4.51) and need not be stored.) By direct analogy with the one-step objective function, the recursion (4.71)–(4.74) is seen to be the solution to the minimization of the objective function (4.61) subject to the model. Most important, assuming consistency of all assumptions, the resulting state vector trajectory $\tilde{\mathbf{x}}(t, +)$ now satisfies the model and no longer displays the jump discontinuities at observation times of the Kalman filter estimate.

As with the Kalman filter, it is possible to examine limiting cases of the RTS smoother. Suppose again that $\mathbf{Q}$ vanishes, and $\mathbf{A}^{-1}$ exists. Then

$$
\mathbf{L}(t + 1) \longrightarrow \mathbf{P}(t)\mathbf{A}^{\mathrm{T}} (\mathbf{A}\mathbf{P}(t)\mathbf{A}^{\mathrm{T}})^{-1} = \mathbf{A}^{-1}, \tag{4.75}
$$

and Eq. (4.71) becomes

$$
\tilde{\mathbf{x}}(t, +) \longrightarrow \mathbf{A}^{-1}\left[\tilde{\mathbf{x}}(t + 1, +) - \mathbf{B}\mathbf{q}(t)\right]. \tag{4.76}
$$

A sensible backward estimate obtained by simply solving

$$
\tilde{\mathbf{x}}(t + 1) = \mathbf{A}\tilde{\mathbf{x}}(t) + \mathbf{B}\mathbf{q}(t), \tag{4.77}
$$

for $\tilde{\mathbf{x}}(t)$. Other limits are also illuminating but are left to the reader.

Example *The smoother result for the straight-line model (4.15) is shown in Figs. 4.4 and 4.5 for both forms of state vector. The time-evolving estimate is now a nearly perfect straight line, whose uncertainty has a terminal value equal to that for the Kalman filter estimate, as it must, and reaches a minimum near the middle of the estimation period, before growing again toward $t = 0$, where the initial uncertainty was very large. In the case where the state vector consisted of* *the constant intercept and slope of the line, both smoothed estimates are seen, in contrast to the filter estimate, to conform very well to the known true behavior. It should be apparent that the best-fitting straight-line solution of Chapter 2 is also the solution to the smoothing problem, but with the data and model handled all at once, in a whole-domain method, rather than sequentially.*

Figures 4.3, 4.6 and 4.7 show the state estimate and its variance for the mass–spring oscillator made from a smoothing computation run backward from $t = 300$. On average, the smoothed estimate is closer to the correct value than is the filtered estimate, as expected. The standard error is also smaller for the smoothed estimate. The figures display the variance, $Q_{11}(t, +)$, of the estimate one can make of the scalar control variable $u(t)$. $\tilde{u}(t)$ does not show the high frequency variability present in practice, because the mass–spring oscillator integrates the one time-step variability in such a way that only an integrated value affects the state vector. But the estimated value is nonetheless always within two standard errors of the correct value.

Example *Consider a problem stimulated by the need to extract information from transient tracers, C, in a fluid, which are assumed to satisfy the equation*

$$\frac{\partial C}{\partial t} + \mathbf{v} \cdot \nabla C - \kappa \nabla^2 C = -\lambda C + q(\mathbf{r}, t), \tag{4.78}$$

where q represents sources/sinks and λ is a decay rate if the tracer is radioactive. To have a simple model that will capture the structure of this problem, the fluid is divided into a set of boxes as depicted in Fig. 4.8. The flow field is represented by exchanges between boxes given by the $J_{ij} \geq 0$. That is, the J_{ij} are a simplified representation of the effects of advection and mixing on a dye C. (A relationship can be obtained between such simple parameterizations and more formal and elaborate finite-difference schemes. Here, it will only be remarked that J_{ij} are chosen to be mass conserving so that the sum over all J_{ij} entering and leaving a box vanishes.) The discrete analogue of (4.78) is taken to be

$$\begin{aligned} C_i(t+1) = {} & C_i(t) - \lambda \Delta t\, C_i(t) \\ & - \frac{\Delta t}{V} \sum_{j \in N(i)} C_i(t) J_{ij} + \frac{\Delta t}{V} \sum_{j \in N(i)} C_j(t) J_{ji}, \end{aligned} \tag{4.79}$$

where the notation $j \in N(i)$ denotes an index sum over the neighboring boxes to box i, V is a volume for the boxes, and Δt is the time step. This model can easily be put into canonical form,

$$\mathbf{x}(t) = \mathbf{A}\mathbf{x}(t-1) + \mathbf{B}\mathbf{q}(t-1) + \boldsymbol{\Gamma}\mathbf{u}(t-1), \qquad \mathbf{Q} = \mathbf{0}, \tag{4.80}$$

with the state vector composed of box concentrations $C_i(t)$, $C_i(t-1)$.

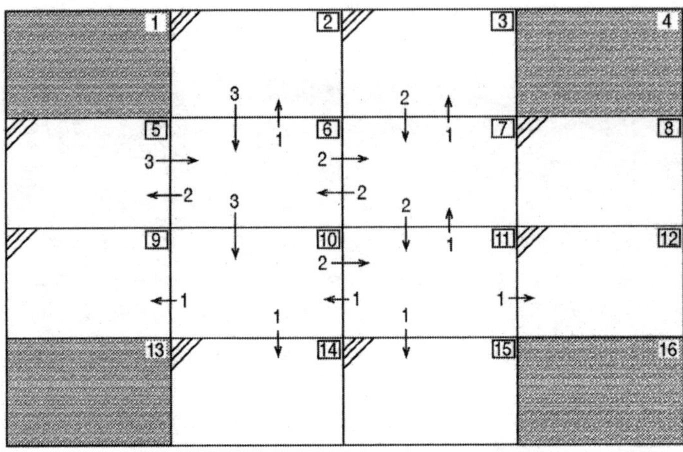

Figure 4.8 Tracer box model where J_{ij} represent fluxes between boxes and are chosen to be mass conserving. Boxes with shaded corners are boundary boxes with externally prescribed concentrations. Numbers in the upper-right corners are used to identify the boxes. Stippled boxes are unconnected and completely passive here. (Source: Wunch, 1988)

A forward computation was run with initial concentrations everywhere of 0, using the boundary conditions depicted in Fig. 4.9, resulting in interior box values as shown. Based upon these correct values, noisy "observations" of the interior boxes only were constructed at times $t = 10,\ 20,\ 30$. The noise variance was 0.01.

An initial estimate of interior tracer concentrations at $t = 0$ was taken (correctly) to be zero, but this estimate was given a large variance (diag $(\mathbf{P}(0)) = 4$). The a-priori boundary box concentrations were set erroneously to $C = 2$ for all t and held at that value. A Kalman filter computation was run as shown in Fig. 4.10. Initially, the interior box concentration estimates rise erroneously (owing to the dye leaking in from the high non-zero concentrations in the boundary boxes). At $t = 10$, the first set of observations becomes available, and the combined estimate is driven much closer to the true values. By the time the last set of observations is used, the estimated and correct concentrations are quite close, although the time history of the interior is somewhat in error. The RTS algorithm was then applied to generate the smoothed histories shown in Fig. 4.10 and to estimate the boundary concentrations (the controls). As expected, the smoothed estimates are closer to the true time history than are the filtered ones. Unless further information is provided, no other estimation procedure could do better, given that the model is the correct one.

Other smoothing algorithms exist. Consider one other approach. Suppose the Kalman filter has been run forward to some time t_c, producing an estimate $\tilde{\mathbf{x}}(t_c)$ with uncertainty $\mathbf{P}(t_c)$. Now suppose, perhaps on the basis of some further observations,

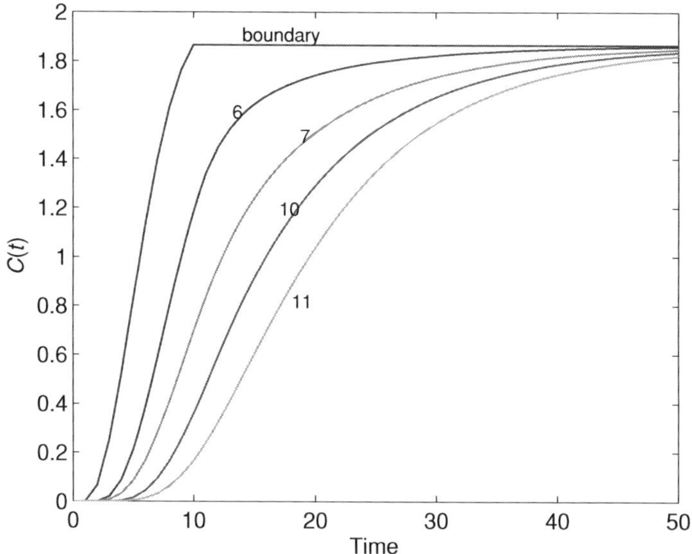

Figure 4.9 Time histories of the forward computation in which boundary concentrations shown were prescribed, and values computed from the forward model for boxes 6, 7, 10, 11. These represent the "truth." Here $\Delta t = 0.05$.

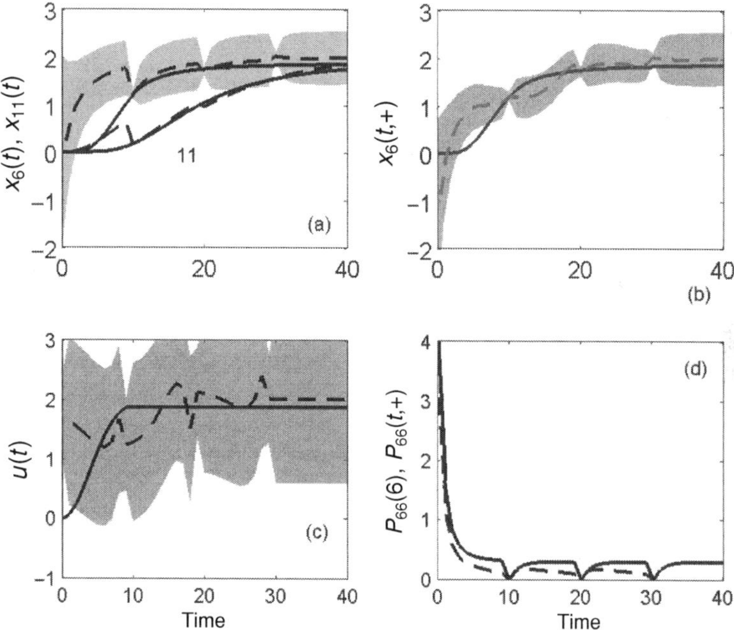

Figure 4.10 (a) Kalman filter estimate (dashed line) with one standard error band, compared to the "true" value in box 6, $x_t(t)$. Filter and true value are shown (without the error band) also in box 11. Observations of all interior values were available at $t = 10, 20, 30$, but the calculations were carried out beyond the last observation. (b) Smoothed estimate $\tilde{x}_6(t, +)$ and one standard error band. (c) Estimated control $\tilde{u}(t, +)$ (dashed) with one standard error, and the correct value (solid). (d) $P_{66}(t)$ (solid) and $P_{66}(t, +)$ (dashed). The smoothed variance is everywhere less than the variance of the filter estimate for all $t < 30$ (the time of the last observation).

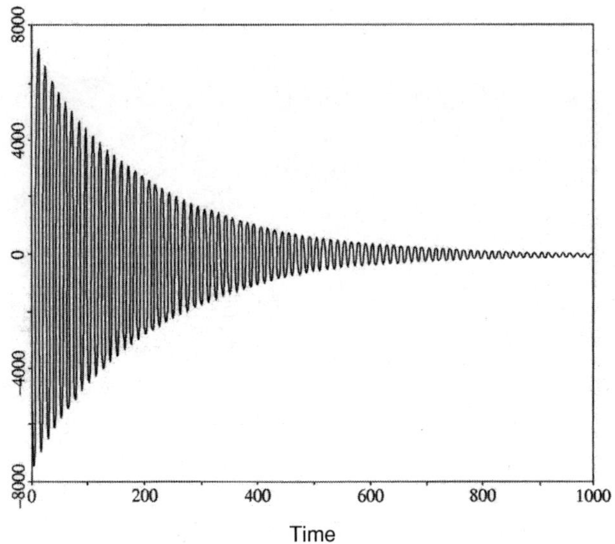

Figure 4.11 Mass–spring oscillator model with friction ($r = 0.01$) run backwards in time from conditions specified at $t = 1000$. The system is unstable, and small uncertainties in the starting conditions would amplify. But the Kalman filter run backwards remains stable because its error estimate also grows – systematically downweighting the model forecast relative to any data that become available at earlier times. A model with unstable elements in the forward direction would behave analogously when integrated in time with growing estimated model forecast error. (Some might prefer to reverse the time scale in such plots.)

that at a *later* time $t_f > t_c$, an independent estimate $\tilde{\mathbf{x}}(t_f)$ has been made, with uncertainty $\mathbf{P}(t_f)$. The independence is crucial – we suppose this latter estimate is made without using any observations at time t_c or earlier so that any errors in $\tilde{\mathbf{x}}(t_c)$ and $\tilde{\mathbf{x}}(t_f)$ are uncorrelated.

Run the model *backwards* in time from t_f to $t_f - 1$:

$$\tilde{\mathbf{x}}_b(t_f - 1) = \mathbf{A}^{-1}\tilde{\mathbf{x}}(t_f) - \mathbf{A}^{-1}\mathbf{B}\mathbf{q}(t_f - 1), \tag{4.81}$$

where the subscript b denotes a backwards-in-time estimate (see Fig. 4.11). The reader may object that running a model backwards in time will often be an unstable operation; this objection needs to be addressed, but ignore it for the moment. The uncertainty of $\tilde{\mathbf{x}}(t_f - 1)$ is

$$\mathbf{P}_b(t_f - 1) = \mathbf{A}^{-1}\mathbf{P}(t_f)\mathbf{A}^{-\mathrm{T}} + \mathbf{A}^{-1}\mathbf{\Gamma}\mathbf{Q}(t_f - 1)\mathbf{\Gamma}^{\mathrm{T}}\mathbf{A}^{-\mathrm{T}}, \tag{4.82}$$

as in the forward-model computation. This backwards computation can be continued to time t_c, at which point we will have an estimate, $\tilde{\mathbf{x}}_b(t_c)$, with uncertainty $\mathbf{P}_b(t_c)$.

The two independent estimates of $\mathbf{x}(t_c)$ can be combined to make an improved estimate using the relations Chapter 2, Eq. (2.444),

$$\tilde{\mathbf{x}}(t_c, +) = \tilde{\mathbf{x}}(t_c) + \mathbf{P}(t_c)(\mathbf{P}(t_c) + \mathbf{P}_b(t_c))^{-1}(\tilde{\mathbf{x}}_b(t_c) - \tilde{\mathbf{x}}(t_c)), \qquad (4.83)$$

and (Eq. 2.446)

$$\begin{aligned}
\mathbf{P}(t_c) &= \langle [\tilde{\mathbf{x}}(t_c, +) - \mathbf{x}(t_c)] \, [\tilde{\mathbf{x}}(t_c, +) - \mathbf{x}(t_c)]^{\mathrm{T}} \rangle \\
&= [\mathbf{P}(t_c)^{-1} + \mathbf{P}_b(t_c)^{-1}]^{-1}.
\end{aligned} \qquad (4.84)$$

This estimate is the same as would be obtained from the RTS algorithm run back to time t_c – because the same objective function, model, and data have been employed. The computation has been organized differently in the two cases. The backwards-running computation can be used at all points of the interval, as long as the data used in the forward and backwards computations are kept disjoint so that the two estimates are uncorrelated.

Running a model backwards in time may indeed be unstable if it contains any dissipative terms. A forward model may be unstable too, if there are unstable elements, either real ones or numerical artifacts. But the expressions in (4.83) and (4.84) are stable, because the computation of $\mathbf{P}_b(t)$ and its use in the updating expression (4.83) automatically downweights unstable elements whose errors will be very large, and which will not carry useful information from the later state concerning the earlier one. The same situation would occur if the forward model had unstable elements – these instabilities would amplify slight errors in the statement of their initial conditions, rendering the initial conditions difficult to estimate from observations at later times. Examination of the covariance propagation equation and the filter gain matrix shows that these elements are suppressed by the Kalman filter, with correspondingly large uncertainties. The filter/smoother formalism properly accounts for unstable, and hence difficult-to-calculate parameters, by estimating their uncertainty as very large, thus handling very general ill-conditioning. In practice, one needs to be careful, for numerical reasons, of the pitfalls in computing and using matrices that may have norms growing exponentially in time. But the conceptual problem is solved. As with the Kalman filter, it is possible to rewrite the RTS smoother expressions (4.71)–(4.74) in various ways for computational efficiency, storage reduction, and improved accuracy.[22]

The dominant computational load in the smoother is again the calculation of the updated covariance matrices, whose size is square of the state-vector dimension, at every time step, leading to efforts to construct simplified algorithms that retain most of the virtues of the filter/smoother combination but with reduced load. For example, it may have already occurred to the reader that in some of the examples displayed, the state vector uncertainties, $\mathbf{P}$, in both the filter and the smoother appear to rapidly approach a steady state. This asymptotic behavior in turn means that the gain matrices, **K**, **L**, **M** will also achieve a steady state, implying that one no longer needs to undertake the updating steps – fixed gains can be used. Such steady-state operators are known as "Wiener filters" and "smoothers" and they represent a potentially very large computational saving. One needs to understand the circumstances under which such steady states can be expected to appear, and we will examine the problem in Section 4.5.

4.3.4 Other smoothers

The RTS algorithm is an example of what is usually called a "fixed-interval" smoother because it assumed that the results are required for a particular interval $t = 0, 1, \ldots, t_f$. Other forms for other purposes are described in the literature, including "fixed-lag" smoothers in which one is interested in an estimate at a fixed time $t_f - t_1$ as t_f advances, usually in real-time. A "fixed-point" smoother addresses the problem of finding a best estimate $\tilde{\mathbf{x}}(t_1)$ with t_1 fixed and t_f continually increasing. When $t_1 = 0$, as data accumulates, the problem of estimating the initial conditions is a special case of the fixed-point smoother problem.

4.4 Control and estimation problems

4.4.1 Lagrange multipliers and adjoints

The results of the last section are recursive schemes for computing first a filtered, and then a smoothed estimate. As with recursive least-squares, the combination of two pieces of information to make an improved estimate demands knowledge of the uncertainty of the information. For static problems, the recursive methods of Chapter 2 may be required, either because all the data were not available initially or because one could not handle them all at once. But, in general, the computational load of the combined least-squares problem introduced in Chapter 2, Eq. (2.424), is less than the recursive one, if one chooses not to compute any of the covariance matrices.

Because the covariance computation will usually dominate, and potentially overwhelm, the filter/smoother algorithms, it is at least superficially very attractive to find algorithms that do not require the covariances – that is, which employ the entire time domain of observations simultaneously – a "whole-domain" or "batch" method. The algorithms that emerge are best known in the context of "control theory." Essentially, there is a more specific focus upon determining the $\mathbf{u}(t)$: the control variables making a system behave as desired. Conventional control engineering has been directed at finding the electrical or physical impulses, e.g., to make a robotic machine tool assemble an automobile, to land an airplane at a specified airfield, or to shift the output of a chemical plant. Because the motion of an airplane is described by a set of dynamical equations, the solution to the problem can equally well be thought of as making a *model* behave as required instead of the actual physical system. Thus if one observes a fluid flow, one whose behavior differs from what one's model said it should, we can seek those controls (e.g., boundary or initial conditions or internal parameters) that will force the model to be consistent with the observed behavior. It may help the reader who further explores these methods to recognize that we are still doing *estimation*, combining observations and models, but sometimes using algorithms best known under the control rubric.

To see the possibilities, consider again the two-point objective function (4.61) where $\mathbf{P}$, etc., are just weight matrices, not necessarily having a statistical significance. We wish to find the minimum of the objective function subject to (4.62). Now append the model equations as done in Chapter 2 (as in Eq. (2.149)), with a vector of Lagrange multipliers, $\boldsymbol{\mu}(1)$, for a new objective function,

$$
\begin{aligned}
J = {}& (\tilde{\mathbf{x}}(0, +) - \tilde{\mathbf{x}}(0))^{\mathrm{T}} \mathbf{P}(0)^{-1} (\tilde{\mathbf{x}}(0, +) - \tilde{\mathbf{x}}(0)) \\
& + (\tilde{\mathbf{u}}(0, +) - \tilde{\mathbf{u}}(0))^{\mathrm{T}} \mathbf{Q}(0)^{-1} (\tilde{\mathbf{u}}(0, +) - \tilde{\mathbf{u}}(0)) \\
& + (\mathbf{y}(1) - \mathbf{E}(1)\tilde{\mathbf{x}}(1))^{\mathrm{T}} \mathbf{R}(1)^{-1} (\mathbf{y}(1) - \mathbf{E}(1)\tilde{\mathbf{x}}(1)) \\
& - 2\boldsymbol{\mu}(1)^{\mathrm{T}} [\tilde{\mathbf{x}}(1) - \mathbf{A}\tilde{\mathbf{x}}(0, +) - \mathbf{B}\mathbf{q}(0) - \boldsymbol{\Gamma}\tilde{\mathbf{u}}(0, +)].
\end{aligned}
\tag{4.85}
$$

As with the filter and smoother, the model is being imposed as a hard constraint, *but with the control term permitting the model to be imperfect*. Tildes have once again been placed over all variables to be estimated. The presence of the Lagrange multiplier now permits treating the differentials as independent; taking the derivatives of J with respect to $\tilde{\mathbf{x}}(0, +)$, $\tilde{\mathbf{x}}(1)$, $\tilde{\mathbf{u}}(0, +)$, $\boldsymbol{\mu}(1)$ and setting them to zero,

$$
\mathbf{P}(0)^{-1} [\tilde{\mathbf{x}}(0, +) - \tilde{\mathbf{x}}(0)] + \mathbf{A}^{\mathrm{T}}\boldsymbol{\mu}(1) = \mathbf{0},
\tag{4.86}
$$

$$
\mathbf{E}(1)^{\mathrm{T}} \mathbf{R}(1)^{-1} [\mathbf{y}(1) - \mathbf{E}(1)\tilde{\mathbf{x}}(1)] + \boldsymbol{\mu}(1) = \mathbf{0},
\tag{4.87}
$$

$$
\mathbf{Q}(0)^{-1} [\tilde{\mathbf{u}}(0, +) - \tilde{\mathbf{u}}(0)] + \boldsymbol{\Gamma}^{\mathrm{T}}\boldsymbol{\mu}(1) = \mathbf{0},
\tag{4.88}
$$

$$
\tilde{\mathbf{x}}(1) - \mathbf{A}\tilde{\mathbf{x}}(0, +) - \mathbf{B}\mathbf{q}(0) - \boldsymbol{\Gamma}\tilde{\mathbf{u}}(0, +) = \mathbf{0}.
\tag{4.89}
$$

Equation (4.86) is the "adjoint model" for $\boldsymbol{\mu}(1)$ involving $\mathbf{A}^{\mathrm{T}}$.

Because the objective function in (4.85) is identical to that used with the smoother for this problem, and because the identical dynamical model has been imposed, Eqs. (4.86)–(4.89) must produce the same solution as that given by the smoother. A demonstration that Eqs. (4.86)–(4.89) can be manipulated into the form of (4.69) and (4.71) is an exercise in matrix identities.[23] As with smoothing algorithms, finding the solution of (4.86)–(4.89) can be done in a number of different ways, trading computation against storage, coding ease, convenience, etc.

Let us show explicitly the identity of smoother and Lagrange multiplier methods for a restricted case – that for which the initial conditions are known exactly, so that $\tilde{x}(0)$ is not modified by the later observations. For the one-term smoother, the result is obtained by dropping (4.86), as $\tilde{x}(0)$ is no longer an adjustable parameter. Without further loss of generality, put $\tilde{u}(0) = 0$, and set $R(1) = R$, reducing the system to

$$\tilde{x}(1) = A\tilde{x}(0) + Bq(0) + \Gamma\tilde{u}(0, +), \tag{4.90}$$

$$\begin{aligned}
\tilde{u}(0, +) &= -Q(0)\Gamma^T\mu(1) \\
&= Q(0)\Gamma^T E^T R^{-1}[y(1) - E(1)\tilde{x}(1)]. \tag{4.91}
\end{aligned}$$

Eliminating $\tilde{u}(0, +)$ from (4.90) produces

$$\tilde{x}(1) = A\tilde{x}(0) + Bq(0) + \Gamma Q(0)\Gamma^T E^T R^{-1}[y(1) - E\tilde{x}(1)]. \tag{4.92}$$

With no initial error in $x(0)$, $P(1, -) = \Gamma Q(0)\Gamma^T$, and with

$$\tilde{x}(1, -) \equiv Ax(0) + Bq(0), \tag{4.93}$$

(4.92) can be written as

$$[I + P(1, -)E^T R^{-1}E]\tilde{x}(1) = \tilde{x}(1, -) + P(1, -)E^T R^{-1}y(1), \tag{4.94}$$

or (factoring $P(1, -)$)

$$\begin{aligned}
\tilde{x}(1) = {}&[P(1, -)^{-1} + E^T R^{-1}E]^{-1}P(1, -)^{-1}\tilde{x}(1, -) \\
&+ [P(1, -)^{-1} + E^T R^{-1}E]^{-1}E^T R^{-1}y(1).
\end{aligned}$$

Applying the matrix inversion lemma in the form (2.35), to the first term on the right, and in the form (2.36) to the second term on the right,

$$\begin{aligned}
\tilde{x}(1) = {}&\{P(1, -) - P(1, -)E^T[EP(1, -)E^T + R]^{-1}EP(1, -)\} \\
&\times P(1, -)^{-1}\tilde{x}(1, -) + P(1, -)E^T[R + EP(1, -)E^T]^{-1}y(1), \tag{4.95}
\end{aligned}$$

or

$$\tilde{x}(1) = \tilde{x}(1, -) + P(1, -)E^T[EP(1, -)E^T + R]^{-1}[y(1) - E\tilde{x}(1, -)]. \tag{4.96}$$

This last result is the ordinary Kalman filter estimate, as it must be, but it results here from the Lagrange multiplier formalism.

Now consider this approach for the entire interval $t = 0, 1, \ldots, t_f$. Start with the objective function (4.41) and append the model consistency demand using Lagrange multipliers:

$$J = [\tilde{\mathbf{x}}(0, +) - \tilde{\mathbf{x}}(0)]^{\mathrm{T}} \mathbf{P}(0)^{-1} [\tilde{\mathbf{x}}(0, +) - \tilde{\mathbf{x}}(0)]$$

$$+ \sum_{t=1}^{t_f} [\mathbf{y}(t) - \mathbf{E}(t)\tilde{\mathbf{x}}(t, +)]^{\mathrm{T}} \mathbf{R}(t)^{-1} [\mathbf{y}(t) - \mathbf{E}(t)\tilde{\mathbf{x}}(t, +)]$$

$$+ \sum_{t=0}^{t_f-1} \tilde{\mathbf{u}}(t, +)^{\mathrm{T}} \mathbf{Q}(t)^{-1} \tilde{\mathbf{u}}(t, +) \tag{4.97}$$

$$- 2 \sum_{t=1}^{t_f} \boldsymbol{\mu}(t)^{\mathrm{T}} [\tilde{\mathbf{x}}(t, +) - \mathbf{A}\tilde{\mathbf{x}}(t - 1, +) - \mathbf{Bq}(t - 1, +) - \boldsymbol{\Gamma}\tilde{\mathbf{u}}(t - 1, +)].$$

Note the differing summation limits.

Note on notation   Equation (4.97) has been written with $\tilde{\mathbf{x}}(t, +)$, $\tilde{\mathbf{u}}(t, +)$ to make it clear that the estimates will be based upon all data, past and future. But unlike the filter/smoother algorithm, there will only be a single estimated value, instead of the multiple estimates previously computed: $\tilde{\mathbf{x}}(t, -)$ (from the model forecast), $\tilde{\mathbf{x}}(t)$ (from the Kalman filter), and $\tilde{\mathbf{x}}(t, +)$ from the smoother, and similarly for $\mathbf{u}(t)$. Of necessity, $\tilde{\mathbf{x}}(t_f, +) = \tilde{\mathbf{x}}(t)$ from the Kalman filter. $\mathbf{x}_0$ is any initial condition estimate with uncertainty $\mathbf{P}(0)$ obtained from any source.

Setting all the derivatives to zero gives the normal equations:

$$\frac{1}{2} \frac{\partial J}{\partial \tilde{\mathbf{u}}(t, +)} = \mathbf{Q}(t)^{-1}\tilde{\mathbf{u}}(t, +) + \boldsymbol{\Gamma}^{\mathrm{T}}\boldsymbol{\mu}(t + 1) = 0, \quad t = 0, 1, \ldots, t_f - 1, \tag{4.98}$$

$$\frac{1}{2} \frac{\partial J}{\partial \boldsymbol{\mu}(t)} = \tilde{\mathbf{x}}(t, +) - \mathbf{A}\tilde{\mathbf{x}}(t - 1, +) - \mathbf{Bq}(t - 1) - \boldsymbol{\Gamma}\tilde{\mathbf{u}}(t - 1, +) = 0,$$

$$t = 0, 1, \ldots, t_f \tag{4.99}$$

$$\frac{1}{2} \frac{\partial J}{\partial \tilde{\mathbf{x}}(0, +)} = \mathbf{P}(0)^{-1}\big(\tilde{\mathbf{x}}(0, +) - \tilde{\mathbf{x}}(0)\big) + \mathbf{A}^{\mathrm{T}}\boldsymbol{\mu}(1) = 0, \tag{4.100}$$

$$\frac{1}{2} \frac{\partial J}{\partial \tilde{\mathbf{x}}(t, +)} = -\mathbf{E}(t)^{\mathrm{T}}\mathbf{R}(t)^{-1}[\mathbf{y}(t) - \mathbf{E}(t)\tilde{\mathbf{x}}(t, +)] - \boldsymbol{\mu}(t) + \mathbf{A}^{\mathrm{T}}\boldsymbol{\mu}(t + 1) = 0,$$

$$t = 1, 2, \ldots, t_f - 1, \tag{4.101}$$

$$\frac{1}{2} \frac{\partial J}{\partial \tilde{\mathbf{x}}(t_f)} = -\mathbf{E}(t_f)^{\mathrm{T}}\mathbf{R}(t_f)^{-1}[\mathbf{y}(t_f) - \mathbf{E}(t_f)\tilde{\mathbf{x}}(t_f)] - \boldsymbol{\mu}(t_f) = 0, \tag{4.102}$$

where the derivatives for $\tilde{\mathbf{x}}(t, +)$, at $t = 0, t = t_f$, have been computed separately for clarity. The so-called adjoint model is now given by (4.101). An equation count shows that the number of equations is exactly equal to the number of unknowns

$[\tilde{\mathbf{x}}(t, +),\ \tilde{\mathbf{u}}(t, +),\ \boldsymbol{\mu}(t)]$. With a large enough computer, we could contemplate solving them all at once. But for real fluid models with large time spans and large state vectors, even the biggest supercomputers are swamped, and one needs to find other methods.

The adjoint model in Eq. (4.101) is

$$\boldsymbol{\mu}(t) = \mathbf{A}^{\mathrm{T}}\boldsymbol{\mu}(t+1) + \mathbf{E}(t)^{\mathrm{T}}\mathbf{R}(t)^{-1}[\mathbf{E}(t)\tilde{\mathbf{x}}(t, +) - \mathbf{y}(t)],$$

in which the model/data misfit appears as a "source term" (compare Eq. (2.355), noting that the $\mathbf{A}$ matrices are defined differently). It is sometimes said that time runs backwards in this equation, with $\boldsymbol{\mu}(t)$ being computed most naturally from $\boldsymbol{\mu}(t+1)$ and the source term, with Eq. (4.102) providing an initial condition. But in fact, time has no particular direction here, as the equations govern a time interval, $t = 1, 2, \ldots, t_f$. Indeed if $\mathbf{A}^{-1}$ exists, there is no problem in rewriting Eq. (4.101) so that $\boldsymbol{\mu}(t+1)$ is given in terms of $\mathbf{A}^{-\mathrm{T}}\boldsymbol{\mu}(t)$.

The Lagrange multipliers – that is, the adjoint solution – have the same interpretation that they did for the steady models described in Chapter 2 – that is, as a measure of the objective function sensitivity to the data,

$$\frac{\partial J'}{\partial \mathbf{Bq}(t)} = 2\boldsymbol{\mu}(t+1). \tag{4.103}$$

The physics of the adjoint model, as in Chapter 2, are again represented by the matrix $\mathbf{A}^{\mathrm{T}}$. For a forward model that is both linear and self-adjoint ($\mathbf{A}^{\mathrm{T}} = \mathbf{A}$), the adjoint solution would have the same physical behavior as the state vector. If the model is not self-adjoint (the usual situation), the evolution of the $\boldsymbol{\mu}(t)$ may have a radically different interpretation than $\mathbf{x}(t)$. Insight into that physics is the road to understanding of information flow in the system. For example, if one employed a large numerical model to compute the flux of heat in a fluid, and wished to understand the extent to which the result was sensitive to the boundary conditions, or to a prescribed flux somewhere, the adjoint solution carries that information. In the future, one expects to see display and discussion of the results of the adjoint model on a nearly equal footing with that of the forward model.

4.4.2 Terminal constraint problem: open-loop control

Consider the adjoint approach in the context of the simple tracer box model already described and depicted in Fig. 4.8. At $t = 0$, the tracer concentrations in the boxes are known to vanish – that is, $\mathbf{x}(0) = \mathbf{x}_0 = \mathbf{0}$ (the initial conditions are supposedly known exactly). At $t = t_f$, a survey is made of the region, and the concentrations $\mathbf{y}(t_f) = \mathbf{E}(t_f)\mathbf{x}(t_f) + \mathbf{n}(t_f)$, $\mathbf{E}(t_f) \equiv \mathbf{I}$, $\langle \mathbf{n}(t) \rangle = \mathbf{0}$, $\langle \mathbf{n}(t_f)\mathbf{n}(t_f)^{\mathrm{T}} \rangle = \mathbf{R}$ are known. No other observations are available. The question posed is: If the boundary conditions are all unknown a priori – that is, $\mathbf{Bq} \equiv \mathbf{0}$, and all boundary conditions are control variables – what boundary conditions would produce the observed values at t_f within the estimated error bars? Write $\mathbf{x}_d = \mathbf{y}(t_f)$ – denoting the "desired" final state.

The problem is an example of a "terminal constraint control problem" – it seeks controls (forces, etc.) able to drive the system from an observed initial state, here zero concentration, to within a given tolerance of a required terminal state, $\mathbf{x}_d$.[24] (The control literature refers to the "Pontryagin Principle.") But in the present context, we interpret the result as an *estimate* of the actual boundary condition with uncertainty $\mathbf{R}(t_f)$. For this special case, take the objective function,

$$
J = [\tilde{\mathbf{x}}(t_f) - \mathbf{x}_d]^{\mathrm{T}} \mathbf{R}(t_f)^{-1} [\tilde{\mathbf{x}}(t_f) - \mathbf{x}_d] + \sum_{t=0}^{t_f-1} \tilde{\mathbf{u}}^{\mathrm{T}}(t) \mathbf{Q}(t)^{-1} \tilde{\mathbf{u}}(t)
$$

$$
- 2 \sum_{1}^{t_f} \boldsymbol{\mu}(t)^{\mathrm{T}} [\tilde{\mathbf{x}}(t) - \mathbf{A}\tilde{\mathbf{x}}(t-1) - \mathbf{Bq}(t-1) - \boldsymbol{\Gamma}\tilde{\mathbf{u}}(t-1)].
$$
(4.104)

From here on, the notation $\tilde{\mathbf{x}}(t, +)$, $\tilde{\mathbf{u}}(t, +)$ in objective functions is suppressed, using $\tilde{\mathbf{x}}(t)$, $\tilde{\mathbf{u}}(t)$ with the understanding that any solution is an estimate, from whatever data are available, past, present, or future. The governing normal equations are

$$
\boldsymbol{\mu}(t-1) = \mathbf{A}^{\mathrm{T}} \boldsymbol{\mu}(t), \quad t = 1, 2, \ldots, t_f,
$$
(4.105)

$$
\boldsymbol{\mu}(t_f) = \mathbf{R}^{-1}(\tilde{\mathbf{x}}(t_f) - \mathbf{x}_d),
$$
(4.106)

$$
\mathbf{Q}(t)^{-1} \tilde{\mathbf{u}}(t) = -\boldsymbol{\Gamma}^{\mathrm{T}} \boldsymbol{\mu}(t+1),
$$
(4.107)

plus the model. Eliminating

$$
\tilde{\mathbf{u}}(t) = -\mathbf{Q}(t) \boldsymbol{\Gamma}^{\mathrm{T}} \boldsymbol{\mu}(t+1),
$$
(4.108)

and substituting into the model, the system to be solved is

$$
\tilde{\mathbf{x}}(t) = \mathbf{A}\tilde{\mathbf{x}}(t-1) - \boldsymbol{\Gamma}\mathbf{Q}(t)\boldsymbol{\Gamma}^{\mathrm{T}}\boldsymbol{\mu}(t), \qquad \tilde{\mathbf{x}}(0) = \mathbf{x}_0 \equiv \mathbf{0},
$$
(4.109)

$$
\boldsymbol{\mu}(t-1) = \mathbf{A}^{\mathrm{T}}\boldsymbol{\mu}(t), \quad t = 1, 2, \ldots, t_f - 1,
$$
(4.110)

$$
\boldsymbol{\mu}(t_f) = \mathbf{R}^{-1}(\tilde{\mathbf{x}}(t_f) - \mathbf{x}_d).
$$
(4.111)

As written, this coupled problem has natural initial conditions for the state vector, $\mathbf{x}(t)$, at $t = 0$, and for $\boldsymbol{\mu}(t)$ at $t = t_f$, but with the latter in terms of the still unknown $\mathbf{x}(t_f)$ – recognizing that the estimated terminal state and the desired one will almost always differ, that is, $\tilde{\mathbf{x}}(t_f) \neq \mathbf{x}_d$.

By exploiting its special structure, this problem can be solved in straightforward fashion without having to deal with the giant set of simultaneous equations.

Using (4.111), step backwards in time from t_f via (4.110) to produce

$$\boldsymbol{\mu}(t_f) = \mathbf{A}^{\mathrm{T}}\mathbf{R}^{-1}(\tilde{\mathbf{x}}(t_f) - \mathbf{x}_d),$$

$$\vdots \tag{4.112}$$

$$\boldsymbol{\mu}(1) = \mathbf{A}^{(t_f)\mathrm{T}}\mathbf{R}^{-1}(\tilde{\mathbf{x}}(t_f) - \mathbf{x}_d),$$

so that $\boldsymbol{\mu}(t)$ is given in terms of the known $\mathbf{x}_d$ and the still unknown $\tilde{\mathbf{x}}(t_f)$. Substituting into (4.109) generates

$$\tilde{\mathbf{x}}(1) = \mathbf{A}\tilde{\mathbf{x}}(0) - \boldsymbol{\Gamma}\mathbf{Q}\boldsymbol{\Gamma}^{\mathrm{T}}\mathbf{A}^{(t_f-1)\mathrm{T}}\mathbf{R}^{-1}(\tilde{\mathbf{x}}(t_f) - \mathbf{x}_d),$$

$$\tilde{\mathbf{x}}(2) = \mathbf{A}\tilde{\mathbf{x}}(1) - \boldsymbol{\Gamma}\mathbf{Q}\boldsymbol{\Gamma}^{\mathrm{T}}\mathbf{A}^{(t_f-2)\mathrm{T}}\mathbf{R}^{-1}(\tilde{\mathbf{x}}(t_f) - \mathbf{x}_d),$$

$$= \mathbf{A}^2\tilde{\mathbf{x}}(0) - \mathbf{A}\mathbf{Q}\boldsymbol{\Gamma}^{\mathrm{T}}\mathbf{A}^{(t_f-1)\mathrm{T}}\mathbf{R}^{-1}(\tilde{\mathbf{x}}(t_f) - \mathbf{x}_d)$$

$$- \boldsymbol{\Gamma}\mathbf{Q}\boldsymbol{\Gamma}^{\mathrm{T}}\mathbf{A}^{(t_f-2)\mathrm{T}}\mathbf{R}^{-1}(\tilde{\mathbf{x}}(t_f) - \mathbf{x}_d),$$

$$\vdots \tag{4.113}$$

$$\tilde{\mathbf{x}}(t_f) = \mathbf{A}^{t_f}\tilde{\mathbf{x}}(0) - \mathbf{A}^{(t_f-1)}\boldsymbol{\Gamma}\mathbf{Q}\boldsymbol{\Gamma}^{\mathrm{T}}\mathbf{A}^{(t_f-1)\mathrm{T}}\mathbf{R}^{-1}(\tilde{\mathbf{x}}(t_f) - \mathbf{x}_d)$$

$$- \mathbf{A}^{(t_f-2)}\boldsymbol{\Gamma}\mathbf{Q}\boldsymbol{\Gamma}^{\mathrm{T}}\mathbf{A}^{(t_f-2)\mathrm{T}}\mathbf{R}^{-1}(\tilde{\mathbf{x}}(t_f) - \mathbf{x}_d)$$

$$- \cdots - \boldsymbol{\Gamma}\mathbf{Q}\boldsymbol{\Gamma}^{\mathrm{T}}\mathbf{R}^{-1}(\tilde{\mathbf{x}}(t_f) - \mathbf{x}_d).$$

The last equation permits us to bring the terms in $\tilde{\mathbf{x}}(t_f)$ over to the left-hand side and solve for $\tilde{\mathbf{x}}(t_f)$ in terms of $\mathbf{x}_d$ and $\tilde{\mathbf{x}}(0)$:

$$\begin{aligned}
\big\{ \mathbf{I} &+ \mathbf{A}^{(t_f-1)}\boldsymbol{\Gamma}\mathbf{Q}\boldsymbol{\Gamma}^{\mathrm{T}}\mathbf{A}^{(t_f-1)\mathrm{T}}\mathbf{R}^{-1} \\
&+ \mathbf{A}^{(t_f-2)}\boldsymbol{\Gamma}\mathbf{Q}\boldsymbol{\Gamma}^{\mathrm{T}}\mathbf{A}^{(t_f-2)\mathrm{T}}\mathbf{R}^{-1} + \cdots + \boldsymbol{\Gamma}\mathbf{Q}\boldsymbol{\Gamma}^{\mathrm{T}}\mathbf{R}^{-1} \big\} \tilde{\mathbf{x}}(t_f) \\
&= \mathbf{A}^{t_f}\tilde{\mathbf{x}}(0) + \big\{ \mathbf{A}^{(t_f-1)}\boldsymbol{\Gamma}\mathbf{Q}\boldsymbol{\Gamma}^{\mathrm{T}}\mathbf{A}^{(t_f-1)\mathrm{T}}\mathbf{R}^{-1} \\
&\qquad + \mathbf{A}^{(t_f-2)}\boldsymbol{\Gamma}\mathbf{Q}\boldsymbol{\Gamma}^{\mathrm{T}}\mathbf{A}^{(t_f-2)\mathrm{T}}\mathbf{R}^{-1} + \cdots + \boldsymbol{\Gamma}\mathbf{Q}\boldsymbol{\Gamma}^{\mathrm{T}}\mathbf{R}^{-1} \big\} \mathbf{x}_d.
\end{aligned} \tag{4.114}$$

With $\tilde{\mathbf{x}}(t_f)$ now known, $\boldsymbol{\mu}(t)$ can be computed for all t from (4.110) and (4.111). Then the control $\tilde{\mathbf{u}}(t)$ is also known from (4.107) and the state vector can be found from (4.109). The resulting solution for $\tilde{\mathbf{u}}(t)$ is in terms of the externally prescribed $\tilde{\mathbf{x}}(0)$, $\mathbf{x}_d$ and is usually known as "open-loop" control.

The canonical form for a terminal constraint problem usually used in the control literature differs slightly; it is specified in terms of a given, non-zero, initial condition $\mathbf{x}(0)$, and the controls are determined so as to come close to a desired zero terminal state. By linearity, the solution to this so-called deadbeat control (driving the system to rest) problem can be used to solve the problem for an arbitrary desired terminal state.

Example *Consider the tracer forward problem in Fig. 4.8 where only boundary box 2 now has a non-zero concentration, fixed at $C = 1$, starting at $t = 1$. A concentration is readily imposed by zeroing the corresponding row of* $\mathbf{A}$*, so that* $\mathbf{B}\mathbf{q}(t)$

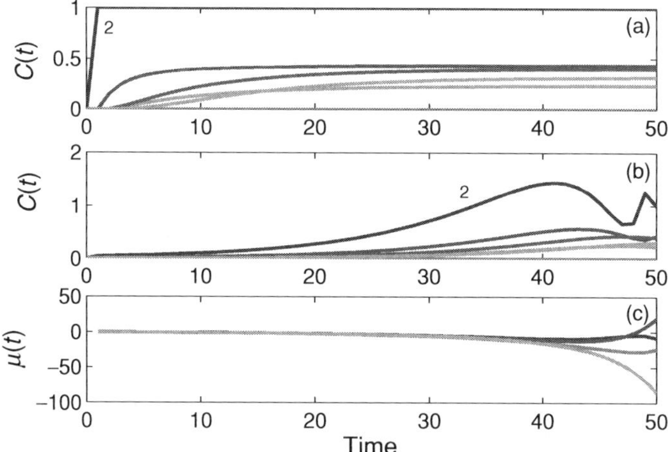

Figure 4.12 Box model example of terminal control. Here the "forward" calculation fixes the concentration in boundary box number 2 as $C = 1$, and all other boundary box concentrations are fixed at zero. (a) The box 2 and interior box concentrations for 50 time-steps with initial condition of zero concentration everywhere. (b) The estimated concentration from the terminal control calculation, in which $\mathbf{R} = 10^{-4}\mathbf{I}$, $\mathbf{Q} = 1$, where the only control value was the box 2 concentration. Thus a slight misfit is permitted to the terminal values $\mathbf{C}(50\Delta t)$, $\Delta t = 0.05$. (c) The Lagrange multipliers (adjoint solution) corresponding to the interior boxes. Having the largest values near the termination point is characteristic, and shows the sensitivity to the near terminal times of the constraints.

or $\mathbf{\Gamma u}(t)$ set the concentration. (An alternative is to put the imposed concentration into the initial conditions and use the corresponding row of $\mathbf{A}$ to force the concentration to be exactly that in the previous time step.) The initial conditions were taken as zero and the forward solution is in Fig. 4.12. Then the same figure shows the solution to the terminal time control problem for the concentration in box 2 giving rise to the terminal values. A misfit was permitted between the desired (observed) and calculated terminal times – with an RMS value of 2×10^{-4}. Clearly the "true" solution is underdetermined by the provision of initial and terminal time tracer concentrations alone. Also shown in the figure are the Lagrange multipliers (adjoint solution) corresponding to the model equations for each box.[25]

In the above formulation, the boundary boxes were contained in the $\mathbf{A}$ matrix, but the corresponding rows were all zero, permitting the $\mathbf{B}$ matrix (here a vector) to control the boundary box concentrations. A variation on this problem is obtained by setting column element j_0 corresponding to boundary box j_0, in $\mathbf{A}$ to unity. $\mathbf{B}$ would then control the time rate of change of the boundary box concentrations. Suppose then that $\mathbf{B}$ is a column vector, vanishing in all elements except for unity

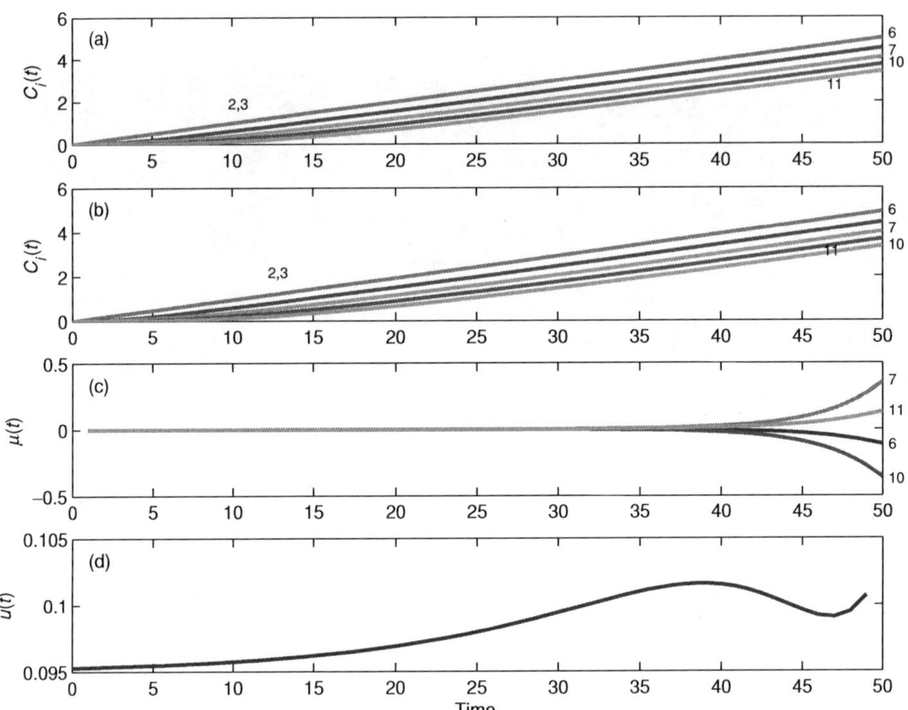

Figure 4.13 The same box model as in Fig. 4.12, except that now $\mathbf{\Gamma u}(t)$ controls the rate of change of concentration rather than concentration itself, and all boundary boxes have a constant rate of change of 0.1. (a) "True" solution. (b) The solution deduced from the terminal state control, with a near-perfect requirement on the terminal values and $\mathbf{Q}$. (c) The Lagrange multipliers for the interior box constraints of the model in Fig. 4.8. (d) Estimated control $\mathbf{\tilde{u}}(t)$. Note the highly compressed amplitude scale.

in all active boundary boxes (the corner boxes are passive here). Then Fig. 4.13 shows the concentration and the result of the terminal control problem in this case.

The smoothing problem has been solved without having to compute the uncertainties, and is the major advantage of the Lagrange multiplier methods over the sequential estimators. Lagrange multiplier methods solve for the entire time domain at once; consequently, there is no weighted averaging of intermediate solutions and no need for the uncertainties. On the other hand, the utility of solutions without uncertainty estimates must be questioned.

In the context of Chapter 1, problems of arbitrary posedness are being solved. The various methods using objective functions, prior statistics, etc., whether in time-evolving or static situations, permit stable, useful estimates to be made under almost any circumstances, using almost any sort of available information. But the reader will by now appreciate that the use of such methods can produce structures in the solution, pleasing or otherwise, that may be present because they are required by (1) the observations, (2) the model, (3) the prior statistics, (4) some norm or smoothness demand on elements of the solution, or (5) all of the preceding in concert. A solution produced in ignorance of these differing sources of structure can hardly be thought very useful, and it is the uncertainty matrices that are usually the key to understanding. Consequently, we will later briefly examine the problem of obtaining the missing covariances. In the meantime, one should note that the covariances of the filter/smoother will also describe the uncertainty of the Lagrange multiplier method solution, because they are the same solution to the same set of equations deriving from the same objective function.

There is one situation where a solution without uncertainty estimates is plainly useful – it is where one simply inquires, "Is there a solution at all?" – that is, when one wants to know if the observations actually contradict the model. In that situation, mere existence of an acceptable solution may be of greatest importance, suggesting, for example, that a model of adequate complexity is already available and that the data errors are understood.

4.4.3 Representers and boundary Green functions

The particular structure of Eqs. (4.105)–(4.107) permits several different methods of solution, of which the version just given is an example. To generalize this problem, assume observations at a set of arbitrary times (not just the terminal time)

$$\mathbf{y}(t) = \mathbf{E}(t)\mathbf{x}(t) + \mathbf{n}(t),$$

and seek a solution in "representers."

Take the objective function to be

$$J = \sum_{t=1}^{t_f} [\mathbf{y}(t) - \mathbf{E}(t)\mathbf{x}(t)]^T \mathbf{R}(t)^{-1} [\mathbf{y}(t) - \mathbf{E}(t)\mathbf{x}(t)]$$

$$+ \sum_{t=0}^{t_f-1} \mathbf{u}(t)^T \mathbf{Q}(t)^{-1} \mathbf{u}(t) - 2 \sum_{t=1}^{t_f} \mu(t)^T \tag{4.115}$$

$$\times [\mathbf{x}(t) - \mathbf{A}\mathbf{x}(t-1) - \mathbf{B}\mathbf{q}(t-1) - \mathbf{\Gamma}(t-1)\mathbf{u}(t-1)],$$

so that the terminal state estimate is subsumed into the first term with $\mathbf{E}(t_f) = \mathbf{I}$, $\mathbf{R}(t_f) = \mathbf{P}(t_f)$. (The tildes are now being omitted.) Let $\mathbf{x}_a(t)$ be the (known) solution to the pure, unconstrained, forward problem,

$$\mathbf{x}_a(t) = \mathbf{A}\mathbf{x}_a(t-1) + \mathbf{B}\mathbf{q}(t-1), \quad \mathbf{x}_a(0) = \mathbf{x}_0. \tag{4.116}$$

Redefine $\mathbf{x}(t)$ to be the difference $\mathbf{x}(t) \rightarrow \mathbf{x}(t) - \mathbf{x}_a(t)$, that is, the deviation from what can be regarded as the a-priori solution. The purpose of this redefinition is to remove any inhomogeneous initial or boundary conditions from the problem – exploiting the system linearity. The normal equations are then

$$\frac{1}{2}\frac{\partial J}{\partial \mathbf{u}(t)} = \mathbf{Q}(t)^{-1}\mathbf{u}(t) + \mathbf{\Gamma}^{\mathrm{T}}\boldsymbol{\mu}(t+1) = \mathbf{0}, \quad t = 0, 1, \ldots, t_f - 1,$$

$$\frac{1}{2}\frac{\partial J}{\partial \mathbf{x}(t)} = \mathbf{E}(t)^{\mathrm{T}}\mathbf{R}(t)^{-1}[\mathbf{E}(t)\mathbf{x}(t) - \mathbf{y}(t)] + \mathbf{A}^{\mathrm{T}}\boldsymbol{\mu}(t+1) - \boldsymbol{\mu}(t) = \mathbf{0},$$

$$t = 1, 2, \ldots, t_f,$$

$$\frac{1}{2}\frac{\partial J}{\partial \boldsymbol{\mu}(t)} = \mathbf{x}(t) - \mathbf{A}\mathbf{x}(t-1) - \mathbf{\Gamma}(t-1)\mathbf{u}(t-1) = \mathbf{0}, \ \mathbf{x}(0) = \mathbf{0},$$

$$t = 1, 2, \ldots, t_f.$$

Eliminating the $\mathbf{u}(t)$ in favor of $\boldsymbol{\mu}(t)$, we have, as before,

$$\mathbf{x}(t) = \mathbf{A}\mathbf{x}(t-1) - \mathbf{\Gamma}\mathbf{Q}(t-1)\mathbf{\Gamma}^{\mathrm{T}}\boldsymbol{\mu}(t), \ \mathbf{x}(0) = \mathbf{0}, \tag{4.117}$$

$$\boldsymbol{\mu}(t) = \mathbf{A}^{\mathrm{T}}\boldsymbol{\mu}(t+1) + \mathbf{E}(t)^{\mathrm{T}}\mathbf{R}(t)^{-1}[\mathbf{E}(t)\mathbf{x}(t) - \mathbf{y}(t)]. \tag{4.118}$$

The system is linear, so we can examine the solution forced by the inhomogeneous term in (4.118) at one time, $t = t_m$. This inhomogeneous term, $\mathbf{E}(t)^{\mathrm{T}}\mathbf{R}(t)^{-1}[\mathbf{E}(t)\mathbf{x}(t) - \mathbf{y}(t)]$, in Eq. (4.118) is, however, unknown until $\mathbf{x}(t)$ has been determined. So to proceed, *first solve the different problem*,

$$\mathbf{M}(t, t_m) = \mathbf{A}^{\mathrm{T}}\mathbf{M}(t+1, t_m) + \mathbf{I}\delta_{t,t_m}, \quad t \le t_m, \tag{4.119}$$

$$\mathbf{M}(t, t_m) = 0, \quad t > t_m, \tag{4.120}$$

where the second argument, t_m, denotes the time of one set of observations (notice that $\mathbf{M}$ is a matrix). Time-step Eq. (4.119) backwards from $t = t_m$. There is then a corresponding solution to (4.117) with these values of $\boldsymbol{\mu}(t)$,

$$\mathbf{G}(t+1, t_m) = \mathbf{A}\mathbf{G}(t, t_m) - \mathbf{\Gamma}\mathbf{Q}\mathbf{\Gamma}^{\mathrm{T}}\mathbf{M}(t+1, t_m), \tag{4.121}$$

which is stepped-forward in time starting with $\mathbf{G}(0, t_m) = \mathbf{0}$, until $t + 1 = t_m$. Both $\mathbf{G}, \mathbf{M}$ are computable independent of the actual data values. Now put

$$\mathbf{m}(t, t_m) = \mathbf{M}(t, t_m)\{\mathbf{E}(t_m)^{\mathrm{T}}\mathbf{R}(t_m)^{-1}[\mathbf{E}(t_m)\mathbf{x}(t_m) - \mathbf{y}(t_m)]\}, \tag{4.122}$$

which is a vector that, by linearity, $\boldsymbol{\mu}(t) = \mathbf{m}(t, t_m)$ is the solution to (4.118) once $\mathbf{x}(t_m)$ is known. Let

$$\boldsymbol{\xi}(t, t_m) = \mathbf{G}(t, t_m)\{\mathbf{E}(t_m)^{\mathrm{T}}\mathbf{R}(t_m)^{-1}[\mathbf{E}(t_m)\boldsymbol{\xi}(t_m, t_m) - \mathbf{y}(t_m)]\}, \tag{4.123}$$

which is another vector, such that $\tilde{\mathbf{x}}(t) = \boldsymbol{\xi}(t, t_m)$ would be the solution sought. Setting $t = t_m$ in Eq. (4.123) and solving,

$$\boldsymbol{\xi}(t_m, t_m) = -[\mathbf{I} - \mathbf{G}(t_m, t_m)\mathbf{E}(t_m)^{\mathrm{T}}\mathbf{R}(t_m)^{-1}\mathbf{E}(t_m)]^{-1} \tag{4.124}$$

$$\times [\mathbf{G}(t_m, t_m)\mathbf{E}(t_m)^{\mathrm{T}}\mathbf{R}(t_m)^{-1}]\mathbf{y}(t_m).$$

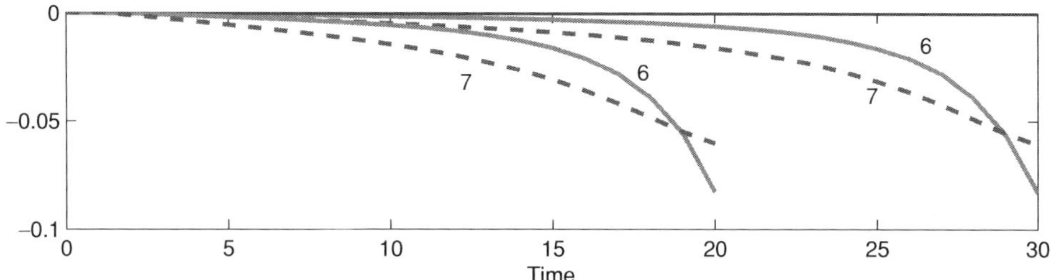

Figure 4.14 Representer (Green function) G for interior box 7, with the columns corresponding to boxes 6, 7 displayed through time. The Green function used numerically is the sum of these two, and displays a near-discontinuity (typical of Green functions) at the data points that are available at $t = 20, 30$.

With $\boldsymbol{\xi}(t_m, t_m)$ known, Eq. (4.123) produces a fully determined $\tilde{\mathbf{x}}(t) = \boldsymbol{\xi}(t, t_m)$ in representer form. This solution is evidently just a variant of Eqs. (4.113) and (4.114). Now suppose that there are multiple observations times, t_m. The problem is a linear one so that solutions can be superimposed,

$$\tilde{\mathbf{x}}(t) = \sum_{t_m} \boldsymbol{\xi}(t, t_m),$$

and after adding $\mathbf{x}_a(t)$ to the result, the entire problem is solved.

The solutions $\mathbf{M}(t, t_m)$ are the Green function for the adjoint model equation; the $\mathbf{G}(t, t_m)$ are "representers,"[26] and they exist independently of the data. *If the data distribution is spatially sparse, one need only compute the subsets of the columns or rows of* $\mathbf{M}$, $\mathbf{G}$ *that correspond to measured elements of* $\mathbf{x}(t)$. That is, in Eq. (4.119) any zero columns in $\mathbf{E}$, representing elements of the state vector not involved in the measurements, multiply the corresponding columns of $\mathbf{M}$, $\mathbf{G}$, and hence one need never compute those columns.

Example *Consider again the 4×4 box model of Fig. 4.8, in the same configuration as used above, with all the boundary boxes having a fixed tracer concentration of $C = 1$, and zero initial condition. Now, it is assumed that observations are available in all interior boxes (6, 7, 10, 11) at time $t = 20, 30$. The representer G is shown in Fig. 4.14.*

The representer emerged naturally from the Lagrange multiplier formulation. Let us re-derive the solution without the use of Lagrange multipliers to demonstrate how the adjoint model appears in unconstrained l_2 norm problems (soft constraints). Introduce the model into the same objective function as above, except we do it by substitution for the control terms; let $\boldsymbol{\Gamma} = \mathbf{I}$, making it possible to solve for $\mathbf{u}(t) = -[\mathbf{x}(t+1) - \mathbf{x}(t)]$ explicitly and producing the simplest results. The

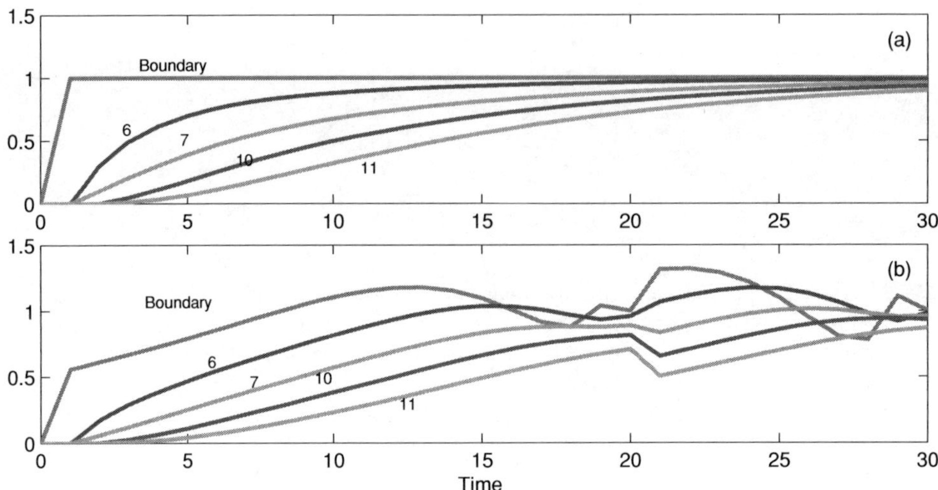

Figure 4.15 (a) The forward "truth" in the box model and (b) the estimated values
from the representer displayed in Fig. 4.14. Data were treated as nearly perfect at
the two observation times.

objective function then is

$$J = \sum_{t=0}^{t_f}[\mathbf{y}(t) - \mathbf{E}(t)\mathbf{x}(t)]^{\mathrm{T}}\mathbf{R}(t)^{-1}[\mathbf{y}(t) - \mathbf{E}(t)\mathbf{x}(t)]$$

$$+ \sum_{t=0}^{t_f-1}[\mathbf{x}(t+1) - \mathbf{A}\mathbf{x}(t)]^{\mathrm{T}}\mathbf{Q}(t)^{-1}[\mathbf{x}(t+1) - \mathbf{A}\mathbf{x}(t)]. \tag{4.125}$$

We again assume that $\mathbf{x}(t)$ is the anomaly relative to the known $\mathbf{x}_a(t)$.

The normal equations include:

$$\frac{1}{2}\frac{\partial J}{\partial \mathbf{x}(t)} = \mathbf{E}(t)^{\mathrm{T}}\mathbf{R}(t)^{-1}[\mathbf{E}(t)\mathbf{x}(t) - \mathbf{y}(t)] - \mathbf{A}^{\mathrm{T}}\mathbf{Q}(t)^{-1}[\mathbf{x}(t+1) - \mathbf{A}\mathbf{x}(t)]$$

$$+ \mathbf{Q}(t)^{-1}[\mathbf{x}(t) - \mathbf{A}\mathbf{x}(t-1)] = 0. \tag{4.126}$$

Define

$$\boldsymbol{\nu}(t+1) = -\mathbf{Q}(t-1)^{-1}[\mathbf{x}(t+1) - \mathbf{A}\mathbf{x}(t)], \tag{4.127}$$

so that the system (4.126) can be written as

$$\boldsymbol{\nu}(t) = \mathbf{A}^{\mathrm{T}}\boldsymbol{\nu}(t+1) + \mathbf{E}(t)^{\mathrm{T}}\mathbf{R}(t)^{-1}[\mathbf{E}(t)\mathbf{x}(t) - \mathbf{y}(t)], \tag{4.128}$$

which, along with (4.127), is precisely the same system of equations (4.117)
and (4.118) that emerged from the Lagrange multiplier approach, if we let
$\boldsymbol{\mu} \to \boldsymbol{\nu}$, $\boldsymbol{\Gamma} = \mathbf{I}$. Representers are again defined as the unit disturbance solution

to the system. As a by-product, we see once again, that l_2-norm least-squares and the adjoint method are simply different algorithmic approaches to the same problem.[27]

4.4.4 The control Riccati equation

Consider yet another solution of the problem. (If the reader is wondering why such a fuss is being made about these equations, the answer, among others, is that it will turn out to be an important route to reducing the computational load required for the Kalman filter and various smoothing algorithms.) We look at the same special case of the objective function (4.104) and the equations that follow from it ((4.105)–(4.107) plus the model). Let $\mathbf{x}_d = \mathbf{0}$, the deadbeat requirement defined above. For this case, the adjoint equation is

$$\boldsymbol{\mu}(t) = \mathbf{A}^T \boldsymbol{\mu}(t+1) + \mathbf{R}(t)^{-1}\mathbf{x}(t), \quad t = 1, 2, \ldots, t_f, \tag{4.129}$$

stipulating that $\mathbf{R}(t)^{-1} = \mathbf{0}$, $t \neq t_f$, if the only requirement is at the terminal time. For simplicity, let $\mathbf{Q}(t) = \mathbf{Q}$.

Take a trial solution, an "ansatz," of the form

$$\boldsymbol{\mu}(t) = \mathbf{S}(t)\mathbf{x}(t), \tag{4.130}$$

where $\mathbf{S}(t)$ is unknown. Then Eq. (4.107) becomes

$$\mathbf{Q}^{-1}\mathbf{u}(t-1) + \boldsymbol{\Gamma}^T\mathbf{S}(t)\mathbf{x}(t) = \mathbf{0}, \tag{4.131}$$

or, using the model,

$$\mathbf{Q}^{-1}\mathbf{u}(t) + \boldsymbol{\Gamma}^T\mathbf{S}(t+1)(\mathbf{A}\mathbf{x}(t) + \boldsymbol{\Gamma}\mathbf{u}(t)) = \mathbf{0}. \tag{4.132}$$

So that

$$\begin{aligned}
\mathbf{u}(t) &= -\{\boldsymbol{\Gamma}^T\mathbf{S}(t+1)\boldsymbol{\Gamma} + \mathbf{Q}^{-1}\}^{-1}\boldsymbol{\Gamma}^T\mathbf{S}(t+1)\mathbf{A}\mathbf{x}(t) \\
&= -\mathbf{L}(t+1)^{-1}\boldsymbol{\Gamma}^T\mathbf{S}(t+1)\mathbf{A}\mathbf{x}(t) \\
\mathbf{L}(t+1) &= \boldsymbol{\Gamma}^T\mathbf{S}(t+1)\boldsymbol{\Gamma} + \mathbf{Q}^{-1}.
\end{aligned} \tag{4.133}$$

Substituting (4.133), and (4.130) for $\boldsymbol{\mu}(t)$, into the adjoint model (4.129),

$$\{\mathbf{A}^T\mathbf{S}(t+1)\mathbf{A} - \mathbf{A}^T\mathbf{S}(t+1)\boldsymbol{\Gamma}\mathbf{L}(t+1)^{-1}\boldsymbol{\Gamma}^T\mathbf{S}(t+1)\mathbf{A} - \mathbf{S}(t) + \mathbf{R}(t)^{-1}\}\,\mathbf{x}(t) = \mathbf{0}. \tag{4.134}$$

Unless $\mathbf{x}(t)$ is to vanish identically,

$$\mathbf{S}(t) = \mathbf{A}^T\mathbf{S}(t+1)\mathbf{A} - \mathbf{A}^T\mathbf{S}(t+1)\boldsymbol{\Gamma}\mathbf{L}(t+1)^{-1}\boldsymbol{\Gamma}^T\mathbf{S}(t+1)\mathbf{A} + \mathbf{R}(t)^{-1}. \tag{4.135}$$

Equation (4.135) is a non-linear difference equation known as the matrix "Riccati equation," and produces a backwards recursion for $\mathbf{S}(t)$. Start the recursion with

$$\mathbf{S}(t_f)\mathbf{x}(t_f) = \mathbf{R}(t_f)^{-1}\mathbf{x}(t_f) \qquad \text{or} \qquad \mathbf{S}(t_f) = \mathbf{R}(t_f)^{-1}, \tag{4.136}$$

(recalling $\mathbf{x}_d = 0$) and step backwards to $t = 0$. The problem has now been solved by what is called the "sweep method."[28] Notice that with $\mathbf{S}(t)$ known, the control is in the form

$$\mathbf{\Gamma}\mathbf{u}(t) = \mathbf{K}_c(t)\mathbf{x}(t). \tag{4.137}$$

This is known as "feedback control" because the values to be applied are determined by the value of the state vector at that time. It contrasts with the open-loop control form derived above, but necessarily produces an identical answer.

With feedback control, the computation of the model update step would now be

$$\mathbf{x}(t) = (\mathbf{A} - \mathbf{K}_c)\mathbf{x}(t - 1) + \mathbf{B}\mathbf{q}(t - 1). \tag{4.138}$$

The structure of the matrix,

$$\mathbf{A}' = \mathbf{A} - \mathbf{K}_c, \tag{4.139}$$

is the center of a discussion of the stability of the scheme, which we will not pursue here.

4.4.5 The initialization problem

Another special case of wide interest is determination of the initial conditions, $\tilde{\mathbf{x}}(0)$, from later observations. For notational simplicity and without loss of generality, assume that the known controls vanish so that the model is

$$\mathbf{x}(t) = \mathbf{A}\mathbf{x}(t - 1) + \mathbf{\Gamma}\mathbf{u}(t - 1), \tag{4.140}$$

that there is an existing estimate of the initial conditions, $\tilde{\mathbf{x}}(0)$, with estimated uncertainty $\mathbf{P}(0)$, and that there is a single terminal observation of the complete state,

$$\mathbf{y}(t_f) = \mathbf{E}\mathbf{x}(t_f) + \mathbf{n}(t_f), \quad \mathbf{E} = \mathbf{I}, \tag{4.141}$$

where the observational noise covariance is again $\mathbf{R}(t_f)$. This problem can now be solved in five different ways:

1. The terminal observations can be written explicitly in terms of the initial conditions as

$$\begin{aligned}
\mathbf{y}(t_f) = \mathbf{A}^{t_f}\tilde{\mathbf{x}}(0) + \mathbf{A}^{t_f-1}\mathbf{\Gamma}\tilde{\mathbf{u}}(0) + \mathbf{A}^{t_f-2}\mathbf{\Gamma}\tilde{\mathbf{u}}(1) + \cdots \\
+ \mathbf{\Gamma}\tilde{\mathbf{u}}(t_f - 1) + \mathbf{n}(t_f),
\end{aligned} \tag{4.142}$$

which, in canonical observation equation form, is

$$\mathbf{y}(t_f) = \mathbf{E}_p \tilde{\mathbf{x}}(0) + \mathbf{n}_p(t_f), \quad \mathbf{E}_p = \mathbf{A}^{t_f},$$
$$\mathbf{n}_p = \mathbf{A}^{t_f-1}\boldsymbol{\Gamma}\tilde{\mathbf{u}}(0) + \cdots + \boldsymbol{\Gamma}\tilde{\mathbf{u}}(t_f - 1) + \mathbf{n}(t_f),$$

and where the covariance of this combined error is

$$\mathbf{R}_p \equiv \langle \mathbf{n}_p \mathbf{n}_p^{\mathrm{T}} \rangle = \mathbf{A}^{t_f-1}\boldsymbol{\Gamma}\mathbf{Q}\boldsymbol{\Gamma}^{\mathrm{T}}\mathbf{A}^{(t_f-1)\mathrm{T}} + \cdots + \boldsymbol{\Gamma}\mathbf{Q}\boldsymbol{\Gamma}^{\mathrm{T}} + \mathbf{R}(t_f). \tag{4.143}$$

Then the least-squares recursive solution leads to

$$\tilde{\mathbf{x}}(0, +) = \tilde{\mathbf{x}}(0) + \mathbf{P}(0)\mathbf{E}_p^{\mathrm{T}}\left[\mathbf{E}_p\mathbf{P}(0)\mathbf{E}_p^{\mathrm{T}} + \mathbf{R}_p\right]^{-1}[\mathbf{y}(t_f) - \mathbf{E}_p\tilde{\mathbf{x}}(0)], \tag{4.144}$$

and the uncertainty estimate follows immediately.

2. A second method (which the reader should confirm produces the same answer) is to run the Kalman filter forward to t_f and then run the smoother backwards to $t = 0$. There is more computation here, but a by-product is an estimate of the intermediate values of the state vectors, of the controls, and their uncertainty.

3. Write the model in backwards form,

$$\mathbf{x}(t) = \mathbf{A}^{-1}\mathbf{x}(t + 1) - \mathbf{A}^{-1}\boldsymbol{\Gamma}\mathbf{u}, \tag{4.145}$$

and use the Kalman filter on this model, with time running backwards. The observation equation (4.141) provides the initial estimate of $\mathbf{x}(t_f)$, and its error covariance becomes the initial estimate covariance $\mathbf{P}(t_f)$. At $t = 0$, the original estimate of $\tilde{\mathbf{x}}(0)$ is treated as an observation, with uncertainty $\mathbf{P}(0)$ taking the place of the usual $\mathbf{R}$. The reader should again confirm that the answer is the same as in (1).

4. The problem has already been solved using the Lagrange multiplier formalism.

5. The Green function representation (4.32) is immediately solvable for $\tilde{\mathbf{x}}(0, +)$.

4.5 Duality and simplification: the steady-state filter and adjoint

For linear models, the Lagrange multiplier method and the filter/smoother algorithms produce identical solutions. In both cases, the computation of the uncertainty remains an issue – in the former case because it is not part of the solution, and in the latter because it can overwhelm the computation. However, if the uncertainty is computed for the sequential estimator solutions, it must also represent the uncertainty derived from the Lagrange multiplier principle. In the interests of gaining insight into both methods, and of ultimately finding uncertainty estimates, consider again the covariance propagation equations for the Kalman filter:

$$\mathbf{P}(t, -) = \mathbf{A}(t - 1)\mathbf{P}(t - 1)\mathbf{A}(t - 1)^{\mathrm{T}} + \boldsymbol{\Gamma}(t - 1)\mathbf{Q}(t - 1)\boldsymbol{\Gamma}(t - 1)^{\mathrm{T}}, \tag{4.146}$$
$$\mathbf{P}(t) = \mathbf{P}(t, -) - \mathbf{P}(t, -)\mathbf{E}(t)^{\mathrm{T}}[\mathbf{E}(t)\mathbf{P}(t, -)\mathbf{E}(t)^{\mathrm{T}} + \mathbf{R}(t)]^{-1}\mathbf{E}(t)\mathbf{P}(t, -),$$
$$\tag{4.147}$$

Table 4.1. *Correspondences between the variables of the control formulation and that of the Kalman filter, which lead to the Riccati equation. Note that time runs backward for control cases and forward for the filter.*

Adjoint/control	Kalman filter
$\mathbf{A}$	$\mathbf{A}^{\mathrm{T}}$
$\mathbf{S}(t, -)$	$\mathbf{P}(t + 1)$
$\mathbf{S}(t + 1)$	$\mathbf{P}(t + 1, -)$
$\mathbf{R}^{-1}$	$\mathbf{\Gamma}\mathbf{Q}\mathbf{\Gamma}^{\mathrm{T}}$
$\mathbf{\Gamma}$	$\mathbf{E}^{\mathrm{T}}$
$\mathbf{Q}^{-1}$	$\mathbf{R}$

where $\mathbf{K}(t)$ has been written out. Make the substitutions shown in Table 4.1; the equations for evolution of the uncertainty of the Kalman filter are identical to those for the control matrix $\mathbf{S}(t)$, given in Eq. (4.135); hence, the Kalman filter covariance *also satisfies a matrix Riccati equation.* To see that, in Eq. (4.135)[29] put

$$\mathbf{S}(t, -) \equiv \mathbf{S}(t + 1) - \mathbf{S}(t + 1)\mathbf{\Gamma}[\mathbf{\Gamma}^{\mathrm{T}}\mathbf{S}(t + 1)\mathbf{\Gamma} + \mathbf{Q}^{-1}]^{-1}\mathbf{\Gamma}^{\mathrm{T}}\mathbf{S}(t + 1), \quad (4.148)$$

and then

$$\mathbf{S}(t) = \mathbf{A}^{\mathrm{T}}\mathbf{S}(t, -)\mathbf{A} + \mathbf{R}(t)^{-1}, \quad (4.149)$$

which correspond to Eqs. (4.146) and (4.147). Time runs backwards in the control formulation and forwards in the estimation problem, but this difference is not fundamental. The significance of this result is that simplifications and insights obtained from one problem can be employed on the other (some software literally makes the substitutions of Table 4.1 to compute the Kalman filter solution from the algorithm for solving the control Riccati equation).

This feature – that both problems produce a matrix Riccati equation – is referred to as the "duality" of estimation and control. It does *not* mean that they are the same problem; in particular, recall that the control problem is equivalent not to filtering, but to smoothing.

Covariances usually dominate the Kalman filter (and smoother) calculations and sometimes lead to the conclusion that the procedures are impractical. But as with all linear least-squares, like estimation problems, the state vector uncertainty does not depend upon the actual data values, only upon the prior error covariances. Thus, the filter and smoother uncertainties (and the filter and smoother gains) can be computed in advance of the actual application to data, and stored. The computation can be done, e.g., by stepping through the recursion in Eqs. (4.146) and (4.147), starting from $t = 0$.

Furthermore, it was pointed out that, in Kalman filter problems, the covariances and Kalman gain can approach a steady state, in which $\mathbf{P}(t)$, $\mathbf{P}(t, -)$, $\mathbf{K}(t)$ become time independent. Physically, the growth in error from the propagation equation (4.146) is then just balanced by the reduction in uncertainty from the incoming data stream (4.147). This simple description supposes the data come in at every time-step; often the data appear only intermittently, but periodically, and the steady-state solution is periodic – errors displaying a characteristic saw-tooth structure between observation times.

If these steady-state values can be found, then the necessity to update the covariances and gain matrix disappears, and the computational load is much reduced, potentially by many orders of magnitude (see also Chapter 5). The equivalent steady state for the control problem is best interpreted in terms of the feedback gain control matrix, $\mathbf{K}_c$, which can also become time independent, meaning that the value of the control to be applied depends only upon the state observed at time t and need not be recomputed at each time step.

The great importance of steady-state estimation and control has led to a large number of methods for obtaining the solution of the various steady-state Riccati equations requiring one of $(\mathbf{S}(t) = \mathbf{S}(t-1), \mathbf{S}(t, -) = \mathbf{S}(t-1, -), \mathbf{P}(t) = \mathbf{P}(t-1),$ or $\mathbf{P}(t, -) = \mathbf{P}(t-1, -))$.[30] The steady-state equation is often known as the "algebraic Riccati equation."[31]

A steady-state solution to the Riccati equation corresponds not only to a determination of the steady-state filter and smoother covariances but also to the steady-state solution of the Lagrange multiplier normal equations – a so-called steady-state control. Generalizations to the steady-state problem exist; an important one is the possibility of a periodic steady state.[32]

Before seeking a steady-state solution, one must determine whether one exists. That no such solution will exist in general is readily seen by considering a physical system in which certain components (elements of the flow) are not readily observed. If these components are initialized with partially erroneous values, then, if they are unstable, they will grow without bound, and there will be no limiting asymptotic value for the uncertainty, which will also have to grow without bound. Alternatively, suppose that there are elements of the state vector whose values cannot be modified by the available control variables. Then no observations of the state vector produce information about the control variables; if the control vector uncertainty is described by $\mathbf{Q}$, then this uncertainty will accumulate from one time-step to another, growing without bound with the number of time steps.

4.6 Controllability and observability

In addition to determining whether there exists a steady-state solution either to the control or estimation Riccati equations, there are many reasons for examining in some detail the existence of many of the matrix operations that have been employed routinely. Matrix inverses occur throughout the developments above, and the issue of whether they exist has been ignored. Ultimately, however, one must face up to questions of whether the computations are actually possible. The questions are intimately connected to some very useful structural descriptions of models and data that we will now examine briefly.

Controllability

Can controls can be found to drive a system from a given initial state $\mathbf{x}(0)$ to an arbitrary $\mathbf{x}(t_f)$? If the answer is "yes," the system is said to be *controllable*. To find an answer, consider for simplicity,[33] a model with $\mathbf{B} = 0$ and with the control, u, a scalar. Then the model time-steps can be written as

$$\mathbf{x}(1) = \mathbf{A}\mathbf{x}(0) + \mathbf{\Gamma}u(0),$$

$$\mathbf{x}(2) = \mathbf{A}\mathbf{x}(1) + \mathbf{\Gamma}u(1),$$

$$= \mathbf{A}^2\mathbf{x}(0) + \mathbf{A}\mathbf{\Gamma}u(0) + \mathbf{\Gamma}u(1),$$

$$\vdots$$

$$\mathbf{x}(t_f) = \mathbf{A}^{t_f}\mathbf{x}(0) + \sum_{j=0}^{t_f-1} \mathbf{A}^{t_f-1-j}\mathbf{\Gamma}u(j),$$

$$= \mathbf{A}^{t_f}\mathbf{x}(0) + [\mathbf{\Gamma}\ \mathbf{A}\mathbf{\Gamma}\cdots\mathbf{A}^{t_f-1}\mathbf{\Gamma}] \begin{bmatrix} u(t_f - 1) \\ \vdots \\ u(0) \end{bmatrix}.$$

To determine $u(t)$, we must be able to solve the system

$$[\mathbf{\Gamma}\ \mathbf{A}\mathbf{\Gamma}\cdots\mathbf{A}^{t_f-1}\mathbf{\Gamma}] \begin{bmatrix} u(t_f - 1) \\ \vdots \\ u(0) \end{bmatrix} = \mathbf{x}(t_f) - \mathbf{A}^{t_f}\mathbf{x}(0), \tag{4.150}$$

or

$$\mathbf{C}\mathbf{u} = \mathbf{x}(t_f) - \mathbf{A}^{t_f}\mathbf{x}(0),$$

for $u(t)$. The state vector dimension is N; therefore, the dimension of $\mathbf{C}$ is N by the number of columns, t_f (a special case – with scalar $u(t)$, $\mathbf{\Gamma}$ is $N \times 1$). Therefore, Eq. (4.150) has no (ordinary) solution if t_f is less than N. If $t_f = N$ and $\mathbf{C}$ is non-singular – that is, of rank N – there is a unique solution, and the system is controllable. If the dimensions of $\mathbf{C}$ are non-square, one could have a discussion, familiar from Chapter 2, of solutions for $u(t)$ with nullspaces present. If $t_f < N$, there is a nullspace of the desired output, and the system would not be controllable. If $t_f > N$, then there will still be a nullspace of the desired output, unless the rank is N, when $t_f = N$, and the system is controllable. The test can therefore be restricted to this last case.

This concept of controllability can be described in a number of interesting and useful ways[34] and generalized to vector controls and time-dependent models. To the extent that a model is found to be uncontrollable, it shows that some elements of the state vector are not connected to the controls, and one might ask why this is so and whether the model cannot then be usefully simplified.

Observability

The concept of "observability" is connected to the question of whether given N perfect observations, it is possible to infer all of the initial conditions. Suppose that the same model is used, and that we have (for simplicity only) a scalar observation sequence,

$$y(t) = \mathbf{E}(t)\mathbf{x}(t) + n(t), \quad t = 0, 1, \ldots, t_f. \tag{4.151}$$

Can we find $\mathbf{x}(0)$? The sequence of observations can be written, with $u(t) \equiv 0$, as

$$y(1) = \mathbf{E}(1)\mathbf{x}(1) = \mathbf{E}(1)\mathbf{A}\mathbf{x}(0),$$

$$\vdots$$

$$y(t_f) = \mathbf{E}(t_f)\mathbf{A}^{t_f}\mathbf{x}(0),$$

which is

$$\mathbf{O}\mathbf{x}(0) = [\, y(1) \ldots y(t_f)\,]^{\mathrm{T}}$$

$$\mathbf{O} = \left\{ \begin{array}{c} \mathbf{E}(1)\mathbf{A} \\ \vdots \\ \mathbf{E}(t_f)\mathbf{A}^{t_f} \end{array} \right\}. \tag{4.152}$$

If the "observability matrix" is square – that is, $t_f = N$ and $\mathbf{O}$ is full rank – there is a unique solution for $\mathbf{x}(0)$, and the system is said to be observable. Should it fail to be observable, it suggests that at least some of the initial conditions are not determinable by an observation sequence and are irrelevant. Determining why that should be would surely shed light on the model one was using. As with controllability, the test (4.152) can be rewritten in a number of ways, and the concept can be extended to more complicated systems. The concepts of "stabilizability," "reachability," "reconstructability," and "detectability" are closely related.[35] A close connection also exists between observability and controllability and the existence of a steady-state solution for the algebraic Riccati equations.

In practice, one must distinguish between mathematical observability and controllability and practical limitations imposed by the realities of observational systems. It is characteristic of fluids that changes occurring in some region at a particular time are ultimately communicated to all locations, no matter how remote, at later times, although the delay may be considerable, and the magnitudes of the signal may be much reduced by dissipation and geometrical spreading. Nonetheless, one anticipates that there is almost no possible element of a fluid flow, no matter how distant from a particular observation, that is not in principle observable. This subject is further discussed in Chapter 5.

4.7 Non-linear models

Fluid flows are non-linear by nature, and one must address the data/model combination problem where the model is non-linear. (There are also, as noted above, instances in which the data are non-linear combinations of the state vector elements.) Nonetheless, the focus here on linear models is hardly wasted effort. As with more conventional systems, there are not many general methods for solving non-linear estimation or control problems; rather, as with forward modeling, each situation has to be analyzed as a special case. Much insight is derived from a thorough understanding of the linear case, and indeed it is difficult to imagine tackling any non-linear problem without a thorough grasp of the linear one. Not unexpectedly, the most accessible approaches to non-linear estimation/control are based upon linearizations.

A complicating factor in the use of non-linear models is that the objective functions need no longer have a unique minimum. There can be many nearby, or distant, minima, and the one chosen by the usual algorithms may depend upon exactly where one starts in the parameter space and how the search for the minimum is conducted. Indeed, the structure of the cost function may come to resemble a chaotic function, filled with hills, plateaus, and valleys into which one may stumble, never to get out again.[36] The combinatorial methods described in Chapter 3 are a partial solution.

4.7.1 The linearized and extended Kalman filter

If one employs a non-linear model,

$$\mathbf{x}(t) = \mathbf{L}(\mathbf{x}(t-1), \mathbf{Bq}(t-1), \mathbf{\Gamma}(t-1)\mathbf{u}(t-1)), \qquad (4.153)$$

then reference to the Kalman filter recursion shows that the forecast step can be taken as before,

$$\tilde{\mathbf{x}}(t, -) = \mathbf{L}(\tilde{\mathbf{x}}(t - 1), \mathbf{Bq}(t - 1), 0), \tag{4.154}$$

but it is far from clear how to propagate the uncertainty from $\mathbf{P}(t - 1)$ to $\mathbf{P}(t, -)$, the previous derivation being based upon the assumption that the error propagates linearly, independently of the true value of $\mathbf{x}(t)$ (or equivalently, that if the initial error is Gaussian, then so is the propagated error). With a non-linear system one cannot simply add the propagated initial condition error to that arising from the unknown controls. A number of approaches exist to finding approximate solutions to this problem, but they can no longer be regarded as strictly optimal, representing different linearizations.

Suppose that we write

$$\mathbf{x}(t) = \mathbf{x}_o(t) + \Delta\mathbf{x}(t), \qquad \mathbf{q} = \mathbf{q}_o(t) + \Delta\mathbf{q}(t), \tag{4.155}$$

Then

$$\mathbf{x}_o(t) = \mathbf{L}_o(\mathbf{x}_o(t - 1), \mathbf{Bq}_o(t - 1), t - 1), \tag{4.156}$$
$$\mathbf{L}_o = \mathbf{L}(\mathbf{x}_o(t - 1), \mathbf{Bq}_o(t - 1), 0, t - 1)$$

defines a nominal solution, or trajectory, $\mathbf{x}_o(t)$.

Non-linear models, in particular, can have trajectories that bifurcate in a number of different ways, so that, subject to slight differences in state, the trajectory can take widely differing pathways as time increases. This sensitivity can be a very serious problem for a Kalman filter forecast, because a linearization may take the incorrect branch, leading to divergences well beyond any formal error estimate. Note, however, that the problem is much less serious in a smoothing problem, as one then has observations available indicating the branch actually taken.

Assuming a nominal solution is available, we have an equation for the solution perturbation:

$$\Delta\mathbf{x}(t) = \mathbf{L}_x(\mathbf{x}_o(t - 1), \mathbf{Bq}_o(t - 1), 0, t)^{\mathrm{T}} \Delta\mathbf{x}(t - 1)$$
$$+ \mathbf{L}_q^{\mathrm{T}} \Delta\mathbf{q}(t - 1) + \mathbf{L}_u^{\mathrm{T}} \mathbf{u}(t - 1), \tag{4.157}$$

where

$$\mathbf{L}_x(\mathbf{x}_o(t), \mathbf{Bq}_o(t), 0, t) = \frac{\partial\mathbf{L}}{\partial\mathbf{x}(t)}, \quad \mathbf{L}_q(\mathbf{x}_o(t), \mathbf{Bq}_o(t), 0, t) = \frac{\partial\mathbf{L}}{\partial\mathbf{q}(t)},$$
$$\mathbf{L}_u(\mathbf{x}_o(t), \mathbf{Bq}_o(t), 0, t) = \frac{\partial\mathbf{L}}{\partial\mathbf{u}(t)},$$

which is linear – called the "tangent linear model," and of the form already used for the Kalman filter, but with redefinitions of the governing matrices. The model is assumed to be differentiable in this manner; discrete models are differentiable, numerically, barring a division by zero somewhere. They are by definition discontinuous, and discrete differentiation automatically accommodates such discontinuities. Numerical models often have switches, typically given by "if xx, then yy" statements. Even these models are differentiable in the sense we need, except at the isolated point where the "if" statement is evaluated; typically the code representing the derivatives will also have a switch at this point. The full solution would be the sum of the nominal solution, $\mathbf{x}_o(t)$, and the perturbation $\Delta\mathbf{x}(t)$. This form of estimate is sometimes known as the "linearized Kalman filter," or the "neighboring optimal estimator." Its usage depends upon the existence of a nominal solution, differentiability of the model, and the presumption that the controls $\Delta\mathbf{q}, \mathbf{u}$ do not drive the system too far from the nominal trajectory.

The so-called "extended Kalman filter" is nearly identical, except that the linearization is taken instead about the most recent estimate $\tilde{\mathbf{x}}(t)$; that is, the partial derivatives in (4.156) are evaluated using not $\mathbf{x}_o(t-1)$, but $\tilde{\mathbf{x}}(t-1)$. This latter form is more prone to instabilities, but if the system drifts very far from the nominal trajectory, it could well be more accurate than the linearized filter. Linearized smoothing algorithms can be developed in analogous ways, and as already noted, the inability to track strong model non-linearities is much less serious with a smoother than with a filter. The references go into these questions in detail. Problems owing to forks in the trajectory[37] can always be overcome by having enough observations to keep the estimates close to the true state. The usual posterior checks of model and data residuals are also a very powerful precaution against a model failing to track the true state adequately.

Example *Suppose, for the mass–spring oscillator on p. 183, there is a non-linear perturbation in the difference equation (4.17) – a non-linear term proportional to $\varepsilon\xi(t)^3$, where ε is very small. Then the governing difference equation becomes, with arbitrary Δt,*

$$\begin{bmatrix} \xi(t) \\ \xi(t-\Delta t) \end{bmatrix} = \left\{ \begin{matrix} 2-\frac{r}{m}\Delta t - \frac{k}{m}(\Delta t)^2 & \frac{r\Delta t}{m}-1 \\ 1 & 0 \end{matrix} \right\} \begin{bmatrix} \xi(t-\Delta t) \\ \xi(t-2\Delta t) \end{bmatrix}$$
$$ + \varepsilon \begin{bmatrix} \xi(t-\Delta t)^3 \\ 0 \end{bmatrix} + \begin{bmatrix} (\Delta t)^2 \frac{q(t-\Delta t)}{m} \\ 0 \end{bmatrix},$$

or,

$$\mathbf{x}(t) = \mathbf{A}\mathbf{x}(t-\Delta t) + \varepsilon\mathbf{L}_1(\mathbf{x}(t-\Delta t)) + \mathbf{B}\mathbf{q}(t), \quad \mathbf{x}(t) = \mathbf{x}(0),$$

a discrete analogue of the so-called hard spring equation. Define a nominal trajectory, $x_0(t)$, satisfying the linearized version of this last equation (there is no general necessity for the nominal trajectory to be linear):

$$\mathbf{x}_0(t) = \mathbf{A}\mathbf{x}_0(t - \Delta t) + \mathbf{B}\mathbf{q}(t), \quad \mathbf{x}_0(0) = \mathbf{x}(0),$$

and let $\mathbf{x}(t) = \mathbf{x}_0(t) + \varepsilon \Delta\mathbf{x}(t)$, so that, to $O(\varepsilon)$,

$$\Delta\mathbf{x}(t) = \mathbf{A}\Delta\mathbf{x}(t - \Delta t) + \varepsilon \mathbf{L}_1(\mathbf{x}_0(t - \Delta t)), \quad \Delta\mathbf{x}(0) = \mathbf{0}, \tag{4.158}$$

which is a special case of (4.157). $\varepsilon \mathbf{L}_1(\mathbf{x}_0(t - \Delta t))$ takes the role of $\mathbf{B}\mathbf{q}(t)$ in the linear problem. Whether $\varepsilon \Delta\mathbf{x}(t)$ remains sufficiently small with time can be determined empirically (the continuous version of this problem is the Duffing equation, about which a great deal is known – the approximation we have made can lead to unbounded perturbations).[38] In general, one would expect to have to re-linearize at finite time. Figure 4.16 shows the full non-linear trajectory of $x_1(t)$, the linear trajectory, $x_{01}(t)$, and the sum of the solution to Eq. (4.158) and that of $x_{01}(t)$. The linear and non-linear solutions ultimately diverge, and in an estimation problem one would almost certainly re-linearize about the estimated trajectory.

It is possible to define physical systems for which sufficiently accurate or useful derivatives of the system cannot be defined[39] so that neither Lagrange multiplier nor linearized sequential methods can be used. Whether such systems occur in practice, or whether they are somewhat like the mathematical pathologies used by mathematicians to demonstrate the limits of conventional mathematical tools (e.g., the failure to exist of the derivative of $\sin(1/t)$, $t \to 0$, or of the existence of space-filling curves) is not so clear. It is clear that all linearization approaches do have limits of utility, but they are and will likely remain, the first choice of practitioners necessarily aware that no universal solution methods exist.

4.7.2 Parameter estimation and adaptive estimation

Often models contain parameters whose values are poorly known. In fluid flow problems, these often concern parameterized turbulent mixing, with empirical parameters which the user is willing to adjust to provide the best fit of the model to the observations. Sometimes, this approach is the only way to determine the parameters.

Suppose that the model is linear in $\mathbf{x}(t)$ and that it contains a vector of parameters, $\mathbf{p}$, whose nominal values, $\mathbf{p}_0$, we wish to improve, while also estimating the state

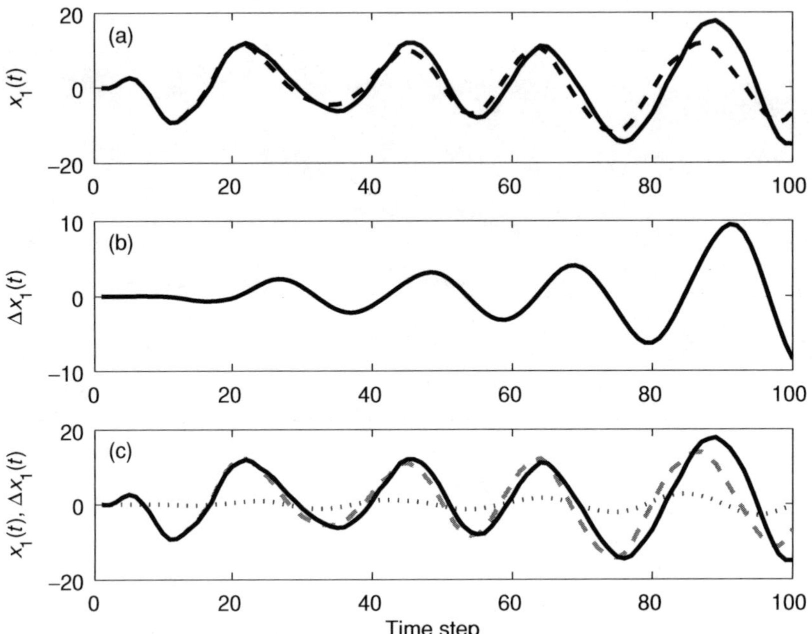

Figure 4.16 (a) The solid curve depicts the time-stepped solution, $x_1(t)$, to the non-linear finite difference oscillator. The dashed line shows the same solution in a linear approximation (here $\varepsilon = 8 \times 10^{-5}$, $\Delta t = 1$, $r = 0.02$, $k = 0.1$). (b) Shows the difference between the linearized and non-linear solutions. (c) The solid curve is the full non-linear solution (as in (a)), the dotted curve is the anomaly solution (to Eq. (4.158)), and the dashed curve is the sum of the linear and anomaly solutions. One would probably re-linearize about the actual solution at a finite time if observational information were available.

vector. Write the model as

$$\mathbf{x}(t) = \mathbf{A}(\mathbf{p}(t-1))\mathbf{x}(t-1) + \mathbf{B}\mathbf{q}(t-1) + \mathbf{\Gamma}\mathbf{u}(t-1), \tag{4.159}$$

where the time dependence in $\mathbf{p}(t)$ arises from the changing estimate of its value rather than a true physical time dependence. A general approach to solving this problem is to augment the state vector. That is,

$$\mathbf{x}_A(t) = \begin{bmatrix} \mathbf{x}(t) \\ \mathbf{p}(t) \end{bmatrix}. \tag{4.160}$$

Then write a model for this augmented state as

$$\mathbf{x}_A(t) = \mathbf{L}_A\left[\mathbf{x}_A(t-1),\ \mathbf{q}(t-1),\ \mathbf{u}(t-1)\right], \tag{4.161}$$

where

$$\mathbf{L}_A = \left\{ \begin{matrix} \mathbf{A}(\mathbf{p}(t-1)) & \mathbf{0} \\ \mathbf{0} & \mathbf{I} \end{matrix} \right\} \mathbf{x}_A(t-1) + \mathbf{B}\mathbf{q}(t-1) + \mathbf{\Gamma}\mathbf{u}(t-1). \tag{4.162}$$

The observation equation is augmented simply as

$$\mathbf{y}_A(t) = \mathbf{E}_A(t)\mathbf{x}_A(t) + \mathbf{n}_A(t),$$
$$\mathbf{E}_A(t) = \{\mathbf{E}(t) \quad \mathbf{0}\}, \quad \mathbf{n}_A(t) = \mathbf{n}(t),$$

assuming that there are no direct measurements of the parameters. The evolution equation for the parameters can be made more complex than indicated here. A solution can be found by using the linearized Kalman filter, for example, linearizing about the nominal parameter values. Parameter estimation is a very large subject.[40]

A major point of concern in estimation procedures based upon Gauss–Markov type methods lies in specification of the various covariance matrices, especially those describing the model error – here lumped into $\mathbf{Q}(t)$. The reader will probably have concluded that there is, however, nothing precluding deduction of the covariance matrices from the model and observations, given that adequate numbers of observations are available. The possibility is briefly discussed on p. 273.

4.7.3 Non-linear adjoint equations: searching for solutions

Consider now a non-linear model in the context of the Lagrange multipliers approach. Let the model be non-linear in either the state vector or model parameters, or both, so that a typical objective function is

$$\begin{aligned} J = {}& [\mathbf{x}(0) - \mathbf{x}_0]^{\mathrm{T}} \mathbf{P}(0)^{-1} [\mathbf{x}(0) - \tilde{\mathbf{x}}_0] \\ & + \sum_{t=1}^{t_f} [\mathbf{E}(t)\mathbf{x}(t) - \mathbf{y}(t)]^{\mathrm{T}} \mathbf{R}(t)^{-1} [\mathbf{E}(t)\mathbf{x}(t) - \mathbf{y}(t)] \\ & + \sum_{t=0}^{t_f-1} \mathbf{u}(t)^{\mathrm{T}} \mathbf{Q}(t)^{-1} \mathbf{u}(t) \\ & - 2 \sum_{t=1}^{t_f} \mu(t)^{\mathrm{T}} [\mathbf{x}(t) - \mathbf{L}[\mathbf{x}(t-1), \mathbf{B}\mathbf{q}(t-1), \mathbf{\Gamma}\mathbf{u}(t-1)]] . \end{aligned} \tag{4.163}$$

Here $\mathbf{x}_o$ is the a-priori estimate of the initial conditions with uncertainty $\mathbf{P}(0)$, and the tildes have been omitted from the remaining variables. The observations continue to be treated as linear in the state vector, but even this assumption can be relaxed. The normal equations are:

$$\frac{1}{2}\frac{\partial J}{\partial \mathbf{u}(t)} = \mathbf{Q}(t)^{-1}\mathbf{u}(t) + \left(\frac{\partial \mathbf{L}(\mathbf{x}(t),\,\mathbf{Bq}(t),\,\mathbf{\Gamma u}(t))}{\partial \mathbf{u}(t)}\right)^{\mathrm{T}}\mathbf{\Gamma}^{\mathrm{T}}\boldsymbol{\mu}(t+1) = \mathbf{0},$$

(4.164)

$$t = 0, 1, \ldots, t_f - 1,$$

$$\frac{1}{2}\frac{\partial J}{\partial \boldsymbol{\mu}(t)} = \mathbf{x}(t) - \mathbf{L}\,[\mathbf{x}(t-1),\,\mathbf{Bq}(t-1),\,\mathbf{\Gamma u}(t-1)] = \mathbf{0}, \quad t = 1, 2, \ldots, t_f,$$

(4.165)

$$\frac{1}{2}\frac{\partial J}{\partial \mathbf{x}(0)} = \mathbf{P}(0)^{-1}\,[\mathbf{x}(0) - \mathbf{x}_0] + \left(\frac{\partial \mathbf{L}(\mathbf{x}(0),\,\mathbf{Bq}(0),\,\mathbf{\Gamma u}(0))}{\partial \mathbf{x}(0)}\right)^{\mathrm{T}}\boldsymbol{\mu}(1) = \mathbf{0},$$

(4.166)

$$\frac{1}{2}\frac{\partial J}{\partial \mathbf{x}(t)} = \mathbf{E}(t)^{\mathrm{T}}\mathbf{R}(t)^{-1}\,[\mathbf{E}(t)\mathbf{x}(t) - \mathbf{y}(t)] - \boldsymbol{\mu}(t)$$

(4.167)

$$+ \left(\frac{\partial \mathbf{L}(\mathbf{x}(t),\,\mathbf{Bq}(t),\,\mathbf{\Gamma u}(t))}{\partial \mathbf{x}(t)}\right)^{\mathrm{T}}\boldsymbol{\mu}(t+1) = \mathbf{0}, \quad t = 1, 2, \ldots, t_f - 1,$$

$$\frac{1}{2}\frac{\partial J}{\partial \mathbf{x}(t_f)} = \mathbf{E}(t_f)^{\mathrm{T}}\mathbf{R}(t_f)^{-1}[\mathbf{E}(t_f)\mathbf{x}(t_f) - \mathbf{y}(t_f)] - \boldsymbol{\mu}(t_f) = \mathbf{0}.$$

(4.168)

These are non-linear because of the non-linear model (4.165) – although the adjoint model (4.167) remains linear in $\boldsymbol{\mu}(t)$ – and the linear methods used thus far will not work directly. The operators that appear in the above equations,

$$\left(\frac{\partial \mathbf{L}(\mathbf{x}(t),\,\mathbf{Bq}(t),\,\mathbf{\Gamma u}(t),\,t)}{\partial \mathbf{u}(t)}\right),\quad \left(\frac{\partial \mathbf{L}(\mathbf{x}(t),\,\mathbf{Bq}(t),\,\mathbf{\Gamma u}(t),\,t)}{\partial \mathbf{x}(t)}\right),$$

(4.169)

are, as in Eq. (4.156), the derivatives of the model with respect to the control and state vectors. Assuming that they exist, they represent a linearization of the model about the state and control vectors and again are the tangent linear model. Their transposes are, in this context, the adjoint model. There is some ambiguity about the terminology: the form of (4.169) or the transposes are definable independent of the form of J. Otherwise, Eq. (4.167) and its boundary condition (4.168) depend upon the actual observations and the details of J; one might call this pair the "adjoint evolution" equation to distinguish it from the adjoint model.

If the non-linearity is not too large, perturbation methods may work. This notion leads to what is usually called "neighboring optimal control."[41] Where the non-linearity is large, the approach to solution is an iterative one. Consider what one is trying to do. At the optimum, if we can find it, J will reach a stationary value in which the terms multiplying the $\boldsymbol{\mu}(t)$ will vanish. Essentially, one uses *search* methods that are able to find a solution (there may well be multiple such solutions, each corresponding to a local minimum of J).

There are many known ways to seek approximate solutions to a set of simultaneous equations, linear or non-linear, using various search procedures. Most such methods are based upon what are usually called "Newton" or "quasi-Newton" methods, or variations on steepest descent. The most popular approach to tackling the set (4.164)–(4.168) has been a form of conjugate gradient algorithm.[42] The iteration cycles are commonly carried out by making a first estimate of the initial conditions and the boundary conditions – for example, setting $\mathbf{u} = \mathbf{0}$. One integrates (4.165) forwards in time to produce a first guess for $\mathbf{x}(t)$. A first guess set of Lagrange multipliers is obtained by integrating (4.167) backwards in time. Normally, (4.164) is not then satisfied, but because the values obtained provide information on the gradient of the objective function with respect to the controls, one knows the sign of the changes to make in the controls to reduce J. Perturbing the original guess for $\mathbf{u}(t)$ in this manner, one does another forward integration of the model and backward integration of the adjoint. Because the Lagrange multipliers provide the partial derivatives of J with respect to the solution (Eq. (4.166) permits calculation of the direction in which to shift the current estimate of $\mathbf{x}(0)$ to decrease J), one can employ a conjugate gradient or steepest descent method to modify $\tilde{\mathbf{x}}(0)$ and carry out another iteration.

In this type of approximate solution, the adjoint solution, $\tilde{\boldsymbol{\mu}}(t)$, is really playing two distinct roles. On the one hand, it is a mathematical device to impose the model constraints; on the other, it is being used as a numerical convenience for determining the direction and step size to best reduce the objective function – as it contains information on the change in J with parameter perturbations. The problem of possibly falling into the wrong minimum of the objective function remains.

In practice, $\mathbf{L}(\mathbf{x}(t-1), \mathbf{Bq}(t-1), \boldsymbol{\Gamma}\mathbf{u}(t-1), t-1)$ is represented as many lines of computer code. Generating the derivatives in Eq. (4.169) can be a major undertaking. Fortunately, and remarkably, the automatic differentiation (AD) tools mentioned above can convert the code for $\mathbf{L}[\mathbf{x}(t), \mathbf{Bq}(t), \boldsymbol{\Gamma}\mathbf{u}(t), t]$ into the appropriate code for the derivatives. While still requiring a degree of manual intervention, this AD software renders Lagrange multiplier methods a practical approach for model codes running to many thousands of lines.[43] The basic ideas are sketched in the next subsection and in the Appendix to this chapter.

4.7.4 Automatic differentiation, linearization, and sensitivity

The linearized and extended filters and smoothers (e.g., Eq. (4.156)) and the normal equations (4.164)–(4.168) involve derivatives such as $\partial \mathbf{L}/\partial \mathbf{x}(t)$. One might wonder how these are to be obtained. Several procedures exist, including the one used above for the weakly non-linear spring, but to motivate what is perhaps the most elegant method, begin with a simple example of a two-dimensional non-linear model.

Example *Let*

$$\mathbf{x}(t) = \mathbf{L}(\mathbf{x}(t-1)) = \begin{bmatrix} a\mathbf{x}^{\mathrm{T}}(t-1)\mathbf{x}(t-1) + c \\ b\mathbf{x}^{\mathrm{T}}(t-1)\mathbf{x}(t-1) + d \end{bmatrix}, \tag{4.170}$$

where a, b, c, d are fixed constants. Time-stepping from $t = 0$,

$$\mathbf{x}(1) = \begin{bmatrix} a\mathbf{x}^{\mathrm{T}}(0)\mathbf{x}(0) + c \\ b\mathbf{x}^{\mathrm{T}}(0)\mathbf{x}(0) + d \end{bmatrix}, \tag{4.171}$$

$$\mathbf{x}(2) = \begin{bmatrix} a\mathbf{x}^{\mathrm{T}}(1)\mathbf{x}(1) + c \\ b\mathbf{x}^{\mathrm{T}}(1)\mathbf{x}(1) + d \end{bmatrix},$$

$$\dots$$

Consider the dependence, $\partial\mathbf{x}(t)/\partial\mathbf{x}(0)$,

$$\frac{\partial\mathbf{x}(t)}{\partial\mathbf{x}(0)} = \left\{ \begin{array}{cc} \dfrac{\partial x_1(t)}{\partial x_1(0)} & \dfrac{\partial x_2(t)}{\partial x_1(0)} \\[2ex] \dfrac{\partial x_1(t)}{\partial x_2(0)} & \dfrac{\partial x_2(t)}{\partial x_2(0)} \end{array} \right\}. \tag{4.172}$$

For $t = 2$, by the definitions and rules of Chapter 2, we have

$$\frac{\partial\mathbf{x}(2)}{\partial\mathbf{x}(0)} = \left\{ \begin{array}{cc} \dfrac{\partial x_1(2)}{\partial x_1(0)} & \dfrac{\partial x_2(2)}{\partial x_1(0)} \\[2ex] \dfrac{\partial x_1(2)}{\partial x_2(0)} & \dfrac{\partial x_2(2)}{\partial x_2(0)} \end{array} \right\} = \frac{\partial\mathbf{L}(\mathbf{L}(\mathbf{x}(0)))}{\partial\mathbf{x}(0)} = \mathbf{L}'(\mathbf{x}(0))\mathbf{L}'(\mathbf{L}(\mathbf{x}(0))). \tag{4.173}$$

Note that

$$\frac{\partial\mathbf{x}(2)}{\partial\mathbf{x}(0)} = \frac{\partial\mathbf{L}(\mathbf{x}(1))}{\partial\mathbf{x}(0)} = \frac{\partial\mathbf{x}(1)}{\partial\mathbf{x}(0)}\frac{\partial\mathbf{L}(\mathbf{x}(1))}{\partial\mathbf{x}(1)} = \frac{\partial\mathbf{L}(\mathbf{x}(0))}{\partial\mathbf{x}(0)}\frac{\partial\mathbf{L}(\mathbf{x}(1))}{\partial\mathbf{x}(1)}, \tag{4.174}$$

where we have used the "chain rule" for differentiation. Substituting into (4.174),

$$\frac{\partial\mathbf{x}(2)}{\partial\mathbf{x}(0)} = \left\{ \begin{array}{cc} 2ax_1(0) & 2bx_1(0) \\ 2ax_2(0) & 2bx_2(0) \end{array} \right\} \left\{ \begin{array}{cc} 2ax_1(1) & 2bx_1(1) \\ 2ax_2(1) & 2bx_2(1) \end{array} \right\}$$

$$= \left\{ \begin{array}{cc} 2ax_1(0) & 2bx_1(0) \\ 2ax_2(0) & 2bx_2(0) \end{array} \right\} \left\{ \begin{array}{cc} 2a^2 & 2ab \\ 2ab & 2b^2 \end{array} \right\} \mathbf{x}^{\mathrm{T}}(0)\mathbf{x}(0).$$

By direct calculation from Eq. (4.171), we have

$$\frac{\partial\mathbf{x}(2)}{\partial\mathbf{x}(0)} = 4(a^2 + b^2)\mathbf{x}(0)^{\mathrm{T}}\mathbf{x}(0) \left\{ \begin{array}{cc} ax_1(0) & bx_1(0) \\ ax_2(0) & bx_2(0) \end{array} \right\}. \tag{4.175}$$

Substituting (4.170) into Eq. (4.173), then

$$\mathbf{L}'(\mathbf{x}(0)) = \left\{ \begin{matrix} 2ax_1(0) & 2bx_1(0) \\ 2ax_2(0) & 2bx_2(0) \end{matrix} \right\},$$

and

$$\mathbf{L}'\left(\mathbf{L}\left(\mathbf{x}\left(0\right)\right)\right) = \mathbf{L}'\left(\mathbf{x}\left(1\right)\right) = \left\{ \begin{matrix} 2ax_1\left(1\right) & 2bx_1(1) \\ 2ax_2\left(1\right) & 2bx_2\left(1\right) \end{matrix} \right\}$$

$$= \left\{ \begin{matrix} 2a^2\mathbf{x}\left(0\right)^{\mathrm{T}}\mathbf{x}\left(0\right) & 2ab\mathbf{x}\left(0\right)^{\mathrm{T}}\mathbf{x}\left(0\right) \\ 2ab\mathbf{x}\left(0\right)^{\mathrm{T}}\mathbf{x}\left(0\right) & 2b^2\mathbf{x}\left(0\right)^{\mathrm{T}}\mathbf{x}\left(0\right) \end{matrix} \right\}.$$

Multiplying, as in (4.173),

$$\left\{ \begin{matrix} 2ax_1\left(0\right) & 2bx_1(0) \\ 2ax_2\left(0\right) & 2bx_2\left(0\right) \end{matrix} \right\} \left\{ \begin{matrix} 2a^2\mathbf{x}\left(0\right)^{\mathrm{T}}\mathbf{x}\left(0\right) & 2ab\mathbf{x}\left(0\right)^{\mathrm{T}}\mathbf{x}\left(0\right) \\ 2ab\mathbf{x}\left(0\right)^{\mathrm{T}}\mathbf{x}\left(0\right) & 2b^2\mathbf{x}\left(0\right)^{\mathrm{T}}\mathbf{x}\left(0\right) \end{matrix} \right\}$$

$$= 4\left(a^2 + b^2\right)\mathbf{x}\left(0\right)^{\mathrm{T}}\mathbf{x}\left(0\right) \left\{ \begin{matrix} ax_1\left(0\right) & bx_1\left(0\right) \\ ax_2\left(0\right) & bx_2\left(0\right) \end{matrix} \right\},$$

which is consistent with (4.175). Hence, as required,

$$d\mathbf{x}\left(2\right) = d\mathbf{x}\left(0\right)^{\mathrm{T}} \left\{ \begin{matrix} 2ax_1\left(0\right) & 2bx_1(0) \\ 2ax_2\left(0\right) & 2bx_2\left(0\right) \end{matrix} \right\} \left\{ \begin{matrix} 2a^2\mathbf{x}\left(0\right)^{\mathrm{T}}\mathbf{x}\left(0\right) & 2ab\mathbf{x}\left(0\right)^{\mathrm{T}}\mathbf{x}\left(0\right) \\ 2ab\mathbf{x}\left(0\right)^{\mathrm{T}}\mathbf{x}\left(0\right) & 2b^2\mathbf{x}\left(0\right)^{\mathrm{T}}\mathbf{x}\left(0\right) \end{matrix} \right\}.$$

As a computational point note that this last equation involves a matrix–matrix multiplication on the right. But if written as a transpose,

$$\mathrm{d}\mathbf{x}\left(2\right)^{\mathrm{T}} = \left\{ \begin{matrix} 2a^2\mathbf{x}\left(0\right)^{\mathrm{T}}\mathbf{x}\left(0\right) & 2ab\mathbf{x}\left(0\right)^{\mathrm{T}}\mathbf{x}\left(0\right) \\ 2ab\mathbf{x}\left(0\right)^{\mathrm{T}}\mathbf{x}\left(0\right) & 2b^2\mathbf{x}\left(0\right)^{\mathrm{T}}\mathbf{x}\left(0\right) \end{matrix} \right\}^{\mathrm{T}} \left\{ \begin{matrix} 2ax_1\left(0\right) & 2bx_1(0) \\ 2ax_2\left(0\right) & 2bx_2\left(0\right) \end{matrix} \right\}^{\mathrm{T}} \mathrm{d}\mathbf{x}\left(0\right),$$

$\mathrm{d}\mathbf{x}\left(2\right)^{\mathrm{T}}$ can be found from matrix–vector multiplications alone, which for large matrices is a vastly reduced computation. This reduction in computational load lies behind the use of so-called reverse mode methods described below.

Going much beyond these simple statements takes us too far into the technical details.[44]

The chain rule can be extended, such that

$$\frac{\partial \mathbf{x}(t)}{\partial \mathbf{x}(0)} = \frac{\partial \mathbf{L}(\mathbf{x}(t-1))}{\partial \mathbf{x}(0)} = \frac{\partial \mathbf{x}(t-1)}{\partial \mathbf{x}(0)}\frac{\partial \mathbf{L}(\mathbf{x}(t-1))}{\partial \mathbf{x}(t-1)} = \frac{\partial \mathbf{L}(\mathbf{x}(t-2))}{\partial \mathbf{x}(0)}\frac{\partial \mathbf{L}(\mathbf{x}(t-1))}{\partial \mathbf{x}(t-1)}$$

$$\tag{4.176}$$

$$= \frac{\partial (\mathbf{x}(t-2))}{\partial \mathbf{x}(0)}\frac{\partial \mathbf{L}(\mathbf{x}(t-2))}{\partial \mathbf{x}(t-2)}\frac{\partial \mathbf{L}(\mathbf{x}(t-1))}{\partial \mathbf{x}(t-1)} = \cdots$$

$$= \frac{\partial \mathbf{L}(\mathbf{x}(0))}{\partial \mathbf{x}(0)}\cdots\frac{\partial \mathbf{L}(\mathbf{x}(t-2))}{\partial \mathbf{x}(t-2)}\frac{\partial \mathbf{L}(\mathbf{x}(t-1))}{\partial \mathbf{x}(t-1)}.$$

Although this result is formally correct, such a calculation could be quite cumbersome to code and carry out for a more complicated model (examples of such codes do exist). An alternative, of course, is to systematically and separately perturb each of the elements of $\mathbf{x}(0)$, and integrate the model forwards from $t = 0$ to t_f, thus numerically evaluating $\partial\mathbf{x}(t)/\partial x_i(0)$, $i = 1, 2, \ldots, N$. The model would thus have to be run N times, and there might be issues of numerical accuracy. (The approach is similar to the determination of numerical Green functions considered above.)

Practical difficulties such as these have given rise to the idea of "automatic (or algorithmic) differentiation" in which one accepts from the beginning that a computer code will be used to define $\mathbf{L}(\mathbf{x}(t), t, \mathbf{q}(t))$ (reintroducing the more general definition of $\mathbf{L}$).[45] One then seeks automatic generation of a second code, capable of evaluating the elements in Eq. (4.176), that is, terms of the form $\partial\mathbf{L}(\mathbf{x}(n))/\partial\mathbf{x}(n)$, for any n. Automatic differentiation (AD) tools take the computer codes (typically in Fortran, C, or Matlab) and generate new codes for the *exact* partial derivatives of the code. Various packages go under names like ADIFOR, TAF, ADiMAT, etc. (see www.autodiff.org). The possibility of using such procedures has already been alluded to, where it was noted that for a linear model, the first derivative would be the state transition matrix $\mathbf{A}$, which may not otherwise be explicitly available. That is,

$$\mathbf{A}(t) = \frac{\partial\mathbf{L}(\mathbf{x}(t), t, \mathbf{q}(t))}{\partial\mathbf{x}(t)}.$$

Actual implementation of AD involves one deeply in the structures of computer coding languages, and is not within the scope of this book. Note that the existing implementations are not restricted to such simple models as we used in the particular example, but deal with the more general $\mathbf{L}(\mathbf{x}(t), t, \mathbf{q}(t))$.

In many cases, one cares primarily about some scalar quantity, $H(\tilde{\mathbf{x}}(t_f))$, e.g., the heat flux or pressure field in a flow, as given by the state vector at the end time, $\mathbf{x}(t_f)$, of a model computation. Suppose[46] one seeks the sensitivity of that quantity to perturbations in the initial conditions (any other control variable could be considered), $\mathbf{x}(0)$. Let $\mathbf{L}$ continue to be the operator defining the time-stepping of the model. Define $\Psi_t = \mathbf{L}(\mathbf{x}(t), t, \mathbf{q}(t))$. Then

$$H = H(\Psi_{t_f}[\Psi_{t_f-1}[\cdots \Psi_1[\mathbf{x}(0)]]]),$$

that is, the function of the final state of interest is a nested set of operators working on the control vector $\mathbf{x}(0)$. Then the derivative of H with respect to $\mathbf{x}(0)$ is again obtained from the chain rule,

$$\frac{\partial H}{\partial\mathbf{x}(0)} = H'(\Psi'_{t_f}[\Psi'_{t_f-1}[\cdots \Psi'_1[\mathbf{x}(0)]]]), \tag{4.177}$$

where the prime denotes the derivative with respect to the argument of the operator $\mathbf{L}$ evaluated at that time,

$$\frac{\partial \Psi_t(\mathbf{p})}{\partial \mathbf{p}}.$$

Notice that these derivatives, are the Jacobians (matrices) of dimension $N \times N$ at each time-step, and are the same derivatives that appear in the operators in (4.169). The nested operator (4.177) can be written as a matrix product,

$$\frac{\partial H}{\partial \mathbf{x}(0)} = \nabla h^{\mathrm{T}} \frac{\partial \Psi_{t_f}(\mathbf{p})}{\partial \mathbf{p}} \frac{\partial \Psi_{t_f}(\mathbf{p})}{\partial \mathbf{p}} \cdots \frac{\partial \Psi_1(\mathbf{p})}{\partial \mathbf{p}}. \tag{4.178}$$

∇h is the vector of derivatives of function H (the gradient) and so (4.178) is a column vector of dimension $N \times 1$. $\mathbf{p}$ represents the state vector at the prior time-step for each Ψ_t. The adjoint compilers described above compute $\partial H / \partial \mathbf{x}(0)$ in what is called the "forward mode," producing an operator that runs from right to left, multiplying $t_f - N \times N$ matrices starting with $\partial \Psi_1(\mathbf{p}) / \partial \mathbf{p}$.

If, however, Eq. (4.178) is transposed, then

$$\left(\frac{\partial H}{\partial \mathbf{x}(0)} \right)^{\mathrm{T}} = \left(\frac{\partial \Psi_1(\mathbf{p})}{\partial \mathbf{p}} \right)^{\mathrm{T}} \left(\frac{\partial \Psi_2(\mathbf{p})}{\partial \mathbf{p}} \right)^{\mathrm{T}} \cdots \left(\frac{\partial \Psi_{t_f}(\mathbf{p})}{\partial \mathbf{p}} \right)^{\mathrm{T}} \nabla h, \tag{4.179}$$

where the first multiplication on the right involves multiplying the column vector ∇h by an $N \times N$ matrix, thus producing another $N \times 1$ vector. More generally, the set of products in (4.179), again taken from right to left, involves only multiplying a vector by a matrix, rather than a matrix by a matrix as in (4.178), with a potentially very large computational saving. Such evaluation is the *reverse* or *adjoint mode* calculation alluded to above (the transposes generate the required adjoint operators, although a formal transpose is not actually formed) and have become available in some automatic differentiation tools only comparatively recently. In comparing the computation in the forward and reverse modes, one must be aware that there is a storage penalty in (4.179) not incurred in (4.178).[47] In practice, the various operators $\partial \Psi_i(\mathbf{p}) / \partial \mathbf{p}$ are not obtained explicitly, but are evaluated.

Historically, the forward mode was developed first, and remains the most common implementation of AD. It permits one to systematically linearize models, and, by repeated application of the AD tool, to develop formal Taylor series for non-linear models. With the rise in fluid state estimation problems of very large dimension, there has recently been a much greater emphasis on the reverse mode.

Many fluid models rely on "if ... then ..." and similar branching statements, such as assignment of a variable to the maximum value of a list. For example, if some region is statically unstable owing to cooling at the surface, a test for instability may lead the model to homogenize the fluid column; otherwise, the stratification is unaffected. Objections to AD are sometimes raised, apparently based on the intuitive belief that such a model cannot be differentiated. In practice, once a branch is chosen, the state vector is well-defined, as is its derivative, the AD code itself then having corresponding branches or, e.g., assignments to maxima or minima of a list. A brief example of this issue is given in the Appendix to this chapter. Our employment so far of the adjoint model and the adjoint evolution equation has been in the context of minimizing an objective function – and, to some degree, the adjoint has been nothing but a numerical convenience for algorithms that find minima. As we have seen repeatedly, however, Lagrange multipliers have a straightforward interpretation as the sensitivity of an objective function, J, to perturbations in problem parameters. This use of the multipliers can be developed independently of the state estimation problem.

Example *Consider the linear time invariant model*

$$\mathbf{x}(n) = \mathbf{A}\mathbf{x}(n-1),$$

such that

$$\mathbf{x}(n) = \mathbf{A}^n\mathbf{x}(0).$$

Suppose we seek the dependence of $H = \mathbf{x}(n)^{\mathrm{T}}\mathbf{x}(n)/2 = (\mathbf{x}(0)^{\mathrm{T}}\mathbf{A}^{n\mathrm{T}}\mathbf{A}^n\mathbf{x}(0))/2$ on the problem parameters. The sensitivity to the initial conditions is straightforward:

$$\frac{\partial H}{\partial \mathbf{x}(0)} = \mathbf{A}^{n\mathrm{T}}\mathbf{A}^n\mathbf{x}(0).$$

Suppose instead that $\mathbf{A}$ depends upon an internal parameter, k, perhaps the spring constant in the example of the discrete mass–spring oscillator, for which H would be an energy. Then

$$\frac{\partial H}{\partial k} = \frac{1}{2}\frac{\partial(\mathbf{x}(n)^{\mathrm{T}}\mathbf{x}(n))}{\partial k} = \frac{1}{2}\left(\frac{\partial(\mathbf{x}(n)^{\mathrm{T}}\mathbf{x}(n))}{\partial \mathbf{x}(n)}\right)^{\mathrm{T}}\frac{\partial \mathbf{x}(n)}{\partial k} = \mathbf{x}(n)^{\mathrm{T}}\frac{\partial \mathbf{x}(n)}{\partial k}.$$

We have, from Eq. (2.32),

$$\frac{d\mathbf{x}(n)}{dk} = \frac{d\mathbf{A}^n}{dk}\mathbf{x}(0) = \left[\frac{d\mathbf{A}}{dk}\mathbf{A}^{n-1} + \mathbf{A}\frac{d\mathbf{A}}{dk}\mathbf{A}^{n-2} + \cdots + \mathbf{A}^{n-1}\frac{d\mathbf{A}}{dk}\right]\mathbf{x}(0),$$

and so,

$$\frac{\partial H}{\partial k} = \mathbf{x}(0)^{\mathrm{T}}\mathbf{A}^{n\mathrm{T}}\left[\frac{d\mathbf{A}}{dk}\mathbf{A}^{n-1} + \mathbf{A}\frac{d\mathbf{A}}{dk}\mathbf{A}^{n-2} + \cdots + \mathbf{A}^{n-1}\frac{d\mathbf{A}}{dk}\right]\mathbf{x}(0).$$

and with evaluation of $d\mathbf{A}/dk$ being straightforward, we are finished.

The Appendix to this chapter describes briefly how computer programs can be generated to carry out these operations.

4.7.5 Approximate methods

All of the inverse problems discussed, whether time-independent or not, were reduced ultimately to finding the minimum of an objective function, either in unconstrained form (e.g., (2.352) or (4.61)) or constrained by exact relationships (e.g., models) (2.354) or (4.97). Once the model has been formulated, the objective function agreed on, and the data obtained in appropriate form (often the most difficult step), the formal solution is reduced to finding the constrained or unconstrained minimum. "Optimization theory" is a very large, very sophisticated subject directed at finding such minima, and the methods we have described here – sequential estimation and Lagrange multiplier methods – are only two of a number of possibilities.

As we have seen, some of the methods stop at the point of finding a minimum and do not readily produce an estimate of the uncertainty of the solution. One can distinguish inverse methods from optimization methods by the requirement of the former for the requisite uncertainty estimates. Nonetheless, as noted before in some problems, mere knowledge that there is at least one solution may be of intense interest, irrespective of whether it is unique or whether its stability to perturbations in the data or model is well understood.

The reader interested in optimization methods generally is referred to the literature on that subject.[48] Geophysical fluid problems often fall into the category of extremely large, non-linear optimization, one that tends to preclude the general use of many methods that are attractive for problems of more modest size.

The continued exploration of ways to reduce the computational load without significantly degrading either the proximity to the true minimum or the information content (the uncertainties of the results) is a very high priority. Several approaches are known. The use of steady-state filters and smoothers has already been discussed. Textbooks discuss a variety of possibilities for simplifying various elements of the solutions. In addition to the steady-state assumption, methods include: (1) "state reduction" – attempting to remove from the model (and thus from the uncertainty calculation) elements of the state vector that are either of no interest or comparatively unchanging;[49] (2) "reduced-order observers,"[50] in which some components of the model are so well observed that they do not need to be calculated; and (3) proving or assuming that the uncertainty matrices (or the corresponding information matrices) are block diagonal or banded, permitting use of a variety of sparse algorithms. This list is not exhaustive.

4.8 Forward models

The focus we have had on the solution of inverse problems has perhaps given the impression that there is some fundamental distinction between forward and inverse modeling. The point was made at the beginning of this book that inverse *methods* are important in solving forward as well as inverse *problems*. Almost all the inverse problems discussed here involved the use of an objective function, and such objective functions do not normally appear in forward modeling. The presence or absence of objective functions thus might be considered a fundamental difference between the problem types.

But numerical models do not produce universal, uniformly accurate solutions to the fluid equations. Any modeler makes a series of decisions about which aspects of the flow are most important for accurate depiction – the energy or vorticity flux, the large-scale velocities, the non-linear cascades, etc. – and which cannot normally be achieved simultaneously with equal fidelity. It is rare that these goals are written explicitly, but they could be, and the modeler could choose the grid and differencing scheme, etc., to minimize a specific objective function. The use of such explicit objective functions would prove beneficial because it would quantify the purpose of the model.

One can also consider the solution of ill-posed forward problems. In view of the discussion throughout this book, the remedy is straightforward: one must introduce an explicit objective function of the now-familiar type, involving state vectors, observations, control, etc., and this approach is precisely that recommended. If a Lagrange multiplier method is adopted, then Eqs. (2.334) and (2.335) show that an over- or underspecified forward model produces a complementary under- or overspecified adjoint model, and it is difficult to sustain a claim that modeling in the forward direction is fundamentally distinct from that in the inverse sense.

Example *Consider the ordinary differential equation*

$$\frac{d^2x(t)}{dt^2} - k^2x(t) = 0. \tag{4.180}$$

Formulated as an initial value problem, it is properly posed with Cauchy conditions $x(0) = x_0$, $x'(0) = x_0'$. *The solution is*

$$x(t) = A\exp(kt) + B\exp(-kt), \tag{4.181}$$

with A, B determined by the initial conditions. If we add another condition – for example, at the end of the interval of interest, $x(t_f) = x_{t_f}$ – the problem is ill-posed because it is now overspecified. To analyze and solve such a problem using the methods of this book, discretize it as

$$x(t+1) - (2+k^2)x(t) + x(t-1) = 0, \tag{4.182}$$

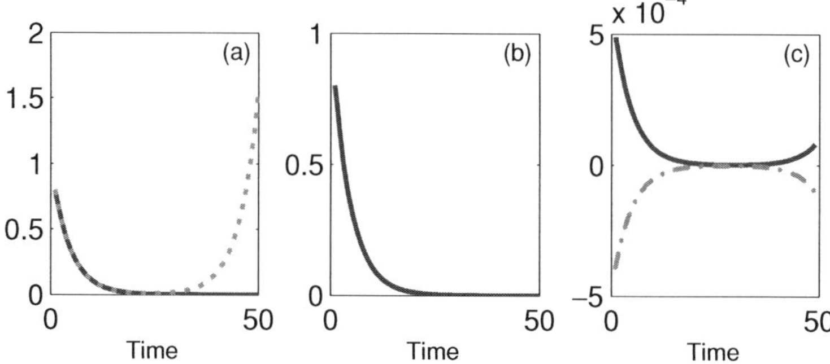

Figure 4.17 (a) Stable solution. $x_1(t)$ (dashed curve) to Eq. (4.182) obtained by setting $k^2 = 0.05$, $\mathbf{x}(0) = [0.800, 1.00]^T$. The dotted line is an unstable solution obtained by modifying the initial condition to $\mathbf{x}(0) = [0.80001, 1.00]^T$. The growing solution takes awhile to emerge, but eventually swamps the stable branch. (b) Solution obtained by overspecification, in which $\tilde{\mathbf{x}}(0) = [0.80001, 1.00]^T$, $\mathbf{P}(0) = .01\mathbf{I}_2$, $\tilde{\mathbf{x}}(50) = [1.4 \times 10^{-5}, 1]$, $\mathbf{P}(50) = \mathrm{diag}([10^{-4}, 10^4])$. (c) Lagrange multiplier values used to impose the initial and final conditions on the model. Solid curve is $\mu_1(t)$, and chain is $\mu_2(t)$.

taking $\Delta t = 1$, with corresponding redefinition of k^2. A canonical form is

$$\mathbf{x}(t) = \mathbf{A}\mathbf{x}(t-1), \quad \mathbf{x}(t) = [x(t), \quad x(t-1)]^T, \quad \mathbf{A} = \begin{Bmatrix} 2+k^2 & -1 \\ 1 & 0 \end{Bmatrix}.$$

The reduced form of equations (4.164)–(4.168) is easily solved (the only "observations" are at the final time) by a backwards sweep of the adjoint model (4.101) to obtain $\boldsymbol{\mu}(1)$, which through (4.114) produces $\tilde{\mathbf{x}}(1)$ in terms of $\mathbf{x}(t_f) - \mathbf{x}_d(t_f)$. A forward sweep of the model, to t_f, produces the numerical value of $\tilde{\mathbf{x}}(t_f)$; the backwards sweep of the adjoint model gives the corresponding numerical value of $\tilde{\mathbf{x}}(1)$, and a final forward sweep of the model completes the solution. The subproblem forward and backwards sweeps are always well-posed. This recipe was run for

$$k^2 = 0.05, \quad \Delta t = 1, \quad \tilde{\mathbf{x}}(1) = [0.805, \ 1.0]^T, \quad \mathbf{P}(1) = 10^{-2}\mathbf{I},$$
$$\tilde{x}(t_f) = 1.427 \times 10^{-5}, \quad \mathbf{P}(t_f) = \mathrm{diag}\{ 10^{-4} \quad 10^4 \}, \quad t_f = 50,$$

with results as shown in Fig. 4.17. (The large subelement uncertainty in $\mathbf{P}(50)$, corresponding to scalar element, $x(49)$, is present because we sought to specify only scalar element $x(50)$, in $\mathbf{x}(50)$.) The solution produces a new estimated value $\tilde{\mathbf{x}}(0) = [0.800, 1.00]^T$, which is exactly the value used in Fig. 4.17 to generate the stable forward computation. Notice that the original ill-posedness in both overspecification and instability of the initial value problem have been dealt with. The Lagrange multipliers (adjoint solution) are also shown in the figure, and imply that *the system sensitivity is greatest at the initial and final times. For a full GCM, the technical details are much more intricate, but the principle is the same.*

This example can be thought of as the solution to a forward problem, albeit ill-posed, or as the solution to a more or less conventional inverse one. The distinction between forward and inverse problems has nearly vanished. Any forward model that is driven by observed conditions is ill-posed in the sense that there can again be no unique solution, only a most probable one, smoothest one, etc. As with an inverse solution, forward calculations no more produce unique solutions in these circumstances than do inverse ones. All problems involving observed parameters, initial or boundary conditions are necessarily ill-posed.

4.9 A summary

Once rendered discrete for placement on a digital computer, time-dependent inverse problems all can be reduced to the minimum variance/least-squares problems already considered in Chapters 2 and 3, depending upon how the weight matrices are chosen. With a large enough and fast enough computer, they could even be solved by the same methods used for the static problems. Given, however, the common need to reduce the computing and storage burdens, a number of algorithms are available for finding solutions by considering the problem either in pieces, as in the sequential methods of filter/smoother, or by iterations as in the Lagrange multiplier methods. Solutions can be either accurate or approximate depending upon one's needs and resources. In the end, however, the main message is that one is still seeking the solution to a minimum variance/least-squares problem, and the differences among the techniques are algorithmic ones, with trade-offs of convenience and cost.

Appendix. Automatic differentiation and adjoints

The utility of automatic differentiation (AD) of computer model codes was alluded to on pp. 189 and 241, both as a way to determine the state transition matrix **A**, when it was only implicit in a code, and as a route to linearizing non-linear models. The construction of software capable of taking (say) a Fortran90 code and automatically generating a second Fortran90 code for the requisite derivatives of the model is a remarkable, if not altogether complete, achievement of computer science. Any serious discussion is beyond the author's expertise, and well outside the scope of this book. But because only AD methods have made the Lagrange multiplier (adjoint) method of state estimation a practical approach for realistic fluid problems, we briefly sketch the possibilities with a few simple examples. The references given in note 43 should be consulted for a proper discussion.

Consider first the problem of finding the state transition matrix. A simple time-stepping code written in Matlab for a 2-vector is

```
function y=lin(x);
y(1)=0.9*x(1)+0.2*x(2);
y(2)=0.2*x(1)+0.8*x(2);
```

Here $\mathbf{x}$ would be the state vector at time $t-1$, and $\mathbf{y}$ would be its value one time-step in the future. A matrix/vector notation is deliberately avoided so that $\mathbf{A}$ is not explicitly specified. When the AD tool ADiMat[51] is used, it writes a new Matlab code:

```
function [g_y, y]= g_lin(g_x, x) %lin.m;
%x is assumed to be a 2-vector
g_lin_0= 0.9* g_x(1);
lin_0= 0.9* x(1);
g_lin_1= 0.2* g_x(2);
lin_1= 0.2* x(2);
g_y(1)= g_lin_0+ g_lin_1;
y(1)= lin_0+ lin_1;
clear lin_0 lin_1 g_lin_0 g_lin_1 ;
g_lin_2= 0.2* g_x(1);
lin_2= 0.2* x(1);
g_lin_3= 0.8* g_x(2);
lin_3= 0.8* x(2);
g_y(2)= g_lin_2+ g_lin_3;
y(2)= lin_2+ lin_3;
clear lin_2 lin_3 g_lin_2 g_lin_3 ;
```

The notation has been cleaned up somewhat to make it more readable. Consider for example, the new variable, $g\_lin\_0 = 0.9 * g\_x(1)$. The numerical value 0.9 is the partial derivative of $y(1)$ with respect to $x(1)$. The variable $g\_x(1)$ would be the partial derivative of $x(1)$ with respect to some other independent variable, permitting the chain rule to operate if desired. Otherwise, one can set it to unity on input. Similarly the notation g_lin_i denotes the corresponding derivative of $y(1)$ with respect to $x(i)$. By simple further coding, one can construct the $\mathbf{A}$ matrix of the values of the partial derivatives. Here, ADiMat has produced the tangent linear model, which is also the exact forward model. More interesting examples can be constructed.

The Matlab code corresponding to a simple switch is

```
function y= switch1(a); if a > 0, y= a; else, y=
  a^2+2*a; end
```

that is, $y = a$ if independent variable a (externally prescribed) is positive, or else $y = a^2 + 2a$. Running this code through AdiMat produces (again after some cleaning up of the notation):

```
function [g_y, y]= g_switch1(g_a, a);
 if a> 0, g_y= g_a;
 y= a;
else, g_tmp_2=2* a^(2- 1)* g_a;
   tmp_0= a^2;
   g_tmp_1= 2* g_a;
   tmp_1= 2* a;
   g_y= g_tmp_0+ g_tmp_1;
   y= tmp_0+ tmp_1;
   end
```

The derivative, `g_y`, is 1 for positive a, otherwise it is `a+2`. `g_a` can be interpreted as the derivative of a with respect to another, arbitrary, independent variable, again permitting use of the chain rule. The derivative is continuous for all values of a except $a = 0$.

Consider now a physical problem of a reservoir as shown in Fig. 4.18.[52] The model is chosen specifically to have discontinuous behavior: there is inflow, storage, and outflow. If the storage capacity is exceeded, there can be overspill, determined by the `max` statement below. A forward code, now written in Fortran, is:

Thresholds: a hydrological reservoir model (I)

```
do t = 1, msteps
```

- get sources and sinks at time t:

  inflow, evaporation, release (read fields)

- calculate water release based on storage:

```
release(t) = 0.8*storage(t-1)**0.7
```

- calculate projected stored water:

  storage = storage + inflow-release-evaporation

```
nominal = storage(t-1) +
   h*( infl(t)-release(t)-evap(t) )
```

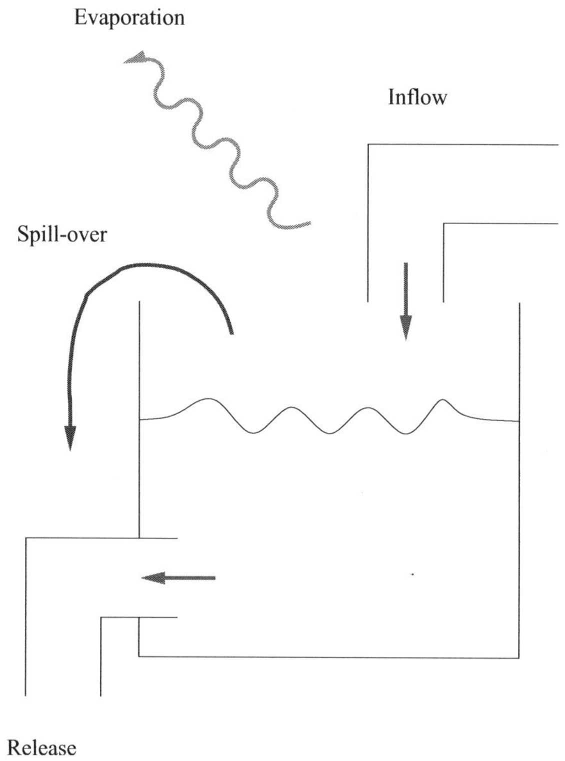

Figure 4.18 Reservoir model used to demonstrate automatic/algorithmic differentiation. The possibility of spill-over is an example of a switch in a computer model. (From P. Heimbach, personal communication, 2004.)

- If threshold capacity is exceeded, spill-over:

```
spill(t) = MAX(nominal-capac , 0.)
```

- re-adjust projected stored water after spill-over:

```
storage(t) = nominal - spill(t)
```

- determine outflow:

```
out(t) = release(t) + spill(t)/h

end do
```

Note the presence of the `max` statement.

When run through the AD tool TAF (a product of FastOpt®), one obtains for the tangent linear model,

Thresholds: a hydrological reservoir model (II)

The tangent linear model

```
do t=1, msteps
g_release(t) =0.56*g_storage(t-1)*storage(t-1)**
   (-0.3)
release(t)=0.8*storage(t-1)**0.7
g_nominal=-g_release(t)*h+g_storage(t-1)
nominal=storage(t-1)+h*(infl(t)-release(t)-evap(t))
g_spill(t)=g_nominal*(0.5+sign(0.5,nominal-
   capac-0.))
spill(t)=max(nominal-capac,0.)
g_storage(t)=g_nominal-g_spill(t)
storage(t)=nominal-spill(t)
g_out(t)=g_release(t)+g_spill(t)/h
out(t)=release(t)+spill(t)/h
end do
```

- `g_release(t)` not defined for `storage(t-1) = 0`
- `g_spill(t)` not defined for `nominal = capac.`

Note how the maximum statement has given rise to the new variable `g_spill(t)`, its corresponding tangent linear variable.

Note that the AD tool can cope with such apparently non-differentiable operators as the maximum of a vector. In practice, it internally replaces the function `max` with a loop of tests for relative sizes of successive elements. Not all AD tools can cope with all language syntaxes, not all all structures are differentiable, and one must be alert to failures owing to incomplete handling of various structures. Nonetheless, the existing tools are a considerable achievement.

TAF and some other AD tools are capable of producing the reverse mode. A major issue in optimization calculations is the ability to restart the computation from intermediate results, in an operation called "checkpointing."[53]

Notes

1　Liebelt (1967), Gelb (1974), Bryson and Ho (1975), Anderson and Moore (1979), and Brown and Hwang (1997) are especially helpful.

2　Daley (1991).

3　Meteorologists have tended to go their own idiosyncratic way – see Ide *et al.* (1997) – with some loss in transparency to other fields.

4　Box *et al.* (1994).

5　Luenberger (1979).

6　Stammer and Wunsch (1996), Menemenlis and Wunsch (1997).

7　von Storch *et al.* (1988).

8　Giering and Kaminski (1998), Marotzke *et al.* (1999).

9　See Bryson and Ho (1975, p. 351).

10　For example, Stengel (1986).

11　Munk *et al.* (1995).

12 A method exploited by Stammer and Wunsch (1996).

13 Kalman (1960). Kalman's derivation was for the discrete case. The continuous case, which was derived later, is known as the "Kalman–Bucy" filter and is a much more complicated object.

14 Stengel (1986, Eq. 4.3–22).

15 For example, Goodwin and Sin (1984, p. 59).

16 Feller (1957).

17 Anderson and Moore (1979) discuss these and other variants of the Kalman filter equations.

18 Some history of the idea of the filter, its origins in the work of Wiener and Kolmogoroff, and a number of applications, can be found in Sorenson (1985).

19 Bryson and Ho (1975, p. 363) or Brown and Hwang (1997, p. 218).

20 Adapted from Bryson and Ho (1975, Chapter 13), whose notation is unfortunately somewhat difficult.

21 For Rauch *et al.* (1965).

22 Gelb (1974), Bryson and Ho (1975), Anderson and Moore (1979), Goodwin and Sin (1984), Sorenson (1985).

23 Some guidance is provided by Bryson and Ho (1975, pp. 390–5) or Liebelt (1967). In particular, Bryson and Ho (1975) introduce the Lagrange multipliers (their equations 13.2.7–13.2.8) simply as an intermediate numerical device for solving the smoother equations.

24 Luenberger (1979).

25 Wunsch (1988) shows a variety of calculations as a function of variations in the terminal constraint accuracies. An example of the use of this type of model is discussed in Chapter 6.

26 Bennett (2002) has a comprehensive discussion, albeit in the continuous time context.

27 The use of the adjoint to solve l_2-norm problems is discussed by Bryson and Ho (1975, Section 13.3), who relax the restriction of full controllability, $\Gamma = \mathbf{I}$. Because of the connection to regulator/control problems, a variety of methods for solution is explored there.

28 Bryson and Ho (1975).

29 Franklin *et al.* (1998).

30 Anderson and Moore (1979), Stengel (1986), Bittanti *et al.* (1991), Fukumori *et al.* (1992), Fu *et al.* (1993), Franklin *et al.* (1998).

31 See Reid (1972) for a discussion of the history of the Riccati equation in general; it is intimately related to Bessel's equation and has been studied in scalar form since the eighteenth century. Bittanti *et al.* (1991) discuss many different aspects of the matrix form.

32 Discussed by Bittanti *et al.* (1991).

33 Franklin *et al.* (1998).

34 For example, Stengel (1986), Franklin *et al.* (1998).

35 Goodwin and Sin (1984) or Stengel (1986).

36 Miller *et al.* (1994) discuss some of the practical difficulties.

37 For example, Miller *et al.* (1994).

38 For example, Nayfeh (1973).

39 For example, Lea *et al.* (2000), Köhl and Willebrand (2002).

40 For example, Anderson and Moore (1979), Goodwin and Sin (1984), Haykin (2002).

41 Stengel (1986, Chapter 5).

42 Gilbert and Lemaréchal (1989).

43 Giering and Kaminski (1998), Marotzke *et al.* (1999), Griewank (2000), Corliss *et al.* (2002), Heimback *et al.* 2005.

44 See Marotzke *et al.* (1999) or Giering (2000).

45 Rall (1981) and Griewank (2000).

46 We follow here, primarily, Marotzke *et al.* (1999).

47 Restrepo *et al.* (1995) discuss some of the considerations.

48 Luenberger (1984), Scales (1985), Gill *et al.* (1986), Tarantola (1987).

49 Gelb (1974).

50 O'Reilly (1983), Luenberger (1984).

51 Available through www.sc.rwth-aachen.de/vehresschild/adimat.

52 Example due to P. Heimbach. For more information see http://hdl.handle.net/1721.1/30598.

53 Restrepo *et al.* (1995).

5

Time-dependent methods – 2

This brief chapter includes a number of extensions of the material in Chapter 4, as well as providing introductions to a number of more specialized topics that would be of interest to anyone attempting to use the methods for large-scale practical problems. References are given for fuller coverage.

5.1 Monte Carlo/ensemble methods

5.1.1 Ensemble methods and particle filters

When a model is non-linear, one of the fundamental computational steps of the Kalman filter (and any related calculation such as a smoother) is no longer possible. Consider a Kalman filter problem in which the initial conditions, $\tilde{\mathbf{x}}(0)$, contain errors characterized by the covariance matrix, $\mathbf{P}(0)$, and the zero-mean disturbance or control, $\mathbf{u}(0)$, is unknown with covariance, $\mathbf{Q}(0)$. The state forecast step, Eq. (4.50) proceeds as before. Computation of the forecast error covariance, Eq. (4.51), however, which sums the error owing to the initial conditions and that of the unknown controls, $\mathbf{A}(0)\mathbf{P}(0)\mathbf{A}(0)^{\mathrm{T}} + \mathbf{\Gamma}\mathbf{Q}\mathbf{\Gamma}^{\mathrm{T}}$, depends directly upon the linearity assumption, and can no longer be carried out rigorously. For weak non-linearities, the extended or linearized Kalman filters and associated smoothers may be adequate. But when linearizing assumptions fail, some other method must be used. A commonly discussed example of a non-linear (but scalar) model is

$$x_t = \frac{1}{2}x_{t-1} + \frac{25x_{t-1}}{1 + x_{t-1}^2} + 8\cos 1.2t + \varepsilon_t,$$

$$y_t = \frac{x_t^2}{20} + \eta_t,$$

which is non-linear in both the evolution equation and in the measurement; ε_t, η_t are Gaussian white noise processes.[1] Extended Kalman filters work badly for this low-dimensional example.

The basic idea behind so-called ensemble or Monte Carlo methods is in some ways even simpler than the use of Eq. (4.51). (In the signal-processing literature, closely related approaches are usually called "sequential Monte Carlo methods," or "particle filtering."[2]) One directly simulates a sufficiently large number of forecasts, $\tilde{\mathbf{x}}^{(i)}(t, -)$, all having the same statistical properties; then $\mathbf{P}(t, -)$ can be estimated by brute force computation from the many simulations. If a sufficiently large ensemble can be generated, one can contemplate estimating not just the second moments, but calculating the empirical frequency function for the forecast step.

To see how this approach might work, generate an ensemble of initial conditions, $\tilde{\mathbf{X}}(0)$, where each column of the $N \times L$, $L < N$ matrix corresponds to a possible initial condition consistent with both $\tilde{\mathbf{x}}(0)$ and $\mathbf{P}(0)$. Form a similar ensemble for $\tilde{\mathbf{u}}(0)$ based upon $\langle \mathbf{u}(0) \rangle = 0$, and $\mathbf{Q}(0)$. (We discuss generation of such ensembles below.) Then one can run the model on each column of $\tilde{\mathbf{X}}(0)$, with a disturbance from the corresponding column of $\tilde{\mathbf{U}}(0)$, and compute the ensemble of forecasts $\tilde{\mathbf{X}}(1, -)$. Assuming that the true mean of $\mathbf{x}(1)$ is zero, estimate

$$\tilde{\mathbf{P}}(1, -) = \frac{1}{L}\tilde{\mathbf{X}}(1, -)\tilde{\mathbf{X}}(1, -)^{\mathrm{T}} = \frac{1}{L}\sum_{j=1}^{L}\tilde{\mathbf{x}}_j(1, -)\tilde{\mathbf{x}}_j(1, -)^{\mathrm{T}}, \qquad (5.1)$$

where $\tilde{\mathbf{x}}_j(1, -)$ is column j of $\tilde{\mathbf{X}}(t, -)$, as an estimate of $\mathbf{P}(1, -)$. Note that if the mean is computed from the columns of $\tilde{\mathbf{X}}$, and subtracted from the estimate, the factor in front becomes $1/(L - 1)$. With $\tilde{\mathbf{P}}(1, -)$ known, the filter averaging step (4.52) can be carried out, although if the probability densities of the model and data errors are different, the average may have little meaning. Because averaging is a linear operation, the conventional filter error covariance calculation (4.54) is still appropriate, and one can continue in this fashion through the filter loop. In essence, this approach characterizes the so-called ensemble Kalman filter method. The main issue here concerns the reliability of the estimates for small ensemble sizes.[3] Because of its structure, if L is less than N, the maximum rank of $\tilde{\mathbf{P}}(1, -)$ is $K = L < N$, and the matrix will be singular. Singularity implies that some structures (those in the nullspace of $\tilde{\mathbf{P}}(1, -)$) are impossible in the initial conditions – a potentially troublesome outcome.

In principle, one can use the ensemble members to produce estimates of the complete probability densities of $\tilde{\mathbf{x}}$, $\tilde{\mathbf{u}}$, no matter how non-linear the model – leading to the use of maximum likelihood methods. These are computationally more demanding, however. Even small ensembles provide at least a qualitative indication of where maximum uncertainty is likely to lie, but their use should not stretch beyond their actual limited information content.

How does one generate ensemble members with zero mean and given spatial covariance, $\mathbf{P}(0)$? Let

$$\mathbf{P}(0) = \mathbf{V}\boldsymbol{\Lambda}\mathbf{V}^{\mathrm{T}},$$

and suppose $\alpha_i^{(p)}$ is white noise from a pseudo-random number generator such that $\langle \alpha_i^{(p)} \rangle = 0$, $\langle \alpha_i^{(p)} a_j^{(p)} \rangle = \delta_{ij}$, where p is the ensemble member label. Form

$$\tilde{\mathbf{x}}_p(0) = \sum_{j=1}^{N} \sqrt{\lambda_j} \alpha_j^{(p)} \mathbf{v}_j.$$

Then it follows that

$$\langle \tilde{\mathbf{x}}^{(p)}(0) \rangle = 0,$$

$$\langle (\tilde{\mathbf{x}}_p(0) - \tilde{\mathbf{x}}(0))(\tilde{\mathbf{x}}_p(0) - \tilde{\mathbf{x}}(0))^{\mathrm{T}} \rangle = \left\langle \left(\sum_{j=1}^{N} \sqrt{\lambda_j} \alpha_j^{(p)} \mathbf{v}_j \right) \left(\sum_{n=1}^{N} \sqrt{\lambda_n} \alpha_n^{(p)} \mathbf{v}_n \right)^{\mathrm{T}} \right\rangle$$

$$= \sum_{j=1}^{N} \lambda_j \langle \alpha_j^{(p)\,2} \rangle \mathbf{v}_j \mathbf{v}_j^{\mathrm{T}} = \mathbf{P}(0),$$

as required. It is readily confirmed too, that the ensemble members are uncorrelated with each other.

The members of the ensemble of initial conditions can have highly non-Gaussian probability densities. One would then select the $\alpha_j^{(p)}$ from populations with whatever is the appropriate probability density.[4] More generally, the initial condition disturbances may have specific structures related to the dynamics. Some of those structures may give rise to particularly rapidly growing disturbances, and which if excited can give an ensemble spread much larger than that obtained from purely random components. A lot of effort in weather forecasting, in particular, has gone into generating small ensembles that are representative of the spread of the true probability density.[5] Unknown model parameters can include initial conditions, including for example, mixing parameterizations, boundary conditions, source/sink components, etc. Ensembles can be generated by calculating solutions from random perturbations to any and all of these problem elements simultaneously.

Example *Let the initial estimate for the mass–spring oscillator of the Example on p. 183, be $\tilde{\mathbf{x}}(0) = [1, 1]^{\mathrm{T}}$, and have error covariance*

$$\mathbf{P}(0) = \begin{Bmatrix} 1 & 0 \\ 0 & 1 \end{Bmatrix}.$$

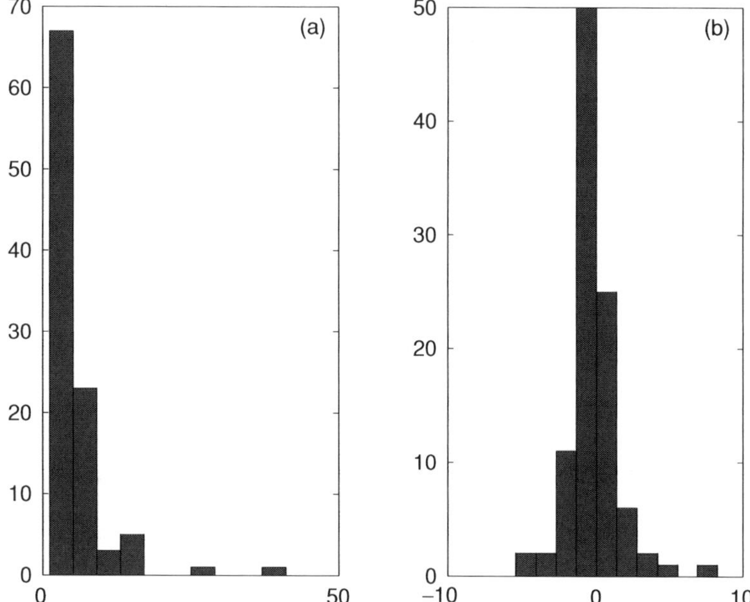

Figure 5.1 (a) Empirical histogram of an ensemble of 100 values of $\tilde{x}_1$ (0) obtained from a log-normal distribution with parameters (1, 1). A similar ensemble of values was generated for $x_2(0)$, with no correlation between the two. The random variables are regarded as different realizations of the noise occurring in the true value $\mathbf{x}(0) = [1, 1]^{\mathrm{T}}$. (b) Histogram of 100 realizations of x_1 ($t = 500$) from the mass–spring oscillator ($k = 0.1$, $\Delta t = 1$, $r = 0.01$). Note the tendency for the frequency function to tend towards Gaussian.

It is thought, however, that the errors follow a log-normal distribution,

$$p(\xi) = \frac{1}{\sqrt{2\pi}\,\xi} \exp(-(\ln \xi - 1)^2/2). \tag{5.2}$$

An ensemble of initial conditions of the form $\tilde{\mathbf{x}}(0) = [1, 1]^{\mathrm{T}} + \boldsymbol{\xi}$ was generated, producing a frequency function (histogram) of values for $\tilde{x}_1(0)$ as shown in Fig. 5.1. After 500 time steps (with $k = 0.1, r = 0.01, \Delta t = 1$) $\tilde{x}_1$ (500) tends to become Gaussian in this linear system, and it would be reasonable to calculate the mean position as $\overline{\tilde{x}_1 (500)}$ where the overbar indicates an average of the ensemble, and the error of the mean position is computed simply from its standard deviation in the ensemble. The latter can be used in the Kalman averaging step. But in a highly non-Gaussian distribution, as seen in the left panel of Fig. 5.1, the sample mean and variance may differ greatly from the true mean and variance owing to the presence of a few strong outliers. (One would be strongly advised to work with the logarithm of $\tilde{x}_1$ (0);[6] such a transformation would not be possible with $\tilde{x}_1$ (500) because it can be negative.)

Calculation of probability densities, or the defining low moments of the probability densities for filter/smoother solutions can also be approached by solving equations for the evolution of the densities or moments. Dynamical evolution equations can be used to write an explicit equation (the so-called Fokker–Planck, or Kolmogorov equation in the physics and mathematics literature, respectively) for the evolution of the state probability density.[7] Because the solution of such equations in high dimensions is forbidding, one can alternatively seek approximate equations for the evolution of the low moments.

This subject is a large and sophisticated one; a growing literature describes applications at various levels of approximation in oceanography and meteorology. But we must leave it to the references to deal with it further.[8]

5.2 Numerical engineering: the search for practicality

Estimation theory is comparatively straightforward in its goals, and in the methods of solution. When it comes to real problems, particularly those involving fluids, the main issues tend to be much less the principle of what one wants to do (it is usually reasonably clear), and more the problems of practicality. Even linear three-dimensional fluid models, particularly those arising in the geophysical world, readily overwhelm the largest available computers and storage devices. Investigators exploiting the special structure of the simultaneous equations represented by time-evolving models may still have serious computational difficulties. The major issues are then primarily those of "numerical engineering" – finding approximate practical methods adequate for a particular goal, while keeping the underlying theory in mind as a guideline. Engineering involves all aspects of the problem, including the forward model, the algorithms for doing minimization, representation and computation of weight matrices, finding adequate estimates of model, and overall system errors. Because of the diversity of the problems that arise, only some very general description of various applications and remedies can be described here.

5.2.1 Meteorological assimilation

"Data assimilation" is a term widely used in numerical weather prediction (NWP) to describe the process of combining a forecast with current observations for the primary purpose of updating a dynamical model – usually in preparation for another forecast. In this book, we use the term "state estimation" for the more general problem of forming model/data combinations, and reserve "assimilation" for the specific meteorological application. For fluid models, forecasting is probably more highly developed in meteorology than in any other field. Astronomers forecasting planetary or cometary positions have a longer history, and ballistic engineers are greatly experienced with a range of trajectory and impact prediction problems. But the meteorological problem is of much greater dimension than any of these, and the economic stakes are so high, that many person-years have been devoted to making and improving weather forecasts. The field is thus a highly developed one and a correspondingly large literature on meteorological assimilation exists and is worth examining.[9] The specialized terminology used by meteorologists can, however, be a major problem. For example, what we are calling the method of Lagrange multipliers (or adjoint method) is known in meteorology as 4DVAR, to imply four-dimensional models and variational techniques. As has been seen, however, the methodology operates in arbitrary dimensions and as used in practice, is least-squares rather than variational in derivation.

Much data assimilation involves simplified forms of objective mapping, in which the model dynamics are used in a primitive fashion to help choose covariances in both time and space for interpolation as in Chapter 2.[10] The formal uncertainties of the forecast are not usually computed – the forecaster learns empirically, and very quickly, how accurate his forecast is. If something works, then one keeps on doing it; if it doesn't work, one changes it. Because of the short timescale, feedback from the public, the military, farmers, the aviation industry, etc., is fast and vehement. Theory often takes a backseat to practical experience. It is important to note that, despite the dense fog of jargon that has come to surround meteorological practice, the methods in actual use remain, almost universally, attempts at the approximate least-squares fitting of a time-evolving atmospheric model to the oncoming observations. The primary goal is forecasting, rather than smoothing. Ensemble methods are used to obtain semi-quantitative understanding of the uncertainty of the forecast. They are semi-quantitative primarily because the ensembles tend to be small in size compared to the model dimension, although useful methods exist for determining the most uncertain elements of short-term forecasts.[11]

5.2.2 Nudging and objective mapping

A number of meteorological schemes can be understood by referring back to the Kalman filter averaging step,

$$\tilde{\mathbf{x}}(t) = \tilde{\mathbf{x}}(t, -) + \mathbf{K}(t)[\mathbf{y}(t) - \mathbf{E}\tilde{\mathbf{x}}(t, -)]. \tag{5.3}$$

This equation has the form of a predictor–corrector – the dynamical forecast of $\tilde{\mathbf{x}}(t, -)$ is compared to the observations and corrected on the basis of the discrepancies. Some assimilation schemes represent guesses for $\mathbf{K}$ rather than the computation of the optimum choice, which we know – for a linear model – is given by the

Kalman gain, replacing (5.3) with

$$\tilde{\mathbf{x}}(t) = \tilde{\mathbf{x}}(t, -) + \mathbf{K}_m[\mathbf{y}(t) - \mathbf{E}\tilde{\mathbf{x}}(t, -)], \tag{5.4}$$

where $\mathbf{K}_m$ is a modified gain matrix. Thus, in "nudging," $\mathbf{K}_m$ is diagonal or nearly so, with elements that are weights that the forecaster assigns to the individual observations.[12] To the extent that the measurements have uncorrelated noise, as might be true of pointwise meteorological instruments like anemometers, and the forecast error is also nearly spatially uncorrelated, pushing the model values pointwise to the data may be very effective. If, in (5.4), the observations $\mathbf{y}(t)$ are direct measurements of state vector elements (e.g., if the state vector includes the density and $\mathbf{y}(t)$ represents observed densities), then $\mathbf{E}(t)$ is very simple – but only if the measurement point coincides with one of the model grid points. If, as is common, the measurements occur between model grid points, $\mathbf{E}$ is an interpolation operator from the model grid to the data location. Usually, there are many more model grid points than data points, and this direction for the interpolation is the most reasonable and accurate. With more data points than model grid points, one might better interchange the direction of the interpolation. Formally, this interchange is readily accomplished by rewriting (5.3) as

$$\tilde{\mathbf{x}}(t) = \tilde{\mathbf{x}}(t, -) + \mathbf{K}_m\mathbf{E}[\mathbf{E}^+\mathbf{y}(t) - \mathbf{x}(t, -)], \tag{5.5}$$

where $\mathbf{E}^+$ is any right inverse of $\mathbf{E}$ in the sense of Chapter 2, for example, the Gauss–Markov interpolator or some plausible approximation to it.

There are potential pitfalls of nudging, however. If the data have spatially correlated errors, as is true of many real observation systems, then the model is being driven toward spatial structures that are erroneous. More generally, the expected great variation in time and space of the relative errors of model forecast and observations cannot be accounted for with a fixed diagonal gain matrix. A great burden is placed upon the skills of the investigator who must choose the weights. Finally, one can calculate the uncertainty of the weighted average (5.4), using this suboptimal gain, but it requires that one specify the true covariances. As noted, however, in NWP formal uncertainty estimates are not of much interest. User feedback is, however, rarely available when the goal is understanding – the estimation problem – rather than forecasting the system for public consumption. When forecasts are made in many contexts, e.g., for decadal climate change, the timescale is often so long as to preclude direct test of the result.

As with the full Kalman filter, in the "analysis step," where the model forecast is averaged with the observations, there is a jump in the state vector as the model is pulled toward the observations. Because the goal is usually forecasting, this state vector discontinuity is not usually of any concern, except to someone instead interested in understanding the fluid physics.

Another more flexible, approximate form of time-dependent estimation can also be understood in terms of the Kalman filter equations. In the filter update equation (4.52), all elements of the state vector are modified to some degree, given any difference between the measurements and the model-prediction of those measurements. The uncertainty of the statevector is *always* modified whenever data become available, even if the model should perfectly predict the observations. As time evolves, information from measurements in one part of the model domain is distributed by the model dynamics over the entire domain, leading to correlations in the uncertainties of all the elements.

One might suppose that some models propagate information in such a way that the error correlations diminish rapidly with increasing spatial and temporal separation. Supposing this to be true (and one must be aware that fluid models are capable of propagating information, be it accurate or erroneous, over long distances and times), static approximations can be found in which the problem is reduced back to the objective mapping methods employed in Chapter 2. The model is used to make an estimate of the field at time t, $\tilde{\mathbf{x}}(t, -)$, and one then finds the prediction error $\Delta\mathbf{y}(t) = \mathbf{y}(t) - \mathbf{E}\tilde{\mathbf{x}}(t, -)$. A best estimate of $\Delta\mathbf{x}(t)$ is sought based upon the covariances of $\Delta\mathbf{y}(t)$, $\Delta\mathbf{x}(t)$, etc. – that is, objective mapping – and the improved estimate is

$$\tilde{\mathbf{x}}(t) = \tilde{\mathbf{x}}(t, -) + \Delta\tilde{\mathbf{x}}(t) = \tilde{\mathbf{x}}(t, -) + \mathbf{R}_{xx}\mathbf{E}^{\mathrm{T}}(\mathbf{E}\mathbf{R}_{xx}\mathbf{E}^{\mathrm{T}} + \mathbf{R}_{nn})^{-1}\Delta\mathbf{y}, \qquad (5.6)$$

which has the form of a Kalman filter update, but in which the state uncertainty matrix, $\mathbf{P}$, is replaced in the gain matrix, $\mathbf{K}$, by $\mathbf{R}_{xx}$ representing the prior covariance of $\Delta\mathbf{x}$. $\mathbf{R}_{xx}$ is fixed, with no dynamical evolution of the gain matrix permitted. Viewed as a generalization of nudging, this approach permits one to specify spatial structure in the noise covariance through choice of a non-diagonal $\mathbf{R}_{nn}$. The weighting of the $\Delta\mathbf{y}$ and the modification for $\tilde{\mathbf{x}}$ is potentially more complex than in pure nudging.

The major issues are the specification of $\mathbf{R}_{xx}$, $\mathbf{R}_{nn}$. Most attempts to use these methods have been simulations by modelers who were content to ignore the problem of determining $\mathbf{R}_{nn}$ or to assume that the noise was purely white. In principle, estimates of $\mathbf{R}_{xx}$ can be found either from observations or from the model itself.

Methods that permit data to be employed from finite-time durations, weighting them inversely with their deviation from some nominal central time, are localized approximations to smoothing algorithms of the Wiener type. Many variations on these methods are possible, including the replacement of $\mathbf{R}_{xx}$ by its eigenvectors (the singular vectors or EOFs), which again can be computed either from the model or from data. Improvements could be made by comparison of the covariance matrices used against the estimates emerging from the calculations of $\tilde{\mathbf{x}}(t)$, $\tilde{\mathbf{n}}(t)$.

All practical linearized assimilation methods are a weighted average of a model estimate of the fluid state with one inferred from the observations. If the model and the observations are physically inconsistent, the forced combination will be impossible to interpret. Thus, the first step in any assimilation procedure has to be to demonstrate that model physics and data represent the same fluid – with disagreement being within the error bounds of both. Following this confirmation of physical consistency, one recognizes that the weighted average of model and data will be useful only if the weights make sense – chosen to at least well-approximate the relative uncertainties of these two. Otherwise, the result of the combination is an average of "apples and oranges."

5.2.3 *Approximate filter/smoother methods*

This book has been primarily devoted to the principles underlying various state estimation methods, rather than to addressing practical issues of implementation. A few methods were introduced to reduce computation (Lagrange multipliers, and ensemble methods), avoiding the calculation of the covariance matrices using the model. Lagrange multiplier methods are attractive because they do not demand the covariance matrices; but their main weakness is that they therefore do not provide them.

Unsurprisingly, numerous approaches have attempted to approximate the full results of the filter/smoother algorithms, both to reduce the burden of the state estimates themselves and of the corresponding error covariances. We examine some examples of such approaches.[13]

Steady-state approximation

Consider, as an example, the Kalman filter, Eqs. (4.50)–(4.54) of Chapter 4. The error covariances, $\mathbf{P}(t, -)$, $\mathbf{P}(t)$ are propagated as

$$\mathbf{P}(t, -) = \mathbf{A}(t - 1)\mathbf{P}(t - 1)\mathbf{A}(t - 1)^{\mathrm{T}} + \mathbf{\Gamma}\mathbf{Q}(t - 1)\mathbf{\Gamma}^{\mathrm{T}}, \tag{5.7}$$

$$\mathbf{P}(t) = \mathbf{P}(t, -) - \mathbf{P}(t, -)\mathbf{E}(t)^{\mathrm{T}}[\mathbf{E}(t)\mathbf{P}(t, -)\mathbf{E}(t)^{\mathrm{T}} + \mathbf{R}(t)]^{-1}\mathbf{E}(t)\mathbf{P}(t, -), \tag{5.8}$$

and do not involve the actual data. These equations can be simply time-stepped from $\mathbf{P}(0)$ to any time t, assuming the availability of $\mathbf{R}(t)$, $\mathbf{E}(t)$ and $\mathbf{P}(0)$. Knowledge of $\mathbf{P}(t)$ then permits the finding of $\mathbf{K}(t)$, and both are determined before any observations actually exist.

Let the model and data stream be time independent, $\mathbf{A}(t) = \mathbf{A}$, $\mathbf{E}(t) = \mathbf{E}$, $\mathbf{Q}(t) = \mathbf{Q}$, $\mathbf{R}(t) = \mathbf{R}$. Substituting for $\mathbf{P}(t, -)$, one has the matrix Riccati

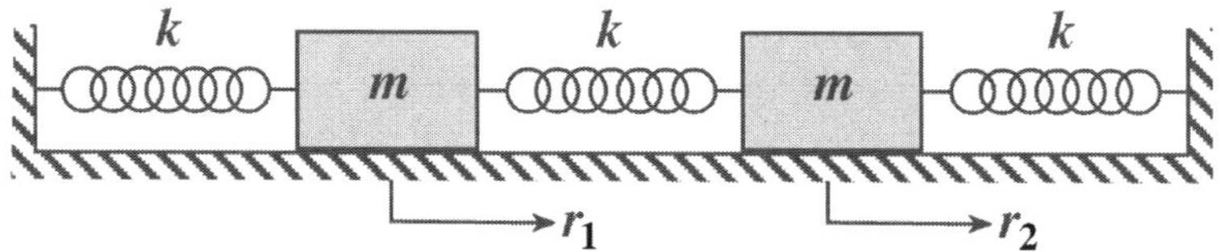

Figure 5.2 Coupled mass-spring oscillator. Rest positions of the two masses define the coordinates $r_{1,2}$. (After McCuskey, 1959.)

equation,

$$\mathbf{P}(t) = \mathbf{A}\mathbf{P}(t-1)\mathbf{A}^{\mathrm{T}} + \mathbf{\Gamma}\mathbf{Q}\mathbf{\Gamma}^{\mathrm{T}} \tag{5.9}$$
$$- [\mathbf{A}\mathbf{P}(t-1)\mathbf{A}^{\mathrm{T}} + \mathbf{\Gamma}\mathbf{Q}\mathbf{\Gamma}^{\mathrm{T}}]\mathbf{E}^{\mathrm{T}}\{\mathbf{E}[\mathbf{A}\mathbf{P}(t-1)\mathbf{A}^{\mathrm{T}} + \mathbf{\Gamma}\mathbf{Q}\mathbf{\Gamma}^{\mathrm{T}}]\mathbf{E}^{\mathrm{T}} + \mathbf{R}\}^{-1}$$
$$\times \mathbf{E}[\mathbf{A}\mathbf{P}(t-1)\mathbf{A}^{\mathrm{T}} + \mathbf{\Gamma}\mathbf{Q}\mathbf{\Gamma}^{\mathrm{T}}], \ t = 0, 1, \ldots .$$

Suppose the difference equation (5.9) approaches a steady state. That is, as $t \to \infty$, $\mathbf{P}(t) = \mathbf{P}(t-1) \equiv \mathbf{P}_\infty$. Then it follows from Eq. (4.53) that $\mathbf{K}(t) = \mathbf{K}_\infty$ also becomes steady. Once $\mathbf{P}$ and $\mathbf{K}$ cease to change, the computational load of the filter is much reduced: the model must be run only once at each time-step. This reduction in load leads one to understand under which circumstances Eq. (5.9) will asymptote to a steady state, and, when it does, to find methods for determining that state. With $\mathbf{K}_\infty$ known, one can, if one chooses, use it in place of $\mathbf{K}(t)$, *even during the period when the steady state is invalid.* To the extent that the system "forgets" its initial conditions, experience suggests that eventually the estimated state will converge to the correct one, even though the initial transient is not properly computed. A steady-Kalman filter is a "Wiener filter"; they are usually applied by fast convolution methods (which we omit).[14] Similar considerations apply to the problem of obtaining steady-state solutions to the evolution equation for the RTS smoother (Wiener smoother); further discussion can be found in the references.

Example *Consider two masses coupled to each other and to the boundaries as indicated in Fig. 5.2. A governing set of differential equations for the position, r_i, of each oscillator is readily shown to be*

$$\mathbf{M}\frac{d^2\mathbf{r}}{dt^2} + \mathbf{D}\frac{d\mathbf{r}}{dt} + \mathbf{L}\mathbf{r} = \mathbf{f}. \tag{5.10}$$

Here, $\mathbf{M}$, $\mathbf{D}$, $\mathbf{L}$ are matrices, $\mathbf{r}(t) = [r_1(t), r_2(t)]^{\mathrm{T}}$ is the non-equilibrium displacement of the masses, and $\mathbf{f}$ is the forcing vector. To generate the simplest case, take $\mathbf{M} = m\mathbf{I}_2$, so that the masses are identical; $\mathbf{D} = \varepsilon\mathbf{I}_2$, so that the dissipation is of ordinary Rayleigh type, and that

$$\mathbf{L} = \left\{ \begin{array}{cc} 2k & -k \\ -k & 2k \end{array} \right\}$$

couples the masses through the connecting springs. Using a one-sided discretization of Eq. (5.10), a canonical state space approximation is

$$\mathbf{x}(n+1) = \mathbf{A}\mathbf{x}(n) + \mathbf{f}_d(n),$$

$$\mathbf{A} = \left\{ \begin{array}{cc} 2\mathbf{I}_2 - \mathbf{L}(\Delta t)^2/m & (-1 + \varepsilon \Delta t)\mathbf{I}_2 \\ \mathbf{I}_2 & 0 \end{array} \right\},$$

$$\mathbf{x}(n) = [\, r_1(n) \, r_2(n) \, r_1(n-1) \, r_2(n-1) \,]^{\mathrm{T}},$$

$$\mathbf{f}_d(n) = (\Delta t)^2 [\, \mathbf{f}(n)^{\mathrm{T}} \quad \mathbf{0} \,]^{\mathrm{T}}.$$

*(**A** includes block sub-matrices.) Taking $k = 1, m = 1, \varepsilon = 0.01, \Delta t = 0.25$, and f to be a unit variance zero mean forcing of r_1 alone (no forcing applied to r_2), a realization of $r_1(t), r_2(t)$ is shown in Fig. 5.3. Now assume that $\mathbf{E} = [\,1\,0\,0\,0\,]$ so that each time step, only $x_1(n)$, that is the position r_1, is measured. Assume $\mathbf{P}(0) = \mathrm{diag}(\,1\,1\,1\,1\,), \mathbf{R} = \mathrm{diag}\,([1, 0]),$ and $\mathbf{Q} = \mathbf{I}_4$. Then time-stepping Eq. (5.9) leads to the results for the diagonal elements of $\mathbf{P}(n)$ as depicted in Fig. 5.3. Both P_{11}, P_{12} (and the off-diagonal elements as well) reach steady-state values before $t = n\Delta t = 10$. At that time, $\mathbf{K}(t)$ has become a constant, and one can cease updating either it or the $\mathbf{P}(t)$. ($\mathbf{P}(t, -)$ would also have reached a steady state.)*

How might one find the steady state of Eq. (5.9) – if it exists? Several methods are known. One of them has been used in the above example: time-step the equation until it asymptotes. Other algorithms exist, including a remarkable one called "doubling." In this algorithm, one time steps the equation from $t = 0$, $\mathbf{P}(0)$, to obtain $\mathbf{P}(1\Delta t)$. One then doubles the time step to compute $\mathbf{P}(3\Delta t)$, doubles again for $\mathbf{P}(6\Delta t)$, etc. With this geometric increase in the time step, convergence, if it occurs, is extremely rapid. A simplified equation is treated this way in the Appendix to this chapter.[15]

When does a steady state exist? In general, uncertainty grows because of errors in initial conditions, and the unknown system perturbations (unknown controls, **u**). Information that reduces uncertainty is provided by the incoming data stream. Under the right circumstances, one can reach an equilibrium where the new information just balances the new uncertainties. A quantitative answer to the question depends directly upon the discussion in Chapter 4 of the observability of the system. Although we omit the formal derivation, one can understand physically why those requirements must be met. Suppose, as in Chapter 4, there is an element of the model which is not observable. Then any error, e.g., in its initial conditions, could grow indefinitely, undetected, without bound. Such growth would mean that the corresponding elements of **P** would have to grow, and there would be no steady state. Suppose to the contrary, that such growth *is* observed. Then if those elements are controllable, one can find controls, $\mathbf{u}(t)$, such that the growth is halted. Note that neither $\mathbf{x}(t)$, nor $\mathbf{u}(t)$ will generally become steady – the state continues to evolve.

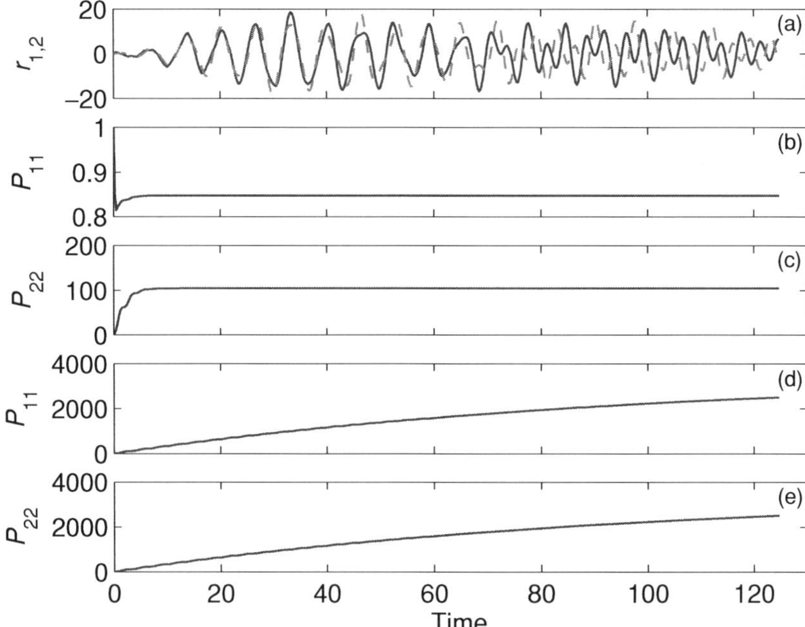

Figure 5.3 (a) Graph of the positions $r_i(t)$, $i = 1, 2$, for the coupled mass–spring oscillator. (b), (c) Graph of $P_{11}(t)$, $P_{22}(t)$ under the assumption that $\mathbf{E} = [1\,0\,0\,0]$, that is, only $r_1(t)$ is measured (with noise). Both $P_{11}(t)$, $P_{22}(t)$ asymptote to a steady state, albeit $P_{11}(t) \ll P_{22}(t)$. (d), (e) Graph of $P_{11}(t)$, $P_{22}(t)$ when the observation matrix is changed to $\mathbf{E} = [-1\,1\,0\,0]$ – that is, the observation is of the relative separation of the two masses. In this case, the uncertainty in the absolute positions continues to grow and a steady state is not reached (there is no dissipation in this example).

In the situation shown in Fig. 5.3, when there is a single measurement of the position, $r_1(t)$, the eigenvalues of the observability matrix, $\mathbf{O}$, range in magnitude from 1.9 down to 0.0067. Albeit there will be considerable uncertainty involved, one can fully determine the initial conditions from the observations. In contrast, when only the relative position, $r_2(t) - r_1(t)$ is measured, two of the eigenvalues of O vanish identically, the system is not completely observable, as seen in Fig. 5.3, and the uncertainties continue to grow without bound. If one were discussing the smoothing algorithm errors, the structure of $\boldsymbol{\Gamma}$ would enter similarly.[16]

5.2.4 Reduced state methods

The computational load of the Kalman filter and smoothers grows approximately as the cube of the state vector dimension. Thus either decoupling the problem into several smaller problems, or removing elements of the state vector, can have a very large payback in terms of the computational load reduction. (If one could solve the problem as two $(n/2)^3$ problems rather than as one n^3 problem the difference in load is a factor of four.) One method for solving large fluid state problems is based upon the assumption that large spatial scales in a fluid flow evolve largely independent of small scales, and that it is the largest scales that are of primary interest.[17] Let $\mathbf{D}$ be a matrix operator that has the effect of averaging a vector spatially, so that $\mathbf{x}'(t) = \mathbf{Dx}(t)$ is a spatial average of $\mathbf{x}(t)$, with an equivalent reduced dimension, N'. (We refer to the "coarse state" and "fine state".) Then if $\mathbf{P}(t)$ is the error covariance of $\mathbf{x}(t)$,

$$\langle (\tilde{\mathbf{x}}' - \mathbf{x}')(\tilde{\mathbf{x}}' - \mathbf{x}')^{\mathrm{T}} \rangle = \mathbf{P}'(t) = \mathbf{DP}(t)\mathbf{D}^{\mathrm{T}}$$

will be of dimension $N' \times N'$ rather than $N \times N$. Now assume further that $\mathbf{D}$ has a left-inverse, $\mathbf{D}^+$, as described in Chapter 2, that would map the coarse state to the finer one. Suppose further that one has a coarse resolution model capable of propagating $\mathbf{x}'$. This model might be obtained from the fine-resolution model:

$$\mathbf{Dx}(t+1) = \mathbf{DA}(t)\mathbf{D}^+\mathbf{x}'(t) + \mathbf{DBD}^+\mathbf{u}'(t) + \mathbf{D\Gamma D}^+\mathbf{q}'(t),$$

or

$$\mathbf{x}'(t+1) = \mathbf{A}'\mathbf{x}'(t) + \mathbf{B}'\mathbf{u}'(t),$$

where $\mathbf{u}'(t) = \mathbf{Du}(t)$, $\mathbf{A}' = \mathbf{DA}(t)\mathbf{D}^+$, $\mathbf{B}' = \mathbf{DBD}^+$, etc. Then the Kalman filter (and smoother) can be applied to $\mathbf{x}'(t)$ and the filtered data, $\mathbf{Dy}(t)$. One can estimate that

$$\tilde{\mathbf{x}}(t) = \mathbf{D}^+\tilde{\mathbf{x}}'(t),$$

and

$$\mathbf{P}(t) = \mathbf{D}^+\mathbf{P}'(t)\mathbf{D}^{+\mathrm{T}}. \tag{5.11}$$

Given $\mathbf{P}(t)$, one has $\mathbf{K}(t)$ for the fine state, under the assumption that Eq. (5.11), based wholly upon the large scales, is adequate. One can put any small scales in the fine-state observations into the data error of the coarse state. A further reduction in computational load can be made by assuming a steady state for $\mathbf{P}'(t)$, $\mathbf{P}(t)$, and finding it using the doubling algorithm. In Chapter 7 we will describe an application of this method. The main issue with its general validity would lie with the assumption that errors in the fine state do not strongly influence the error budget of the coarse state. This assumption cannot in general be correct: spatially averaged equations of fluid motion are not proper representations of the equations governing the averaged fields. One must carefully assess the behavior of the algorithm as it evolves.

Determination of $\mathbf{D}$, $\mathbf{D}^+$ is important. In principle, the Gauss–Markov mapping procedures, as described in Chapter 2, would be appropriate (and would include error estimates should one choose to use them). Various strategies for reducing storage and computation are available.[18]

Other approaches to state reduction

The Eckart–Young–Mirsky theorem, described in Chapter 2, shows that sometimes a comparatively small number of singular vectors can represent a field with considerable accuracy. Here "small" is measured relative to the number of grid points or basis functions used by the underlying model.[19] Suppose that the state vector $\mathbf{x}(t) = \mathbf{V}\alpha(t)$, where $\mathbf{V}$ is the matrix of $\mathbf{v}_i$, the singular vectors of a large span of model – that is, the matrix to which the Eckart–Young–Mirsky theorem is applied is $\{\mathbf{x}(0) \quad \mathbf{x}(2) \quad \ldots \quad \mathbf{x}(t_N)\}$ – is then truncated to some acceptable sub-set,

$$\mathbf{x}(t) \approx \mathbf{V}_K \alpha(t).$$

Taking the canonical, full, model,

$$\mathbf{V}_K \mathbf{x}(t+1) = \mathbf{V}_K \mathbf{A}(t)\mathbf{V}_K \alpha(t) + \mathbf{V}_K \mathbf{B}\mathbf{V}_K \mathbf{u}(t) + \mathbf{V}_K \mathbf{\Gamma}\mathbf{V}_K \mathbf{q}(t),$$

or

$$\alpha(t+1) = \mathbf{A}'(t)\alpha(t) + \mathbf{B}'\mathbf{u}(t) + \mathbf{\Gamma}'\mathbf{q}(t),$$

where $\mathbf{A}'(t) = \mathbf{V}_K \mathbf{A}(t)\mathbf{V}_K$, $\mathbf{B}' = \mathbf{V}_K \mathbf{B}\mathbf{V}_K$, etc., is an evolution equation for the new state vector $\alpha(t)$ whose dimension is $K \ll N$. (If $\mathbf{A}$ is time-independent, an alternative is to diagonalize it in the canonical equation by using its singular vector decomposition.[20]) Then each mode can be handled independently. As with the coarse-to-fine resolution transformation, one is assuming that the suppressed singular vectors (those banished to the nullspace are being assumed zero) do not significantly affect the errors of those retained. One must test the original assumptions against the solution obtained.

5.3 Uncertainty in Lagrange multiplier method

When using the Lagrange multiplier approach, the issue remains of estimating the uncertainty of the solution, even if the system is linear. One approach is to calculate it from the covariance evolution equation of the filter/smoother. When one wishes to avoid that computational load, some limited information about it can be obtained from the Hessian of the cost function at the solution point.[21]

Understanding of the Hessian is central to quadratic norm optimization problems in general. Let $\boldsymbol{\xi}$ represent all of the variables being optimized, including $\mathbf{x}(t)$, $\mathbf{u}(t)$

for all t. Let $\boldsymbol{\xi}^*$ be the optimal value that is sought. Then, as with the static problems of Chapter 2, if we are close enough to $\boldsymbol{\xi}^*$ in the search process, the objective function is locally

$$J = \text{constant} + (\boldsymbol{\xi} - \boldsymbol{\xi}^*)^{\mathrm{T}} \cdot \mathcal{H} \cdot (\boldsymbol{\xi} - \boldsymbol{\xi}^*) + \Delta J,$$

where $\mathcal{H}$ is the Hessian (see Eq. (2.31)) and ΔJ is a higher-order correction. The discussion of the behavior of the solution in the vicinity of the estimated optimal value proceeds then, exactly as before, with row and column scaling being relevant, and issues of ill-conditioning, solution variances, etc., all depending upon the eigenvalues and eigenvectors of $\mathcal{H}$.[22] One of the important future applications of automatic differentiation (AD) tools as described in Chapter 4 would be to calculate the second derivatives of objective functions to produce the Hessian.

The only problem, albeit a difficult one, is that the dimensions of $\mathcal{H}$ are square of the dimensions of $\mathbf{x}(t)$ plus $\mathbf{u}(t)$ over the entire time history of the model and data. Finding ways to understand the solution structure and uncertainty with realistic fluids and large-scale datasets remains as one of the most important immediate challenges to results such as those described in Chapter 7.

5.4 Non-normal systems

Consider a forward model,

$$\mathbf{x}(t) = \mathbf{A}\mathbf{x}(t - 1), \tag{5.12}$$

with t again an integer. In general, the underlying physics will fail to be self-adjoint and hence $\mathbf{A}$ will be non-normal, that is, $\mathbf{A} \neq \mathbf{A}^{\mathrm{T}}$. We suppose the system is unforced, but is started with initial conditions $\mathbf{x}(0)$ which are a realization of white noise with variance σ^2. Thus, at time t,

$$\mathbf{x}(t) = \mathbf{A}^t \mathbf{x}(0), \tag{5.13}$$

Recalling the discussion on p. 152, it follows immediately that the eigenvalues and right eigenvectors of $\mathbf{A}^t$ satisfy

$$\mathbf{A}^t \mathbf{g}_i = \lambda_i^t \mathbf{g}_i. \tag{5.14}$$

Expanding,

$$\mathbf{x}(0) = \sum_{i=1}^{N} \alpha_i(0) \mathbf{g}_i, \tag{5.15}$$

and

$$\mathbf{x}(t) = \sum_{i=1}^{N} \lambda_i^t \alpha_i(0) \mathbf{g}_i. \tag{5.16}$$

Stability of the model demands that all $|\lambda_i| \le 1$. But the lack of orthogonality of the $\mathbf{g}_i$ means that some of the α_i may be very large, despite the white noise properties of $\mathbf{x}(0)$. This result implies that elements of $\mathbf{x}(t)$ can become very large, even though the limit $\lambda_i^t \to 0$, $t \to \infty$ means that they are actually transients. To an onlooker, the large response of the system to a bounded initial disturbance may make the system look unstable. Furthermore, the disturbance may become so large that the system becomes non-linear, and possibly non-linearly unstable.[23] That is, stable fluid systems may well appear to be unstable owing to the rapid growth of transients, or linearly stable systems may become unstable in the finite amplitude sense if the transients of the linearized system become large enough.

Now consider the forced situation with time-independent $\mathbf{A}$,

$$\mathbf{x}(t) = \mathbf{A}\mathbf{x}(t-1) + \mathbf{q}(t-1). \tag{5.17}$$

Take the Fourier transform of the difference equation (5.17), using the result[24] that if the transform of $\mathbf{x}(t)$ is

$$\hat{\mathbf{x}}(s) = \sum_{t=0}^{\infty} \mathbf{x}(t)\,e^{-2\pi i s t}, \tag{5.18}$$

then the transform of $\mathbf{x}(t-1)$ is $e^{-2\pi i s}\hat{\mathbf{x}}(s)$. Solving for $\hat{\mathbf{x}}(s)$,

$$\hat{\mathbf{x}}(s) = (e^{-2\pi i s}\mathbf{I} - \mathbf{A})^{-1}\hat{\mathbf{q}}(s). \tag{5.19}$$

We call $(e^{-2\pi i s}\mathbf{I} - \mathbf{A})^{-1}$ the "resolvent" of $\mathbf{A}$, in analogy to the continuous case terminology of functional analysis.[25] If the resolvent is infinite for real values of $s = s_i$ it implies $\hat{\mathbf{x}}(s_i)$ is an eigenvector of $\mathbf{A}$ and an ordinary resonance is possible. For the mass–spring oscillator of Chapter 2, the complex eigenvalues of $\mathbf{A}$ produce $s_{1,2} = \pm 0.0507 + 0.0008i$, and the damped oscillator has no true resonance. Should any eigenvalue have a negative imaginary part, leading to $|e^{-2\pi i s_i t}| > 1$, the system would be unstable.

Define $z = e^{-2\pi i s}$, to be interpreted as an analytic continuation of s into the complex plane. The unit circle $|z| = 1$ defines the locus of real frequencies. The gist of the discussion of what are called "pseudo-spectra" is the possibility that the norm of the resolvent $\|(z\mathbf{I} - \mathbf{A})^{-1}\|$ may become very large, but still finite, on $|z| = 1$ without there being either instability or resonance, giving the illusion of linear instability.

5.4.1 POPs and optimal modes

For any linear model in canonical form, the right eigenvectors of $\mathbf{A}$ can be used directly to represent fluid motions,[26] as an alternative, e.g., to the singular vectors

(EOFs). These eigenvectors were called "principal oscillation patterns," or POPs, by K. Hasselmann. Because $\mathbf{A}$ is usually not symmetric (not self-adjoint), the eigenvalues are usually complex, and there is no guarantee that the eigenvectors are a basis. But assuming that they provide an adequate description – usually tested by trying them – the right eigenvectors are used in pairs when there are complex conjugate eigenvalues. The expansion coefficients of the time-evolving field are readily shown to be the eigenvectors of $\mathbf{A}^{\mathrm{T}}$ – that is, the eigenvectors of the adjoint model. Assuming that the eigenvectors are not grossly deficient as a basis, and/or one is interested in only a few dominant modes of motion, the POP approach gives a reasonably efficient representation of the field.

Alternatively, $\mathbf{A}$ always has an SVD and one can try to use the singular vectors of $\mathbf{A}$ – directly – to represent the time evolving field. The complication is that successive multiplications by non-symmetric $\mathbf{A}$ transfers the projection from the $\mathbf{U}$ vectors to the $\mathbf{V}$ vectors and back again. Write $\mathbf{A} = \mathbf{U}\mathbf{\Lambda}\mathbf{V}^{\mathrm{T}}$ and assume, as, is normally true of a model, that it is full rank $K = N$ and $\mathbf{\Lambda}$ is square. Using Eqs. (4.99) and (4.101), in the absence of observations,

$$\mathbf{x}(t) = \mathbf{A}\mathbf{x}(t - 1), \tag{5.20}$$

$$\boldsymbol{\mu}(t - 1) = \mathbf{A}^{\mathrm{T}}\boldsymbol{\mu}(t), \tag{5.21}$$

one can always write

$$\mathbf{x}(t) = \mathbf{V}\boldsymbol{\alpha}(t), \tag{5.22}$$

where $\boldsymbol{\alpha}(t)$ is a set of vector coefficients. Write the adjoint solution as

$$\boldsymbol{\mu}(t) = \mathbf{U}\boldsymbol{\beta}(t). \tag{5.23}$$

Multiply (5.20) by $\boldsymbol{\mu}(t - 1)^{\mathrm{T}}$ and (5.21) by $\mathbf{x}(t)^{\mathrm{T}}$, and then subtract,

$$\boldsymbol{\mu}(t - 1)^{\mathrm{T}}\mathbf{A}\mathbf{x}(t - 1) = \mathbf{x}(t)^{\mathrm{T}}\mathbf{A}^{\mathrm{T}}\boldsymbol{\mu}(t) = \boldsymbol{\mu}(t)^{\mathrm{T}}\mathbf{A}\mathbf{x}(t), \tag{5.24}$$

or, using (5.22) and (5.23),

$$\boldsymbol{\beta}(t - 1)^{\mathrm{T}}\mathbf{\Lambda}\boldsymbol{\alpha}(t - 1) = \boldsymbol{\alpha}(t)^{\mathrm{T}}\mathbf{\Lambda}\boldsymbol{\beta}(t)^{\mathrm{T}}, \tag{5.25}$$

which can be interpreted as an energy conservation principle, summed over modes.

Assume $\|\mathbf{A}\| < 1$ so that the system is fully stable. We can ask what disturbance of unit magnitude at time $t - 1$, say, would lead to the largest magnitude of $\mathbf{x}(t)$? That is, we maximize $\|\mathbf{A}\mathbf{q}(t - 1)\|$ subject to $\|\mathbf{q}(t - 1)\| = 1$. This requirement is equivalent to solving the constrained maximization problem for the stationary values of

$$J = \mathbf{q}(t - 1)^{\mathrm{T}}\mathbf{A}^{\mathrm{T}}\mathbf{A}\mathbf{q}(t - 1) - 2\mu(\mathbf{q}(t - 1)^{\mathrm{T}}\mathbf{q}(t - 1) - 1), \tag{5.26}$$

where μ is a scalar Lagrange multiplier, and which leads to the normal equations,

$$\mathbf{A}^{\mathrm{T}}\mathbf{A}\mathbf{q}\,(t-1) = \mu\mathbf{q}(t-1), \tag{5.27}$$

$$\mathbf{q}\,(t-1)^{\mathrm{T}}\,\mathbf{q}\,(t-1) = 1. \tag{5.28}$$

Equation (5.27) shows that the solution is $\mathbf{q}\,(t-1) = \mathbf{v}_1$, $\mu = \lambda_1^2$, which is the first singular vector and value of $\mathbf{A}$ with (5.28) automatically satisfied. The particular choice of μ ensures that we obtain a maximum rather than a minimum. With $\mathbf{q}\,(t-1)$ proportional to the $\mathbf{v}_1$ singular vector of $\mathbf{A}$, the growth rate of $\mathbf{x}(t)$ is maximized.[27] The initial response would be just $\mathbf{u}_1$, the corresponding singular vector. If the time-step is very small compared to the growth rates of model structures, the analysis can be applied instead to $\mathbf{A}^{t_1}$, that is, the transition matrix after t_1 time-steps. The next largest singular value will give the second fastest growing mode, etc.

5.5 Adaptive problems

A major point of concern in estimation procedures based upon Gauss–Markov type methods lies in the specification of the various covariance matrices, especially those describing the model error (here included in $\mathbf{Q}(t)$). Nothing precludes deduction of the covariance matrices from the model and observations, given that adequate numbers of observations are available. For example, it is straightforward to show that if a Kalman filter is operating properly, then the so-called innovation, $\mathbf{y}(t) - \mathbf{E}\tilde{\mathbf{x}}(t, -)$, should be uncorrelated with all previous measurements (recall Chapter 2, Eq. (2.433)):

$$\langle \mathbf{y}(t') \, [\mathbf{y}(t) - \mathbf{E}\tilde{\mathbf{x}}(t, -)] \rangle = 0, \quad t' < t. \tag{5.29}$$

To the extent that (5.29) is not satisfied, the covariances need to be modified, and algorithms can be formulated for driving the system toward this condition. The possibilities for such procedures are known under the title "adaptive estimation."[28]

The major issues here are that accurate determination of a covariance matrix of a field, $\langle \mathbf{z}(t)\mathbf{z}(t') \rangle$, requires a vast volume of data. Note in particular that if the mean of the field $\mathbf{z}(t) \neq \mathbf{0}$, and it is inaccurately removed from the estimates, then major errors can creep into the estimated second moments. This bias problem is a very serious one in adaptive methods.

In practical use of adaptive methods, it is common to reduce the problem dimensionality by *modeling the error covariance matrices*, that is, by assuming a particular, simplified structure described by only a number of parameters much less than the number of matrix elements (accounting for the matrix symmetry). We must leave this subject to the references.[29]

Appendix. Doubling

To make the doubling algorithm plausible,[30] we consider the matrix equation

$$\mathbf{B}_{k+1} = \mathbf{F}\mathbf{B}_k\mathbf{F}^{\mathrm{T}} + \mathbf{C}, \tag{5.30}$$

and seek to time-step it. Starting with $\mathbf{B}_1$, one has, time-stepping as far as $k = 3$,

$$\mathbf{B}_2 = \mathbf{F}\mathbf{B}_1\mathbf{F}^{\mathrm{T}} + \mathbf{C},$$
$$\mathbf{B}_3 = \mathbf{F}\mathbf{B}_2\mathbf{F}^{\mathrm{T}} + \mathbf{C} = \mathbf{F}^2\mathbf{B}_1\mathbf{F}^{2\mathrm{T}} + \mathbf{F}\mathbf{C}\mathbf{F}^{\mathrm{T}} + \mathbf{C},$$
$$\mathbf{B}_4 = \mathbf{F}\mathbf{B}_3\mathbf{F}^{\mathrm{T}} + \mathbf{C}$$
$$= \mathbf{F}^2\mathbf{B}_2\mathbf{F}^{2\mathrm{T}} + \mathbf{F}\mathbf{C}\mathbf{F}^{\mathrm{T}} + \mathbf{C}$$
$$= \mathbf{F}^2\mathbf{B}_2\mathbf{F}^{2\mathrm{T}} + \mathbf{B}_2,$$

that is, $\mathbf{B}_4$ is given in terms of $\mathbf{B}_2$. More generally, putting $\mathbf{M}_{k+1} = \mathbf{M}_k^2$, $\mathbf{N}_{k+1} = \mathbf{M}_k\mathbf{N}_k\mathbf{M}_k^{\mathrm{T}} + \mathbf{N}_k$, with $\mathbf{M}_1 = \mathbf{F}$, $\mathbf{N}_1 = \mathbf{Q}$, then $\mathbf{M}_{2k} = \mathbf{F}^{2^k}$, $\mathbf{N}_{k+1} = \mathbf{B}_{2^k}$, and one is solving Eq. (5.30) so that the time-step doubles at each iteration. An extension of this idea underlies the doubling algorithm used for the Riccati equation.

Notes

1 See Kitagawa and Sato (2001) for references.
2 See Arulampalam *et al.* (2002). Their development relies on a straightforward Bayesian approach.
3 See Daley (1991) and Evensen and VanLeeuwen (1996) and the references therein for a more complete discussion.
4 See Press *et al.* (1996) for detailed help concerning generating values from known probability distributions.
5 Kalnay (2003) discusses, e.g., "breeding" vectors, which are selected to display fastest growth in the model.
6 See Aitchison and Brown (1957).
7 Gardiner (1985).
8 Evensen (1994) and Evensen and VanLeeuwen (1996) are good starting points for practical applications, insofar as problem dimensions have permitted. See Daley (1991) and Kalnay (2003) for a broad discussion of the specific numerical weather forecasting problem.
9 Reviews are by Lorenc (1986), Daley (1991), Ghil and Malanotte-Rizzoli (1991), or Kalnay (2003).
10 Usually called "3DVAR," by meteorologists, although like "4DVAR" it is neither variational nor restricted to three dimensions.
11 For example, "breeding"; see Kalnay (2003).
12 Anthes (1974).
13 Gelb (1974, Chapters 7 and 8) has a general discussion of the computation reduction problem, primarily in the continuous time context, but the principles are identical.
14 Kalman's (1960) filter derivation was specifically directed at extending the Wiener theory to the transient situation, and it reduces to the Wiener theory when a steady-state is appropriate.)
15 Anderson and Moore (1979) should be consulted for a complete discussion.
16 Fukumori *et al.* (1992) discuss this problem in greater generality for a fluid flow.
17 Fukumori (1995), who interchanges the roles of $\mathbf{D}$, $\mathbf{D}^{+}$. See also Fukumori (2002).
18 A general discussion of various options for carrying out the transformations between fine and coarse states is provided by Fieguth *et al.* (2003).

19 Used, for example, by Cane *et al.* (1996).
20 For example, Brogan (1991).
21 Thacker (1989), Marotzke and Wunsch (1993).
22 Tziperman *et al.* (1992) grapple with ill-conditioning in their results; the ill-conditioning is interpretable as arising from a nullspace in the Hessian.
23 This potential confusion is the essence of the conclusions drawn by Farrell (1989) and Farrell and Moore (1993), and leads to the discussion by Trefethen (1997, 1999) of pseudo-spectra.
24 Bracewell (2000).
25 Trefethen (1997).
26 Hasselmann (1988), von Storch *et al.* (1988), von Storch and Zwiers (1999).
27 The meteorological literature, e.g., Farrell and Moore (1993), renamed this singular vector as the "optimal" vector.
28 Among textbooks that discuss this subject are those of Goodwin and Sin (1984), Haykin (1986), and Ljung (1987).
29 See Menemenlis and Chechelnitsky (2000) and the references given there.
30 Following Anderson and Moore (1979, p. 67).

Part II

Applications

6

Applications to steady problems

The focus will now shift away from discussion of estimation methods in a somewhat abstract context, to more specific applications, primarily for large-scale fluid flows, and to the ocean in particular. When the first edition of this book (OCIP[1]) was written, oceanographic uses of the methods described here were still extremely unfamiliar to many, and they retained an aura of controversy. Controversy arose for two reasons: determining the ocean circulation was a classical problem that had been discussed with ideas and methods that had hardly changed in 100 years; the introduction of algebraic and computer methods seemed to many to be an unwelcome alien graft onto an old and familiar problem. Second, some of the results of the use of these methods were so at odds with "what everyone knew," that those results were rejected out of hand as being obviously wrong – with the methods being assumed flawed.

In the intervening 25+ years, both the methodology and the inferences drawn have become more familiar and less threatening. This change in outlook permits the present chapter to focus much less on the why and how of such methods in the oceanographic context, and much more on specific examples of how they have been used.[2] Time-dependent problems and methods will be discussed in Chapter 7.

This chapter is not intended to be an oceanographic textbook, but does outline the so-called geostrophic inverse problem. The chapter should be accessible even to those without any oceanographic knowledge: the problems are primarily those of making inferences from tracers in a large-scale fluid flow. Anyone seeking understanding of the oceanographic details should consult one of the references provided in the text.

Inverse methods were introduced into oceanography by the need to determine the ocean circulation from large-scale hydrographic data. At that time (the middle 1970s), the focus was still on determining a "mean" circulation based on data obtained from ships over timescales ranging from weeks to decades. Ironically, much of the interest in this problem has waned in the intervening years, as it became

279

obvious that the ocean is highly time-dependent. The fiction that the flow could be regarded as steady over arbitrarily long periods became increasingly difficult to sustain in the teeth of increasing evidence to the contrary. It is still useful, nonetheless, to study this problem in some depth: (1) A very large literature is focussed on it, and the steady-state approximation is sometimes sufficiently accurate to provide useful information about the circulation. (2) Interpretation of the time-dependent problem, both in terms of the methodology, and of the physics, depends upon understanding the steady-flow results. As with most approximations in science, however, one must always bear in mind the limitations of the result.

We begin with a review of the steady tracer problem, one applicable to a wide class of fluid-flow problems, and then turn to the ocean circulation problem.

6.1 Steady-state tracer distributions

A generic distribution equation for tracer C in steady state is

$$\mathbf{v} \cdot \nabla C - \nabla (\mathbf{K} \nabla C) = -\lambda C + m_C(\mathbf{r}). \tag{6.1}$$

Here the velocity field, $\mathbf{v}(\mathbf{r})$, can be one- to three-dimensional (in one dimension it is a "pipeflow"), and will in general be a function of the spatial coordinate $\mathbf{r}$. The mixing tensor, $\mathbf{K}$, can also be dependent upon position and is often anisotropic because it represents a parameterization of subgridscale (eddy) mixing processes sometimes greatly influenced by fluid stratification or geometry. ($\mathbf{K}$ could be the molecular value, which would thus be a special case with its value known nearly perfectly.) λC is present to permit the study of tracers (carbon-14 is an example) undergoing radioactive decay. Finally, $m_C(\mathbf{r})$ represents any interior sources and sinks of C other than radio-decay. In the most general situation, m_C can be a function of C, permitting representation of chemical reaction terms.

Equation (6.1) is bilinear in $([\mathbf{v}, \mathbf{K}], C)$ – that is, with $[\mathbf{v}, \mathbf{K}]$ known, it is linear in C, and if C is known, it is linear in $[\mathbf{v}, \mathbf{K}]$. It thus lends itself to two distinct classes of linear inverse problem, depending upon which of $[\mathbf{v}, \mathbf{K}]$, or C is regarded as known/unknown. Usually, however, one has imperfect knowledge of both sets and any resulting inverse problem will be non-linear to some degree (m_C, if partly unknown, does not usually render the problem non-linear).

The most common application of steady tracer distributions has been in so-called box models, where integral versions of Eq. (6.1) have been used, as in the example on p. 10, to describe *net* exchanges between large reservoirs, without distinguishing between tracer movement by way of advection, $\mathbf{v}$, and by diffusion, $\mathbf{K}$. Most such models have been used in forward mode alone (recall, from Chapter 1, that "forward" or "direct" mode refers to the conventional well-posed initial/boundary value problem).

Box models are powerful tools for obtaining scale analyses for order of magnitude estimates of exchange rates. They often suffer from an underparameterization of the possible flow paths, and, consequently, can lead to grossly incorrect inferences. Consider the minimal reservoir problem (p. 59), in which a volume V_0 with tracer concentration C_0 is supposed fed from two reservoirs of concentration C_1, C_2, such that $C_1 \le C_0 \le C_2$. It was found that assuming two feeding reservoirs produced a definitive answer. If, however, one was suspicious of the existence of a third (or more) feeding reservoir, it was found that formulating the problem so as to deal with the indeterminacy was a better strategy than simply pretending it was not possible.

Another common approach to inferences from steady tracers is exemplified by the calculation of Munk (1966), for a one-dimensional pipeflow, using a representation of two tracers in vertical balance given by

$$w\frac{\partial C_1}{\partial z} - k\frac{\partial^2 C_1}{\partial z^2} = 0, \tag{6.2}$$

$$w\frac{\partial C_2}{\partial z} - k\frac{\partial^2 C_2}{\partial z^2} = -\lambda C_2, \tag{6.3}$$

where C_1 is temperature and C_2 is radiocarbon, ^{14}C. Because a one-dimensional balance is assumed, the flow, w, must be a constant, and the problem was further simplified by assuming the turbulent mixing coefficient $k = $ constant. By fitting observations of $C_1(z)$, $C_2(z)$ at one location in the central Pacific Ocean, and representing $\partial C_i/\partial z$, $\partial^2 C_i/\partial z^2$ as fixed, known, constants with depth, Munk (1966) reduced Eqs. (6.2) and (6.3) to two equations in two unknowns, $[w, k]$, that is, as

$$\mathbf{Ex} + \mathbf{n} = \mathbf{y},$$

although no formal uncertainty estimate was provided. The missing error analysis of this solution was later provided by Olbers and Wenzel (1989) who found that the calculation was nearly singular.

Wunsch (1985) and McIntosh and Veronis (1993) discuss a finite difference form of Eq. (6.1) in two dimensions. On a grid with Δx, Δy, and using indices i, j, one has

$$\begin{aligned}
(C_{i+1,j} + C_{i,j})&u_{i,j} - (C_{i,j} + C_{i-1,j})u_{i-1,j} \\
&+ \delta[(C_{i,j+1} + C_{i,j})v_{i,j} - (C_{i,j} + C_{i,j-1})v_{i,j-1}] \\
= 2K/\Delta x[C_{i+1,j} &+ C_{i-1,j} - 2C_{i,j} + \delta^2(C_{i,j+1} + C_{i,j-1} - 2C_{i,j})],
\end{aligned} \tag{6.4}$$

subject to the continuity relation

$$u_{i,j} - u_{i-1,j} + \delta(v_{i,j} - v_{i,j-1}) = 0, \qquad \delta = \Delta y / \Delta x. \tag{6.5}$$

Solution of this set of simultaneous equations for various assumptions concerning C_{ij} for u_{ij}, v_{ij} are discussed by Wunsch (1985) and McIntosh and Veronis (1993). The latter authors employ a streamfunction for the velocity, $u = \partial \psi / \partial y$, $v = -\partial \psi / \partial x$, and explore the use of weighting functions in least-squares that attempt (recall Eq. (2.134)) to find the smoothest possible solutions. As noted previously, with least-squares one can invoke any desired aesthetic principle for the solution, as long as it can be expressed through a weight matrix. It is for this reason that least-squares is best regarded as a curve-fitting procedure, rather than as an estimation one, unless the weight matrices are chosen specifically to produce the same solution as does the Gauss–Markov result.

6.2 The steady ocean circulation inverse problem

6.2.1 Equations of motion

The ocean circulation is governed by Newton's laws of motion plus those of the thermodynamics of a heat- and salt-stratified fluid. That these physics govern the system is a concise statement that a great deal is known about it. One seeks to exploit this information to the fullest extent possible, combining this theoretical knowledge with whatever observations are available.

The resulting equations of motion describing the ocean are the so-called Navier–Stokes equations for a thin shell of temperature- and salinity-stratified fluid on a bumpy, near-spheroidal body undergoing rapid rotation, augmented with an equation of state for density. Appropriate boundary conditions are those of no flow of fluid into the bottom and sides, statements about the stress exerted on these boundaries, and those representing exchange of momentum, heat, and moisture with the atmosphere at the surface. So-called tracer advection equations govern the distributions of temperature, salt, oxygen, and other scalar properties. The science of oceanography is in large part occupied with finding simplifications of this complicated system of equations adequate to provide description, understanding, and, ultimately, forecasting where possible.

6.2.2 Geostrophy

Local Cartesian approximations to the full equations are sufficient for many purposes. Both theory and decades of ocean observation lead to the inference that local "geostrophic, hydrostatic" balance is an excellent approximation:

$$-\rho(x, y, z, t) f(y) v(x, y, z, t) = -\frac{\partial p(x, y, z, t)}{\partial x}, \tag{6.6}$$

$$\rho(x, y, z, t) f(y) u(x, y, z, t) = -\frac{\partial p(x, y, z, t)}{\partial y}, \tag{6.7}$$

$$0 = -\frac{\partial p(x, y, z, t)}{\partial z} - g\rho(x, y, z, t), \tag{6.8}$$

where p is the pressure (Eq. (6.8) asserts that it is hydrostatic); u, v are the zonal (x-direction), and meridional (y-direction) flow fields; $\rho(x, y, z)$ is the fluid density; g is local gravity; and $f = 2\Omega \sin \phi$ is the Coriolis parameter, where ϕ is the local latitude and Ω is the Earth's rotation rate (radians/second). One commonly writes $f = f_0 + \beta y$ to account for its local dependence on latitude. *The reader is cautioned that x, y are here spatial coordinates and should not be confused with $\mathbf{x}$, the state vector, or $\mathbf{y}$, the observation vector, used previously.* This convention for spatial coordinates is so deeply embedded in the literature that it seems quixotic to not use it here. (The time, t, has been placed into the arguments in the above equation as a reminder that these balances are often excellent ones, even in a time-evolving field.) The above equations are not exact, but they are so close to balancing that observational tests of their differences almost universally fail. On the other hand, the small deviations from precise equality are essential to determining oceanic physics.[3]

Defining w to be the local vertical velocity, approximate mass conservation is readily shown to be equivalent to

$$\frac{\partial u}{\partial x} + \frac{\partial v}{\partial y} + \frac{\partial w}{\partial z} = 0, \tag{6.9}$$

and conservation of density is

$$u\frac{\partial \rho}{\partial x} + v\frac{\partial \rho}{\partial y} + w\frac{\partial \rho}{\partial z} = 0, \tag{6.10}$$

(a special case of Eq. (6.1)) or, if combined with (6.9),

$$\frac{\partial(\rho u)}{\partial x} + \frac{\partial(\rho v)}{\partial y} + \frac{\partial(\rho w)}{\partial z} = 0. \tag{6.11}$$

The equation of state,

$$\rho(x, y, z, t) = \varrho(T(x, y, z, t), S(x, y, z, T), p(x, y, z, t)), \tag{6.12}$$

is an empirical relationship for density ρ in terms of local temperature, T, salinity, S, and pressure, p.

Without diffusion, tracer conservation can be written as

$$u\frac{\partial(\rho C)}{\partial x} + v\frac{\partial(\rho C)}{\partial y} + w\frac{\partial(\rho C)}{\partial z} = m_C(x, y, z, t), \qquad (6.13)$$

or, if combined with (6.9), as

$$\frac{\partial(\rho C u)}{\partial x} + \frac{\partial(\rho C v)}{\partial y} + \frac{\partial(\rho C w)}{\partial z} = m_C(x, y, z, t), \qquad (6.14)$$

where m_C is a generic trace source or sink. For readers unfamiliar with, or uninterested in, the ocean, the equations (6.6)–(6.14) should simply be taken as axiomatic relations among physical variables.

The classical dynamic method

The "thermal wind equations" are obtained from (6.6)–(6.8) by cross-differentiation to eliminate the pressure,

$$-f\frac{\partial(\rho v)}{\partial z} = g\frac{\partial \rho}{\partial x}, \qquad (6.15)$$

$$f\frac{\partial(\rho u)}{\partial z} = g\frac{\partial \rho}{\partial y}, \qquad (6.16)$$

and can be integrated in the vertical to produce

$$\rho u(x, y, z, t) = \frac{g}{f}\int_{z_0}^{z} \frac{\partial \rho}{\partial y}dz + \rho c(x, y, t, z_0)$$
$$\equiv \rho(x, y, z, t)\left[u_R(x, y, z, t) + c(x, y, z_0, t)\right], \qquad (6.17)$$

$$\rho v(x, y, z, t) = -\frac{g}{f}\int_{z_0}^{z} \frac{\partial \rho}{\partial x}dz + \rho b(x, y, t, z_0)$$
$$\equiv \rho(x, y, z, t)[v_R(x, y, z, t) + b(x, y, z_0, t)]. \qquad (6.18)$$

Depth, z_0, is the "reference depth" and is arbitrary. Note that the two integration constants b, c do not depend upon z, but only upon the fixed value z_0. u_R, v_R are the "relative velocities," and the integration constants b, c are the "reference-level velocities."

From the middle of the nineteenth century, scientists learned to measure T, S as functions of depth (usually determined from the pressure via the hydrostatic equation) at sea permitting the computation of $\rho(x, y, z, t)$. From Eqs. (6.17) and (6.18) calculation of the velocity follows – except for the contribution from b, c. Apart from that portion, the geostrophic flow can then be calculated from shipboard measurements of the horizontal derivatives of ρ alone. The integration constants – that is, the reference level velocities – are mathematically simple, but the inability to determine them plagued oceanography for almost 100 years, and has been one of the major obstacles to understanding the ocean circulation.

In practice, most investigators assumed values for b, c by asserting that there was some depth, z_0 (or equivalently, a temperature or pressure surface), where $u(x, y, z = z_0(x, y)) = v(x, y, z = z_0(x, y)) = 0$, implying $b(x, y) = c(x, y) = 0$ and then proceeded to discuss the implications of the resulting flow field. z_0 was commonly referred to as the "level-of-no-motion." This approach is the classical "dynamical method." (The t-dependence is generally omitted from here on.) Although there is definite evidence that at great depths the flow becomes comparatively small compared to its near-surface values, unhappily there is neither quantitative evidence nor theory suggesting that a true depth of no flow, even one that might be a complicated function, $z_0(x, y)$, exists anywhere. Even very small velocities, when integrated over large depths, can produce substantial water, and property, transports.

On the other hand, from Eqs. (6.10)–(6.13), it *is* possible to find an expression for the three components of velocity, as (Needler, 1985)

$$\rho(x, y, z)[u, v, w] \qquad (6.19)$$

$$= g\frac{\hat{\mathbf{k}} \cdot (\nabla\rho \times \nabla q)}{\nabla(f\partial q/\partial z) \cdot (\nabla\rho \times \nabla q)} \nabla\rho \times \nabla q, \quad q = f(y)\frac{\partial\rho}{\partial z}$$

Alternate derivations can be found in Pedlosky, 1987; and in Wunsch, 1996. Here ∇ is the three-dimensional gradient operator, and $\hat{\mathbf{k}}$ is the vertical unit normal vertical vector. Equation (6.19) shows that, up to the approximations implicit in the equations used, *a knowledge of the density field alone is adequate to completely determine the absolute oceanic flow field.* In practice, expression (6.19) is never used – for two reasons: (1) it involves the third derivative of the density field, in the element $\nabla(f\partial q/\partial z)$, rendering the calculation hopelessly noisy and, (2) exploration of the sea by ships most commonly takes place along linear tracks, precluding the calculation of density gradients normal to the track.

Because inverse methods were first introduced into oceanography to deal with the determination of absolute geostrophic velocities, we now review that application. Fundamentally, it involves using the same equations that lead to (6.19), but in an approximate, linearized, form that is less noise sensitive than the exact expression, and that will reduce to the canonical $\mathbf{Ex} + \mathbf{n} = \mathbf{y}$.

Consider the transport of properties by a geostrophic flow. Let the concentration of a scalar property (temperature, carbon, oxygen, etc.) be given by $C(x, y, z)$ per unit mass (suppressing t). Then if the geostrophic flow is known, the integrated flux in the meridional direction of C between $x = x_1$ and $x = x_2$ lying between two depths $z_1(x)$ and $z_2(x)$ is

$$V_c(y) \equiv \int_{x_1}^{x_2} \int_{z_1(x)}^{z_2(x)} \rho C(x, y, z)[v_R(x, y, z) + b(x, y)]\, dz dx, \qquad (6.20)$$

which is made up of a part owing to the thermal wind and a part owing to the unknown reference-level velocity. The dependence upon z_0 is now implicit. A corresponding expression exists for the flux in the zonal direction. From here on, we will use v, b to denote the two velocities normal to the section, irrespective of the true direction of flow or of the x, y coordinates, as the Cartesian approximation does not distinguish between x, y in the thermal wind. When the property is mass itself, $C = 1$. Transport calculations are of very great importance both because they underlie all large-scale property conservation statements, but also because the overall movement of properties, C, which can include temperature (enthalpy), freshwater, oxygen, and carbon, among other quantities, are determinants of biological productivity and of climate. If the term involving $b(x, y)$ in Eq. (6.20) is arbitrarily set to zero, then an error occurs in the computation of V_c. Suppose $b \approx 1$ mm/s, a small oceanographic velocity, and that $C = 1$, so that the transport of mass is being calculated. Then in a 5000 m deep ocean, with b extending over 3000 km, its mass flux contribution is

$$\int_{x_1}^{x_2} \int_{z_1(x)}^{z_2(x)} \rho b(x, y = y_c)\,\mathrm{d}z\mathrm{d}x \approx 3000 \text{ km} \times 5000 \text{ m} \times 10^3 \text{ kg/m}^3 \times 10^{-3} \text{ m/s}$$

$$= 15 \times 10^9 \text{ kg/s},$$

(y_c is constant) or about 50% of the mass transport of the Gulf Stream near Florida. So, as asserted above, small velocities at depth can generate unacceptable errors in transport calculations.

How can one determine the missing integration constants, using the two equations omitted in the classical dynamical method, but without having to resort to using third derivatives of the calculated density field? (The example shown in Fig. 1.5 described the seemingly artificial problem of determining an absolute flow field when only its vertical derivatives were available. For many decades it was the central problem of physical oceanography – through equations (6.15) and (6.16).)

It is useful first to slightly complicate this relatively simple story in two ways:

1. The direct action of the wind blowing on the ocean drives a local flow, usually called the "Ekman transport."[4] This flow is confined to roughly the upper 100 m of the ocean, and does not produce any readily measured influence on the density field – it is thus invisible to observers who only measure temperature and salinity on board a ship. In practice, Ekman transports are estimated from climatological wind fields provided by meteorologists. At any given location, define the Ekman mass flux as F_E, in the direction normal to the section or station pair. Any property C is advected by the Ekman layer, as $F_C = \bar{C}_E(x, y, z_E)F_E$, where $\bar{C}_E$ is the mean of C over the Ekman layer depth. We will write the horizontally integrated Ekman flux of C as $\overline{F_C} = \overline{C_E F_E}$. This contribution has to be accounted for in any flux or conservation statements.

2. The ocean is stably stratified in density, so that any vertical motion, w, carries fluid of one density into a region of lower ($w > 0$), or higher ($w < 0$) density. If the stratification

is to be maintained in the presence of a time average w, then the fluid with density anomalous compared to its surroundings must be mixed with fluid of a compensating density so as to maintain the steady state. This process of mixing is often represented as a pseudo-diffusive mass flux in the vertical direction using an empirical coefficient, K_v, multiplying the vertical density gradient, so that the vertical density flux becomes approximately

$$w\rho - K_\mathrm{v}\frac{\partial\rho}{\partial z}, \tag{6.21}$$

and Eq. (6.10) is modified to

$$u\frac{\partial\rho}{\partial x} + v\frac{\partial\rho}{\partial y} + w\frac{\partial\rho}{\partial z} - \frac{\partial}{\partial z}\left(K_v\frac{\partial\rho}{\partial z}\right) = 0. \tag{6.22}$$

It is common to define z as normal to the surfaces of constant density, rather than as strictly vertical. For many purposes, the near-vertical oceanic stratification renders the difference immaterial. Sometimes it is either not necessary, or it is infeasible, to separate the effects of w and K_v, so a combined variable w^* is defined so that (6.22) becomes

$$u\frac{\partial\rho}{\partial x} + v\frac{\partial\rho}{\partial y} + w^*\frac{\partial\rho}{\partial z} = 0,$$

where

$$w^* = \left[w\frac{\partial\rho}{\partial z} - \frac{\partial}{\partial z}\left(K_v\frac{\partial\rho}{\partial z}\right)\right] \Big/ \frac{\partial\rho}{\partial z},$$

and analogously for any property C.

6.2.3 Integral version

The equations above are in differential form. In practical use, an integral representation is preferred (although an exception is outlined later). Note that the Navier–Stokes equations are commonly derived from the integral form, letting the volumes involved become differentially small; we now reverse that process. Consider Eq. (6.20), the flux across a line, when written generally,

$$V_C = \iint \rho C\mathbf{v} \cdot \hat{\mathbf{n}}\mathrm{d}z\mathrm{d}s + \overline{F_C}, \tag{6.23}$$

where the Ekman flux appears only if the vertical integration extends to the sea surface. s is now being used as the horizontal coordinate in place of either x or y and is just a horizontal line element.

Consider any closed volume of ocean, made up of horizontal line segments (Fig. 6.1). Flux of C in and out involves the flow through vertical boundaries plus anything entering or leaving across horizontal ones. Invoking Gauss's and Stokes's

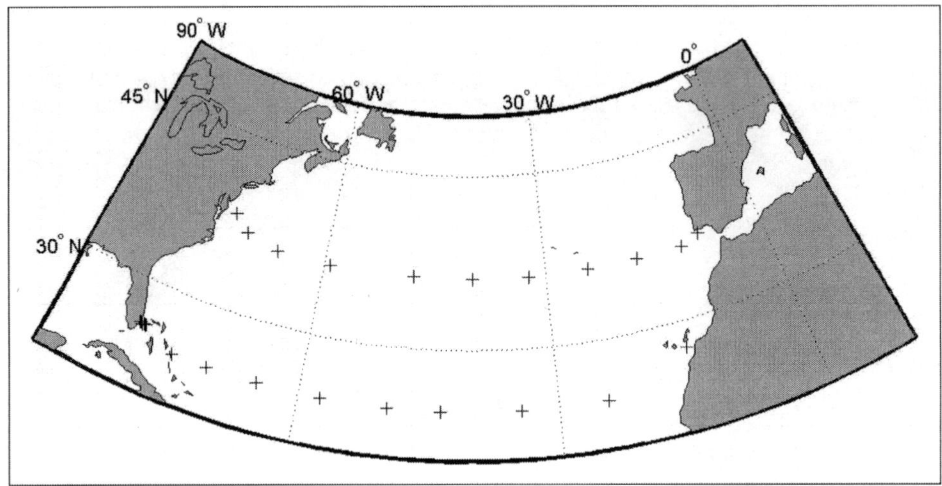

Figure 6.1 Station positions used in prototypical box inversion. There are six stations (five pairs) in the Florida Straits, not visible at this map resolution. A total of 23 station pairs defines the flow into the box, and there is a w_i^* at each of nine isopycnal interfaces. Nominal section names of "Florida Straits," 24° N, and 36° N are used to label the groups of stations. Another box lying between 36° N and the northern boundary of the North Atlantic is also defined, but is not used here.

theorems, Eq. (6.13) becomes

$$
\iint \rho C \mathbf{v} \cdot \hat{\mathbf{n}} \, dA - \iint \left(w\rho C - K_v \frac{\partial \rho C}{\partial z} \right)\bigg|_{z_{i+1}} dA \tag{6.24}
$$

$$
+ \iint \left(w\rho C - K_v \frac{\partial \rho C}{\partial z} \right)\bigg|_{z_i} dA + \sum_i \overline{F_C}^{(i)} = \iiint m_C \, dV,
$$

where $z_i\,(x, y)$ are bounding surfaces, in the vertical coordinate; $\hat{\mathbf{n}}$ is the unit normal to lateral bounding surfaces, e.g., the zonal or meridional sections; dV is the volume differential; and $\sum_i \overline{F_C}^{(i)}$ is the sum of the Ekman fluxes (if the top layer is included), over the different lateral segments. We will make the convention here that inwardly directed flows are positive (the reverse of the usual convention). In the combined form, Eq. (6.24) becomes

$$
\underbrace{\iint \rho C \mathbf{v} \cdot \hat{\mathbf{n}} \, dA}_{\text{horiz advection}} - \underbrace{\iint w^*\rho C\big|_{z_{i+1}} dA}_{\text{vert. advection}} + \underbrace{\iint w^*\rho C\big|_{z_i} dA}_{\text{vert. advection}} + \underbrace{\sum_i \overline{F_C}^{(i)}}_{\text{Ekman flux}} = \underbrace{\iiint m_C \, dV}_{\text{mass source}}.
$$

$$\tag{6.25}$$

These equations are algebraic statements that what flows in, must flow out, except for that amount of C produced or consumed within the volume by the source/sink m_C.

6.2.4 Discrete version

Although the equations have been written, as is conventional, in continuous form, they are solved almost universally using computers, and are necessarily discretized. In the thermal wind equations (6.17) and (6.18), density is computed from the equation of state as $\rho(x_j, y_j, z_k, t)$ at horizontal positions, x_j, y_j, at vertical positions z_k and times t (typically about every 2 m in the vertical, 30–50 km in the horizontal, and separated laterally by times of a few hours). The measurement "stations" are used in pairs for computing the horizontal derivatives $\Delta\rho/\Delta x$, $\Delta\rho/\Delta y$, written here as $\Delta\rho/\Delta s$. Integrals in Eqs. (6.18), (6.20) and (6.25) are calculated from discrete approximations to the integrals (often with a trapezoidal rule), the summations running upward and downward, starting from $z_i = z_0$, the reference depth. (Depth is most commonly measured in pressure units, but that is a technical detail.) Thus an estimate from the thermal wind of the relative velocity perpendicular to any pair of stations is:

$$\rho v_R(x_j, y_j, z_q) = \rho v_R(j, q) \approx \frac{g}{f} \sum_{k=0}^{q} \frac{\Delta\rho(j, k)}{\Delta x(j)} \Delta z_k. \tag{6.26}$$

Here $\Delta\rho(j, k) = \rho(x_{j+1}, y_{j+1}, z_k) - \rho(x_j, y_{j}, z_k)$, or some similar horizontal difference, and it is assumed that x_j lies along the section (t is being suppressed). Note the substitution $x_j \to j$, etc.

The conservation equation for mass (6.25), when integrated (summed) from top-to-bottom, becomes

$$V_C \approx \sum_{j \in J} \sum_q \rho_j(q) C(j, q)[v_R(j, q) + b_j] \Delta a(j, q) + \overline{F_C}, \tag{6.27}$$

which is obtained from (6.26). The notation $j \in J$ is used to denote the summation over the station pairs lying along the section in question. $\Delta a(j, q)$ is the area occupied in the section between depths z_q, z_{q+1}, over pair separation $\Delta s(j)$. That is, $\Delta a(j, q) = \Delta s(j)(z_{q+1} - z_q)$. An approximation sign has been used in Eq. (6.27) both because of the numerical errors incurred in the discretization, but also acknowledging the reality that we are using an approximate physics (of which the assumption of a steady state is the most stringent), and that because ρ and F_C are obtained from measurements, they are necessarily somewhat inaccurate. In recognition of the inaccuracy, and because equalities are far easier to work with, re-write the equation as

$$V_C = \sum_{j \in J} \sum_q \rho(j, q) C(j, q)[v_R(j, q) + b_j] \Delta a(j, q) + \overline{F_C} + n_1, \tag{6.28}$$

where n_1 is the noise.

Using w^*, the net flux of C into a closed region becomes

$$\sum_{j\in J}\sum_{q}\rho(j,q)C(j,q)\delta(j)\Delta a(j,q)[v_R(j,q)+b_j] \tag{6.29}$$

$$-\rho_{k+1}C_{k+1}w_{k+1}^*A_{k+1}+\rho_k w_k^*C_k A_k+\sum_i \overline{F_C}^{(i)}+n_2=m_C,$$

where n_2 represents whatever noise is required to render the equation an equality. Here $\delta(j)=\pm 1$ depending upon whether a positive velocity represents a net flow into or out of the volume. A_k is the horizontal area of density ρ_k multiplying the pseudo-vertical velocity w_k^*, and C_k is the corresponding horizontal average property on the surface. q is the index in the vertical used to sum the properties between interfaces $k,k+1$. As before, the Ekman fluxes only appear if the top layer(s) is included in the particular equation. m_C now represents any volume integrated source or sink. So, for mass, it would represent (in the top layer) any net evaporation or precipitation; for oxygen it would involve photosynthetic production, gas exchange with the atmosphere, and oxidation at depth.

Equations such as (6.28) and (6.29) represent constraints on the flow field, and thus upon the unknown integration constants b_j, and unknown vertical exchanges w_i^*. The solution to the venerable problem of determining the geostrophic flow in the ocean is reduced to writing down enough such constraints that one can make useful inferences about the b_j, w_i^*. In practice, the number of knowns, b_j, can be very large, even thousands. It should be clear, however, defining $\mathbf{x}^T=[[b_j]^T,[w_j^*]^T]$, $\mathbf{n}=[n_j]$, or $\mathbf{x}^T=[[b_j]^T,[w_j]^T,[K_{vj}]^T]$, that they all are in the conventional form

$$\mathbf{Ex}+\mathbf{n}=\mathbf{y}, \tag{6.30}$$

and can be addressed by the methods of Chapters 2 and 3.

This discussion is far from exhaustive. Within the context of linear models, one can incorporate complex structures describing whatever physics, chemistry, or biology that is reasonably believed to depict the observed fields. But if the model (6.30) fails to represent important governing physics, etc., one may be building in errors that no mathematical machinery can overcome.

6.2.5 A specific example

Consider an intermediate scale problem. Three hydrographic sections in the North Atlantic (Fig. 6.1) are used to define a closed volume of ocean within which various fluid properties are likely conserved. Along the lines, each pair of stations defines the relative velocity. The original trans-Atlantic sections (Roemmich and Wunsch, 1985) had a total of 215 stations, but for reasons of discussion and display, the number is here reduced to 25 (6 in the Florida Straits, 8 from the Bahama Bank to

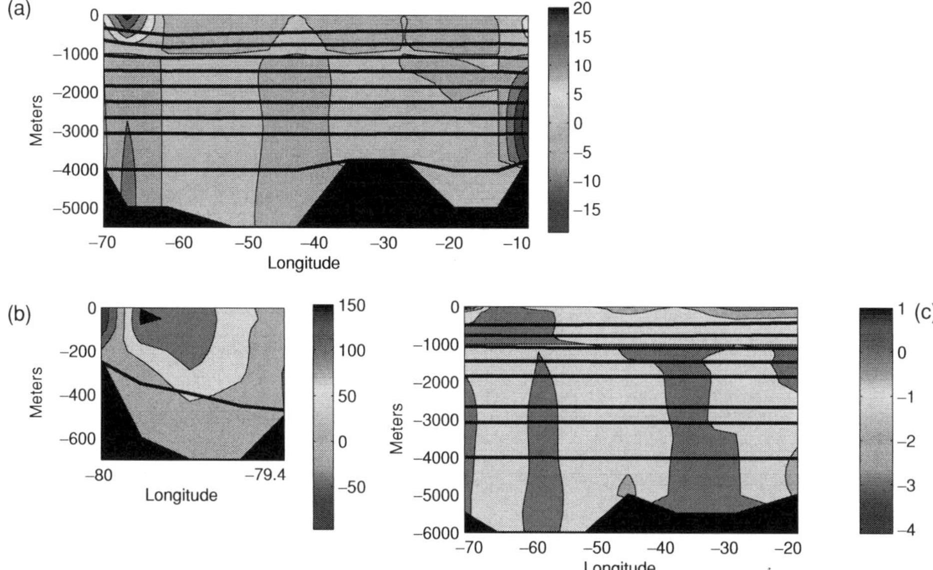

Figure 6.2 Geostrophic velocity (colors, in cm/s) relative to a 1000-decibar reference level, or the bottom, whichever is shallower, and isopycnal surfaces (thick lines) used to define ten layers in the constraints. Part (a) is for 36° N, (b) for the Florida Straits, and (c) for 24° N east of the Florida Straits. Levels-of-no-motion corresponding to the initial calculation are visible at 1000 m in the deep sections, but lie at the bottom in the Florida Straits. Note the greatly varying longitude, depth, and velocity scales. (See color figs.)

Africa along nominally 24° N, and 11 along the nominal 36° N section) producing 23 usable pairs for computing the thermal wind. The water column is divided into 10 layers bounded by isopycnals and/or the top and bottom. (The box model on p. 111 is a simplified analogue of this one.)

For the present primarily pedagogical purposes, the reference level z_0 was fixed at 1000 m (actually measured as a pressure of 1000 decibars, where a decibar is the hydrostatic pressure exerted by one meter of sea water) and which shows up in a zero-line along the sections shown in Fig. 6.2. In the Florida Straits section, which is shallower than 1000 m, z_0 was defined to be at the bottom. In practice, one would try to choose an initial reference level to reflect whatever knowledge one had of the circulation prior to the computation. In particular, note that there is a strong tendency for the sense of the geostrophic flow to reverse across the reference level. It would be sensible, should one have reason to believe that such a reversal occurred in practice, to place the reference level at the transition depth. If the data being used prove too noisy to add useful information, one would find that $\tilde{\mathbf{x}} = \mathbf{0}$, and the resulting best estimate of the ocean would revert to the statement that it was geostrophic with no flow at $z = z_0$. (We repeat: no such depth following any simple rule exists – the ancient idea that there is some approximately constant depth where $v = 0$ is unsupported by any observations or theory; it is a convenient numerical device for defining and calculating a relative velocity. In the absence of information, however, $v = 0$ may be the best available estimate.)

The spatial resolution being used here is so poor that no sensible oceanographer would expect the result to be very realistic. For example, the Gulf Stream velocities visible in Fig. 6.2 are generally much too low, as are the temperatures being assigned there. Nonetheless, the orders of magnitude are roughly correct and the aim is to be demonstrative rather than to be definitive.

The procedure is to write as many equations as we can involving the unknown b_i, w_j^*, representing whatever is thought true about the field, with some estimate of how accurate the constraints are. One reasonable equation asserts that in a steady state, the total mass flowing into the volume must vanish, up to the errors present; evaporation and precipitation are too small to matter in such a small area. Estimates derived from the windfield suggest that there is an Ekman flow to the south across $36°$ N of about 3×10^9 kg/s, and about 6.5×10^9 kg/s to the north across the $24°$ N line (ignoring the very small Ekman contribution in the Florida Straits).[5]

A net Ekman flux into the box of $(3 + 6.5) \times 10^9$ kg/s $= 9.5 \times 10^9$ kg/s can be sustained in a steady state only if it is exported by the geostrophic flow. We can write an equation insisting that the total geostrophic flow across $36°$ N has to be (approximately) equal and opposite to the southward-going Ekman flux across that line, and that the sum of the mass fluxes across the Florida Straits and $24°$ N has to balance the 6.5×10^9 kg/s driven northward by the wind in the surface layer. Thus total mass balance provides one equation, and the three individual section flux constraints provide three more; note, however, that these equations are redundant, in that the sum of the net flux equations (taking account of the sign of the flux) must sum to the total mass balance equation. No particular harm is done by carrying redundant equations, and there is some gain in obtaining solution diagnostics, as we shall see.

Many years of direct measurement in the Florida Current (the name for the Gulf Stream in that region), suggest that its time average mass flux is approximately 31×10^9 kg/s to the north. Using the bottom reference level, the relative velocity produces only 22×10^9 kg/s to the north, and some non-zero b_j are expected in any solution producing a transport of about 31×10^9 kg/s there.

We now have four equations (only three of which are independent) in 32 unknowns. A number of possibilities exist for additional constraints. Oceanographers have evidence that the rate of mixing of fluid across surfaces of constant density (isopycnals) tends to be quantitatively small.[6] In other words, fluid entering the ocean volume defined in Fig. 6.1 between two different densities, $\rho_i \le \rho \le \rho_{i+1}$, can be expected to be leaving. Given, however, that some exchange (mixing) of different density types can occur, we may wish to account for it. Using the reduced form, Eq. (6.29), add, with ten layers and ten equations,

$$
\sum_j \sum_q \rho(j, q) b_j \delta_j \Delta a(j, q) - \rho_i w_i^* A_i + \rho_{i+1} w_{i+1}^* A_{i+1} + n_i
$$
$$
= - \sum_j \sum_q \rho(j, q) v_R(j, q) \delta(j) \Delta a(j, q). \tag{6.31}
$$

All unknowns, b_j, w_i^*, n_i have been placed on the left-hand side. In the top layer, the upper boundary is at the seasurface, and the corresponding w_i^* is zero; similarly the lowest layer lower boundary is at the seafloor and $w_i^* = 0$. For the moment, Ekman fluxes are being deliberately omitted, to demonstrate the appearance of model error. Evaporation and precipitation are again ignored as being very small.

With ten layers, we obtain ten additional equations of the form (6.31). In addition to the new equations, we have introduced nine new unknowns w_i^* (the number of interior isopycnals defining ten layers) exactly equal to the maximum number of independent new equations (whether all the layer equations are really independent needs to be determined), as well as an additional nine new noise unknowns, n_i. The top-to-bottom sum of these ten layers must be the same as the equation previously written for total mass conservation. In that sum, all of the w_i^* drop out. The main reason for using isopycnals as the layer boundaries, as opposed, e.g., to constant depths, is to render $w_i^* = 0$ as a reasonable starting solution – that is, one that would be acceptable if the new equations do not produce useful new information. The number of independent equations thus cannot exceed 12. That adding new equations generates twice as many new unknowns, w_i^*, n_i, is nonetheless advantageous, follows from the ability to specify information about the new unknowns in the form of their statistics (specifically their covariances).

Note that in Eq. (6.31) and elsewhere, the elements of the coefficient matrix, $\mathbf{E}$, involve observed quantities (e.g., $\rho(j, q)$, $\Delta a(j, q)$), and thus the elements of $\mathbf{E}$ contain errors of observation, $\Delta \mathbf{E}$. The system of equations, rigorously, instead of being of the canonical form (6.30), should really be written as

$$
(\mathbf{E} + \Delta \mathbf{E}) \mathbf{x} + \mathbf{n} = \mathbf{y}.
$$

Neglect of $\Delta \mathbf{E} \mathbf{x}$ relative to $\mathbf{E} \mathbf{x}$, represents a linearization of the system, whose validity must always be a concern. This linearization of the problem is the price being paid to avoid using the high spatial derivatives appearing in the exact expression Eq. (6.19). We here neglect the non-linear term, assuming its influence to be negligible. Let us now analyze the existing 14 equations in 32 unknowns.

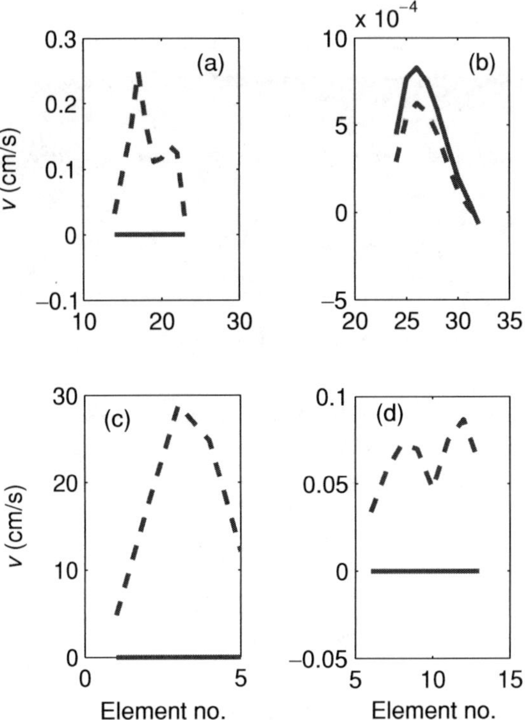

Figure 6.3 The solution to the box model at ranks $K = 9$ (solid line), 12 (dashed line), corrresponding to Fig. 6.4. (a) Results shown for $x_i = b_i$ in the $36°$ N section, (c) and (d) are the corresponding values for the Florida Straits and $24°$ N sections, respectively. The values of w_i^* are displayed in (b). Note the weaker values of w_i^* for $K = 12$ relative to $K = 9$. Abcissae is the value of i in x_i.

6.2.6 Solution by SVD

Proceeding to use least-squares, by invoking the singular value decomposition on the equations written down in "raw" (unscaled) form, two solutions, for ranks $K = 9$, 12 are displayed in Fig. 6.3. The singular values are displayed in Fig. 6.4. Two of the values are extremely small (not precisely zero, but so small relative to the others that a reasonable inference is that they represent roundoff errors). Thus the rank of this system is at best $K = 12$ as anticipated above. When plotted on a linear scale (also in Fig. 6.4), it is seen that an additional three singular values are significantly smaller than the others, suggesting some information in the equations which is likely relatively noise sensitive.

To proceed, form the data resolution matrices, $\mathbf{T}_u(K)$ for $K = 9$, 12, whose diagonals are depicted in Fig. 6.4. With $K = 9$, the total mass transport equations are completely unresolved – a perhaps surprising result. (The figure caption lists the order of equations.) If one examines the corresponding $\mathbf{T}_v(9)$, one finds that *all*

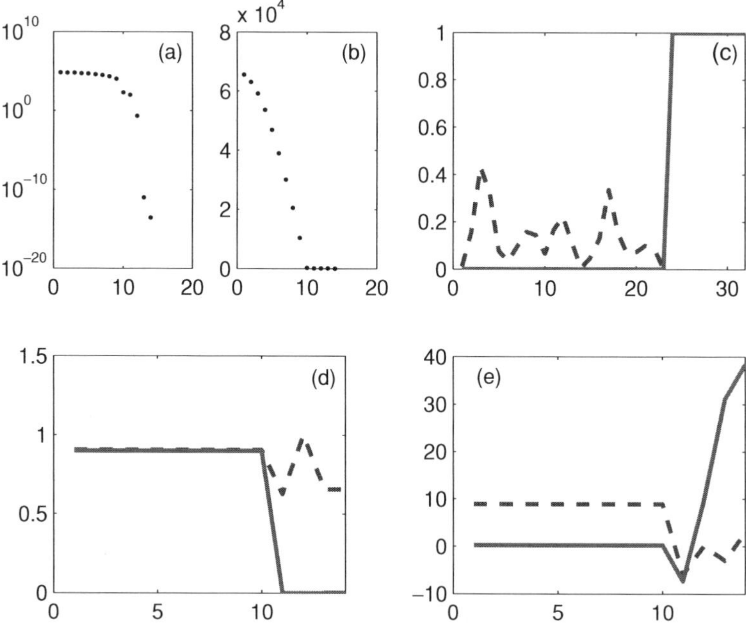

Figure 6.4 (a) and (b) Singular values, λ_i, shown as both logarithmic and linear values for the simple mass balance equations defined by the stations shown in Fig. 6.1. Two λ_i are zero, corresponding to the known redundancy built in, and three are distinctly smaller than the others. (c) $\mathbf{T}_v$ at rank 9 (solid) and at rank 12 (dashed). At rank 9, only the interior layer equations are active, and only the w_i^* are resolved. (d) The diagonal elements of $\mathbf{T}_u$ at ranks 9 (solid) and 12 (dashed). Large residuals (e) are left in the top-to-bottom flux equations when $K = 9$ (the corresponding $\mathbf{T}_u$ shows that they are inactive). At rank 12, there is partial resolution of the b_i, and all equations are active in the solution. The $K = 12$ residuals have been multiplied by 10 for visibility. Residual units are 10^9 kg/s. Note that the first ten elements are all positive – a symptom of the inherent contradiction left present by the omission of the Ekman flux in the layer equations. Order of equations: (1)–(10) are mass balance in each of the individual layers; (11) is total box mass balance (sum of Eqs. (1)–(10)); (12) sets the Florida Straits transport; (13) specifies the 24° N Ekman flux; (14) sets 36° N Ekman flux.

of the w_i^* are fully resolved, and *none* of the reference level velocities b_i is resolved at all, consistent with the zero values for the b_i with $K = 9$. Why should this be? Examination of the columns of $\mathbf{E}$ shows that the interface areas, A_k, multiplying the w_k^* are of order 3×10^4 times larger than the vertical areas multiplying the b_i. Or, alternatively, the column norms of $\mathbf{E}$ corresponding to the w_i^* are much larger than for the b_i. Recalling that a singular vector solution solves the least-squares problem of minimizing the solution norm for the smallest possible residual norm, it is unsurprising that the system would seek to make the w_i^* finite relative to the b_i, because small values of w_k^* have a greatly amplified effect on the residuals.

The rank 9 solution is then found to be $\tilde{b}_i \approx 0$, and the $\tilde{w}_i^*$ as shown in Fig. 6.4(b), corresponding to an upwelling throughout the water column. Although the w_i^* are somewhat large, the real reason for rejecting this solution is the behavior of the residuals. Strong violation of the overall mass balance constraints is occurring – vertical velocities alone cannot produce the import and export of fluid required to balance the overall mass budget, and the net flux constraint through the Florida Straits.

To invoke the constraints on overall (top-to-bottom) mass balance, increase the rank to $K = 12$, with diagonals of $\mathbf{T}_u$ shown in Fig. 6.4 demonstrating the new importance of the equations for overall balance. The rank 12 solution (Fig. 6.3) and its residuals (Fig. 6.4e) are also displayed. Now the residuals in all layers are 0.9×10^9 kg/s, but the top to bottom mass imbalance is -0.6×10^9 kg/s and the two zonal flux integrals fail to balance by $\pm 0.3 \times 10^9$ kg/s.

In looking at the solution, we see that the reference level velocities in the Florida Straits (bottom velocities) are of order 10–20 cm/s, while elsewhere, they range from 0.01–0.1 cm/s. The $\tilde{w}_i^*$ are all positive, increasing to a mid-depth maximum from 1×10^{-4} cm/s at the lowest interface to 7×10^{-4} cm/s in the middle of the water column.

This solution might be an acceptable one, but it is to some degree an accident of the various row and column norms appearing in matrix $\mathbf{E}$. For example, some of the layers are thinner than others, and they are given less weight in the solution than thicker ones. Whether this outcome is desirable depends upon what the investigator thinks the appropriate weighting is. The $\tilde{w}_k^*$ values are perhaps a bit large by conventional standards, but before attempting to control them, note the main reason for solution rejection: a contradiction has been left in the system (for demonstration purposes). The total top-to-bottom mass flux was required to vanish so that the geostrophic flow exported the net incoming Ekman flux. But in the individual layer equations, the Ekman flux was omitted altogether. Because the sum of the ten layer equations should be the same as the total top-to-bottom mass balance, there is a contradiction. The present solution has resolved this contradiction by leaving mass residuals in each of the layers – corresponding to a weak, unexplained, mass source in them. It is this mass source, when summed vertically, that permits the geostrophic outflow to balance the Ekman inflow.

The system has apportioned the erroneous ageostrophic flow over all layers – it had no information to the contrary. That there is a systematic error is shown to the user by the fact that at $K = 12$, *all of the layer imbalances have the same sign* (Fig. 6.4(e)) – if the residuals were pure noise, they should have a much more random character. This example demonstrates a truism of such methods: most of the information about solution acceptability or shortcomings is obtained from the residuals. To account for the systematic error, one can "inform" the

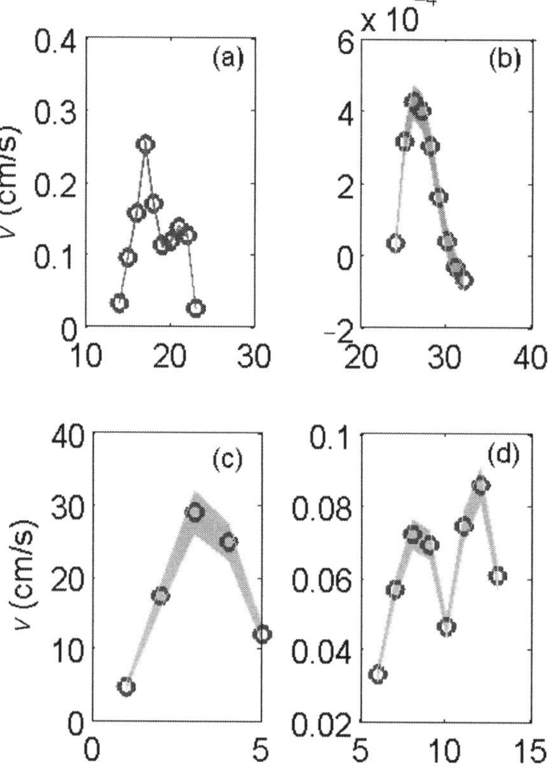

Figure 6.5 Inverse solution by SVD, now consistently accounting for the Ekman flux, for $K = 12$. As before, but now the very small standard error is also displayed (shading) as determined from the diagonal elements of $\mathbf{C}$. The $K = 9$ solution is not detectably changed from that shown in Fig. 6.4. (a) 36° N values of $\tilde{b}_i$; (b) $\tilde{w}_i^*$, (c) and (d) $\tilde{b}_i$ for the Florida Straits and 24° N sections, respectively.

system, that the Ekman flux is probably confined to the uppermost layer by requiring that the net geostrophic outflow there should be equal and opposite to the Ekman inflow.

When that is done, the $K = 12$ solution changes very slightly overall (Fig. 6.5), but the layer mass residuals are very much smaller and fluctuating in sign (Fig. 6.6). The main point here is that the slightly subtle error in the equations (the model), is detected because the residuals do not behave as expected. In the present case, in the consistent system, the residuals are now too small – they are inconsistent with the prior estimates of their magnitude. In a more realistic situation, this outcome would be cause for suspecting a further error. Here, the explanation is immediate: the y_i were artificially constructed so that all equations are now fully consistent and there are no errors present other than roundoff noise.

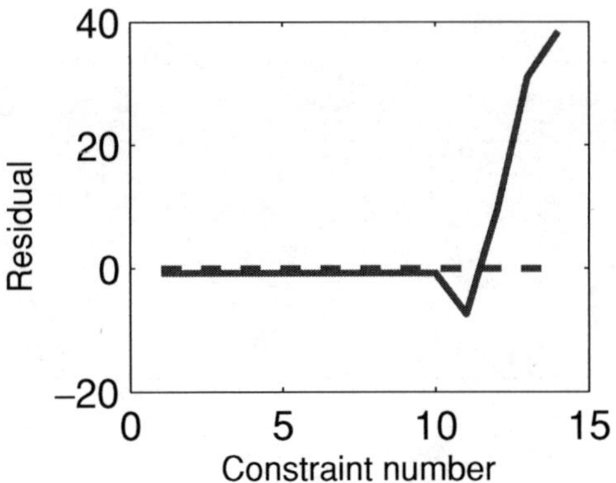

Figure 6.6 Residuals for the consistent mass conservation, at $K = 9$ (solid), and $K = 12$ (dashed). The latter are vanishingly small by construction of the artificial "observations."

The SVD solution permits a full description of the relationship of the equations and the structure of the solution. Figure 6.7 shows the singular vectors of the mass-conserving system. Consider the first term in the SVD solution:

$$\tilde{\mathbf{x}} = \frac{\left(\mathbf{u}_1^T \mathbf{y}\right)}{\lambda_1} \mathbf{v}_1.$$

The orthonormal structure, proportional to $\mathbf{u}_1$ in $\mathbf{y}$, produces an orthonormal structure $\mathbf{v}_1$ in the solution. These two vectors are displayed in Fig. 6.7. The dot product of $\mathbf{u}_1$ with $\mathbf{y}$, forms the successive differences on the right-side of the equations, that is, approximately, $-y_1/2 + y_2 - y_3 + \cdots + y_{10}/2$, with some extra emphasis on the equations for layers at intermediate depths in the water column. The corresponding structure, $\mathbf{v}_1$, is approximately zero for all of the b_i, but is non-zero in the w_i^*. Because $w_{1,10}^*$ each appears in only one equation, whereas all the others appear in two of them – the flow into one layer being the flow out of its neighbor – the two end layer equations are given roughly $1/2$ the weight of the interior ones. When dotted with $\mathbf{x}$, $\mathbf{v}_1$ produces, approximately, the sum $(-w_1^* + w_2^* - w_3^* + \cdots)$ again with some extra weight assigned in the middle of the water column. The strongest information present in the set of equations concerns the differences of the vertical transfers, which are proportional to $-w_1^* + w_2^* - w_3^* + \cdots$, that is the vertical convergences, layer-by-layer in the vertical. The second most important information available concerns a different set of differences of the w_i^*. The first nine $\mathbf{u}_i$, $\mathbf{v}_i$ pairs shown in Fig. 6.7 pertain to the w_i^* – evidently information about the unknown b_i is weak in these equations, as we already knew. The $\mathbf{u}_{10}$, $\mathbf{v}_{10}$ pair involving the equations for total mass, and the two meridional flux equations, evidently carry the most important information about the b_i: the (weighted) differences of the flows

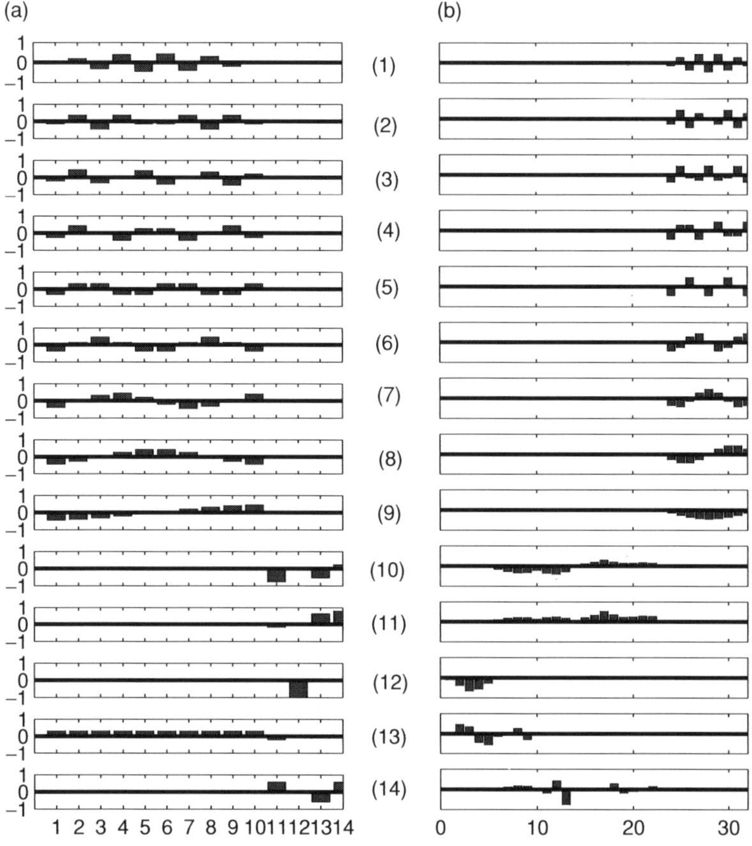

Figure 6.7 (a) First 14 of the $\mathbf{u}_i$, $\mathbf{v}_i$ pairs. The first ten equations describe layer-by-layer mass balance, Eq. (11) is top-to-bottom mass balance (redundantly with the sum of Eqs. (1)–(10); Eq. (12) enforces the observed mass flux through the Florida Straits, and Eqs. (13) and (14) impose a top-to-bottom geostrophic flow equal and opposite the 24° N and 36° N Ekman fluxes. (b) The first five unknowns are the Florida Straits reference level velocities; (6)–(13) are the reference level velocities in the 24° N section, and (13)–(33) are in the 36° N section. As discussed in the text, for $\mathbf{u}_1$, $\mathbf{v}_1$, successive differences of the layer equations resolve sums of w_i^*. Note that $\mathbf{u}_{12}$, which brings in only the Florida Straits mass flux constraint, corresponds to a weighted mean of the bottom velocities there. $\mathbf{u}_{13}$, when dotted with the equations, asserts that the sum of the ten-layer equations, with overall box mass balance plus and minus the 24 and 36° N mass balances, should add to zero (they are redundant). The 13th and 14th $\mathbf{u}_i$, $\mathbf{v}_i$ are in the nullspace. A dot product with $\mathbf{u}_{14}$ asserts that the 11th equation (box mass balance), is redundant with the difference of the 24 and 36° N mass balances, which we know to be true. The corresponding $\mathbf{v}_{13,14}$ are linear combinations of b_i about which there is no information.

across 24 and 26° N. Pair 12 is particularly simple: it involves only the equation that sets the Florida Straits mass flux to 31×10^9 kg/s; only the Florida Straits velocities appear in it.

Controlling the solution

The solutions shown are determined in large part by the various row and column norms appearing in $\mathbf{E}$. For example, the very large horizontal areas multiplying w_i^* give them high resolution, and values at depth that some might regard as too high. Without entering into a debate as to the reality of the outcome, we can nonetheless gain some control over the solution – *to the extent that it does not actually conflict with the equations themselves.*

Following the principles outlined in Chapter 2, column-normalize by the square root of the column lengths – putting all elements on an equal footing. A reasonable estimate of the rank is still $K = 12$. A solution (not shown) to this column-normalized form does *not* produce equal elements – the equations do not permit it. For example, in the Florida Straits, the horizontal velocities remain of order 20 cm/s, and the w_i^* are $O(10^{-7}$ cm/s). Indeed, no solution of uniform magnitude could simultaneously satisfy the Florida Straits flux conditions, and the interior mass balances.

In the spirit of least-squares then, we remain free to impose any prior weights that we wish. Having column-normalized, we can impose variances on the x_i. The b_i variances in the Florida Straits will be estimated as 100 (cm/s)^2, outside that region, they are about 1 (cm/s), and $(10^{-5}$ cm/s)^2 for the w_i^*. These values are meant to roughly reflect prior knowledge. Thus,

$$\mathbf{R}_{xx} = \mathrm{diag}([100\,(\mathrm{cm/s})^2, \ldots, 1\,(\mathrm{cm/s}), \ldots, (10^{-5}\mathrm{cm/s})^2]). \tag{6.32}$$

Now let us try to gain control over the residuals left in each equation. A plausible analysis suggests the Florida Straits mass flux of 31×10^9 kg/s is accurate to about $\pm 2 \times 10^9$ kg/s, and that the various Ekman fluxes are accurate to about $\pm 1 \times 10^9$ kg/s. Total mass balance is required to $\pm 0.5 \times 10^9$ kg/s, and in the individual layers to $\pm 1 \times 10^9$ kg/s. No information is available about the covariances of the errors among the equations (not strictly true). Because of the different layer thicknesses being used in the constraints, the equations have very different contributions to the solution for any finite rank. We thus opt to first divide each equation by its corresponding row norm, and then weight it by the square root of

$$\mathbf{R}_{nn} = \mathrm{diag}\{1\,1\,1\,1\,1\,1\,1\,1\,1\,1\,1\,1/(0.5)^2\,1/2^2\,1\,1\}, \tag{6.33}$$

in units of $(10^9$ kg/s)$^{-2}$. Note that we are making the a-priori inference that the noise estimates apply to the row-normalized equations; determining whether this inference is the best one becomes a discussion of the specifics of noise in oceanographic

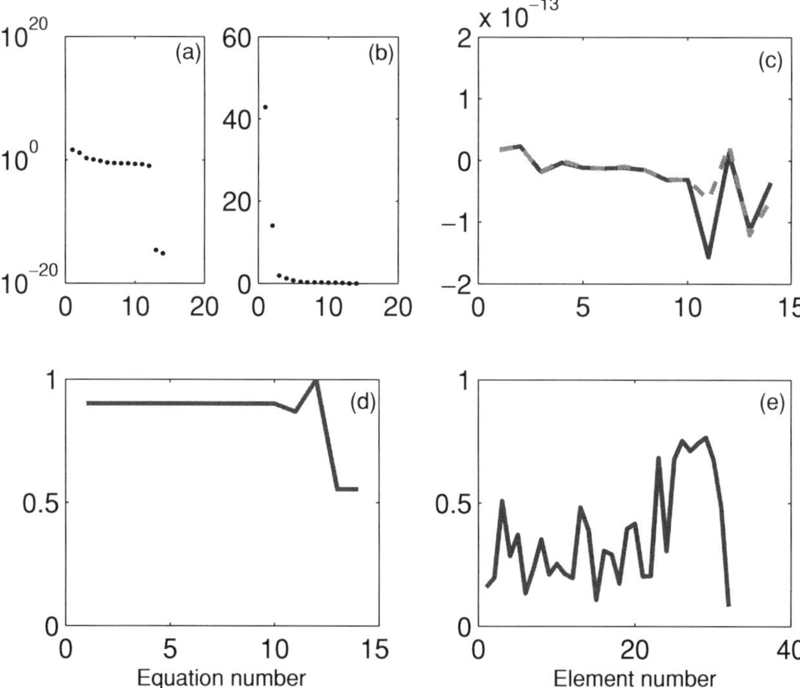

Figure 6.8 Singular values on a logarithmic (a) and linear (b) scale, for the row and column normalized mass-conserving system; (c) the non-dimensional (solid) and dimensional residuals (dashed) when $K = 12$; (d) $\mathbf{T}_u$; and (e) $\mathbf{T}_v$ for $K = 12$.

measurements and computation. For the present, it is a reasonable choice. The fully scaled system is then (Eq. (2.131)),

$$\mathbf{E}'\mathbf{x}' + \mathbf{n}' = \mathbf{y}'.$$

Again use the SVD (in the scaled spaces), and take rank $K = 12$ (see Fig. 6.8). Now the resolution of the w_i^* is lower than before, much of the information having gone instead to determining the b_i. Equation (11) (top-to-bottom conservation) is given little weight (but is redundant with the highly weighted sum of Eqs. (1)–(10)). Figure 6.8 displays the singular values, the diagonals at $K = 12$ of $\mathbf{T}_u$, $\mathbf{T}_v$, and the rank 12 residual of the system. The solution itself (in the original, unscaled, space) can be seen in Fig. 6.9 along with its standard errors.

The $K = 12$ rows (or columns) of the dimensional resolution matrices $\mathbf{T}_u$, $\mathbf{T}_v$ are displayed in Fig. 6.10 for this case. The first 12 equations are almost uniformly used in determining the solution. But Eqs. (11) (the overall box mass balance), (13) (the $24°$ N net mass flux), and (14) (the $36°$ N net mass flux) are used only in linear combination with each other, which makes sense because they are highly

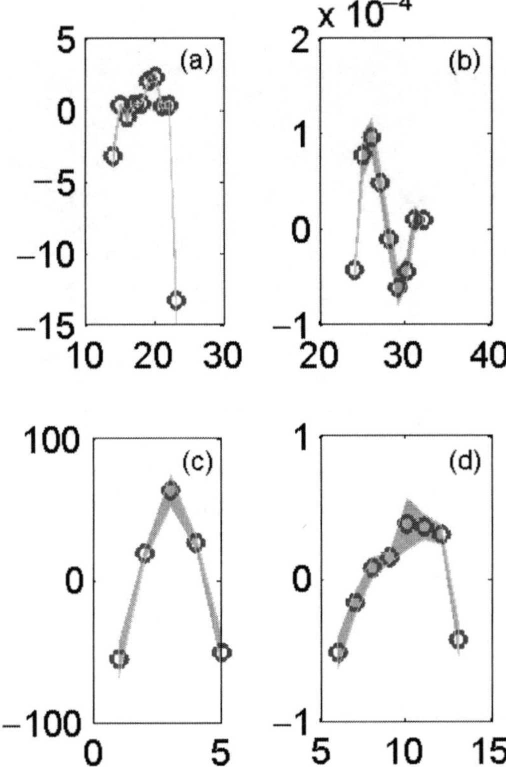

Figure 6.9 Solutions for the row and column normalized mass-conserving system. Graphs (a), (c) and (d) show the $\tilde{b}_i$ in the 36° N, Florida Straits and 24° N sections, respectively, while (d) shows the corresponding w_i^*. Shading is a one-standard-deviation uncertainty estimate.

dependent. Only the equation (12) that fixes the mass transport through the Florida Straits is fully resolved. Otherwise, all equations have some partial projection onto the others, giving rise to dependencies, and produce the two smallest non-zero singular values. The horizontal velocities are resolved in various weighted average groups whose structure can be inferred directly from $\mathbf{T}_v$.

6.2.7 Solution by Gauss–Markov estimate

The great power of the SVD solution is that it provides a nearly exhaustive method to understand exactly why the solution comes out the way it does, permits ordering the equations (data) in terms of importance, and provides estimates of the uncertainty and its structure. As the solution dimension grows, the amount of available information becomes overwhelming, and one may prefer a somewhat more opaque, but nonetheless still very useful, approach. Consider, therefore, the solution by the

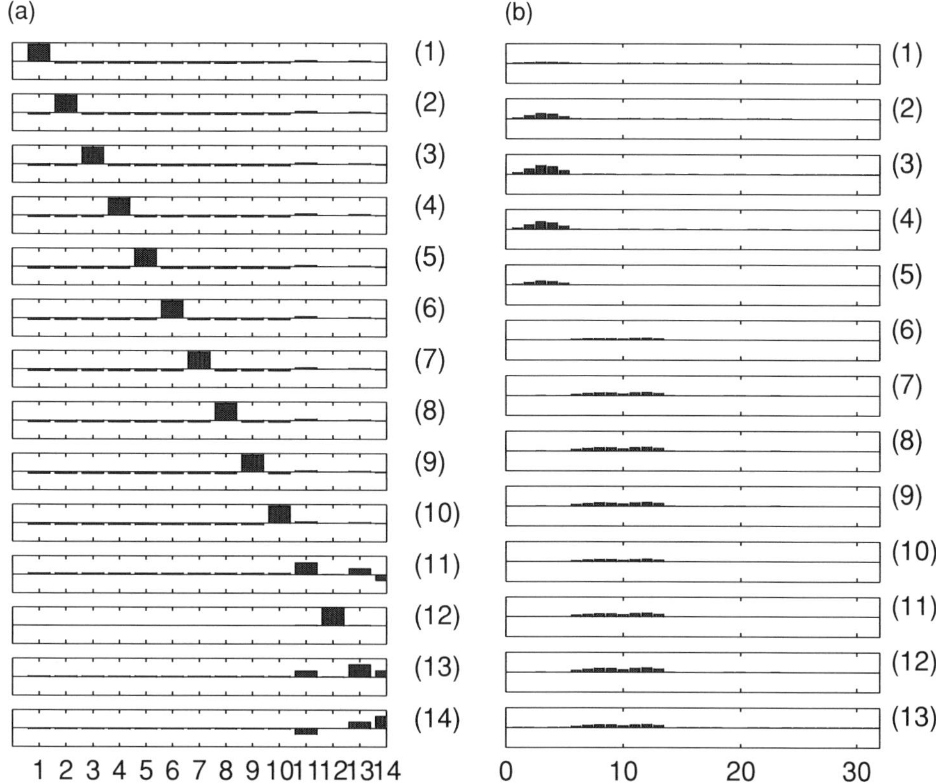

Figure 6.10 (a) The successive rows (or columns) of $\mathbf{T}_u$ for $K = 12$ for the row and column normalized mass-conservation equations. As discussed in the text, Eqs. (11), (13) and (14) show strong dependencies. The order of equations is listed in the caption to Fig. 6.4. (b) The first 13 columns of $\mathbf{T}_v$ (the remaining columns correspond to the w_i^*, which are nearly completely resolved). Full scale in all cases is ± 1. Note that many of the estimated b_i exhibit compact resolution (a local average is determined), as in columns 2–6, where others (e.g., column 8) are determined only in combination with spatially more remote values. None is fully resolved by itself.

Gauss–Markov estimator, Eq. (2.403). In that notation, and consistent with the numbers already used, put $\mathbf{R}_{nn}$ as defined in Eq. (6.33). An a-priori estimate of the solution covariance was taken to be the column weights used above: (Eq. 6.32) for the Florida Straits, open ocean b_i, and w_k^*, respectively. The resulting solution is displayed in Fig. 6.11 along with the formal uncertainty, $\pm\sqrt{\operatorname{diag}(\mathbf{P})}$. The solution closely resembles that from the SVD approach. The x_i are seen to be generally consistent with the values in Eq. (6.32), although a few of the w_i^* become somewhat large (but not enough to be truly suspicious). The estimates of $\mathbf{n}$ (not shown) are again, generally too small to be consistent with $\mathbf{R}_{nn}$ – a result that would lead one

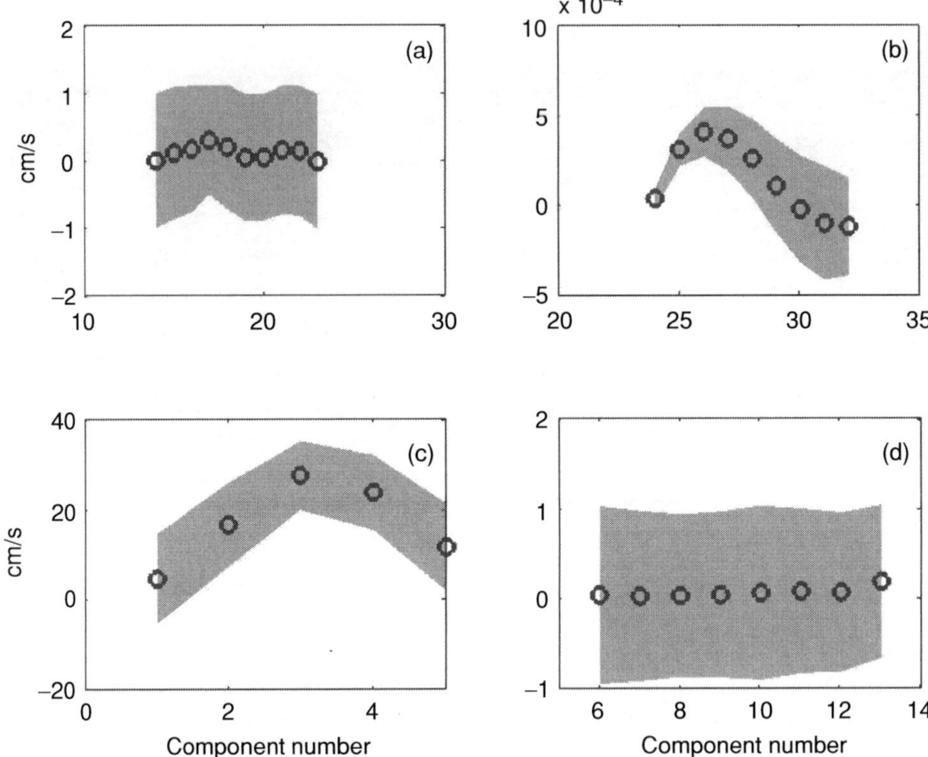

Figure 6.11 Gauss–Markov solution, with standard errors, for the geostrophic box model. Only some of the solution elements are distinguishable from zero within one standard deviation of the error. (a) Results shown for $\tilde{b}_i$ in the 36° N section, (c) Florida Straits, and (d) for 24° N, while (b) is $\tilde{w}_i^*$.

to conclude that the system was inconsistent with the prior covariance, which we know is true because of the artificially low error.

6.2.8 Adding further properties

This approach to estimating b_i, w_k^* can be extended indefinitely by adding conservation requirements on as many properties, C, as have been measured. Among the properties that have been used are salinity, S, silica, and oxygen concentration. The most commonly available property is S, and so let us examine the consequences of using it. Salt is not carried by the atmosphere, and no measurable internal sources or sinks of S exist; in the assumed steady state, however, much salt enters a volume of ocean, so an equivalent amount must exit. Salt conservation equations can be written in the form of (6.25), completely analogous to those already written for mass, including Ekman contributions $\overline{F_S}$, and putting $C = S$, $M_S = 0$. (Note

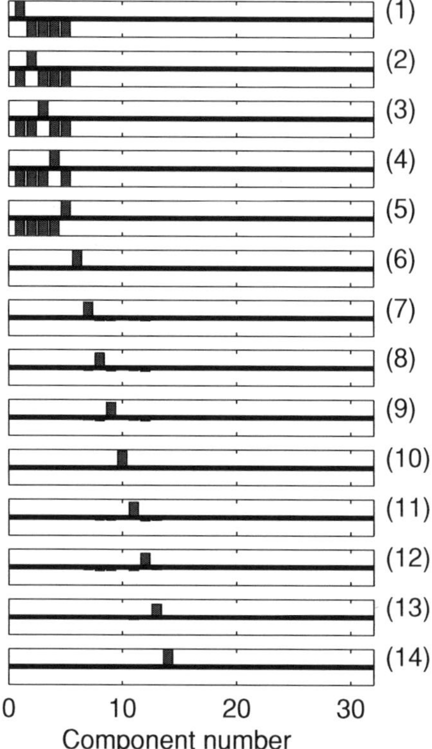

Figure 6.12 First 14 columns (or rows) of the **P** matrix for the solution in Fig. 6.11 corresponding to the $\tilde{b}_i$ elements of the solution, and showing the covariances among the solution elements. All columns are scaled to ± 1, but differ dimensionally by orders of magnitude. Most conspicuous is a strong negative correlation among the errors of the different b_i.

that modern salinity values are dimensionless, and measured on what is called the "practical salinity scale"; despite widespread use, there is no such thing as a "practical salinity unit.") Adding ten equations for salt conservation in the layers to the present system, plus top-to-bottom salt conservation (again redundant), produces 11 additional equations, for a total of 25 equations in the same 32 unknowns. At most ten independent equations have been added. To illustrate, the SVD solution will be used.

Oceanic salinities hardly differ from $S = 35$, so, following the column normalization, row normalize the 13 new equations by 35. The singular values of the new, larger, system are displayed in Fig. 6.13. The new rank could be at most $K = 22$, and indeed there are 22 non-zero singular values, but the range of values is now quite large – over four orders of magnitude. Clearly a choice of the practical rank

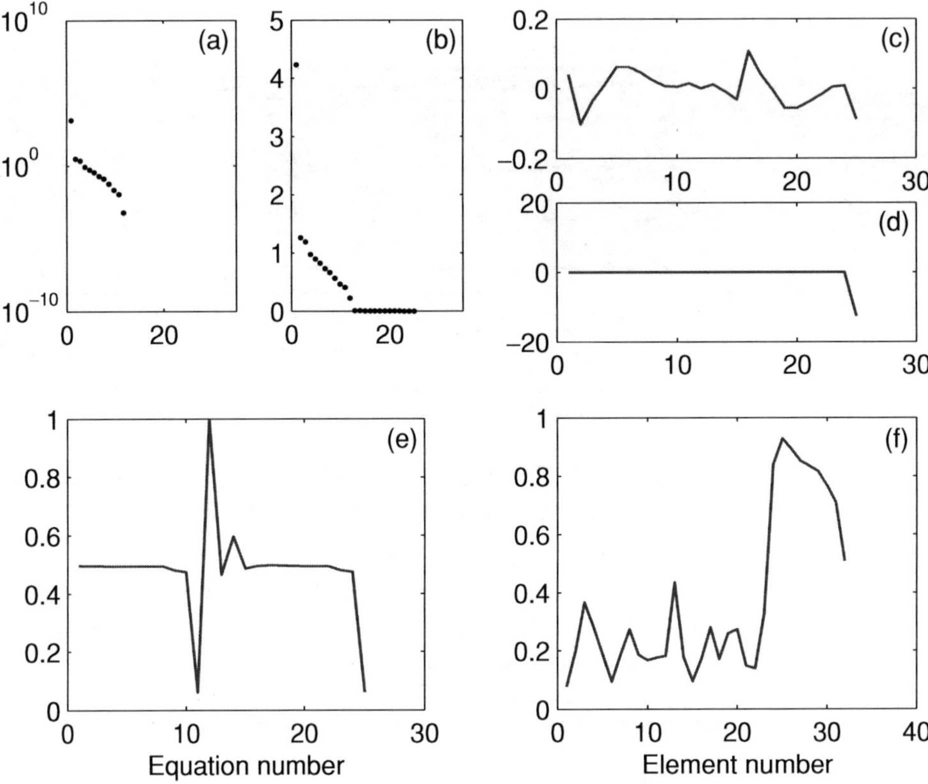

Figure 6.13 Combined mass and salt equations, showing the singular values (a), (b), and the rank 12 diagonals of $\mathbf{T}_u$, $\mathbf{T}_v$ (e), (f). Non-dimensional and dimensional equation residuals are in (c), (d).

depends directly on how small a λ_i can be tolerated. Taking again $K = 12$, the results are shown in Figs. 6.13–6.15.

Why doesn't salt balance prove more useful? If one examines the $\mathbf{u}_i$, $\mathbf{v}_i$ (Fig. 6.15) or the rank 22 resolution matrices (not shown), one sees that the mass and salt equations are seen to have a strong dependency (redundancy). The reason is not difficult to find. The equation of state (Eq. (6.12)) is, in practice, linearizable:

$$\rho \approx \rho_0(1 - \alpha_1 T + \beta_1 S).$$

Thus there is a near linear dependency between ρ (mass) and S. Whether the measurement and model accuracies are sufficient to render the small singular values that result useful is a question that can only be answered in quantitative terms, employing the actual noise levels.

The row normalization by 35, before anything else has been done, renders the resulting values of $C/35 \approx 1$ with little variation. For this reason, some authors

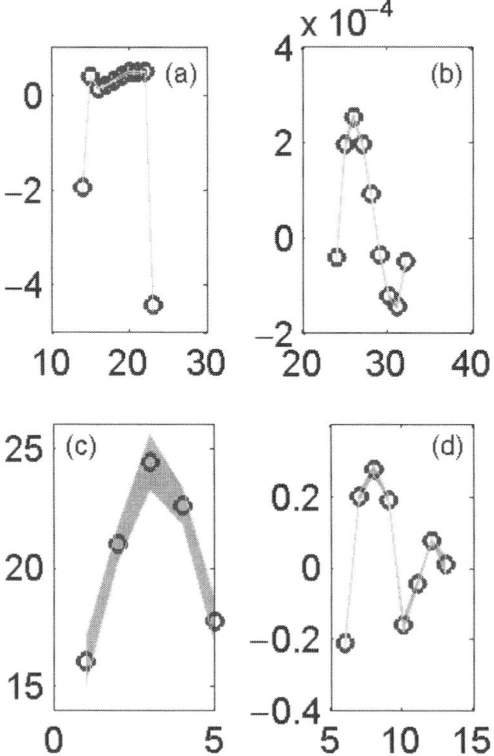

Figure 6.14 SVD solution for the combined mass and salt system. (a) Estimated b_i for the $36°\,$N section, (c), (d) b_i for the Florida Straits and $24°\,$N sections, respectively; (b) shows the estimated w_i^*.

have advocated working instead with the salinity anomaly. Consider a salt balance equation (either flux or overall box conservation), e.g.,

$$\sum_{j\notin J}\sum_{j}\rho(j,q)S(j,q)\delta(j)\Delta a(j,q)[v_R(j,q)+b_j] \tag{6.34}$$

$$-\rho_{k+1}S_{k+1}w_{k+1}^*A_{k+1}+\rho_k S_k w_k^* A_k+\sum_i \overline{F_S}^{(i)}+n_2=0,$$

and the corresponding mass conservation equation,

$$\sum_{j\in J}\sum_{q}\rho(j,q)\delta(j)\Delta a(j,q)[v_j(j,q)+b_j]-\rho_{k+1}w_{k+1}^*A_{k+1} \tag{6.35}$$

$$+\rho_k w_k^* A_k+\sum_i \overline{F_E}^{(i)}+n_1=m_\rho.$$

m_ρ has been retained in the mass balance equation to permit a very slight net evaporation or precipitation if the integration extends to the seasurface. $\mathbf{v}$ is the

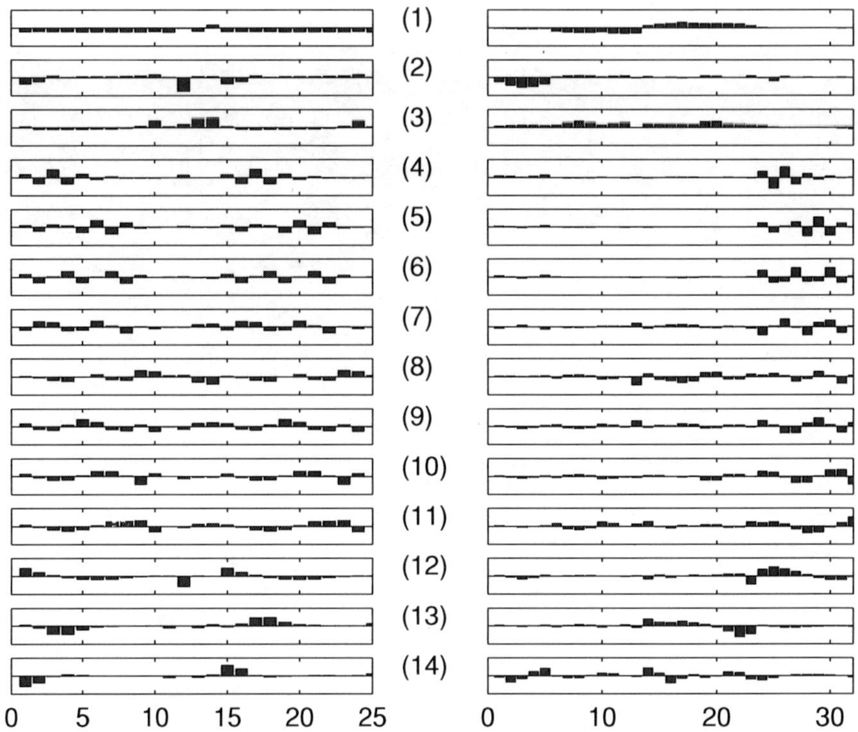

Figure 6.15 First 14 $\mathbf{u}_i$, $\mathbf{v}_i$ for the mass- and salt-conserving system. $\mathbf{u}_1$, $\mathbf{v}_1$ involve a linear combination of all of the equations, that is, primarily determining the b_i in the 24° N and 36° N sections. Elements 1 to 6 in the Florida Straits are determined by a complex combination of equations. Full scale is ± 1 for all.

horizontal velocity. Multiply the second of these equations by $\bar{S}$, and subtract from the first:

$$\sum_{j \notin J} \sum_q \rho(j, q)[S(j, q) - \bar{S}]\delta(j)\Delta a(j, q)[v_R(j, q) + b_j] \qquad (6.36)$$

$$- \rho_{k+1} w_{k+1}^*[S_{k+1} - \bar{S}]A_{k+1} + \rho_k w_k^*[S_k - \bar{S}]A_k$$

$$+ \sum_i \overline{F_S}^{(i)} - \bar{S}\overline{F_E}^{(i)} + n_2 - \bar{S}n_1 = \bar{S}m_\rho.$$

This result is a new equation in $S' = S - \bar{S}$. The argument is commonly made that this subtraction improves the numerical conditioning of the system of equations – as the relative variation of S' is much greater than that of S. In practice, this reason is *not* why the system appears to be more useful: modern computers have such large word lengths, and with the accuracies of the input data hardly exceeding five significant figures, ordinary software has no difficulty in exploiting the small variations in S. The real reason salinity anomaly equations prove useful is quite different; the noise element in the new equation is $n_2 - \bar{S}n_1$, and it is usually assumed that this difference nearly vanishes, producing a strong noise suppression. Although the noise removal will not, in practice, be exact, noise does appear, empirically, to be much reduced. One could obtain the same result by using a non-diagonal $\mathbf{R}_{nn}$, in which a very high correlation was imposed on the noise elements of corresponding mass- and salt-conservation equations. The only reason why that would be preferable is that it makes clearer what is actually going on – it is not the numerical conditioning that matters here, but the assumption about the noise structure. Equation (6.36), when integrated to the surface, cannot normally be used to constrain $\mathbf{x}$, as m_ρ is too poorly known. It is more commonly used to infer m_ρ, from estimates of $\tilde{\mathbf{x}}$ made using the other equations. If the noise does not effectively vanish, the resulting estimate will not be meaningful: m_ρ typically has a magnitude much smaller than any plausible error in the original equations.

6.3 Property fluxes

Mass and salt fluxes have been used in the equations employed to determine $\mathbf{x}$. One of the most important reasons for studying the ocean circulation, however, is the need to compute the fluxes of other scalar properties, physical and chemical, which are central to the climate system. Often, from the estimate of the resulting flow and mixing parameters, one seeks to calculate the fluxes of other important properties that were not necessarily employed as constraints. One might not use them as constraints if, for example, sources and sinks were so uncertain that little useful information would be obtained if realistic errors were assigned. An example would be the top-to-bottom oxygen balance, in the presence of highly uncertain air–sea gas exchange, photosynthesis, and remineralization at depth, all of which might be regarded as essentially unknown. In that case, the m_C appearing on the right-hand side of equations such as Eq. (6.29) would be best obtained by calculating them as residuals of the equations.

If some m_C is so poorly known that one wishes to compute it from $\tilde{\mathbf{x}}$, there are two equivalent approaches. One can include the equation in the set to be solved, but giving it such small weight that it has no influence on the solution. Evaluation of the system residuals then automatically evaluates m_C. The alternative is to omit the equation completely, and then to evaluate it after the fact. In the first approach, the equation system is slightly larger than is strictly necessary, but unless there are many such downweighted constraints, the consequences are slight.

Consider any property transport equation, e.g., for temperature across a line:

$$\tilde{H} = \sum_{j \in J} \sum_{q} \rho(j,q) C(j,q) \delta(j) \Delta a(j,q) \tilde{b}_j \tag{6.37}$$

$$+ \sum_{j \in J} \sum_{q} \rho(j,q) C(j,q) \delta(j) \Delta a(j,q) v_R(j,q) + \overline{F_C}$$

$$= \mathbf{d}^{\mathrm{T}} \tilde{\mathbf{x}} + \sum_{j \in J} \sum_{q} \rho(j,q) C(j,q) \delta(j) \Delta a(j,q) v_R(j,q) + \overline{F_C},$$

$$d_j = \left[\sum_{q} \rho(j,q) C(j,q) \delta(j) \Delta a(j,q) \right].$$

The first term on the right of $\tilde{H}$ is determined from an inversion for b_i^*; the remaining two terms are known. Apart from the numerical estimate of $\tilde{H}$ itself, the most important piece of information is its accuracy. So, for example, estimates of the heat flux across $36°$ N in the North Atlantic are approximately 1×10^{15} W. For studies of climate and of climate change, one must quantify the expected error in the value. There will be three contributions to that error, from uncertainties in $\tilde{b}_j$, from $v_R(j, m)$, and from F_C. The first error is the easiest to estimate. Suppose one has $\tilde{\mathbf{x}}$, and its uncertainty $\mathbf{P}$. Let the elements of $\tilde{\mathbf{x}}$ corresponding to the b_i be the vector of elements $\mathbf{x}_b$, with corresponding error covariance $\mathbf{P}_b$ (that is, omitting the w_i^*, or other parameters, not appearing in Eq. (6.37)). Then the uncertainty of the flux owing to the errors in $\tilde{b}_i$ is

$$P_{H_1} = \mathbf{d}^{\mathrm{T}} \mathbf{P}_b \mathbf{d}. \tag{6.38}$$

One might attempt then to add the uncertainty owing to the variances of $\overline{F_C}$, and v_{Rj}. The difficulty is that the $\tilde{b}_j$ are determined from $\overline{F_C}$, v_{Rj}, and the errors cannot be considered independently. For this reason, these other errors are usually omitted. A computation of the uncertainty owing to time variations in $\overline{F_C}$, and v_{Rj} is done separately, and is described later on in this chapter.

Consider the heat flux estimated from the Gauss–Markov method used above. Property C is then the temperature, T, times the heat capacity, $h_p = 4 \times 10^3$ J kg^{-1} $°$C^{-1}. Then, $\tilde{H} = (4.3 \pm 2.2) \times 10^{14}$ W (a number that, owing to the coarse spatial resolution, is too low by about a factor of about 2). The uncertainty was obtained directly from Eq. (6.38). The flux divergence and its error estimate is obtained in the same way, except using instead, the equation for the difference of the $24°$ and $36°$ N fluxes. Because of the summation implied by $\mathbf{d}^{\mathrm{T}} \mathbf{P}_b \mathbf{d}$, the error for the flux can be much lower than for individual elements of the velocity field, if error cancellation, owing to spatial correlation structures, can take place. In other circumstances, the errors will amplify through the summation.

An alternative approach is to write $\tilde{\mathbf{x}} = \tilde{\mathbf{x}}_{SVD} + \mathbf{q}$, where $\tilde{\mathbf{x}}_{SVD}$ is the particular SVD solution as defined in Chapter 2, and $\mathbf{q}$ is the nullspace contribution, which is unknown, and which may dominate $\mathbf{P}$, $\mathbf{P}_b$. If the vector $\mathbf{d}$ in Eq. (6.37) is such that $\mathbf{d}^T\mathbf{q} \approx 0$, $\tilde{H} = \mathbf{d}^T\tilde{\mathbf{x}}_{SVD}$, then P_{H_1} will tend to be small. That is, the missing nullspace will have little effect on the heat flux estimate because it is nearly orthogonal to the vector of station heat contents ($\mathbf{d}$). This situation seems to obtain, approximately, over much of the ocean, and thus oceanic heat flux calculations are often more accurate than might have been anticipated, given the large remaining uncertainty in the mass flux owing to its nullspace. Much of that nullspace lies in small spatial scales (eddies).

6.4 Application to real oceanographic problems

The estimation machinery discussed here has been applied in a great variety of physical oceanographic settings, ranging from comparatively small regions involving state vectors $\mathbf{x}$ of modest dimension (tens to hundreds of elements), to global calculations in which $\mathbf{x}$ has thousands of elements. Here we survey some of these calculations, primarily to give the flavor of real problems. It is a truism of most estimation problems that the construction of the model (defining $\mathbf{E}$, $\mathbf{R}_{xx}$, $\mathbf{R}_{nn}$, etc.) and interpreting the result is 90% of the labor, with the actual inversion being comparatively straightforward. Here we omit most of the detail involved in editing real data (temperature, salinity, oxygen, nutrients, etc.) and getting it into a form whereby the station-pair sums in equations like (6.34) and (6.35) can be carried out. The reader is referred to OCIP (Wunsch, 1996) and the references cited there for some of this detail. Software for carrying out practical inversions with real hydrographic data has been developed at a number of oceanographic organizations, and is available over the Internet. But given the ephemeral nature of websites, I will refrain from providing specific sources here.

6.4.1 Regional applications

Numerous applications of these ideas to particular regions exist and only a few will be described. Consider the hydrographic stations in Fig. 6.16 (from Joyce *et al.*, 2001). This single section defines a closed, if complicated, region of ocean to the west, involving much of the western North Atlantic as well as a major part of the Caribbean Sea. In addition to temperature, salinity, and various conventional scalar fields such as nitrate, silicate concentration, the investigators had available direct, if noisy, velocity measurements from a so-called lowered acoustic Doppler current profiler (LADCP). They also had "transient tracer" data in the form of chlorofluorocarbons (CFC-11,12). The density field (computed from temperature,

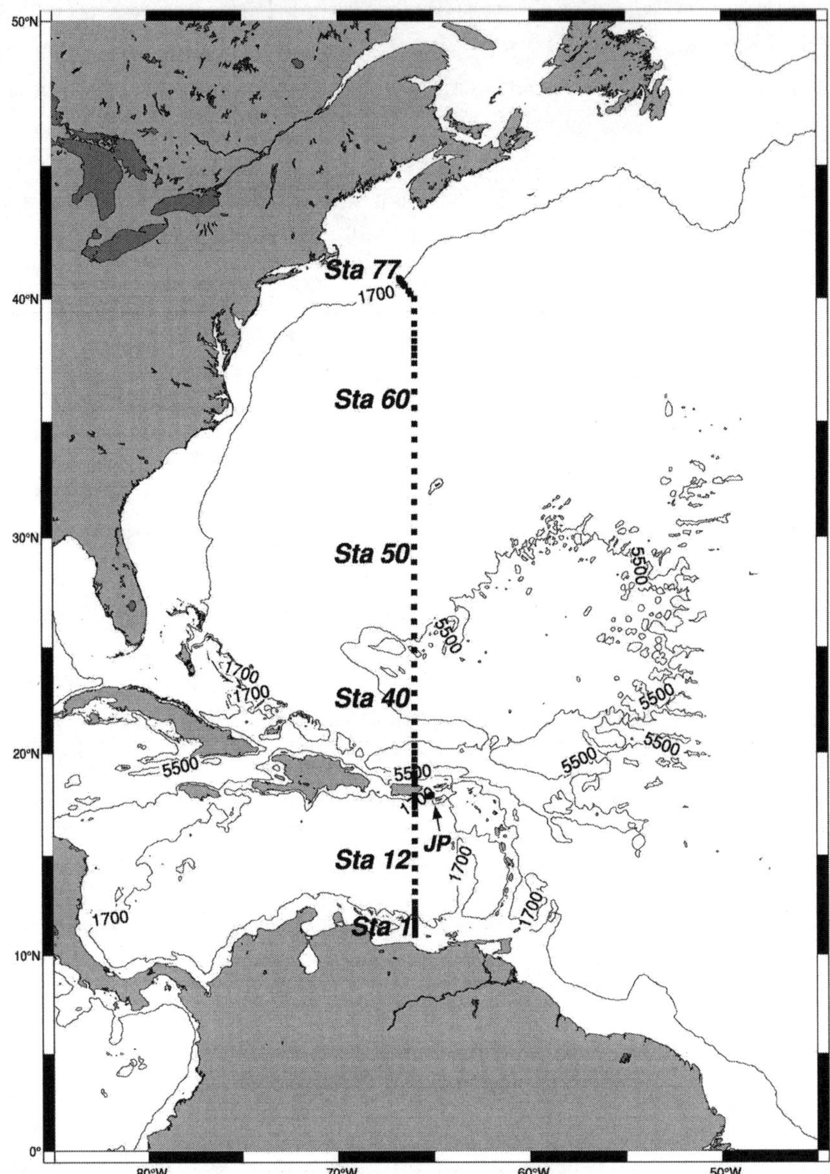

Figure 6.16 Positions of the hydrographic and LADCP measurements taken by Joyce *et al.* (2001). The region to the west of the line represents a closed volume of ocean for which various approximate conservation requirements are written in the form of standard simultaneous linear equations. (Source: Joyce *et al.*, 2001)

salinity, and pressure) and silica concentration are displayed in Figs. 6.17 and 6.18. The goal was to combine all of the data so as to determine the flow field, and the major property fluxes across the line. (Note that one might regard the section as also defining a closed volume consisting of the entire remainder of the world ocean, to which the net inflow/outflow constraints equally well apply. Internal balances involving isopycnal areas would be different, however, and distant outcrops to the surface would be very important.)

Joyce *et al.* (2001) used the LADCP velocities to provide an initial, non-zero, estimate of the reference level velocity; these are most easily employed by adding equations of the form $x_i = v_i$, where the v_i are the LADCP values at the reference level, and the indices i correspond to the reference level velocities, so that $x_i = b_i$. An estimate is made of the accuracy of these new constraints (this procedure is only approximately that used by Joyce *et al.*, 2001). In addition, to mass constraints overall, and in each of the layers (accounting for the Ekman velocity), near-conservation of silica within the system leads to a canonical problem of 44 equations in 72 unknowns. Error variances for the mass flux equations varied from $(2 \times 10^9 \text{ kg/s})^2$ to $(0.5 \times 10^9 \text{ kg/s})^2$ for the mass equations; silica equations had error variances ranging from $(5\text{--}20 \text{ kmol/s})^2$. In addition, a-priori estimates were made of the reference level variances (see Fig. 6.19). Note that no use was made of vertical exchanges, w^*, and thus any mass transfers across the interfaces will appear in the noise residuals, $\tilde{n}_i$. An inversion was done using the Gauss–Markov estimator; the flow field obtained (see Fig. 6.20) is the best estimate of the combined data sets with the assumptions of geostrophy, a steady state, and an Ekman flux in the surface layer. The resulting flow fields were then used to calculate zonal heat, freshwater, and CFC.

This solution and its employment depicts the general ability to combine data of very different types, to test the adequacy of the model assumptions (residuals are all acceptable), and its application to finding the fluid fluxes of important scalar properties.

The preferred strategy is to write as many constraints on the system as is consistent with prior knowledge, and then to use them to make the best possible estimate of the circulation. Other observation types have been employed by many authors, including direct velocities from mid-depth floats (Mercier *et al.*, 1993); current meters (Matear, 1993); and altimetric measurements of the seasurface height (Martel and Wunsch, 1993a). Gille (1999) used a combination of surface Doppler acoustic measurements of velocity (ADCP), mid-depth floats, and velocities determined from satellite altimeters in addition to the hydrographic data. As always, the only requirement for the use of any kind of data is that there should be an algebraic relationship between the unknowns, x_i, and the observations, and that estimates of the expected errors should be available.

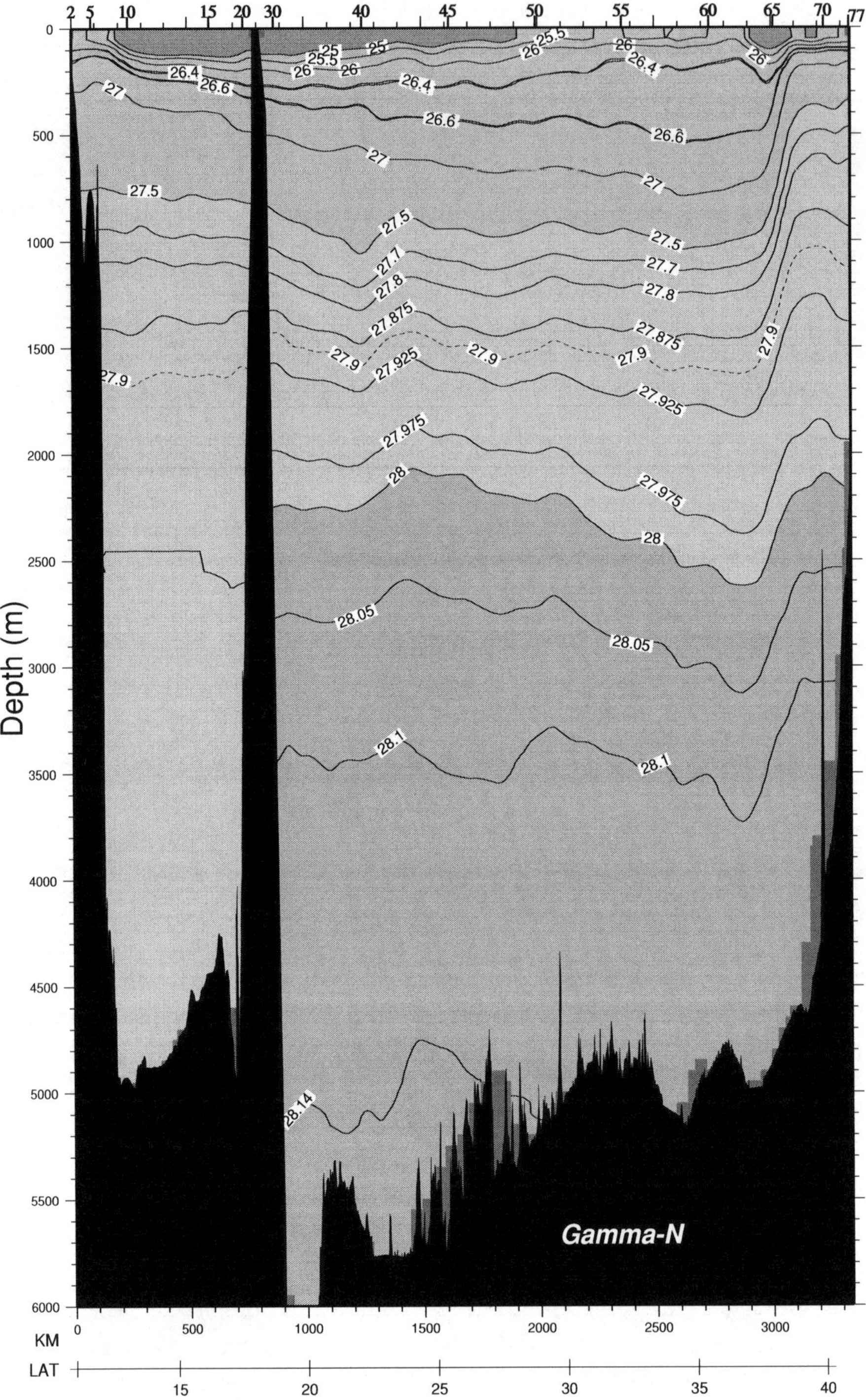

Figure 6.17 Density field constructed from the measurements at the positions in Fig. 6.16. Technically, these contours are of so-called neutral density, but for present purposes they are indistinguishable from the potential density. (See color figs.)

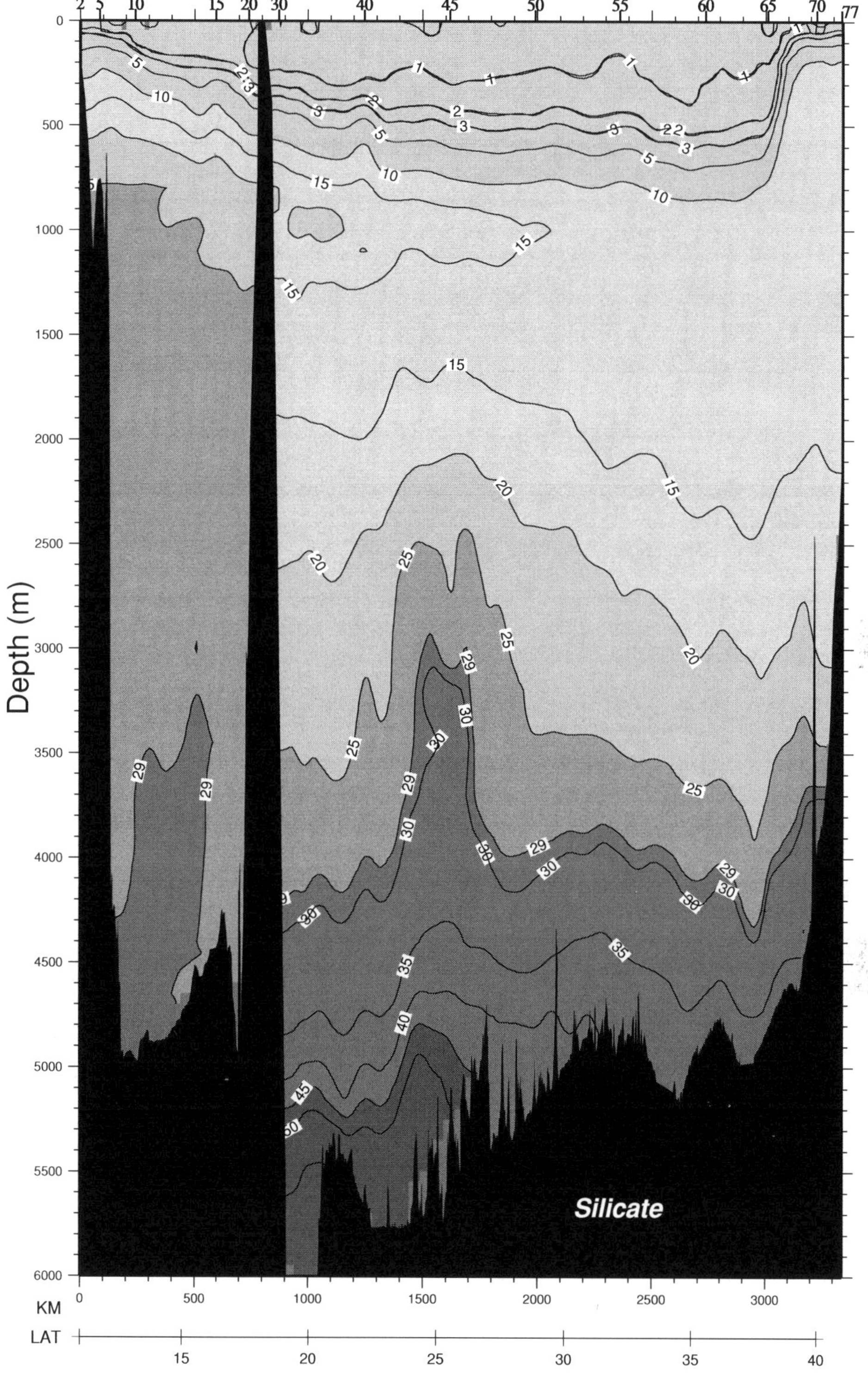

Figure 6.18 Same as Fig. 6.17, except showing the silicate concentration – a passive tracer – in μ moles/kg. (See color figs.)

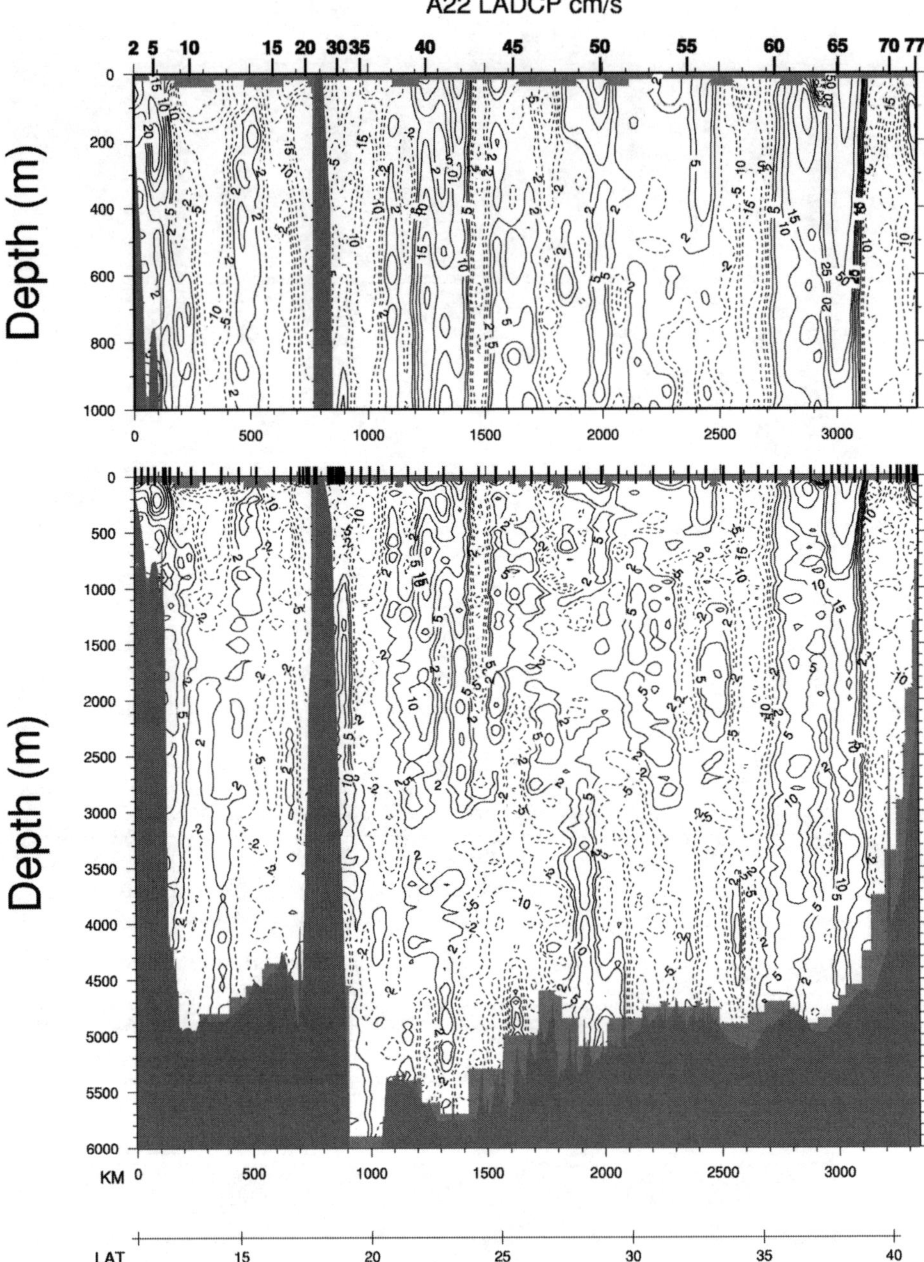

Figure 6.19 Initial velocity field from the hydrographic and LADCP data, as described by Joyce *et al.* (2001). Note the very great noisiness of the data. An expanded version of the upper 1000 m is also shown. (Source: Joyce *et al.*, 2001)

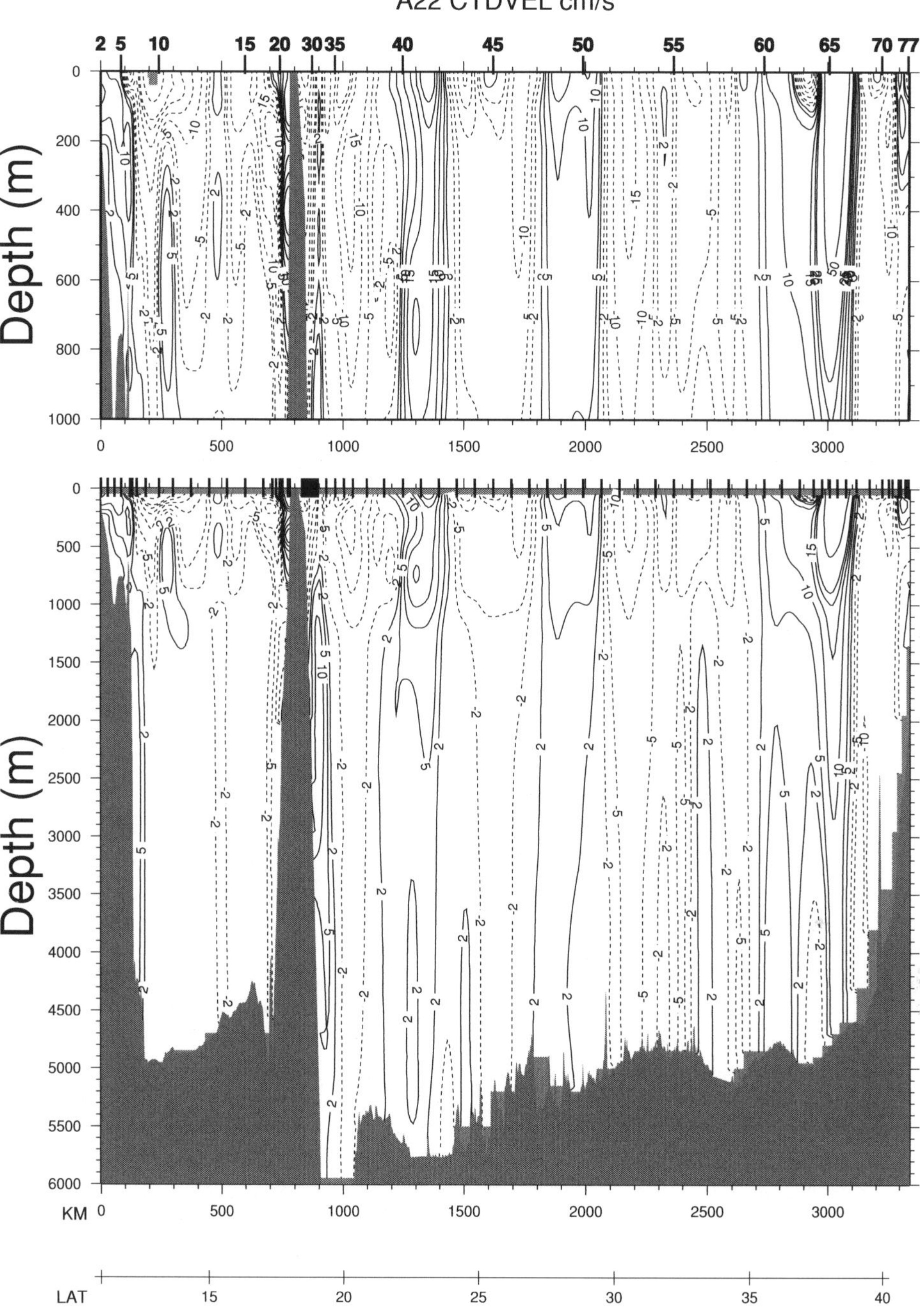

Figure 6.20 Final estimated velocity from the inversion using the constraints of Joyce *et al.* (2001). See their paper for details. Compare to Fig. 6.19, which was the starting point. Note the very strong vertical columnar nature of the estimated flow.

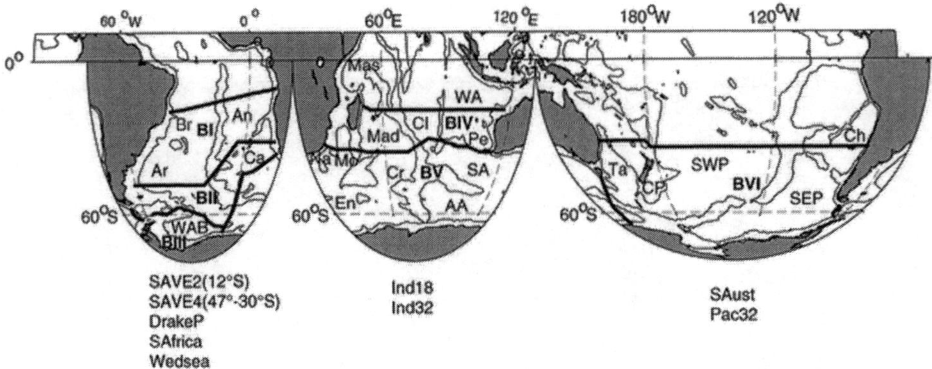

Figure 6.21 Positions of the sections used by Sloyan and Rintoul (2001) to determine the high latitude southern hemisphere oceanic circulation. (Source: Sloyan and Rintoul, 2001)

In another, larger, region Sloyan and Rintoul (2001) used the hydrographic lines depicted in Fig. 6.21. This region is sufficiently large that the results are probably best compared to the global-scale inversions described later. The authors employed 23 layers bounded by neutral surfaces, and the top and bottom. Conservation equations were written for mass, heat, salt, and silica, and a number of section mass transports, as well as a requirement of heat loss over the Weddell Sea, were imposed. A distinguishing feature of this calculation was that Sloyan and Rintoul (2001) permitted the vertical exchange coefficients w^* to be different for different properties (heat, silica, etc.) in contrast to other inversions where they were forced to be identical for all properties. Their solution produced a much more vigorous circulation than is conventionally accepted (e.g., the production rate of Antarctic Bottom Water exceeded 50×10^9 kg/s, which is much larger than conventional wisdom suggests). The extra degrees-of-freedom from the property-variable mixing do not by themselves necessitate a stronger circulation. Rather, one can trace the strong solution back to the choices of a-priori solution and residual values. To the degree that their solution is consistent with all of the imposed information, and in the absence of additional information, it cannot be rejected. One interpretation is that the information employed is inadequate to invalidate a 50×10^9 kg/s bottom water formation rate hypothesis.

Another interesting regional solution is that of Naveira Garabato *et al.* (2003) who focussed on determining the mixing coefficients, K, in the Scotia Sea region of the Southern Ocean, finding remarkably strong abyssal vertical diffusion coefficients.

6.4.2 The columnar result

When inverse methods were initially applied to the ocean circulation problem (Wunsch, 1977, 1978), they elicited a highly skeptical, generally negative, response.

Prominent among the reasons for this reaction was the vertical, or columnar, structures visible in the solutions, e.g., as seen in Fig. 6.20. That the lines of vanishing flow tended to be oriented more vertically, rather than horizontally – as a simple level-of-no-motion would imply – was regarded as strong evidence that the method was producing erroneous results. The level-of-no-motion assumption and related inferences about the ocean, had led to a widespread presumption that the true flows were layer-like. As it gradually became clear that these features were commonplace, and robust, and it was more widely appreciated that there was no rationale for quasi-horizontal levels-of-no-motion, the negative chorus gradually diminished, but has not yet entirely vanished.

It does remain possible that very long time-averages of the solutions would produce time-mean velocities that had a more layered character. A full discussion of this issue more properly belongs in an oceanographic book. But as one piece of evidence, Fig. 6.22 shows the directly measured two-year mean velocity (that is, not inferred from an inverse method) from a moored array of current meters in the western South Pacific Ocean. Note the marked near-vertical orientation of the zero lines. Of course, two years is still an interval short compared to the longest time scales present in the ocean, and a 100-year average might well have a different character, although there is no evidence for it. There would also be serious questions about the interpretation of the time-averaged equations governing such a hypothetical flow – they would not, in general, be the same linear equations we have been using. (The data all lie below 2000 m, but the accompanying hydrography suggests, in fact, that these columns do extend close to the sea surface as in Fig. 6.20.)

6.4.3 Global-scale applications

There have been two attempts at estimating the global ocean circulation with these methods (Macdonald, 1998; Ganachaud, 2003a), and the latter is briefly examined as representative of the state-of-the-art. The sections shown in Fig. 6.23 were used to write approximate conservation equations for the volumes bounded by sections and continents (note that all sections terminate at both ends in shallow water). Properties used as constraints were mass, temperature, salinity, silica, and "PO." The last is a combination of phosphate and oxygen concentrations, "PO" $= 170[PO_4] + [O_2]$, for which biological arguments exist that it should be nearly conservative below the zone of biological productivity (the brackets denote "concentration of"). For present purposes, it is just one more property carried passively by the fluid. Not all properties were required to be conserved top-to-bottom, but some only in interior layers (temperature, "PO"), and silica was conserved only top-to-bottom and not in individual layers. The layers were defined by constant density surfaces (technically, neutral surfaces), and a very large number of auxiliary requirements were written

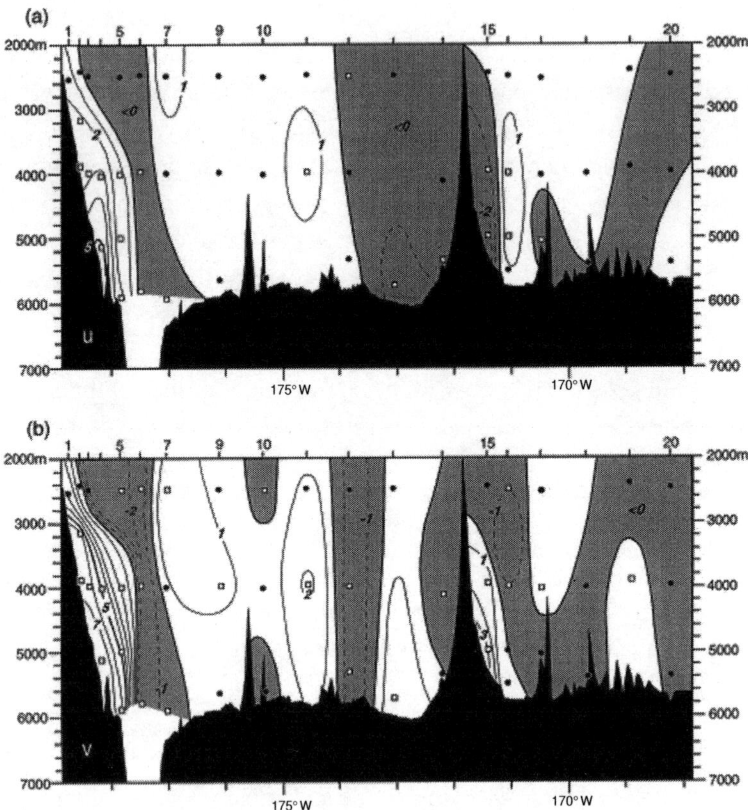

Figure 6.22 (a) The mean zonal velocity from two years of data, and (b) the mean meridional (normal to the section) velocity from two years of direct measurement. Instrument depths are shown as dots. Although there were no measurements above 2000 m, the cellular nature of the mean flow is nonetheless plain. (Source: Whitworth *et al.*, 1999)

for different regions. Ekman fluxes were represented as $\bar{F}_C + \Delta F_C$, where ΔF_C are a new set of unknown adjustments relative to the initial estimates of $\bar{F}_C$. Vertical exchanges were written in the full separated form, Eq. (6.21), but with w^* interpreted as being normal to the isopycnals.

A description of all of the regional constraints is not attempted here because that becomes a discussion of everything that was known, a-priori, about the general circulation, globally. But to give some of the flavor of the procedure, consider only the Pacific Ocean. There, the net northward flux across 17° S and 32° S was constrained to $(7.8 \pm 10) \times 10^9$ kg/s to represent the mass flux previous measurements had suggested were passing into the Bering Sea and into the Indian Ocean north of Australia. At 32° S, current meter measurements were interpreted to demand $(16 \pm 5) \times 10^9$ kg/s moving northward between 179 and 168° W between two of

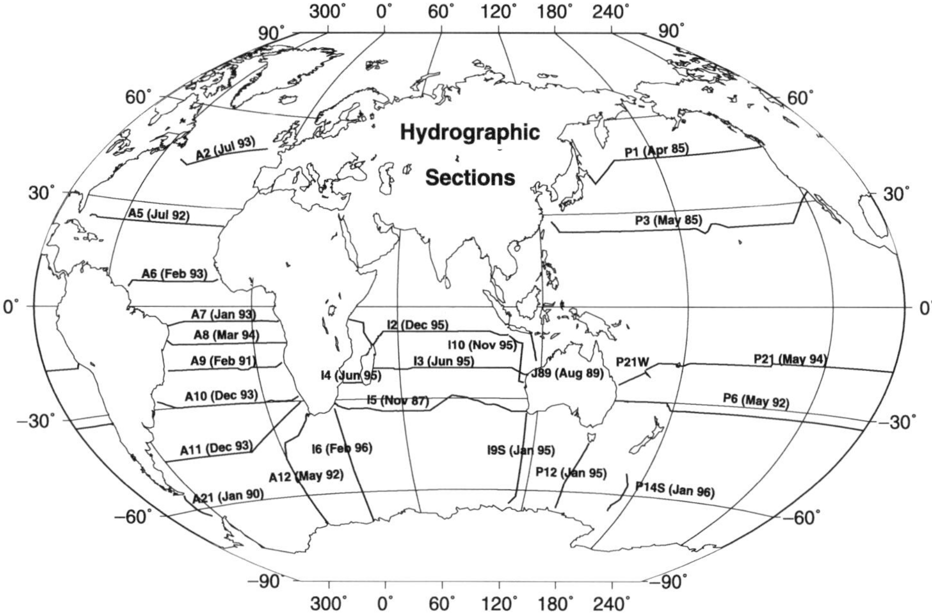

Figure 6.23 Sections defining the volumes of ocean for which property conservation equations were written. (Source: Ganachaud, 2003a)

the isopycnals. Further current meter measurements in the Samoa Passage were used to enforce a mass flux of very cold water to the north of $(12 \pm 4) \times 10^9$ kg/s there. Finally, the reference level, z_0, was chosen to correspond with different isopycnals at different sections, in such a way that a reasonable a-priori estimate of the reference level velocities, b_i, would be zero (see Table 6 of Ganachaud, 2003a).

Similar constraints were written in all ocean basins. As with the regional inversions, one seeks as many constraints, whose accuracy can be quantified, as is possible. The final system was approximately 1200 equations in 3000 unknowns, and solution was by the Gauss–Markov method using the prior error estimates discussed in detail by Ganachaud (2003b). That is, a total of 4200 solution and noise variances were assigned. (An SVD analysis, which is very desirable, involves, e.g., solution resolution matrices, which are 3000×3000 – a size that renders analysis and understanding very difficult.)

A schematic of the final solution is shown in Figs. 6.24 and 6.25. Figure 6.24 displays the total flows (arrows) as integrated across the sections in three layers (sums of the layers used in the inversion) spanning the entire water column. Also shown on the figure are vertical exchanges across the defining interfaces. Figure 6.25 displays the same solution, but showing the horizontal distribution of the flow integrated from west to east (north to south) in the three layers. The interpretation

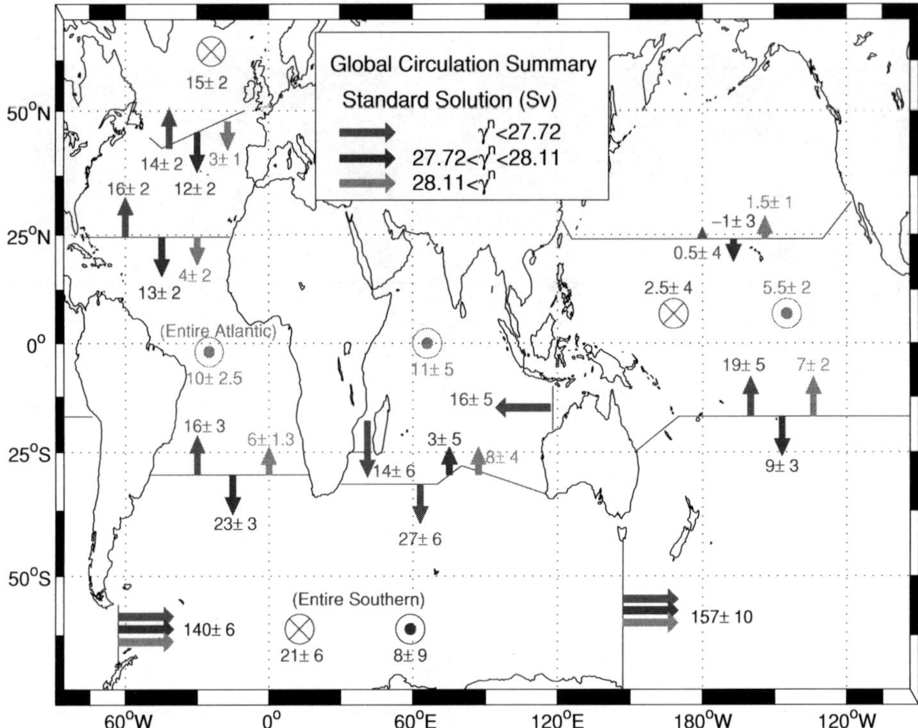

Figure 6.24 Mass flux in the Ganachaud (2003a) solution. Red, blue, and green arrows depict the vertically and horizontally averaged mass flux between the neutral surfaces noted. (See color figs.) (Source: Ganachaud, 2003a)

of these results would take us too deeply into physical oceanography, and the reader is urged to consult the references for further information.

It is worth, however, producing one example of the use of the solution. Figure 6.26 shows the estimated flux of dissolved nitrate in the ocean. Nitrate is a vital biogeochemical field in the ocean, important for both biology and climate. No nitrate constraints were employed in the above solution for the flow and mixing fields and these fluxes were simply computed from the inferred flow field and the observed nitrate distribution. These estimated fluxes (and more important, their divergences) are probably the best available at the present time.

6.4.4 Error estimates

Ganachaud (2003b) revisited the question of specifying the error covariances to be used in large-area box models such as the ones just described. Many sources of error must be accounted for, ranging from internal waves disturbing the density field, to

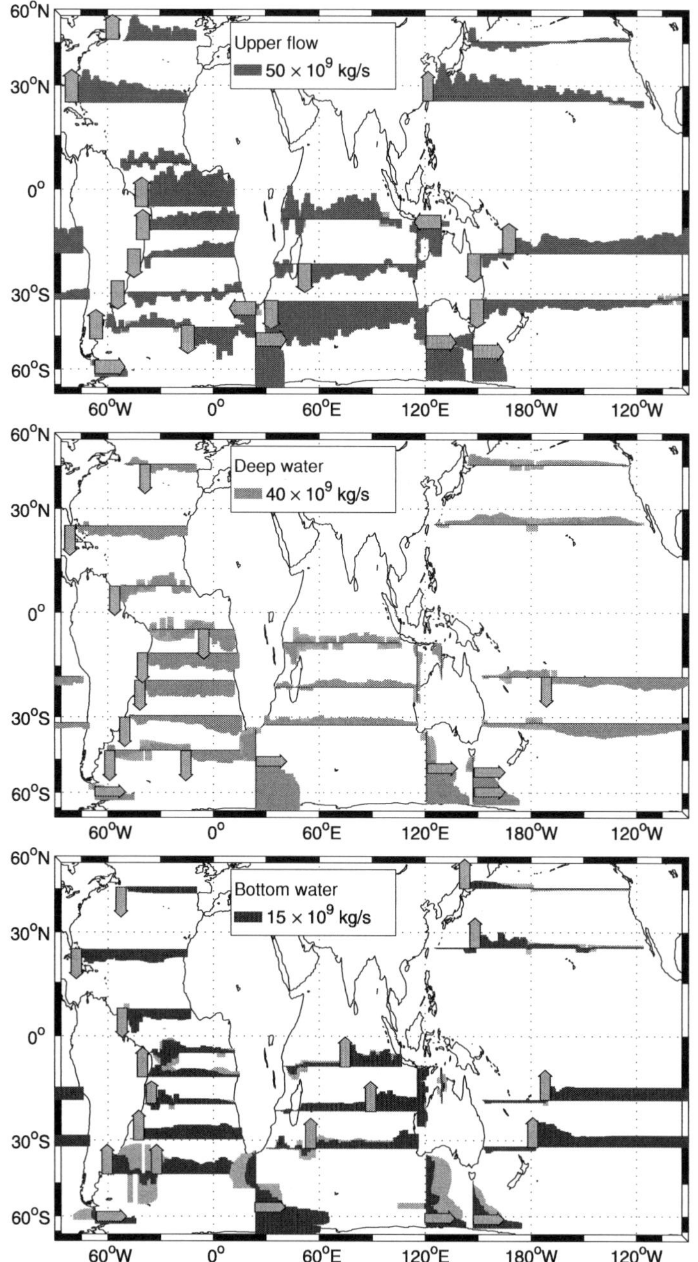

Figure 6.25 Mass transports integrated from west to east and north to south for the solution displayed in Fig. 6.24. Light shading shows the estimated standard errors. Arrows denote the major currents of the system. (See color figs.) (Source: Ganachaud, 2003a)

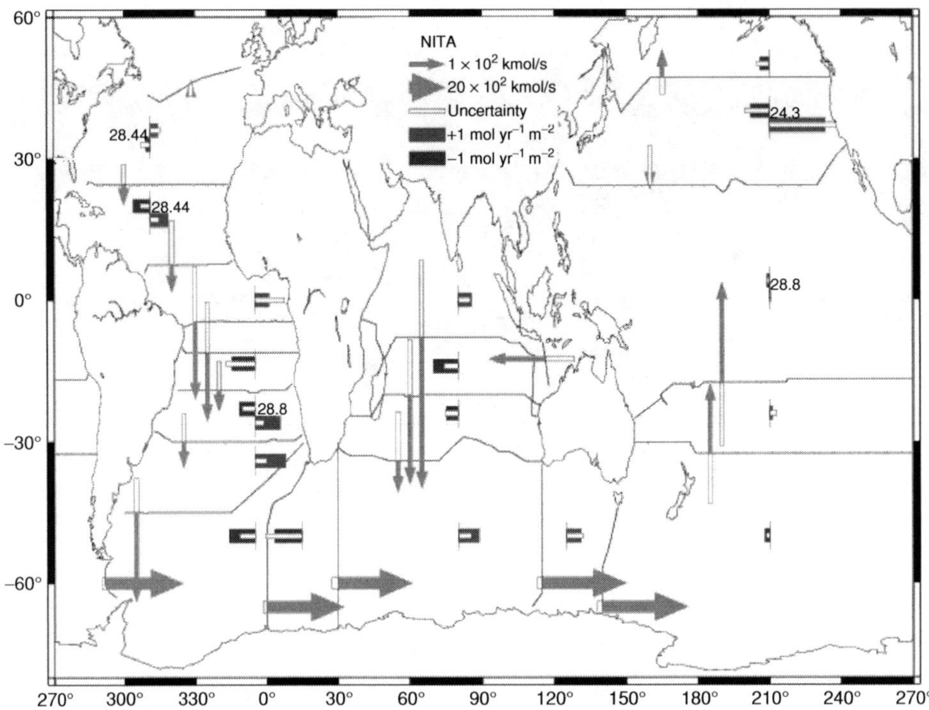

Figure 6.26 Integrated nitrate flux corresponding to the mass flux in Figs. 6.24 and 6.25. Although there is some resemblance to the mass fluxes, significant differences in the net movement of nitrate occur – owing to the spatially varying concentration of nitrate. (See color figs.)

the extrapolation of temperature, salinity, and other properties into regions without data, to instrumental noise, to the finite duration required by a ship to sample the ocean. He relied heavily on the results of a so-called "eddy-permitting" general circulation model, and reference is made to OCIP (Wunsch, 1996) and Ganachaud's paper for further discussion. Perhaps the most significant error, thought to dominate the solutions, is the difference between a flow computed from a quasi-instantaneous ocean sample (a "synoptic" section) and one computed from a hypothetical time average of the flows from sections made over a long period of time. It has been argued (Wunsch, 1978, 1996), that the large-scale spatial integrations of the fluxes involved in the constraint equations might, under an ergodic hypothesis, closely resemble their long-term time average. This hypothesis fails quantitatively, and one cannot, without supplying an error estimate, equate the property fluxes obtained from the box balances to those in a hypothetical time-average ocean. Two separate issues arise: the errors incurred when synoptic sections measured at different times are combined to write regional box balances for any property, and the extent to which the result then can be interpreted as equal to the long-term average. Both have errors, and they must be carefully evaluated as suggested in the references.

6.4.5 Finite-difference models

Any system that can be put into the standard form $\mathbf{E}\mathbf{x} + \mathbf{n} = \mathbf{y}$ can be solved using the methods of this book. Although the geostrophic box inverse models have usually been written for integrals over large volumes of ocean, the same procedures can be used for the equations in differential form, including, e.g., linear numerical models of the ocean. A simple version was discussed above for generic tracer problems.

As a complement to the above discussion of very large-scale constraints, consider (Martel and Wunsch, 1993b) the use of these methods on a system of equations based upon a discretized generic, steady conservation equation with Laplacian diffusion. That is, consider the partial differential system

$$\mathbf{u} \cdot \nabla(\rho C) - \nabla_{\mathrm{H}}(K_{\mathrm{H}} \nabla(\rho C)) - \frac{\partial}{\partial z}\left(K_{\mathrm{V}} \frac{\partial (\rho C)}{\partial z}\right) = 0, \tag{6.39}$$

subject to boundary conditions (a special case of Eq. (6.1)); K_{H} is the horizontal dimension analogue of K_{V}. To solve such equations in practice, one commonly resorts to numerical methods, discretizing them, e.g., as

$$\frac{u(i+1, j, k+1/2)\rho C(i+1, j, k+1/2) - u(i, j, k+1/2)\rho C(i, j, k+1/2)}{\Delta x} + \cdots$$

$$+ \frac{w(i+1/2, j+1/2, k+1)\rho C(i+1/2, j+1/2, k+1) - w(i+1/2, j+1/2, k)\rho C(i+1/2, j+1/2, k)}{\Delta z} + \cdots$$

$$- K_{\mathrm{H}}(i+1, j, k+1/2)\frac{(\rho C(i+1, j, k+1/2) - 2\rho C(i, j, k+1/2) + \rho C(i-1, j, k+1/2))}{(\Delta x)^2} + \cdots$$

$$+ n(i, j, k) = 0, \tag{6.40}$$

where i, j are horizontal indices, k is a vertical one (x is again here a coordinate, not to be confused with the state vector, $\mathbf{x}$). Compare these to Eqs. (6.4) and (6.5). The $1/2$ index is for a variable evaluated halfway between grid points. ρC_{ijk} is a shorthand for the product of the discretized density and tracer. A discretized form of the mass-conservation equation (6.9) was also written for each unit cell. $n(i, j, k)$ has been introduced to permit the model, Eq. (6.39), to be regarded as imperfect. If $C(i, j, k+1/2)$ is known at each grid point, the collection of (6.40) plus mass conservation is a set of simultaneous equations for $u(i, j, k+1/2)$ and K_{xx}, etc. One writes a second finite difference equation corresponding to Eq. (6.9) for mass conservation involving $u(i+1, j, k+1/2)$, $v(i, j+1, k+1/2)$, $w(i+1/2, j+1/2, k+1), \ldots$.

If the velocity field is calculated using the thermal wind equations on the grid, then at each horizontal grid position, i, j, there is a corresponding pair of unknown reference level velocities $c(i, j)$, $b(i, j)$ as above. This would then be an inverse problem for a state vector consisting of c, b (and/or K_H, if it is unknown). The equations are in the canonical form, $\mathbf{E}\mathbf{x} + \mathbf{n} = \mathbf{y}$. (In conventional forward modeling, one would specify the u, v, K on the grid and solve the linear equation set (6.40) for $C(i, j, k)$, etc. Here, the problem is inverse because C is known and elements (b, c, K) are the unknowns.) After some rearrangement for numerical convenience, there were about 9000 equations in 29 000 unknowns (the horizontal mixing unknowns were suppressed by giving them zero column weights and are not counted), a system that was solvable by tapered least-squares, taking advantage of the very great sparsity of the equations. The main outcome of this calculation is a demonstration of its feasibility. As the spatial resolution becomes finer, the time-independence assumption fails, and a steady model becomes untenable; that is, it fails on physical grounds. One is driven to employ the models and methods described in Chapter 7.

An interesting study by Tziperman and Hecht (1987) considered a repeated data set that permitted a comparison of solutions from averaged data, as opposed to those from averaging a series of solutions. Unsurprisingly, the results depend upon the order of averaging and calculation, and the differences lead into a discussion of the meaning of the eddy-coefficients, K, not pursued here.

6.5 Linear programming solutions

Linear programming, as outlined in Chapter 3, has been comparatively little used in fluid problems as compared to methods based upon least-squares. The reasons probably lie with its relative unfamiliarity in the physics community and the need to specify rigid upper and lower bounds. The latter appear more demanding than the soft constraints of least-squares, which can be violated. Advantages of linear programming include the very efficient and sophisticated software developed because of widespread economics and management applications; the easy ability to employ semi-qualitative constraints expressing, e.g., the requirement that some field or parameter should be positive or less than some large upper bound; the general insensitivity of 1-norms to data outliers; and the ease of finding absolute upper and lower bounds on interesting properties such as the flux of a scalar across a boundary (Wunsch, 1984). Many software packages automatically produce the dual (or adjoint) solutions, thus providing crucial sensitivity information.

Wunsch (1984) wrote a system of equations like those used in earlier chapters for the geostrophic box inversions but for multiple boxes spanning the North Atlantic. A form suitable for linear programming was used: The soft constraints for mass, salt, etc., conservation were replaced by hard inequalities representing absolute maximum and minimum bounds. Individual bounds were set on the reference-level velocities. The hard bounds did not have any simple statistical

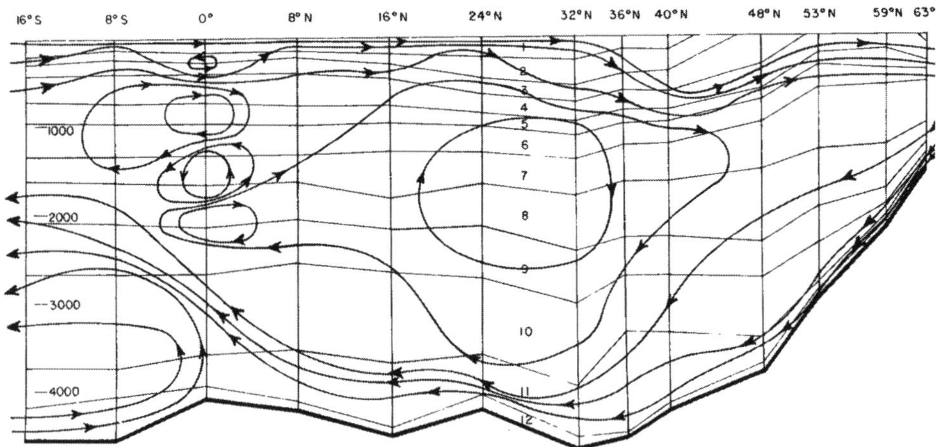

Figure 6.27 Zonal integral of a solution (it is not unique) that maximizes the flux of heat across 24° N in the North Atlantic. The approximate meridional overturning stream function is displayed. (Source: Wunsch, 1984)

interpretation – unlike the soft bounds of Gauss–Markov estimation or least-squares. Rather, they represented partially subjective views, based upon experience, of what would be extreme acceptable values. The objective functions consisted of the heat fluxes across each of the ocean-spanning sections in the form (6.37). As the bounds on x_i are not in linear programming canonical form (they permit both positive and negative values) one introduces new variables, $\mathbf{x}^+$, $\mathbf{x}^-$, which are non-negative, and defines $\mathbf{x} = \mathbf{x}^+ - \mathbf{x}^-$. The result was about 500 constraints in approximately 900 unknown parameters, each with a set of upper and lower bounds (see Eqs. (3.41)–(3.44)).

The problem was solved by a simplex method. The bounding heat fluxes were sought in order to understand their range of uncertainty. One solution that maximizes the net heat flux (including the Ekman component) subject to all of the constraints is depicted in Fig. 6.27.

Such methods, in which bounds are sought, are especially useful in problems that from the point of view of the basic SVD are grossly underdetermined, and for which determining "*the*" value of quantities such as the heat flux is less important than understanding their possible range. If the resulting range is sufficiently small, the remaining uncertainty may be unimportant. For anyone interested in the large-scale properties of the circulation – for example, its chemical and biological flux properties – detailed determination of the flow at any given point may be beyond reach, and not very important, whereas the integrated property extrema may well be well and readily determined. Wunsch (1984) called this approach "eclectic modeling" because inequality constraints are particularly flexible in accounting for the wide variety and inhomogeneous distribution of most oceanographic observations. These methods (Wunsch and Minster, 1982) are very powerful and, because of the ease with which positivity constraints are imposed, are the natural mathematical tool for handling tracers. Schlitzer (1988, 1989) used linear programming to discuss the carbon budget of the North Atlantic Ocean.

Non-linear extensions of these methods called, generically, "mathematical programming" are also available. One oceanographic example, which involves the adjoint to a general circulation model, is discussed by Schröter and Wunsch (1986), but, thus far, little further use has been made of such techniques.

6.6 The β-spiral and variant methods

6.6.1 The β-spiral

Combinations of kinematic and dynamic equations with observations can be made in many different ways. Take the same governing equations (6.6)–(6.10), but now employ them in a local, differential, mode. The so-called β-spiral method of Stommel and Schott (1977) is a rearrangement of the equations used for the geostrophic box balances, written for a point balance. There are several ways to derive the resulting system.

Ignoring some technical details (chiefly the dependence of density on the pressure), the density equation (6.10) is solved for w, and then differentiated by z,

$$
\left(\frac{\partial \rho}{\partial z}\right)^2 \frac{\partial w}{\partial z} = -\left(\frac{\partial \rho}{\partial z}\right)\left(\frac{\partial u}{\partial z}\frac{\partial \rho}{\partial x} + u\frac{\partial^2 \rho}{\partial x \partial z} + \frac{\partial v}{\partial z}\frac{\partial \rho}{\partial y} + v\frac{\partial^2 \rho}{\partial z \partial y}\right)
$$
$$
+ \left(u\frac{\partial \rho}{\partial x} + v\frac{\partial \rho}{\partial y}\right)\frac{\partial^2 \rho}{\partial z^2} .
\tag{6.41}
$$

Now recognize that the Coriolis parameter f, is a function of latitude. This dependence can be made explicit by writing it in local Cartesian approximation as $f = f_0 + \beta y$, where f_0 and β are constants. Then cross-differentiate the two momentum equations (6.6) and (6.7), which removes the large pressure terms from the system, and, using (6.9),

$$
\beta v = f\frac{\partial w}{\partial z},
\tag{6.42}
$$

called the "geostrophic vorticity balance." Equation (6.42) permits us to eliminate w from Eq. (6.41). The thermal wind equations (6.17) and (6.18) can be used to write $u(x, y, z) = u_R(x, y, z) + c(x, y)$, $v(x, y, z) = v_R(x, y, z) + b(x, y)$, where u_R, v_R are again assumed known from the density field. Substituting into (6.41) produces a partial differential equation in ρ involving the integration constants c, b,

$$
(u_R + c)\left[\frac{\partial \rho}{\partial x} - \left(\frac{\partial \rho}{\partial z}\right)\frac{\partial^2 \rho}{\partial x \partial z}\right]
$$
$$
+ (v_R + b)\left[\frac{\partial \rho}{\partial y} - \left(\frac{\partial \rho}{\partial z}\right)\frac{\partial^2 \rho}{\partial z \partial y} - \frac{\beta}{f}\left(\frac{\partial^2 \rho}{\partial z^2}\right)^2\right] = 0.
$$

$$(6.43)$$

If b, c were known, Eq. (6.43) would be an equation for the forward problem determining ρ subject to boundary conditions. Instead, one treats ρ as known and attempts to find b, c such that Eq. (6.43) is satisfied at each point x, y, as best possible. If discretized, the result is a set of simultaneous equations for $b(x, y)$, $c(x, y)$ written for different values of z_j, which can be solved by any one of the methods available to us. Suppose that $(x, y) = (x_i, y_i)$ are fixed and that (6.43) is applied at a series of depths z_i, $i = 1, 2, \ldots, M$. There are then M equations in the two unknown $b(x_i, y_i)$, $c(x_i, y_i)$, and a solution and its uncertainty can be found easily. Note again that the coefficient matrix here is constructed from observations, and inevitably contains errors; the full estimation problem would require simultaneously estimating the errors in ρ, and the suppression of that calculation is again a linearization of the full problem implicit in Eq. (6.19).

Figures 6.28–6.30 show the data positions and results of one of the earliest attempts, by Schott and Stommel (1978), to employ the β-spiral ideas in practice. They re-wrote Eq. (6.43) in terms of the isopycnal depth, rather than in terms of the value of the density field at fixed depths, in effect using $z(\rho)$, rather than $\rho(z)$. There are technical reasons for this change having to do with the dependence of ρ on the pressure, but it does not change the character of the equation or of the problem. Note that the greatly underdetermined geostrophic box balance model has been somehow converted into one with apparently arbitrary overdetermination, dependent only upon the number of depths z_i for which one chooses to write Eq. (6.43). How is this possible? By computing the horizontal derivatives in Eq. (6.43) from linear fits over large latitude and longitude ranges of the sections shown in Fig. 6.28, one is, in effect, asserting that b, c are constant over those long distances. An equivalent reduction in the nullspace of the geostrophic box balances could be made either by assigning a very large-scale spatial covariance via $\mathbf{R}_{xx}$ or by adding equations of the form $b_1 = b_2$, $b_2 = b_3$, etc., thus providing equivalent large-scale assumptions (*cf.* Davis, 1978).

When would one use the β-spiral approach as opposed to the large-scale box balances? The answer is chiefly dictated by the available data: the β-spiral requires the ability to determine both the x and y derivatives of the density field, at a central point. The volume integral approach requires derivatives only along sections, as is more commonly available in large-scale ship surveys, but which must define closed

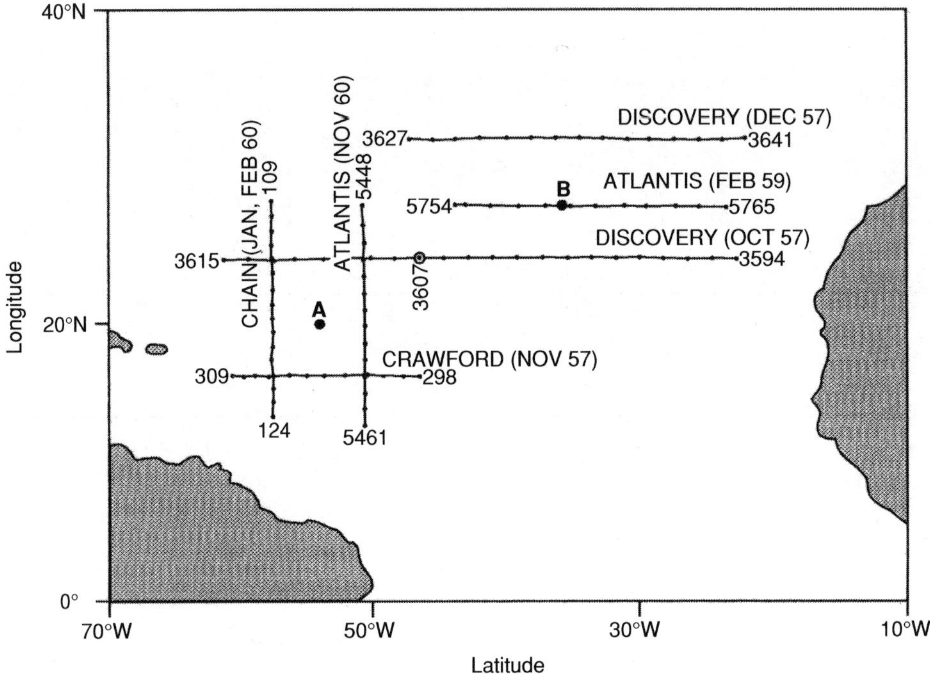

Figure 6.28 Hydrographic sections used by Schott and Stommel (1978) to infer absolute currents at points A,B using the β-spiral constraints. Labels are ship names and dates of measurements; the numbers indicate the ship station number defining the section ends. Isopycnal slopes along the sections are used to estimate the different terms of the β-spiral equations. (Source: after Schott and Stommel, 1978)

volumes. As already noted, one must be very careful about the spatial covariances that may be only implicitly, but nonetheless importantly, present in any attempt to calculate horizontal derivatives. A whole North Atlantic Ocean result for the estimated β-spiral by Olbers *et al.* (1985), based upon a 1° of latitude and longitude climatology, is shown in Fig. 6.31.

In some practical situations, the large-scale box balance constraints have been combined with the point-wise β-spiral balances into one solution (Fukumori, 1991; Ueno and Yasuda, 2003). The non-linear β-spiral problem, accounting for errors, $\Delta \mathbf{E}$, was discussed by Wunsch (1994), and is taken up below.

Many variations on this method are possible. Examples are in Welander (1983), Killworth (1986), Zhang and Hogg (1992), and Chu (1995). These methods might be thought of as hybrids of the straightforward finite difference representations, and the original β-spiral. They have not often been used in practice. Time-dependence in the observed fields remains a major error source, particularly as the regions under study diminish in size.

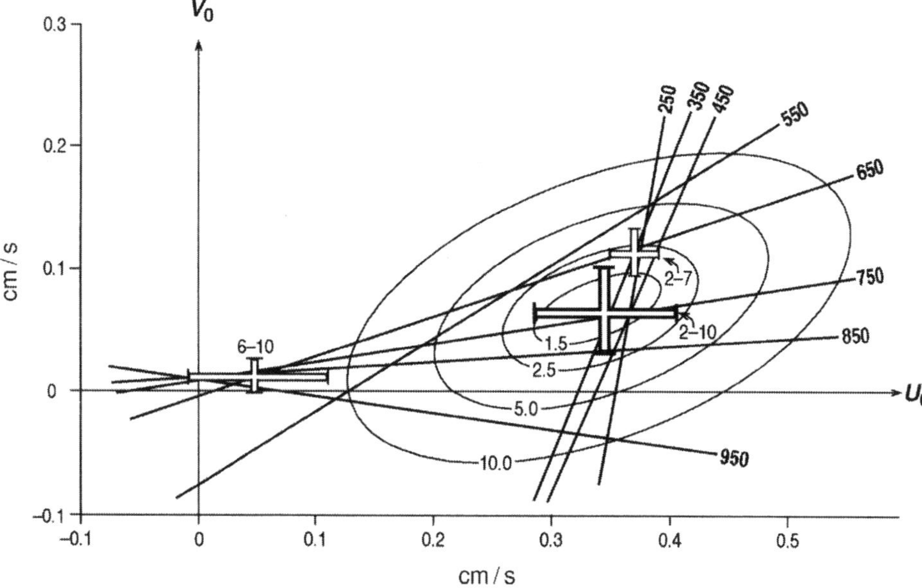

Figure 6.29 β-spiral constraint equations at position B in Fig. 6.28 as a function of depth (numbers on the straight lines). u_0, v_0 correspond to c, b in Eq. (6.43). In the absence of noise, and with a perfect model, the lines should all intersect at a point. Heavy crosses show the best solution over different depth ranges. Error bars are from the equivalent of $\sqrt{\mathrm{diag}\,(\mathbf{P})}$. That the solutions do not agree within error bars suggests that some other physics must also be operative, most probably in the upper ocean depth range.

6.7 Alleged failure of inverse methods

There is a small literature from the early days of the application of inverse methods in physical oceanography, in which various authors claimed to show their failure when applied to the ocean circulation. Unfortunately, the methodologies cannot magically compensate for missing data, or if the model used fails to be consistent with the correct one. The simplest analogy is the one in Chapter 2, attempting to fit a straight line to data actually governed by a higher-order polynomial, or if one grossly misrepresented the noise covariance. The solution is then doomed to be incorrect, not because least-squares or Gauss–Markov methods have failed, but because the problem was incorrectly formulated. Incorrect error covariances can drive a formally correct model away from the correct answer; again one should blame the practitioner, not the tools.

A failure of an inverse method could be claimed if the model relating the observations to the unknowns, $\mathbf{x}$, were fully correct, the prior error covariances were known to be correct, and the estimated solution, $\tilde{\mathbf{x}}$, were then shown to differ from $\mathbf{x}$ beyond the error bars (including the variance of the nullspace, if appropriate).

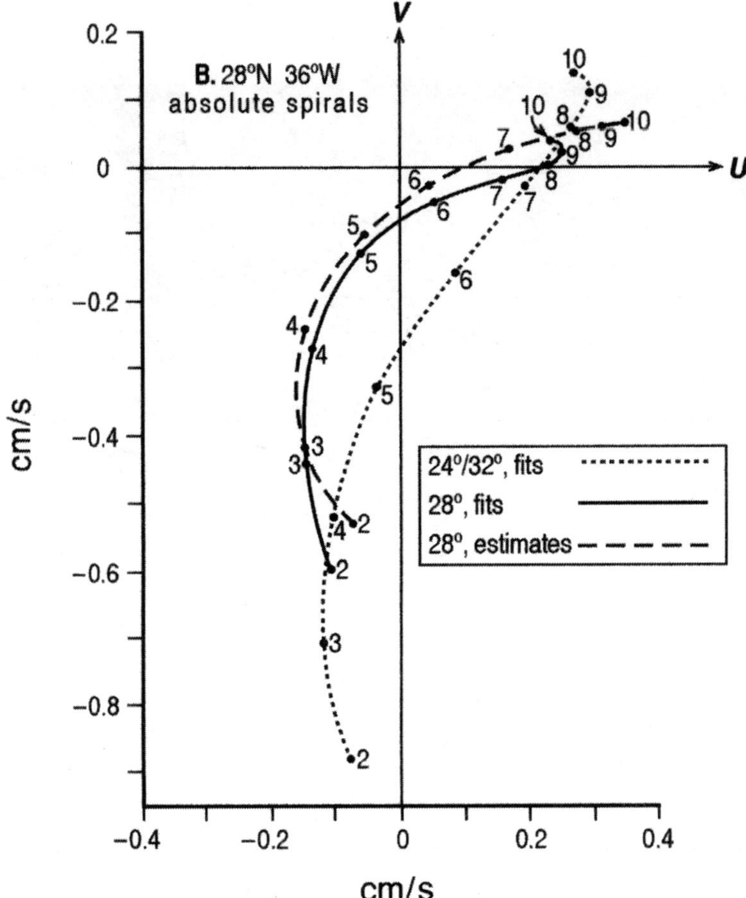

Figure 6.30 Diagram showing the velocity spirals resulting from the solutions to β-spiral equations. Depths are in hundreds of meters. The dashed curve labeled "estimates" is from a visual determination of the isopycnal slopes. The other two curves differ in the central latitude assigned to the Coriolis parameter. (Source: Schott and Stommel, 1978)

Such a failure would be surprising. All of the published claims of failure actually mis-specify a part of the problem in some way, and/or fail to use the uncertainty estimates any inverse method provides.

It sometimes happens that insufficient information is available to resolve some part of the solution of intense interest to a scientist (perhaps he or she seeks the vertical mixing coefficient, K_v); the solution shows it to be poorly resolved (or indistinguishable from zero). Such outcomes have been blamed on the method; but they are the result of *successful* use of the method: one is provided with the specific, and practical information, that more data of certain types would be required to obtain a non-zero estimate.

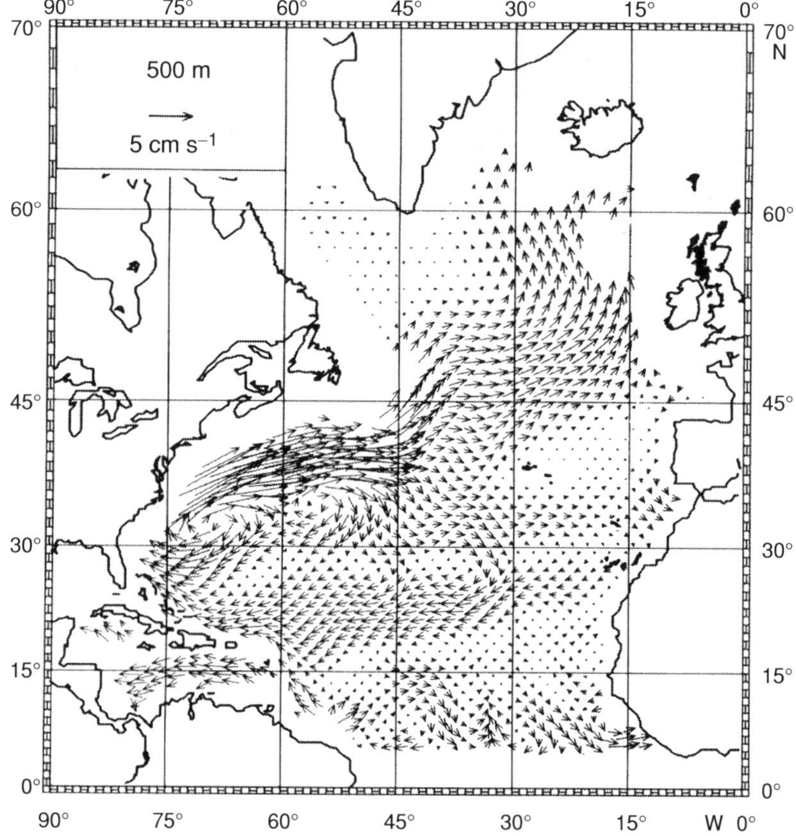

Figure 6.31 Absolute velocity field at 500 m determined from the β-spiral least-squares solution. Straight lines were fit over 1° squares and used a long time-space average to define ρ. (Source: Olbers *et al.*, 1985)

6.8 Applications of empirical orthogonal functions (EOFs) (singular vectors)

Suppose a fluid carries with it a variety of tracer properties. That is, suppose it is tagged by a point in an n-dimensional space of properties, such as temperature, salinity, chlorophyll content, etc. Write the collection of properties at a point as an n-tuple (temperature, depth, horizontal positions (r_x, r_y), salinity, velocity, ...). Oceanographers, with only a two-dimensional display possible on a page, have employed a variety of methods to depict their data. A common one is the section, in which the axes are depth and horizontal distance steamed by a ship, with one property, e.g., temperature, contoured. Another two-dimensional rendering can be obtained by selecting two physical attributes, e.g., temperature and silica concentration, defining those as the graph axes, and plotting each observation on a graph of temperature versus silica, either as a scatter of points, or sometimes contoured as a histogram. The very large number of possible two-dimensional projections onto planes cutting through the n-space renders it nearly impossible to visualize the properties. There have been a few attempts directed at representations and syntheses more convenient and powerful than property-property diagrams or sections. If one is attempting to understand the degree to which a model is consistent with the real ocean, a measure of distance in one particular two- or three-dimensional subspace of the n-space may give a very distorted picture of the true distance to all of the properties.

Some of the fluid properties that oceanographers measure are nearly redundant, appearing sometimes in geographical space as coincident tongues of, say, salinity and silica, or in property–property space as linear or other simple functional relationships (e.g., the conclusion that salinity is a nearly linear, local, function of temperature). A quantitative description of the relationship becomes useful in several ways: finding an explanation; reducing the dimension of the n-dimensional space; using the relationships to produce estimates of properties of a poorly measured variable from measurements of a better measured one; reducing costs by substituting a cheap-to-measure variable for a functionally connected expensive-to-measure one. Efforts in this direction include those of Mackas *et al.* (1987), Fukumori and Wunsch (1991), Hamann and Swift (1991), Tomczak and Large (1989) and the summary in Bennett (1992). You (2002) describes an application.

Mackas *et al.* (1987), working solely within the physical property space (i.e., the geographical coordinates are ignored), calculated the mixing required to form a particular water type in terms of a number of originating, extreme properties. For example, let the water at some particular place be defined by an n-tuple of properties like temperature, salt, oxygen, etc. (θ, S, O, ...). It is supposed that this particular set of water properties is formed by mixing a number, N, of parent water masses, for example, "pure" Labrador Sea Water, Red Sea Intermediate Sea Water, etc.,[7] each defined as $[\theta_i, S_i, O_i, \ldots]$. Mathematically,

$$\theta = m_1\theta_1 + m_2\theta_2 + \cdots + m_N\theta_N,$$
$$S = m_1 S_1 + m_2 S_2 + \cdots + m_N S_N,$$
$$O = m_1 O_1 + m_2 O_2 + \cdots + m_N O_N,$$
$$\cdots$$
$$m_1 + m_2 + m_3 + \cdots + m_N = 1,$$

and one seeks the fractions, m_i, of each water type making up the local water mass. Because the fractions are necessarily positive, $m_i \geq 0$, the resulting system must be solved either with a linear programming algorithm or one for non-negative least-squares. The equations can as always be rotated, scaled, etc., to reflect prior statistical hypotheses about noise, solution magnitude, etc. Mackas *et al.* (1987) and Bennett (1992) discuss this problem in detail.

The empirical orthogonal functions as employed by Fukumori and Wunsch (1991) were based directly upon the SVD, using the spatial dimensions. They define a hydrographic station as a column vector of all the properties measured at a station in sequence. That is, let $\boldsymbol{\theta}_i$ be the vector of potential temperatures at station i; let $\mathbf{S}_i$, $\mathbf{N}_i$, etc., be the corresponding $n \times 1$ vector of salinities, nitrate, etc., at that station. Form an extended column vector,

$$\mathbf{s}_i = [[\boldsymbol{\theta}_i]^{\mathrm{T}}, \ [\mathbf{S}_i]^{\mathrm{T}}, \ [\mathbf{N}_i]^{\mathrm{T}}, \ldots]^{\mathrm{T}}$$

(i.e., first n elements are temperatures, second n are salinities, etc.), and make up a matrix of all available hydrographic stations (p of them):

$$\mathbf{M}_2 = \{\mathbf{s}_1, \ \mathbf{s}_2, \ldots, \mathbf{s}_p\}, \tag{6.44}$$

which is a projection of the $N \times n \times p$-dimensional space onto two dimensions. If the SVD is applied to $\mathbf{M}_2$, then, as discussed in Chapter 3, the singular values and $\mathbf{u}_i$, $\mathbf{v}_i$ produce the most efficient possible representation of the matrix. In this context, then either the $\mathbf{u}_i$ or $\mathbf{v}_i$ are empirical orthogonal functions, or principle components. Fukumori and Wunsch (1991) called these the "form-2" modes. If the matrix is ordered instead, as

$$\mathbf{M}_1 = \{\{\boldsymbol{\theta}_i\}, \ \{\mathbf{S}_i\}, \ \{\mathbf{N}_i\} \ldots\}, \tag{6.45}$$

the first n columns of the matrix are the temperatures, the second n are the salinities, etc. These are "form-1" modes. The representation is not unique – it cannot be because a higher-dimensional space is being projected onto a two-dimensional one. More generally, the column and row indices can be time, space, or any ordering or bookkeeping variable. As with the matrices dealt with in earlier chapters, one commonly wishes to introduce various column and row weights before applying the SVD to $\mathbf{M}_1$ or $\mathbf{M}_2$.

6.9 Non-linear problems

The equations of fluid dynamics are non-linear, and sometimes even the most determined linearization cannot adequately represent the physics under consideration. In the geostrophic box models, for example, the coefficient matrix, $\mathbf{E}$, is constructed from observations, and therefore contains a variety of errors, including internal wave noise, ship navigational errors, large-scale time variations, and others. Whether these, and the non-linearities owing to the dynamical simplifications, are truly negligible can only be determined on a case-by-case basis. That the system is non-linear

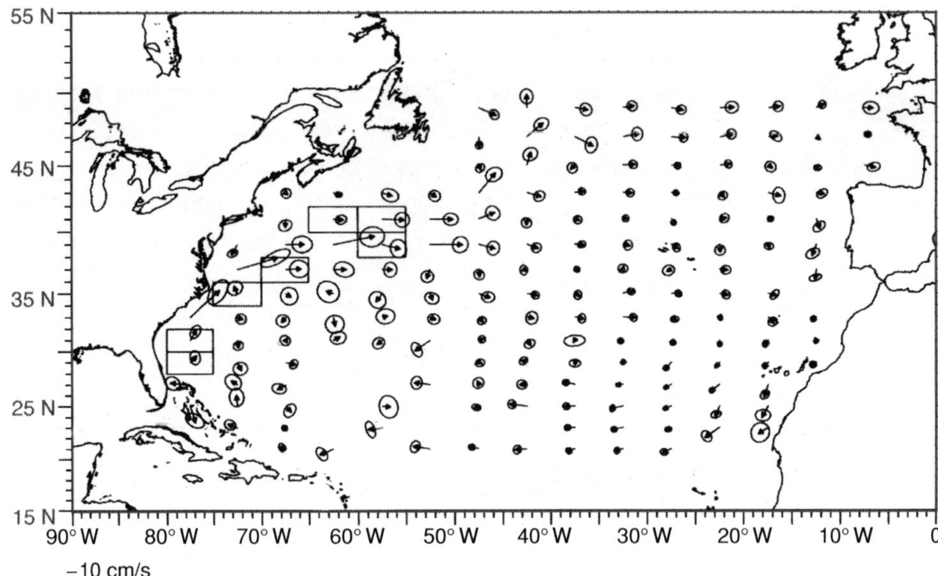

Figure 6.32 Gridded velocity data between about 0 and 100 m depth from neutrally buoyant floats along with an estimated one-standard-deviation error bar, as used in the inverse calculation of Mercier *et al.* (1993). (Adapted from: Mercier *et al.*, 1993)

is implicit in Needler's formula for the absolute velocity. A full representation of the problem of determining the circulation would permit estimation of the density field, simultaneously with the flow and mixing coefficients. This problem has been treated as a non-linear one by Mercier *et al.* (1993) for the North Atlantic time-mean flow.

They first constructed the density field, ρ, as a function of depth and position, and formed the matrix $\mathbf{M} = \{\rho(z, \mathbf{r}_i)\}$, where $\mathbf{r}_i$ is an arbitrary ordering of the station latitudes and longitudes. $\mathbf{M}$ was then approximately represented through Eq. (3.54), the Eckart–Young–Mirsky theorem, in terms of its first ten singular vectors as described in the last section. Each of the EOFs was mapped onto a regular grid using objective mapping and a plausible spatial covariance function. The mapped coefficients of the singular vectors become subject to adjustment. From the mapped density field, one can, using a finite difference scheme, compute the thermal wind (Eqs. (6.15) and (6.16)), and the vertical velocity, w^*, from the thermal wind velocities plus the unknown reference level velocities. In addition to the hydrographic fields, Mercier *et al.* (1993) employed neutrally buoyant float velocities at a variety of levels in the North Atlantic; an example of the gridded values is shown in Fig. 6.32. They wrote an objective function equivalent to

$$J = (\mathbf{d} - \mathbf{d}_0)^{\mathrm{T}} \mathbf{W}_{dd}^{-1} (\mathbf{d} - \mathbf{d}_0) + \mathbf{b}^{\mathrm{T}} \mathbf{W}_{RR}^{-1} \mathbf{b} + (\mathbf{v} - \mathbf{v}_f)^{\mathrm{T}} \mathbf{W}_{ff}^{-1} (\mathbf{v} - \mathbf{v}_f) + \mathbf{m} \mathbf{W}_{mm}^{-1} \mathbf{m},$$
$$(6.46)$$

where $\mathbf{d}$ is the vector of coefficients of the hydrographic singular vectors for all vectors at all locations, with initial estimate $\mathbf{d}_0$; $\mathbf{b}$ is the reference level velocity from the thermal wind equations (assumed, initially, as zero); $\mathbf{v}_f$ are the float velocities as gridded; $\mathbf{v}$ is the geostrophic velocity at that depth and location, expressed in terms of the thermal wind, which is a function of $\mathbf{d}$ and the reference level velocity; and $\mathbf{m}$ is the misfit to the dynamical constraint of geostrophic balance and to any other requisite balance (e.g., that surface velocities include an estimated Ekman transport). The $\mathbf{W}_{ii}$ are the usual weight factors, here all diagonal.

The next step is to seek values of $\mathbf{d}, \mathbf{v}_R$ that render all terms of J in Eq. (6.46) statistically indistinguishable from 1. $\mathbf{d}$ occurs, implicitly, in $\mathbf{m}$. Because the density is being modified, the problem is fully non-linear. To proceed, Mercier *et al.* (1993) used a quasi-Newton descent method suggested by Tarantola and Valette (1982), but any good minimization algorithm would do (see, e.g., Press *et al.*, 1996). They succeeded in finding a generally acceptable solution (there are inevitable discrepancies). The resulting flow, away from the Ekman layer, is consistent with all the data used and with geostrophic hydrostatic balance.

A somewhat different non-linear hydrographic inversion was described by Wunsch (1994) for a generalization of the β-spiral problem discussed above. In the β-spiral, Eq. (6.43) was solved in a simple overdetermined, least-squares form, leaving residuals in each layer of the water column. Suppose, however, that small adjustments to the observed density field could be found such that the residuals were reduced to zero at all depths. The result would then be an estimate not only of the velocity field but of a density field that was completely consistent with it, in the sense that Eq. (6.6)–(6.10) would be exactly satisfied.

To proceed, Wunsch (1994) wrote the β-spiral equation using the pressure field rather than isopycnal layer depths, and in spherical coordinates. Put

$$P \equiv \frac{p}{\rho_0} = -g \int_{z_0}^{z} \frac{\rho}{\rho_0} \mathrm{d}z + P_0(\phi, \lambda), \tag{6.47}$$

where P_0 is an (unknown) reference pressure, and then Eqs. (6.6)–(6.10) can be written as

$$\frac{\partial P}{\partial \phi} \left(\frac{\partial^3 P}{\partial z^3} \frac{\partial^2 P}{\partial \lambda \partial z} - \frac{\partial^2 P}{\partial z^2} \frac{\partial^3 P}{\partial \lambda \partial z^2} \right)$$

$$+ \frac{\partial P}{\partial \lambda} \left(\frac{\partial^2 P}{\partial z^2} \frac{\partial^3 P}{\partial \phi \partial z^2} - \frac{\partial^3 P}{\partial z^3} \frac{\partial^2 P}{\partial \phi \partial z} + \cot \phi \left(\frac{\partial^2 P}{\partial z^2} \right)^2 \right) \tag{6.48}$$

$$= K 2\Omega a^2 \sin \phi \cos \phi \left(\frac{\partial^2 P}{\partial z^2} \frac{\partial^4 P}{\partial z^4} - \left(\frac{\partial^3 P}{\partial z^3} \right)^2 \right),$$

where ϕ is latitude, λ is longitude, and K is a vertical-eddy coefficient, assumed to be constant in the vertical in deriving (6.48). Equation (6.48) was derived by Needler (1967) and is sometimes known, along with $K = 0$, as the "P-equation." Its correspondence with the β-spiral equation was not noticed for a long time.

Substitution of an observed density field and a plausible initial reference pressure, $\tilde{P}_0(0, \phi, \lambda)$, into (6.48) results in large values for K, which is here regarded as a model/data residual rather than as a manifestation of real mixing. Choosing P_0 is the same as having to choose an initial reference level in conventional dynamic computations. The problem was converted to an optimization one by asserting that K appears finite only because the initial reference-level velocity of zero is not correct and because there are also errors in the observed density field, called $\tilde{\rho}(0, \phi, \lambda)$.

Both $\tilde{P}_0$ and $\tilde{\rho}$ were expanded in a three-dimensional set of Chebyschev polynomials, whose coefficients a_{nm}, α_{nml} respectively, become the parameters $\mathbf{x}$. The objective function, J, was formed by computing the sum of squares of K calculated on a three-dimensional grid:

$$J = \sum_{ijk}(\tilde{K}_{ijk} \sin\phi_{ij} \cos\phi_{ij})^2 + R_r^{-2} \sum_{ijk}(\tilde{\rho}_{ijk}(1) - \tilde{\rho}_{ijk}(0))^2. \tag{6.49}$$

The indices, ijk, refer to the horizontal and vertical grid positions where the expression (6.49) is evaluated after substitution of the Chebyschev expansion. The terms in $\tilde{\rho}$ penalize deviations from the initial estimates with a weight given by R_r. The index, 1, in the argument denotes the modified estimate.

If the coefficients of $\tilde{\rho}$ are held fixed at their initial values as determined from the observations, then only the reference pressure is being optimized. The normal equations in that case are linear, and one is simply solving the ordinary β-spiral problem. The search for an improved density/pressure estimate was carried out using a so-called Levenburg–Marquardt method (e.g., Gill *et al.*, 1986), and a minimum was found. The uncertainty of the final solution was then estimated using the Hessian evaluated in its vicinity. The final residuals are not zero, but are sufficiently small that the solution was deemed acceptable. Because slight changes in the prior density field are capable of producing a solution that implies conservation of potential vorticity, density, etc., within uncertainty limits, the result suggests that estimates of vertical mixing coefficients in the ocean differing from zero may well be nothing but artifacts of sampling errors. Davis (1994) draws a similar conclusion from entirely different reasoning.

Notes

1 *The Ocean Circulation Inverse Problem* (Wunsch, 1996).
2 Brief histories of the use of inverse methods for determing the oceanic general circulation can be found in OCIP (Wunsch, 1996), and in Wunsch (1989).

3 See for example, Pedlosky (1987).

4 The geostrophic flow is also largely due to the wind driving, but much more indirectly than is the Ekman component.

5 We will also ignore, for simplicity, the estimated 0.8×10^9 kg/s flowing southward across all sections and representing the inflow from the Arctic (and ultimately from the North Pacific). Including it raises no new issues.

6 In a more exact configuration, one would use so-called neutral surfaces (Jackett and McDougall, 1997), but the practical difference in a region such as this is slight.

7 Attaching pet names to water types is not necessary, but it is a tradition and gives many oceanographers a sense of intimacy with their subject, similar to the way physicists define quarks as having charm, and it does provide some geographic context.

7

Applications to time-dependent fluid problems

So-called "inverse methods," a terminology sometimes restricted to steady-state situations, and "state estimation," are really one and the same. Either label can defensibly be applied to both time-dependent or steady physical problems, with the choice among the differing methods being mainly dictated by problem size and convenience. Thus, in principle, observations and their uncertainties could be combined, e.g., with a governing discretized partial differential equation such as the familiar advection/diffusion equation,

$$\frac{\partial C}{\partial t} + \mathbf{v} \cdot \boldsymbol{\nabla} \mathbf{C} - \nabla \left(\mathbf{K} \boldsymbol{\nabla} C \right) = m_C, \tag{7.1}$$

with known $\mathbf{v}$, $\mathbf{K}$, plus boundary/initial condition information, some estimate of the accuracy of the difference equation, the entire set put into the canonical form, and the collection solved for C, and any other imperfectly known elements. The principle is clear, but for large spatial domains and long time intervals, the practicalities of calculating, storing, and using the enormous resulting matrices drive one, even for completely linear problems, toward numerical algorithms that require less storage or computer time, or do not burden one with the possibility of a complete resolution analysis, or all of these things.

The methods described in this book for solving time-dependent least-squares, or inverse problems, or state estimation problems, have focussed on so-called sequential methods (Kalman filter plus the RTS smoother algorithm), and so-called whole-domain or iterative methods involving Lagrange multipliers (adjoint method). These approaches are not exhaustive; in particular, Monte Carlo methods (genetic algorithms, simulated annealing, etc.) hold considerable promise. If we restrict ourselves to the methods we have treated, however, it can be recognized that they are widely discussed and used in the general context of "control theory" with both pure mathematical and highly practical engineering limits. What distinguishes large-scale fluid problems from those encountered in most of the control literature is primarily

340

the enormous dimensionality of the fluid problem (the "curse of dimensionality," to use R. Bellman's colorful phrase). That one can, in principle, find minima or stationary points of a convex objective function, is hardly in doubt (enough computer power overcomes many numerical difficulties). At the time the first edition of this book was written, the practical application of the ideas in Chapters 3 and 4 to time-dependent geophysical flow problems was primarily one of conjecture: a few investigators, with the computers available then, set out to explore the problems that would arise. Their results suggested that the methods could be used on problems of practical concern. But few results, other than symbolic ones, were available. In the intervening decade, however, computer power continued its inexorable growth, better numerical methods were developed, models improved, and vast new numbers of observations appeared, to the extent that we can now focus on describing actual, scientifically useful results, rather than speculating on the possibility of their existence.

Important global fluid inverse problems, however, nonetheless still outstrip the largest existing, or even foreseeable, computers. Approximations to the estimation algorithms thus remain extremely important, and will be so indefinitely. After examining some of the actual state estimates made, we will describe some of the more important methods being employed to reduce computational load. So-called identical twin results are not discussed here. Numerical experiments with artificial, often noise-free or with artificially simplified noise, data are a necessary step toward testing any estimation method. But as described in much of the published literature, they become an end in themselves, reducing to uninteresting tests of least-squares, a method whose efficacy is hardly in doubt.

7.1 Time-dependent tracers

A two-dimensional version of Eq. (6.1) was used by Kelly (1989) to determine the flow field, and is a regional, comparatively simple example. She expanded satellite measurements of surface temperature $T(\mathbf{r}, t)$ in a set of basis functions (sines and cosines) and minimized the data misfit in time and space, solving for the horizontal flow $\mathbf{v}$. An example of her results is shown in Fig. 7.1 for a region off the coast of California. The solution nullspace is important, and Kelly (1989) discusses the reliability of the estimates.

Observed transient tracers and the extent to which they could be used for determining the flow field were described by Mémery and Wunsch (1990). A very general formalism was described by Haine and Hall (2002), but the method is so demanding of data that the application appears primarily to diagnostics in general circulation models – where 100% "data" coverage is available.

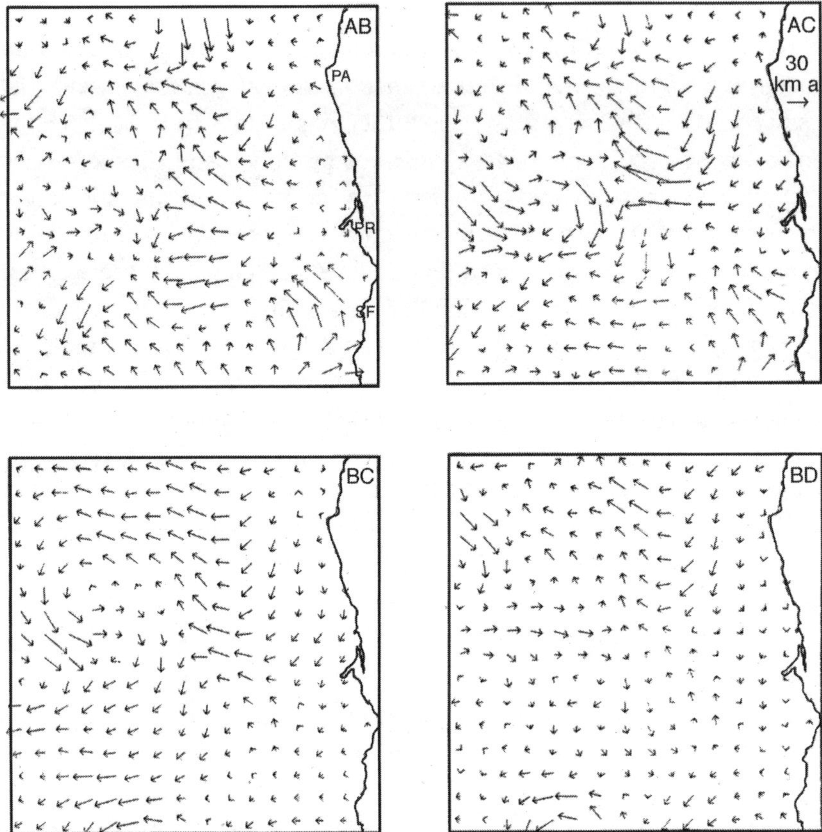

Figure 7.1 Flow fields inferred from the difference of sea surface temperatures at four times, subject to Eq. (7.1). The region depicted is off San Francisco, California. (Source: Kelly, 1989)

7.2 Global ocean states by Lagrange multiplier methods

As has been discussed in Chapter 4, the computational advantage of the Lagrange multiplier methods, as compared to the sequential ones, is the absence of a requirement for producing the error covariances for the state and control vectors as part of the computation. Because the Lagrange multiplier method finds the stationary value of the objective function, J, iteratively, over the entire time-span of observations, no average of two independent state estimates is required. Without the covariances, one can greatly reduce both the computational load and storage requirements (the storage requirement for the RTS smoother is formidable in large problems). Of course, as pointed out earlier, a solution without error covariances is likely to be of limited utility; nonetheless, we can postpone confronting the problem of finding them.

As an example of what is now possible, again focus on the ocean, and, in particular, the problem of determining the three-dimensional, time-varying circulation of the ocean over various durations in the period 1993–2002 (Stammer *et al.*, 2002, 2003, 2004). The model, thought to realistically describe the oceanic flow, is based upon the Navier–Stokes equations for a thin spherical shell using a fully realistic continental geometry and topography. For those interested, the governing system of underlying numerical equations are discussed in textbooks (e.g., Haidvogel and Beckmann, 1999; Griffies, 2004). It suffices here to define the state vector on a model grid (i, j, k) at time $t = n\Delta t$. The state vector consists of three components of velocity, $(u(i, j, k, t), v(i, j, k, t), w(i, j, k, t))$, a hydrostatic pressure $(p(i, j, t))$, temperature $(T(i, j, k, t))$, and salinity $(S(i, j, k, t))$ (some of the variables are defined at positions displaced by $1/2$ of the grid spacing from the others, a distinction we ignore for notational tidiness). A numerical approximation is written (many thousands of lines of Fortran code) to time-step this model approximation, which we abbreviate in the standard form,

$$\mathbf{x}(t) = \mathcal{L}(\mathbf{x}(t-1), \mathbf{Bq}(t-1), \mathbf{\Gamma u}(t-1), \mathbf{p}, t-1, \mathbf{r}), \qquad (7.2)$$

with t discrete, and taking $\Delta t = 1$. Here, as in Chapter 4, $\mathbf{q}(t)$ are the known boundary conditions, sources/sinks (if any), and the $\mathbf{u}(t)$ are the unknown boundary conditions, sources/sinks (controls). The $\mathbf{u}(t)$ generally represent adjustments to the boundary conditions (taken up momentarily), as well as, e.g., corrections to the estimated initial conditions $\tilde{\mathbf{x}}(0)$, and model errors. $\mathbf{r}$ is a three-dimensional position vector $(i\Delta x, j\Delta y, k\Delta z)$. $\Delta x, \Delta y, \Delta z$ are the spatial grid increments (not to be confused with the state or observation vectors, $\mathbf{x}, \mathbf{y}$). The grid increments do not need to be constant in the domain, but again for notational simplicity we will ignore that detail. $\mathbf{p}$ is a vector of internal model parameters, also regarded as unknown, and which could be included in $\mathbf{u}$. These are adjustments to, e.g., empirical friction or diffusion coefficients (eddy-coefficients).

In conventional ocean (forward) modeling, the system is driven by atmospheric winds, and estimated exchanges of heat (enthalpy) and fresh water between the ocean and atmosphere. These appear in the model as numerical approximations, for example, the wind stress might be imposed as a boundary condition,

$$A_v \left[\frac{\Delta u(i, j, k = 0, t)}{\Delta z}, \frac{\Delta v(i, j, k = 0, t)}{\Delta z} \right]\Bigg|_{z=0} = [\tau_x(i, j, t), \tau_y(i, j, t)].$$

Here $\tau = (\tau_x, \tau_y)$ is the vector wind stress evaluated nominally at the sea surface $(z = k\Delta z = 0)$, and A_v is an empirical coefficient. τ and the related fluxes of heat and moisture are taken as given, perfectly. Some initial conditions $\tilde{\mathbf{x}}(0)$ (perhaps, but not necessarily, a state of rest) are used, and Eq. (7.2) is time-stepped forward subject to the time-varying meteorological fields. In the most realistic simulations, wind and air – sea exchanges are taken from meteorological analyses on global grids and available, typically, two or four times per day and then linearly interpolated to the model time-step. As formulated, the resulting problem is well-posed in the sense that enough information is available to carry out the time-stepping of Eq. (7.2). The system can be unstable – either because of numerical approximations or, more interestingly, because the fluid flow is physically unstable. Many such simulations exist in the published oceanographic literature (and in the even larger analogous meteorological literature).

From the oceanographer's point of view, the simulation is only the beginning: one anticipates, and easily confirms, that the resulting model state does not agree (sometimes well, sometimes terribly), with observations of the actual ocean. For someone trying to understand the ocean, the problem then becomes one of (a) understanding why the model state disagrees with what was observed, and (b) improving the model state so that, within error estimates, the model comes to consistency with the observations and one can begin to understand the physics. In the spirit of this book, the latter would then be regarded as the best estimate one could make of the time-evolving ocean state. With such a best estimate one can, as in the previous chapter, calculate property fluxes (heat, carbon, etc.) and analyze the model in detail to understand why the system behaves as it does.

The specific example we discuss is based upon Stammer *et al.* (2002, 2003, 2004), but displaying later estimates, believed to be improved over those previously published. State estimation is open-ended, as models can always be improved, and error covariances made more accurate. The specific examples are based upon a numerical representation of Eq. (7.2) written in a finite difference form on a grid of $2°$ in both latitude and longitude, and with 23 layers in the vertical. The total number of grid points (the number of equations in Eq. 7.2) was thus approximately 360 000. With three components of velocity, pressure, temperature, and salinity at each grid point making up the state vector, the dimension of $\mathbf{x}\,(t)$ is approximately 5.3 million. The model was time-stepped at intervals $\Delta t = 1$ hour. The total state vector in the estimates of Stammer *et al.* (2002, 2003, 2004), $[\mathbf{x}(0)^{\mathrm{T}}, \mathbf{x}(1)^{\mathrm{T}}, \ldots, \mathbf{x}(t_f)^{\mathrm{T}}]^{\mathrm{T}}$, had dimensions over a decade of about 10^{11} elements. An error covariance matrix for such a result would be a forbidding object to compute, store, or understand. (In the interim, state estimates of even larger dimension have been carried out – see Köhl *et al.*, 2006; Wunsch and Heimbach, 2006.)

A great variety of oceanic observations were employed. A proper discussion of these data would take us far afield into oceanography. In highly summary form, the data consisted of, among others:

1. Sea surface elevation observations from altimetry (approximately one/second at a succession of points) over the entire period. These were employed as two separate data sets, one defined by the time mean at each point over the time interval, and the second being the deviation from that mean. (The reason for the separation concerns the error covariances, taken up below.)
2. Observations of sea surface temperature, composited as monthly means, over the entire ocean.
3. Time average climatologies of temperature and salinity at each grid point, i, j, k. That is, the computational-duration temperature, etc., is constrained in its average to observations, as in

$$\sum_j E_{ij} \left[\frac{1}{t_f + 1} \sum_{t=0}^{t_f} x_j(t) \right] + n_i = y_i, \tag{7.3}$$

where E_{ij} selects the variables and grid points being averaged.
4. "Synoptic" temperatures and salinities from shipboard instantaneous observations.
5. Various constraints from observed mass transports of major currents.
6. Model-trend suppression by requiring temperatures and salinities at the beginning, $t = 0$, and at the end, $t = t_f$, to be within an acceptable range (where the meaning of "acceptable" requires study of observed large-scale oceanic temperature changes).

The known controls, $\mathbf{q}(t)$, were also based upon observations, and separating them from state vector observations is an arbitrary choice. They included:

1. twice-daily estimates of gridded vector wind stress, τ;
2. daily estimates of gridded air–sea enthalpy (heat) exchange, H_E;
3. daily estimates of air–sea moisture exchange, H_f.

The control vector $\mathbf{u}(t)$ included:

1. adjustments to the vector wind, $\Delta\tau$;
2. adjustments to the enthalpy flux, ΔH_E;
3. adjustments to the moisture flux, ΔH_f.

The control matrix, Γ, vanished everywhere except at the sea surface, where it was unity at each boundary grid point. That is, each of the controls $\Delta\tau$, etc., could vary independently at each surface grid point. Adjustments to the initial conditions, $\tilde{\mathbf{x}}(0)$, can be regarded as part of the controls; here we retain them as part of the state vector.

A count of the observations produces approximately 6.3×10^6 equations of the form $\mathbf{E}(t)\mathbf{x}(t) + \mathbf{n}(t) = \mathbf{y}(t)$. These were used to form an objective function, J, as in Eq. (4.41). The control terms, $\mathbf{u}(t)\mathbf{Q}(t)^{-1}\mathbf{u}(t)$, provide another 8.1×10^7 elements of J.

Weights must be specified for every term of the objective function (that is, for every term of Eq. (4.41), including expressions such as Eq. (7.3)). Unsurprisingly, obtaining realistic values is a major undertaking. In practice, all such matrices were diagonal, with one exception, with information about the covariance of the observational errors (including that of the control terms) being unavailable.

J should include additional control-like terms representing the model error. (Errors in the wind-field and other boundary conditions can be labeled arbitrarily as being part of the model or data or controls; here we generally regard them as control variables.) But no conventional fluid-flow model has ever been produced with a statement of the quantitative skill expected in reproducing a quantity of interest: for example, the skill with which the temperature at point $i, j, k, n\Delta t$ would be determined if the boundary and initial conditions were perfect. Deviations from perfection would arise from limited spatial and temporal resolution, mis-parameterized process, missing physics, incorrect bottom topography, etc. Such errors are unlikely to be Gaussian, and are in general not known. But they ought to be included in J, J'. In practice, what was done was to permit the model to misfit the data in some regions by much more than the formal data error would allow. Such regions include, for example, the equator, and the areas of the intense boundary currents, where model resolution is expected, a priori, to be inadequate to reproduce what is observed. One is redefining, empirically, the data error to include structures the model cannot reproduce. This situation is not very satisfactory, and a major goal has to be to find ways of formally specifying model error. For the time being, one must regard the solution as being one of least-squares – curve-fitting, rather than as a true minimum variance solution.

The very large number of terms in J, and of the state vector dimension, suggest the practical difficulties facing anyone attempting rigorous state estimation with global-scale fluids. Fortunately, modern computing power is rising to the challenge.

These large dimensions, however, dictate the use of Lagrange multipliers in order to avoid the need to compute even approximate error covariances. Formally then, the model (7.2) was appended to J as in Eq. (4.97) to form a new objective function, J'. The dimension of the Lagrange multipliers $\boldsymbol{\mu}(t)$ is equal to the number of model equations over the whole spatial domain, or 2×10^6 at each time-step or 2×10^{10} over the whole time domain. To find the stationary point of J' (at $\tilde{\mathbf{x}}(t), \tilde{\mathbf{u}}(t), \tilde{\mu}(t)$) the iterative linearized search process described on p. 241, was used. Note, in particular, that the $(\partial \mathbf{L}/\mathbf{x}(t))^{\mathrm{T}}$ matrix used in the equivalent normal equations corresponding to the stationary point ((4.168) is not available. It was instead computed implicitly ("on the fly") as described in Chapter 4, using an automatic differentiation tool. The general procedure employed to determine the best-estimate state followed the iterative scheme outlined there, iterating through a series of trial solutions, $\tilde{\mathbf{x}}^{(i)}(t), \tilde{\mu}^{(i)}(t), \tilde{\mathbf{u}}^{(i)}(t)$, until convergence was obtained (more precisely, until the misfit measured in J was deemed acceptable.) In particular, $\tilde{\mu}^{(i)}(t)$ was used to define the directions of the line search to minimize J. Hundreds of iterations were required before J reached an acceptable level, and its individual

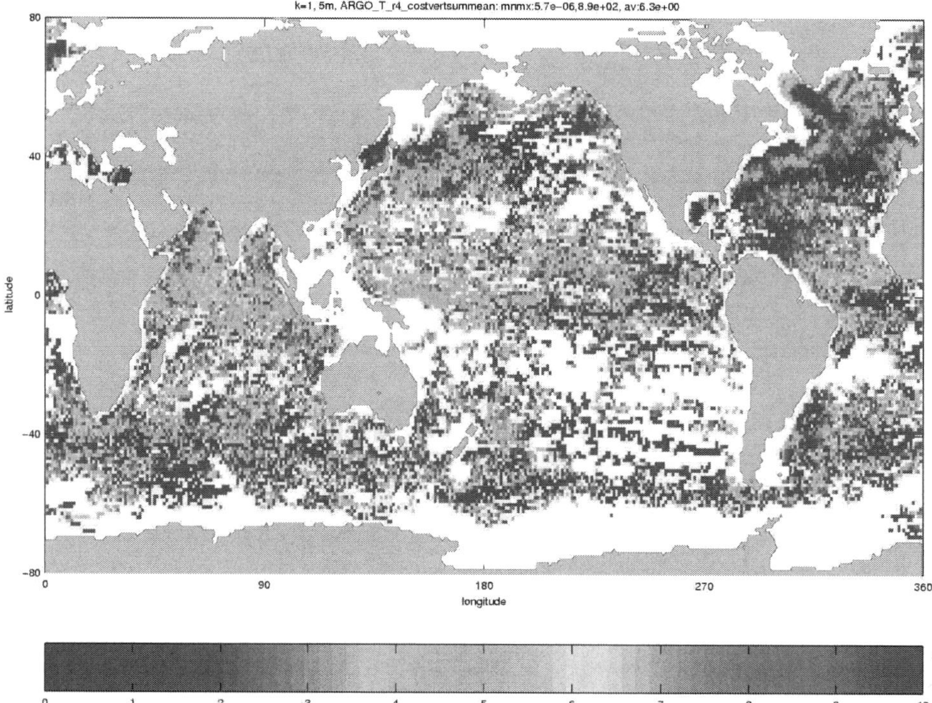

Figure 7.2 Misfit to so-called ARGO float temperature profiles. These instruments produce a vertical profile of temperature in the ocean above about 2000 m at pre-determined intervals (of order 10 days). The misfits shown here are weighted by the estimated errors in both model and data, and then averaged over the entire water column. (See color figs.)

terms were sufficiently close to white noise that the solution appeared to make sense. Non-linear optimization has to be regarded as in part an art – as descent algorithms can stall for many numerical reasons or otherwise fail.

Only a few representative results of these computations are displayed here. Much of the information about the results, and the focus of the scientific effort, once the model has been deemed acceptable, lies with the model-data misfits. These misfits reflect the estimated errors in the data and in the ability of the model to reproduce the data – if the latter were perfect. An example is shown in Fig. 7.2 for vertical profiles of temperature over a period of about eight years ending in 2004. The figure is not untypical, showing near-global coverage (most of these data, however, exist only towards the end of the 13-year period of estimation), which is a generally acceptable order of magnitude, but also patterns of larger misfit that ultimately have to be understood. This calculation is ongoing, and some of the misfits are likely to disappear as the optimization proceeds.

Figures 7.3 and 7.4 display the time average of part of the control vector, $\langle \tilde{\mathbf{u}}(t) \rangle$, as determined by the optimization at one stage (Stammer *et al.*, 2002). A

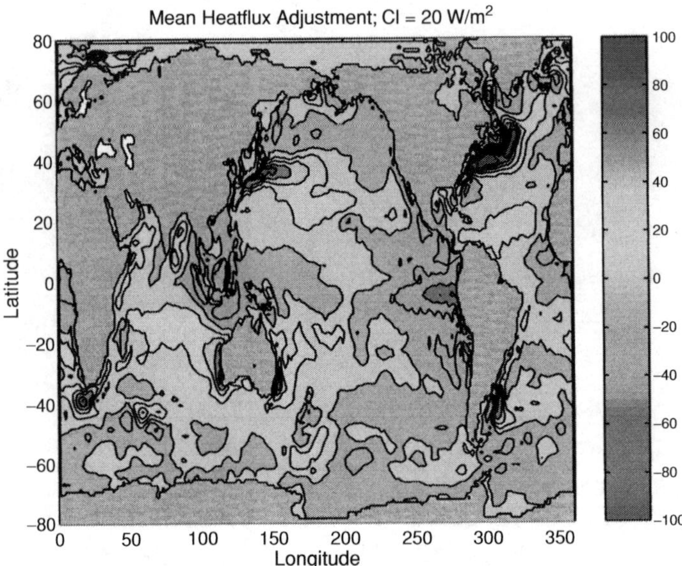

Figure 7.3 Mean changes in (a) net surface heat exchange (W/m^2), determined from six years of meteorological estimates and an ocean model. (See color figs.) (Source: Stammer *et al.*, 2002)

complicated spatial pattern emerges – one that is rationalizable in terms of known problems with the meteorological analyses that produced the "known" controls $\mathbf{q}(t)$ to which these are adjustments. None of the values displayed violates general understanding of the accuracy of the meteorological fields. On the other hand, the comments made above about model errors have to be kept in mind, and the very large values of $\langle \tilde{\mathbf{u}}(t) \rangle$ inferred in some of those regions should be viewed with suspicion.

Some of the flavor of the state vector can be seen in Fig. 7.5, where mean velocities $\langle \tilde{u}(t) \rangle$, $\langle \tilde{v}(t) \rangle$ are displayed for one model depth ($\tilde{u}(t)$ is the scalar velocity component, not the control vector). The reader is reminded that the spatial resolution of this model is $2°$ of latitude and longitude, and physical arguments exist that it would need to approach $1/16°$ over much of the ocean to be adequately realistic. Dimension remains a major challenge. Figure 7.6 gives another view of the time average flow field, in the vertical dimension, across $26°$ N in the North Atlantic, from a later calculation with $1°$ spatial resolution.

The full flavor of model time dependence is best seen in animations of the results that are available on the web. To show something of the great temporal variability in the model, Fig. 7.7 displays the meridional flux of heat across $21°$ S in the South Atlantic, showing values that can actually reverse (to the south), and vary very rapidly from day-to-day.

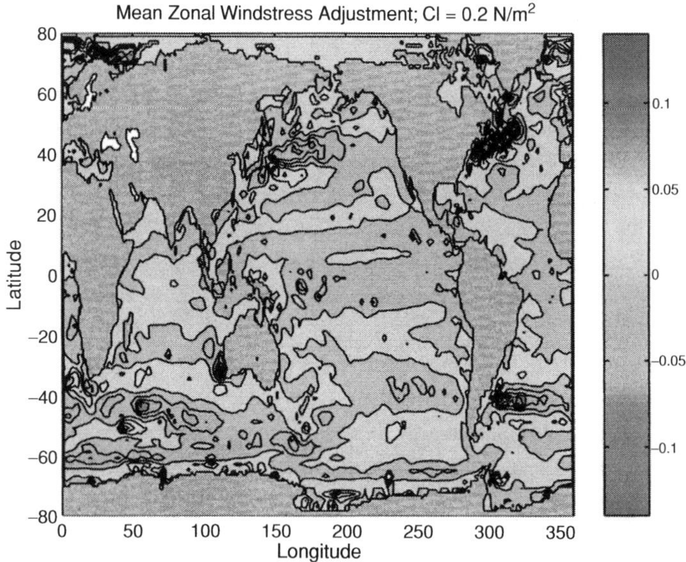

Figure 7.4 Adjusted wind stress (part of the control vector) in N/m^2. (See color figs.) (Source: Stammer *et al.*, 2002)

Because the model calculates the meridional transport of temperature at every grid point at every time-step, it can be integrated zonally and in time to produce the curves displayed in Fig. 7.8. These are compared directly with the values inferred from the static calculation described in Chapter 6. An explicit error bar is available for the latter, as shown, but not for the former. These two types of calculation are clearly two sides of the same coin – one taking explicit account of the temporal variation, the other treating the system as momentarily static.

In the immediate future, one can anticipate exploitation and extension of this type of state estimation, as the approximations made in building numerical models come under increasing scrutiny and question. Numerical models representing any of the important physical partial differential equation systems of physics, chemistry, and biology (e.g., Schrödinger, Navier–Stokes, Maxwell, etc.) are necessarily an approximation. Generally speaking, the modeler makes a whole series of choices about simplified representation of both kinematics and dynamics. Some of these simplifications are so commonplace that they are not normally regarded as parameterizations. In the Navier–Stokes equation example, the geometry of the container is commonly smoothed to a model grid spacing and little further thought given to it. But it is clear that flow fields can be greatly influenced by exactly where the boundary is placed and represented (the classical example is the problem of making the Gulf Stream leave the coastline at Cape Hatteras – if done incorrectly, the Gulf Stream will follow the wrong pathway, with disastrous results for heat and other

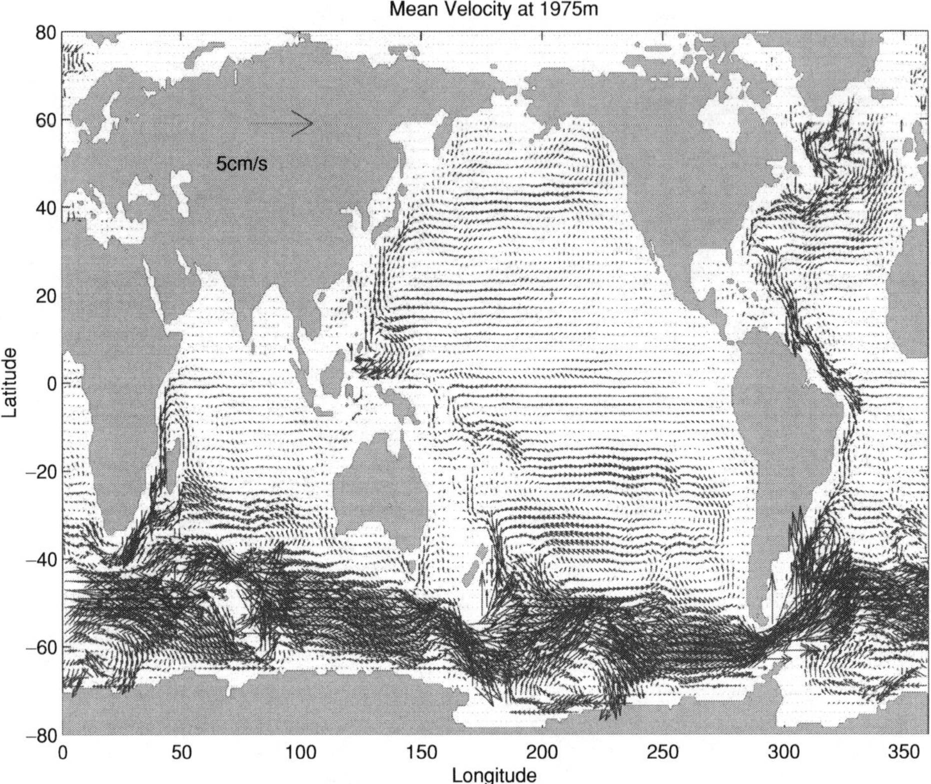

Figure 7.5 Estimated mean velocity field at 1975 m depth, in cm/s, from a six-year estimation period. (Source: Stammer *et al.*, 2002)

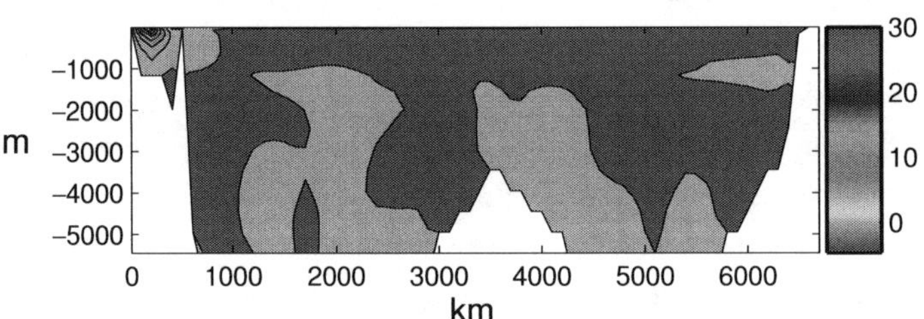

Figure 7.6 Twelve-year mean velocity across the North Atlantic Ocean (cm/s) at 26° N from a 13-year optimization of a 1 degree horizontal resolution (23 vertical layers) general circulation model and a large global data set. Red region is moving southward, remaining regions are all moving northward. Note the absence (recall Chapter 6) of any obvious level-of-no-motion, although there is a region of weak mean meridional flow near 1100 m depth. The northward-flowing Gulf Stream is visible on the extreme western side. (See color figs.)

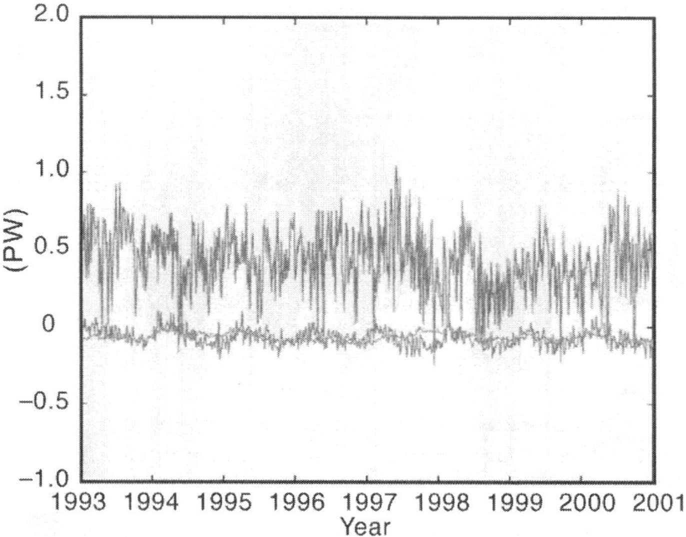

Figure 7.7 Time dependence in components of the meridional flux of heat across $21°$ S, in this case from eight years of analysis. The different colors correspond to different spatial integrals of the heat budget. (See color figs.) (Source: Stammer *et al.*, 2003, Fig. 14)

budgets). The estimation machinery developed here is readily extended to include, e.g., bottom topography in a model (Losch and Heimbach, 2005), and, presumably, such other arbitrary representations as slip/no-slip/hyper-slip boundary conditions that have large elements of ad hoc approximation buried in them.

7.3 Global ocean states by sequential methods

The solution to state estimation problems by filter/smoother methods is highly desirable because one obtains both the estimated state and controls, but also their complete error covariances. In a fluid model, having perhaps 10^7 state elements at each time-step, and a similar order of magnitude of control vector elements, the construction, e.g., in a Kalman filter step, of the error covariance of the state alone requires the equivalent of running the model $10^7 + 1$ times at each time-step. A time-reverse step to, e.g., implement the RTS smoother, involves a similar calculation. This computational load is incurred even if the system is entirely linear. Non-linear systems will usually require many more computations.

With modern computers, dealing with problems with dimensions of thousands or even hundreds of thousands may be feasible. But global scale fluid problems, if tackled even remotely rigorously, are beyond our current capability. Yet the goal of a rigorous (for linear systems) implementation of the sequential algorithms is so

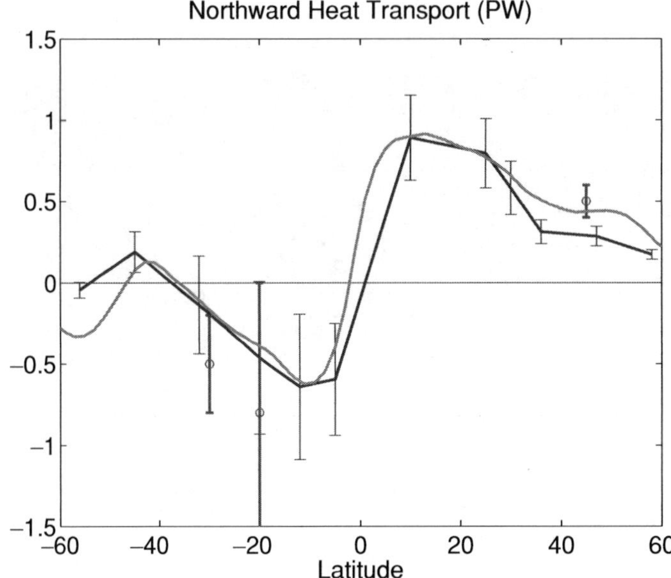

Figure 7.8 The blue curve shows the time and zonally integrated global meridional flux of heat (enthalpy) in the constrained model. The green curve shows the same field, estimated instead by integrating the fluxes through the surface. The two curves differ because heat entering from the surface can be stored rather than necessarily simply transported. Red bars are from the calculation of Ganachaud and Wunsch (2000) described in Chapter 6. Blue bars are the temporal standard deviation of the model heat flux and are an incomplete rendering of the uncertainty of the model result. (See color figs.) (Source: Stammer *et al.*, 2003)

attractive that one is led to explore useful approximations to the full filter/smoother algorithms.

Consider the calculations of Fukumori *et al.* (1999), who used the same numerical model as in the Lagrange multiplier method calculation, but employing the reduced state approximation described in Chapter 5, p. 267. The grid was spaced at $2°$ of longitude and $1°$ of latitude, with 12 vertical layers, and again with the six elements of velocity, pressure, temperature, and salinity for an approximate state vector dimension at each time-step of 3×10^5 over a total time interval of three years, 1992–5. This state vector defines the fine-scale model. Apart from the data used to define the initial temperature and salinity conditions, the only observations were the time-varying sea surface pressure averaged over $2.5°$ sub-satellite arcs.

The general procedure was to both reduce the state vector dimension and to use the Kalman filter/RTS smoother algorithms in their steady-state asymptotes. They reduced the modeled vertical structure to two degrees of freedom by employing a truncated vertical normal mode representation. Dominant horizontal scales of

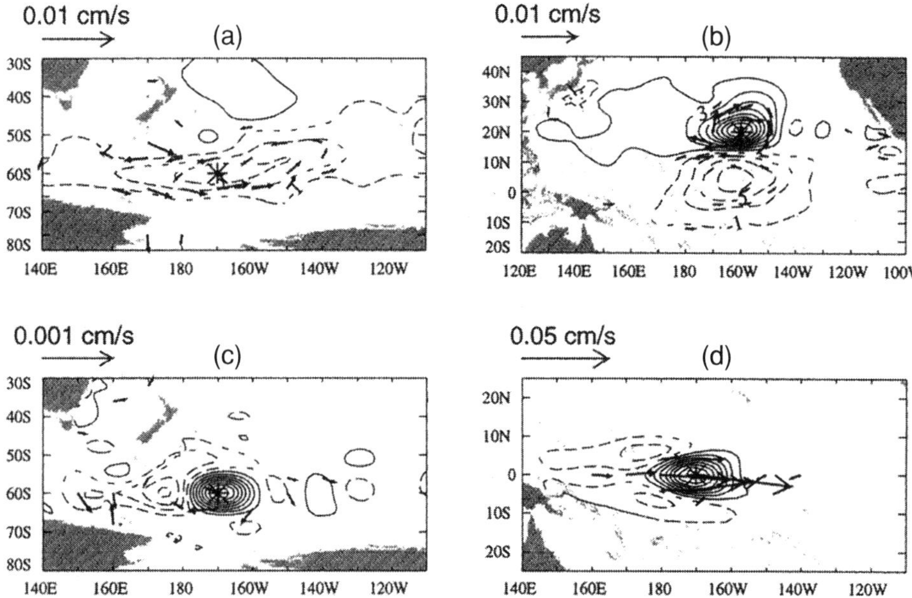

Figure 7.9 Horizontal structure imposed by the approximate Kalman filter corresponding to a 1 cm difference in predicted and observed sea levels at the position of the asterisk: (a) mass transport, (b) stream function, (c) temperature at 1700 m, and (d) temperature at 175 m. (Source: Fukumori *et al.*, 1999)

model variability were identified by running the fine-scale model in a forward computation. The resulting fields were reconstructed using the Eckart–Young–Mirsky representation in the singular vectors, and truncating at $K = 300$, accounting for 99% of the correlation structure of the model (not of the variance – the singular value decomposition was applied to the correlation rather than the covariance matrix). A study of the spatial structure of this reduced representation led to a coarse-scale model defined on a grid that was $10°$ in longitude, $5°$ in latitude. The coarse-to-fine-scale transformation $\mathbf{D}^+$ was carried out using simple objective (but sub-optimal) mapping using a fixed spatial covariance function dependent only upon grid-point separation (neither orientation nor proximity to boundaries was accounted for) and $\mathbf{D}$ was the pseudo-inverse of $\mathbf{D}^+$ (rather than as we defined them in Chapter 5). Special methods were used to assure that lateral boundary conditions were satisfied after a field was mapped from the fine to the coarse grid.

The Kalman gain, computed from the asymptotic steady-state of $\mathbf{P}(t)$, is shown in Fig. 7.9 for temperature elements of the reduced state vector, and for the stream function at different depths.

The effectiveness of the system used is tested in part by its ability to reproduce observations not included as constraints. An example for this computation is shown

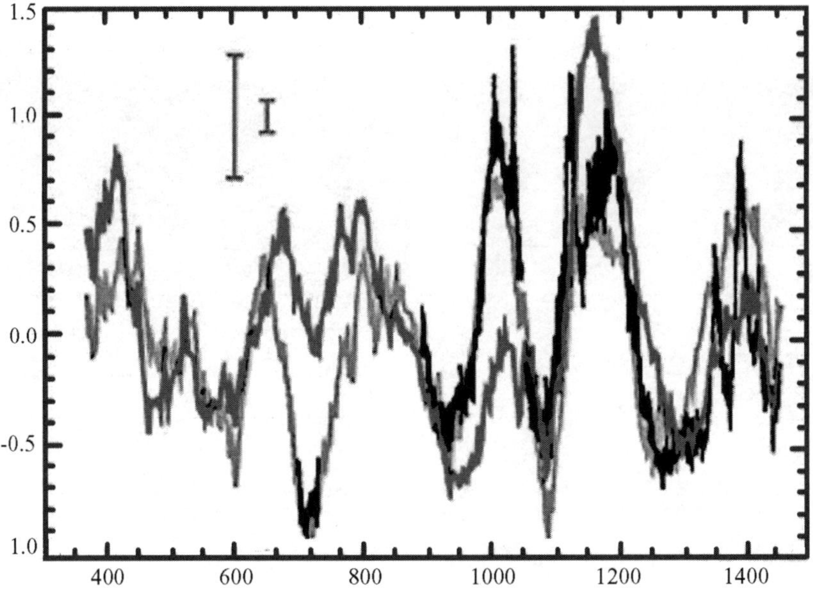

Figure 7.10 Comparison of temperature at 200 m depth at 8° N, 180° E of data (black), unconstrained model (red) and model constrained using altimetric data (blue). See Fukumori *et al.* (1999) for details and further comparisons. An approximate filter/smoother combination was used. (See color figs.) (Source: Fukumori *et al.*, 1999)

in Fig. 7.10 for temperature variations on the equator. Error bars shown represent the estimated value from the asymptotic RTS smoother.

7.4 Miscellaneous approximations and applications

Tomographic applications in the ocean have been discussed in detail by Munk *et al.* (1995). A number of interesting applications of tomographic inversions have been carried out in the years since then (e.g., ATOC Consortium, 1998; Worcester *et al.*, 1999; Worcester and Spindel, 2005). In tomography, very large-scale coverage of the ocean can be obtained almost instantaneously (at the speed of acoustic waves), and hence most calculations have been in the context of an ocean state that is momentarily static.

A very large number of approximate estimates of the ocean using highly reduced forms of Kalman filter has been published. One example is described by Carton *et al.* (2000). In many applications, the need for speed and efficiency outweighs the need for dynamical or statistical rigor. It is not possible to prescribe any method that is uniformly applicable to all problems.

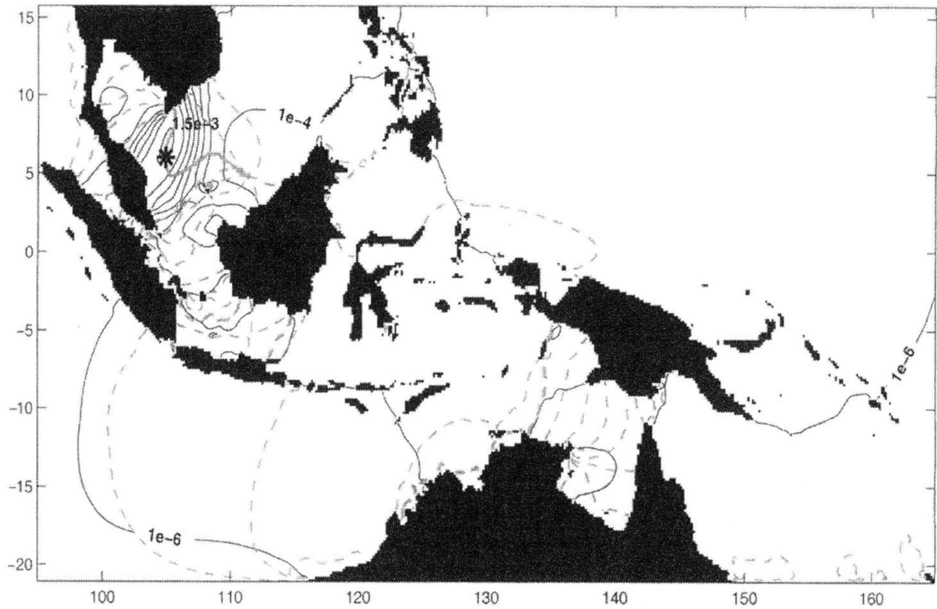

Figure 7.11 Amplitude and phase for a tide modeling representer (Green function), as calculated by Egbert and Erofeeva (2002). Solid contours are the amplitude of influence of a data point at the asterisk. Phases (dashed) correspond to tidal propagation phase changes relative to that point. (Source: Egbert and Erofeeva, 2002)

The mathematical machinery of state estimation has many other applications. Examples are the study of model sensitivity to parameter or other perturbations (e.g., Galanti and Tziperman, 2003; Hill *et al.*, 2004; Li and Wunsch, 2004) by using the adjoint solution directly. Farrell and Moore (1993) have used the SVD of the state transition matrix, **A**, to find the most rapidly growing perturbations in the atmosphere. Marchal (2005) applied the RTS smoother to estimate the variations in atmospheric carbon-14 over the last 10 000 years. The list of applications is now too long to be discussed here.

The representer form of inversion (Chapter 5) has had considerable use, particularly in models that calculate tides and employ altimetric sealevel measurements as constraints. See Egbert and Erofeeva (2002) for a representative discussion. One of their representers is depicted in Fig. 7.11. The methodology has been used in several global calculations.

We have seen the close connection between Lagrange multiplier methods and Green functions. Menemenlis *et al.* (2005) apply Green functions to a variety of parameter estimation problems with a global GCM. Fukumori (2001) discusses this connection, as well as published applications to observability and controllability in ocean models.

7.5 Meteorological applications

This subject was touched on in Chapter 5. Numerical weather prediction (NWP) represents the most highly developed, in a numerical engineering sense, of the methods discussed here, and the terminology "data assimilation" originated there. The main focus of this book has, however, been deliberately placed elsewhere because of the highly specialized nature of the weather forecasting problem. Governmental, aviation, military, agricultural, and wide public interests demand predictions over time spans ranging from minutes (in tornado-prone regions) to a few days, and on a continuous basis. To the extent that one is able to make useful forecasts on any of these timescales, the question of whether either the model or the methodology is completely understood scientifically is secondary. That methods such as extended Kalman filters represent highly desirable goals for weather forecasting has long been known (e.g., Ghil *et al.*, 1981), but NWP has faced such enormous data sets and complex models, that rigorous implementation of the Kalman filter has remained beyond the realm of practice. The smoothing problem has received scant meteorological attention.

A very large number of approximate methods have been used in NWP, with its own textbooks (Daley, 1991; Kalnay, 2003), and review articles continue to appear (e.g., Lorenc, 1986; Ghil and Malanotte-Rizzoli, 1991; Talagrand, 1997). A major issue for anyone attempting to make use of this important literature is the proliferation of ad hoc terminology in the meteorological application. Thus what this book calls the adjoint or method of Lagrange multipliers is widely known as 4DVAR; objective mapping is commonly 3DVAR; and many of the physical terms in fluid mechanics have their own particular meteorological definition (e.g., "mixing ratio"). Most of the meteorological methods can be recognized as variations on methods described here, albeit the development usually begins with the continuous time/space formulation. Chapter 5 discussed some of the approximations invoked, following discretization, including so-called nudging, where the Kalman gain is replaced by a plausible, but guessed, fixed matrix; state reduction; the use of approximate rather than exact adjoints; and many others that have proven of practical use.

For a discussion of the application of methods very similar to those discussed here to the problems of atmospheric trace gas movement, see Enting (2002). The collection of papers by Kasibhatla *et al.* (2000) applies the methods to biogeochemical modeling.

Because meteorological forecasters have recognized the importance of giving their users an idea of reliability, a very large meteorological effort has gone into so-called ensemble forecasting methods, which were briefly discussed in Chapter 5; Kalnay (2003) is a good starting point in what is now a very large literature.

References

Aitchison, J. and Brown, J. A. C. (1957). *The Lognormal Distribution with Special Reference to its Use in Economics.* Cambridge: Cambridge University Press.

Aki, K. and Richards, P. G. (1980). *Quantitative Seismology.* San Francisco: W. H. Freeman, 2 vols.

Anderson, B. D. O. and Moore, J. B. (1979). *Optimal Filtering.* Englewood Cliffs, NJ: Prentice-Hall.

Anthes, R. A. (1974). Data assimilation and initialization of hurricane prediction models. *J. Atm. Sci.*, **31**, 701–19.

Armstrong, M. (1989). *Geostatistics*, 2 vols. Dordrecht: Kluwer.

Arthanari, T. S. and Dodge, Y. (1993). *Mathematical Programming in Statistics.* New York: Wiley.

Arulampalam, M. S., Maskell, S., Gordon, N. and Clapp, T. (2002). A tutorial on particle filters for online nonlinear/non-Gaussian Bayesian tracking. *IEEE Trans. Signal Processing*, **50**, 174–88.

ATOC Consortium (1998). Ocean climate change: comparison of acoustic tomography, satellite altimetry, and modeling. *Science*, **281**, 1327–32.

Barth, N. H. and Wunsch, C. (1990). Oceanographic experiment design by simulated annealing. *J. Phys. Oc.*, **20**, 1249–63.

Bennett, A. F. (1992). *Inverse Methods in Physical Oceanography.* Cambridge: Cambridge University Press.

Bennett, A. F. (2002). *Inverse Modeling of the Ocean and Atmosphere.* Cambridge: Cambridge University Press.

Berkooz, G. P. Holmes and Lumley, J. L. (1993). The proper orthogonal decomposition in the analysis of turbulent flows. *Ann. Revs. Fl. Mech.*, **25**, 539–75.

Bittanti, S., Colaneri, P. and De Nicolao, G. (1991). The periodic Riccati equation. In *The Riccati Equation*, ed. S. Bittanti, A. J. Laub and J. C. Willems. Berlin: Springer-Verlag, pp. 127–62.

Bittanti, S., Laub, A. J. and Willems, J. C., eds (1991). *The Riccati Equation.* New York: Springer.

Box, G. E. P., Jenkins, G. M. and Reinsel, G. C. (1994). *Time Series Analysis. Forecasting and Control.* Englewood Cliffs, NJ: Prentice-Hall.

Bracewell, R. N. (2000). *The Fourier Transform and its Applications*, 3rd edn. Boston: McGraw-Hill.

Bradley, S. P., Hax, A. C. and Magnanti, T. L. (1977). *Applied Mathematical Programming.* Reading: Addison-Wesley.

Bretherton, F. P., Davis, R. E. and Fandry, C. B. (1976). Technique for objective analysis and design of oceanographic experiments applied to MODE-73. *Deep-Sea Res.*, **23**, 559–82.

Brogan, W. L. (1991). *Modern Control Theory*, 3rd edn. Englewood Cliffs, NJ: Prentice-Hall/Quantum.

Brown, R. G. and Hwang, P. Y. C. (1997). *Introduction to Random Signal Analysis and Applied Kalman Filtering: With MATLAB Exercises and Solutions*, 3rd edn. New York: Wiley.

Bryson, A. E., Jr. and Ho, Y.-C. (1975). *Applied Optimal Control*, revised printing. New York: Hemisphere.

Butzer, P. L. and Stens, R. L. (1992). Sampling theory for not necessarily band-limited functions: a historical overview. *SIAM Review*, **34**, 40–53.

Cacuci, D. G. (1981). Sensitivity theory for nonlinear systems. I. Nonlinear functional analysis approach. *J. Math. Phys.*, **22**, 2794–802.

Cane, M., Kaplan, A., Miller, R. N. *et al.* (1996). Mapping tropical Pacific sea level: data assimilation via a reduced state space Kalman filter. *J. Geophys. Res.*, **101**, 22599–617.

Carton, J. A., Chepurin, C., Cao, X. H. and Giese, B. (2000). A simple ocean data assimilation analysis of the global upper ocean 1950–95. Part 1: Methodology. *J. Phys. Oc.*, **30**, 294–309.

Chu, P. C. (1995). P-vector method for determining absolute velocity from hydrographic data. *Mar. Tech. Soc. J.*, **29**, 3–14.

Claerbout, J. C. (2001). *Basic Earth Imaging*. Self-published on web. Stanford University.

Corliss, G., Faure, C., Griewank, A., Hascöet, L. and Naumann, U., eds (2002). *Automatic Differentiation of Algorithms. From Simulation to Optimization*. New York: Springer.

Cramér, H. (1946). *Mathematical Methods of Statistics*. Princeton: Princeton University Press.

Daley, R. (1991). *Atmospheric Data Analysis*. Cambridge: Cambridge University Press.

Dantzig, G. B. (1963). *Linear Programming and Extensions*. Princeton: Princeton University Press.

Davenport, W. B., Jr. and Root, W. L. (1958). *An Introduction to the Theory of Random Signals and Noise*. New York: McGraw-Hill.

David, M. (1988). *Handbook of Applied Advanced Geostatistical Ore Reserve Estimation*. Amsterdam: Elsevier.

Davis, P. J. and Polonsky, I. (1965). Numerical interpolation, differentiation and integration. In *Handbook of Mathematical Functions*, ed. M. Abramowitz and I. A. Stegun. New York: Dover, pp. 875–924.

Davis, R. E. (1978). Predictability of sea surface temperature and sea level pressure anomalies over the North Pacific Ocean. *J. Phys. Oc.*, **8**, 233–46.

Davis, R. E. (1994). Diapycnal mixing in the ocean: the Osborn–Cox model. *J. Phys. Oc.*, **24**, 2560–76.

Denning, P. J. (1992). Genetic algorithms. *Amer. Sci.*, **80**, short review tutorial, 12–14.

Draper, N. R. and Smith, H. (1998). *Applied Regression Analysis*, 3rd edn. New York: Wiley.

Eckart, C. and Young, G. (1939). A principal axis transformation for non-Hermitian matrices. *Bull. Amer. Math. Soc.*, **45**, 118–21.

Egbert, G. D. and Erofeeva, S. Y. (2002). Efficient inverse modeling of barotropic ocean tides. *J. Atm. Oceanic Tech.*, **19**, 183–204.

Enting, I. G. (2002). *Inverse Problems in Atmospheric Constituent Transport*. New York: Cambridge University Press.

Evensen, G. (1994). Sequential data assimilation with a nonlinear quasi-geostrophic model using Monte Carlo methods to forecast error statistics. *J. Geophys. Res.*, **99**, 10143–62.

Evensen, G. and VanLeeuwen, P. J. (1996). Assimilation of Geosat altimeter data for the Agulhas Current using the ensemble Kalman filter with a quasigeostrophic model. *Mon. Wea. Rev.*, **14**, 85–96.

Farrell, B. F. (1989). Optimal excitation of baroclinic waves. *J. Atmos. Sci.*, **46**, 1193–206.

Farrell, B. F. and Moore, A. M. (1993). An adjoint method for obtaining the most rapidly growing perturbations to oceanic flows. *J. Phys. Oc.*, **22**, 338–49.

Feller, W. (1957). *An Introduction to Probability Theory and Its Applications*, 2nd edn. New York: Wiley.

Fiacco, A. V. and McCormick, G. P. (1968). *Nonlinear Programming: Sequential Unconstrained Minimization Techniques*. New York: John Wiley. Reprinted by SIAM (1992).

Fieguth, P. W., Menemenlis, D. and Fukumori, I. (2003). Mapping and pseudoinverse algorithms for ocean data assimilation. *IEEE Trans. Geosci. Remote Sensing*, **41**, 43–51.

Franklin, G. F., Powell, J. D. and Workman, M. L. (1998). *Digital Control of Dynamic Systems*. Menlo Park: Addison-Wesley.

Freeman, H. (1965). *Discrete-Time Systems. An Introduction to the Theory*. New York: Wiley.

Fu, L.-L. (1981). Observations and models of inertial waves in the deep ocean. *Revs. Geophys. and Space Phys.*, **19**, 141–70.

Fu, L.-L., Fukumori, I. and Miller, R. N. (1993). Fitting dynamical models to the Geosat sea level observations in the tropical Pacific Ocean. Part II: A linear wind-driven model. *J. Phys. Oc.*, **23**, 2162–89.

Fukumori, I. (1991). Circulation about the Mediteranean Tongue: an analysis of an EOF-based ocean. *Prog. Oceanog.*, **27**, 197–224.

Fukumori, I. (1995). Assimilation of TOPEX sea level measurements with a reduced-gravity, shallow water model of the tropical Pacific Ocean. *J. Geophys. Res.*, **100**, 25027–39.

Fukumori, I. (2001). Data assimilation by models. In *Satellite Altimetry and Earth Sciences*, ed. L.-L. Fu and A. Cazenave. San Diego: Academic, pp. 237–66.

Fukumori, I. (2002). A partitioned Kalman filter and smoother. *Mon. Wea. Rev.*, **130**, 1370–83.

Fukumori, I., Benveniste, J., Wunsch, C. and Haidvogel, D. B. (1992). Assimilation of sea surface topography into an ocean circulation model using a steady-state smoother. *J. Phys. Oc.*, **23**, 1831–55.

Fukumori, I., Martel, F., Wunsch, C. (1991). The hydrography of the North Atlantic in the early 1980s. An atlas. *Prog. Oceanog.*, **27**, 1–110.

Fukumori, I. and Wunsch, C. (1991). Efficient representation of the North Atlantic hydrographic and chemical distributions. *Prog. Oceanog.*, **27**, 111–95.

Fukumori, I., Raghunath, R., Fu, L.-L. and Chao, Y. (1999). Assimilation of TOPEX/Poseidon altimeter data into a global ocean circulation model: how good are the results? *J. Geophys. Res.*, **104**, 25647–65.

Galanti, E. and Tziperman, E. (2003). A midlatitude ENSO teleconnection mechanism via baroclinically unstable long Rossby waves. *J. Phys. Oc.*, **33**, 1877–88.

Ganachaud, A. (2003a). Large-scale mass transports, water massformation, and diffusivities estimated from World Ocean Circulation Experiment (WOCE) hydrographic data. *J. Geophys. Res.*, **108**, 321310.1029/2002JC001565.

Ganachaud, A. (2003b). Error budget of inverse box models: the North Atlantic. *J. Atm. Oc. Tech.*, **20**, 1641–55.

Ganachaud, A. and Wunsch, C. (2000). Improved estimates of global ocean circulation, heat transport and mixing from hydrographic data. *Nature*, **408**, 453–7.

Gardiner, C. W. (1985). *Handbook of Stochastic Methods for Physics, Chemistry, and the Natural Sciences*. Berlin: Springer-Verlag.

Gauch, H. G. (2003). *Scientific Method in Practice*. Cambridge: Cambridge University Press.

Gelb, A., ed. (1974). *Applied Optimal Estimation*. Cambridge, MA: The MIT Press.

Ghil, M., Cohn, S., Tavantzis, J., Bube, K. and Isaacson, E. (1981). Applications of estimation theory to numerical weather prediction. In *Dynamic Meteorology. Data Assimilation Methods*, ed. L. Bengtsson, M. Ghil and E. Kallen. New York: Springer-Verlag, pp. 139–224.

Ghil, M. and Malanotte-Rizzoli, P. (1991). Data assimilation in meteorology and oceanography. *Adv. Geophys.*, **33**, 141–266.

Giering, R. (2000). Tangent linear and adjoint biogeochemical models. In *Inverse Methods in Global Biogeochemical Cycles*, ed. P. Kasibhatla *et al.* Geophysical Monog. 114, Washington, DC: American Geophysical Union, pp. 33–48.

Giering, R. and Kaminski, T. (1998). Recipes for adjoint code construction. *ACM Trans. Math. Software*, **24**, 437–74.

Gilbert, J. C. and Lemaréchal, C. (1989). Some numerical experiments with variable-storage quasi-Newton algorithms. *Math. Prog.*, **45**(B), 407–35.

Gill, P. E., Murray, W. and Wright, M. H. (1986). *Practical Optimization*. New York: Academic.

Gille, S. T. (1999). Mass, heat, and salt transport in the southeastern Pacific: a Circumpolar Current inverse model. *J. Geophys. Res.*, **104**, 5191–209.

Goldberg, D. E. (1989). *Genetic Algorithms in Search, Optimization and Machine Learning*. Reading, MA: Addison-Wesley.

Golub, G. H. and van Loan, C. F. (1996). *Matrix Computation*, 3rd edn. Baltimore: Johns Hopkins University Press.

Goodwin, G. C. and Sin, K. S. (1984). *Adaptive Filtering Prediction and Control*. Englewood Cliffs, NJ: Prentice-Hall.

Gordon, C. and Webb, D. (1996). You can't hear the shape of a drum. *Am. Sci.*, **84**, 46–55.

Griewank, A. (2000). *Evaluating Derivatives. Principles and Techniques of Algorithmic Differentiation*. Philadelphia: SIAM.

Griffies, S. M. (2004). *Fundamentals of Ocean Climate Models*. Princeton: Princeton University Press.

Haidvogel, D. B. and Beckmann, A. (1999). *Numerical Ocean Circulation Modeling*. River Edge, NJ: Imperial College Press.

Haine, T. W. N. and Hall, T. M. (2002). A generalized transport theory: water-mass composition and age. *J. Phys. Oc.*, **32**, 1932–46.

Hall, M. C. G. and Cacuci, D. G. (1984). Systematic analysis of climatic model sensitivity to parameters and processes. In *Climate Processes and Climate Sensitivity*, ed. J. E. Hansen and T. Takahashi. Washington, DC: American Geophysical Union, pp. 171–9.

Hamann, I. M. and Swift, J. H. (1991). A consistent inventory of water mass factors in the intermediate and deep Pacific Ocean derived from conservative tracers. *Deep-Sea Res.*, **38**(Suppl. J. L. Reid Volume), S129–70.

Hamming, R. W. (1973). *Numerical Methods for Scientists and Engineers*. New York: Dover.

Hansen, P. C. (1992). Analysis of discrete ill-posed problems by means of the L-curve. *SIAM Rev.*, **34**, 561–80.

Hasselmann, K. (1988). PIPs and POPs: the reduction of complex dynamical systems using principal interactions and oscillation pattern. *J. Geophys. Res.*, **93**, 11015–21.

Haykin, S. (2002). *Adaptive Filter Theory*, 4th edn. Englewood Cliffs, NJ: Prentice-Hall.

Heimbach, P., Hill, C. and Giering, R. (2005). Efficient exact adjoint of the parallel MIT general circulation model, generated via automatic differentiation. *Future Gen. Comput. Sys.*, **2**, 1356–71.

Herman, G. T. (1980). *Image Reconstruction From Projections: The Foundations of Computerized Tomography*. New York: Academic.

Hill, C., Bugnion, V., Follows, M. and Marshall, J. (2004). Evaluating carbon sequestration efficiency in an ocean circulation model by adjoint sensitivity analysis. *J. Geophys. Res.*, **109**, doi:(10).1029/2002JC001598.

Holland, J. H. (1992). *Adaptation in Natural and Artificial Systems: An Introductory Analysis with Applications to Biology, Control and Artificial Intelligence*. Cambridge, MA: The MIT Press.

Ide, K., Courtier, P., Ghil, M. and Lorenc, A. C. (1997). Unified notation for data assimilation: operational, sequential and variational. *J. Met. Soc. Japan*, **75**, 181–9.

Jackett, D. R. and McDougall, T. J. (1997). A neutral density variable for the world's oceans. *J. Phys. Oc.*, **27**, 237–63.

Jackson, J. E. (2003). *A User's Guide to Principal Components*. New York: Wiley.

Jaynes, E. T. (2003). *Probability Theory: The Logic of Science*. Cambridge: Cambridge University Press.

Jeffreys, H. (1961). *Theory of Probability*, 3rd edn. Oxford: Oxford University Press.

Jerri, A. J. (1977). The Shannon sampling theorem – its various extensions and applications: a tutorial review. *Proc. IEEE*, **65**, November, 1565–96.

Jolliffe, I. T. (2002). *Principal Component Analysis*, 2nd edn. New York: Springer-Verlag.

Joyce, T. M., Hernandez-Guerra, A. and Smethie, W. M. (2001). Zonal circulation in the NW Atlantic and Caribbean from a meridional World Ocean Circulation Experiment hydrographic section at 66 degrees W. *J. Geophys. Res.*, **106**, 22095–113.

Kač, M. (1966). Can one hear the shape of a drum? *Am. Math. Monthly*, **73**, 1–27.

Kalman, R. E. (1960). A new approach to linear filtering and prediction problems. *J. Basic. Eng.*, **82D**, 35–45.

Kalnay, E. (2003). *Atmospheric Modeling, Data Assimilation, and Predictability*. Cambridge: Cambridge University Press.

Kasibhatla, P., Heimann, M., Rayner, P. *et al.*, eds (2000). *Inverse Methods in Global Biogeochemical Cycles*, Geophys. Monog. (114). Washington, DC: American Geophysical Union.

Kelly, K. A. (1989). An inverse model for near-surface velocity from infrared images. *J. Phys. Oc.*, **19**, 1845–64.

Killworth, P. D. (1986). A Bernoulli inverse method for determining the ocean circulation. *J. Phys. Oc.*, **16**, 2031–51.

Kirkpatrick, S., Gelatt, C. D. and Vecchi, M. P. (1983). Optimization by simulated annealing. *Science*, **220**, 671–80.

Kitigawa, G. and Sato, S. (2001). Monte Carlo smoothing and self-organising state-space model. *Sequential Monte Carlo Methods in Practice*, ed. A. Doucet, N. de Freitas and N. Gordon. New York: Springer-Verlag, pp. 177–95.

Klema, V. C. and Laub, A. J. (1980). The singular value decomposition: its computation and some applications. *IEEE Trans. Automatic Control*, **AC-25**, 164–76.

Köhl, A., Stammer, D. and Cornuelle, B. (2006). Interannual to decadal changes in the ECCO global synthesis. Submitted for publication.

Köhl, A. and Willebrand, J. (2002). An adjoint method for the assimilation of statistical characteristics into eddy resolving ocean models. *Tellus*, **54**, 406–25.

Lanczos, C. (1960). *The Variational Principles of Dynamics*. New York: Dover.

Lanczos, C. (1961). *Linear Differential Operators*. Princeton: Van Nostrand.

Landau, H. J. and Pollak, H. O. (1962). Prolate spheroidal wave functions, Fourier analysis and uncertainty-III: the dimensions of the space of essentially time and bandlimited signals. *Bell System Tech. J.*, **41**, 1295–336.

Lawson, C. L. and Hanson, R. J. (1995). *Solving Least Squares Problems*. Philadelphia: SIAM.

Lea, D. J., Allen, M. R. and Haine, T. W. N. (2000). Sensitivity analysis of the climate of a chaotic system. *Tellus A*, **52**, 523–32.

Li, X. and Wunsch, C. (2004). An adjoint sensitivity study of chlorofluorocarbons in the North Atlantic. *J. Geophys. Res.*, **109**(C1), (10).1029/2003JC002014.

Liebelt, P. B. (1967). *An Introduction to Optimal Estimation*. Reading, MA: Addison-Wesley.

Ljung, L. (1999). *System Identification: Theory for the User*, 2nd edn. Englewood Cliffs, NJ: Prentice-Hall.

Lorenc, A. C. (1986). Analysis methods for numerical weather prediction. *Q. J. Royal Met. Soc.*, **112**, 1177–94.

Losch, M. and Heimbach, P. (2005). Adjoint sensitivity of an ocean general circulation model to bottom topography. Submitted for publication.

Luenberger, D. G. (1969). *Optimization by Vector Space Methods*. New York: Wiley.

Luenberger, D. G. (1979). *Introduction to Dynamic Systems. Theory, Models and Applications*. New York: Wiley.

Luenberger, D. G. (2003). *Linear and Non-Linear Programming*, 2nd edn. Reading, MA: Addison-Wesley.

Macdonald, A. (1998). The global ocean circulation: a hydrographic estimate and regional analysis. *Prog. Oceanog.*, **41**, 281–382.

Mackas, D. L., Denman, K. L. and Bennett, A. F. (1987). Least squares multiple tracer analysis of water mass composition. *J. Geophys. Res.*, **92**, 2907–18.

Magnus, J. R. and Neudecker, H. (1988). *Matrix Differential Calculus with Applications in Statistics and Econometrics*. Chichester: Wiley.

Marchal, O. (2005). Optimal estimation of atmospheric C-14 production over the Holocene: paleoclimate implications. *Clim. Dyn.*, **24**, 71–88.

Marotzke, J., Giering, R., Zhang, K. Q. *et al.* (1999). Construction of the adjoint MIT ocean general circulation model and application to Atlantic heat transport sensitivity. *J. Geophys. Res.*, **104**, 29529–47.

Marotzke, J. and Wunsch, C. (1993). Finding the steady state of a general circulation model through data assimilation: application to the North Atlantic Ocean. *J. Geophys. Res.*, **98**, 20149–67.

Martel, F. and Wunsch, C. (1993a). Combined inversion of hydrography, current meter data and altimetric elevations for the North Atlantic circulation. *Manus. Geodaetica*, **18**, 219–26.

Martel, F. and Wunsch, C. (1993b). The North Atlantic circulation in the early 1980s – an estimate from inversion of a finite difference model. *J. Phys. Oc.*, **23**, 898–924.

Matear, R. J. (1993). Circulation in the Ocean Storms area located in the Northeast Pacific Ocean determined by inverse methods. *J. Phys. Oc.*, **23**, 648–58.

McCuskey, S. W. (1959). *An Introduction to Advanced Dynamics*. Reading, MA: Addison-Wesley.

McIntosh, P. C. and Veronis, G. (1993). Solving underdetermined tracer inverse problems by spatial smoothing and cross validation. *J. Phys. Oc.*, **23**, 716–30.

Mémery, L. and Wunsch, C. (1990). Constraining the North Atlantic circulation with tritium data. *J. Geophys. Res.*, **95**, 5229–56.

Menemenlis, D. and Chechelnitsky, M. (2000). Error estimates for an ocean general circulation model from altimeter and acoustic tomography data. *Monthly Weather Rev.*, **128**, 763–78.

Menemenlis, D., Fukumori, I. and Lee, T. (2005). Using Green's functions to calibrate an ocean general circulation model. *Monthly Weather Rev.*, **133**, 1224–40.

Menemenlis, D. and Wunsch, C. (1997). Linearization of an oceanic general circulation model for data assimilation and climate studies. *J. Atm. Oc. Tech.*, **14**, 1420–43.

Menke, W. (1989). *Geophysical Data Analysis: Discrete Inverse Theory*, 2nd edn. New York: Academic.

Mercier, H., Ollitrault, M. and Le Traon, P. Y. (1993). An inverse model of the North Atlantic general circulation using Lagrangian float data. *J. Phys. Oc.*, **23**, 689–715.

Miller, R. N., Ghil, M. and Gauthiez, F. (1994). Advanced data assimilation in strongly nonlinear dynamical models. *J. Atmos. Sci.*, **51**, 1037–56.

Morse, P. M. and Feshbach, H. (1953). *Methods of Theoretical Physics.* 2 vols. New York: McGraw-Hill.

Munk, W. H. (1966). Abyssal recipes. *Deep-Sea Res.*, **13**, 707–30.

Munk, W., Worcester, P. and Wunsch, C. (1995). *Ocean Acoustic Tomography.* Cambridge: Cambridge University Press.

Munk, W. and Wunsch, C. (1982). Up/down resolution in ocean acoustic tomography. *Deep-Sea Res.*, **29**, 1415–36.

Naveira Garabato, A. C., Stevens, D. P. and Heywood, K. J. (2003). Water mass conversion, fluxes, and mixing in the Scotia Sea diagnosed by an inverse model. *J. Phys. Oc.*, **33**, 2565–87.

Nayfeh, A. H. (1973). *Perturbation Methods.* New York: Wiley.

Needler, G. (1967). A model for the thermohaline circulation in an ocean of finite depth. *J. Mar. Res.*, **25**, 329–42.

Needler, G. T. (1985). The absolute velocity as a function of conserved measurable quantities. *Prog. Oceanog.*, **14**, 421–9.

Noble, B. and Daniel, J. W. (1977). *Applied Linear Algebra*, 2nd edn. Englewood Cliffs, NJ: Prentice-Hall.

Numerical Algorithms Group (2005). *The NAG Fortran Library Manual, Mark 21* (pdf). www.nag.co.uk/numeric/FL/FLdocumentation.asp.

O'Reilly, J. (1983). *Observers for Linear Systems.* London: Academic.

Olbers, D. and Wenzel, M. (1989). Determining diffusivities from hydrographic data by inverse methods with application to the circumpolar current. In *Oceanic Circulation Models: Combining Data and Dynamics*, ed. D. L. T. Anderson and J. Willebrand, Boston: Kluwer, pp. 95–140.

Olbers, D., Wenzel, J. M. and Willebrand, J. (1985). The inference of North Atlantic circulation patterns from climatological hydrographic data. *Rev. Geophys.*, **23**, 313–56.

Oreskes, N., Shrader-Frechette, K. and Belitz, K. (1994). Verification, validation, and confirmation of numerical models in the earth sciences. *Science*, **263**, 641–6.

Paige, C. C. and Saunders, M. A. (1982). Algorithm 583 LSQR: sparse linear equations and least-squares problems. *ACM Trans. Software, Math.*, **8**, 195–209.

Pedlosky, J. (1987). *Geophysical Fluid Dynamics*, 2nd edn. New York: Springer-Verlag.

Percival, D. B. and Rothrock, D. A. (2005). "Eyeballing" trends in climate time series: a cautionary note. *J. Clim.*, **18**, 886–91.

Petersen, D. P. and Middleton, D. (1962). Sampling and reconstruction of wave-number-limited functions in N-dimensional Euclidean space. *Inform. Control*, **5**, 279–323.

Pincus, M. (1968). A closed form solution of certain programming problems. *Oper. Res.*, **16**, 690–4.

Preisendorfer, R. W. (1988). *Principal Component Analysis in Meteorology and Oceanography*. Posthumously compiled and edited by C. D. Mobley. Amsterdam: Elsevier.

Press, W. H., Flannery, B. P., Teukolsky, S. A. and Vetterling, W. T. (1996). *FORTRAN Numerical Recipes*, 2nd edn, corrected. Cambridge: Cambridge University Press.

Priestley, M. B. (1982). *Spectral Analysis and Time Series. Volume 1: Univariate Series. Volume 2: Multivariate Series, Prediction and Control*. Combined edition. London: Academic.

Rall, L. B. (1981). *Automatic Differentiation: Techniques and Applications*. Berlin: Springer-Verlag.

Rauch, H. E., Tung, F. and Striebel, C. T. (1965). Maximum likelihood estimates of linear dynamic systems. *AIAA J.*, **3**, 1445–50 (reprinted in Sorenson, 1985).

Reid, W. T. (1972). *Riccati Differential Equations*. New York: Academic.

Restrepo, J. M., Leaf, G. K. and Griewank, A. (1995). Circumventing storage limitations in variational data assimilation studies. *Siam J. Sci. Computing*, **19**, 1586–605.

Ripley, B. D. (2004). *Spatial Statistics*. New York: Wiley.

Rockafellar, R. T. (1993). Lagrange multipliers and optimality. *SIAM Rev.*, **35**, 183–238.

Roemmich, D. and Wunsch, C. (1985). Two transatlantic sections: meridional circulation and heat flux in the subtropical North Atlantic Ocean. *Deep-Sea Res.*, **32**, 619–64.

Rogers, G. S. (1980). *Matrix Derivatives*. New York: Marcel Dekker.

Sasaki, Y. (1970). Some basic formalisms in numerical variational analysis. *Monthly Weather Rev.*, **98**, 875–83.

Scales, L. E. (1985). *Introduction to Non-Linear Optimization*. New York: Springer-Verlag.

Schlitzer, R. (1988). Modeling the nutrient and carbon cycles of the North Atlantic, 1. Circulation, mixing coefficients, and heat fluxes. *J. Geophys. Res.*, **93**, 10699–723.

Schlitzer, R. (1989). Modeling the nutrient and carbon cycles of the North Atlantic, 2. New production, particle fluxes, CO2, gas exchange, and the role of organic nutrients. *J. Geophys. Res.*, **94**, 12781–94.

Schott, F. and Stommel, H. (1978). Beta spirals and absolute velocities in different oceans. *Deep-Sea Res.*, **25**, 961–1010.

Schröter, J. and Wunsch, C. (1986). Solution of non-linear finite difference ocean models by optimization methods with sensitivity and observational strategy analysis. *J. Phys. Oc.*, **16**, 1855–74.

Seber, G. A. F. and Lee, A. J. (2003). *Linear Regression Analysis*, 2nd edn. Hoboken, NJ: Wiley-Interscience.

Sewell, M. J. (1987). *Maximum and Minimum Principles. A Unified Approach with Applications*. Cambridge: Cambridge University Press.

Sloyan, B. M. and Rintoul, S. R. (2001). The Southern Ocean limb of the global deep overturning circulation. *J. Phys. Oc.*, **31**, 143–73.

Sorenson, H. W., ed. (1985). *Kalman Filtering: Theory and Application*. New York: IEEE Press.

Stammer, D., Ueyoshi, K., Large, W. B., Josey, S. and Wunsch, C. (2004). Global sea surface flux estimates obtained through ocean data assimilation. *J. Geophys. Res.*, **109**, C05023, doi:(10).1029/2003JC(002082).

Stammer, D. and Wunsch, C. (1996). The determination of the large-scale circulation of the Pacific Ocean from satellite altimetry using model Green's functions. *J. Geophys. Res.*, **101**, 18409–32.

Stammer, D., Wunsch, C., Giering, R. *et al.* (2002). Global ocean state during 1992–1997, estimated from ocean observations and a general circulation model. *J. Geophys. Res.*, DOI: (10).1029/2001JC000888.

Stammer, D., Wunsch, C., Giering, R. *et al.* (2003). Volume, heat and freshwater transports of the global ocean circulation 1992–1997, estimated from a general circulation model constrained by WOCE data. *J. Geophys. Res.*, DOI: 10.1029/2001JC001115.

Stengel, R. F. (1986). *Stochastic Optimal Control*. New York: Wiley-Interscience.

Stewart, G. W. (1993). On the early history of the singular value decomposition. *SIAM Rev.*, **35**, 551–66.

Stommel, H. and Schott, F. (1977). The beta spiral and the determination of the absolute velocity field from hydrographic station data. *Deep-Sea Res.*, **24**, 325–9.

Strang, G. (1988). *Linear Algebra and its Applications*, 3rd edn. San Diego: Harcourt, Brace Jovanovich.

Talagrand, O. (1997). Assimilation of observations, an introduction. *J. Meteor. Soc. Japan*, **75**, 191–209.

Tarantola, A. (1987). *Inverse Problem Theory. Methods for Data Fitting and Model Parameter Estimation*. Amsterdam: Elsevier.

Tarantola, A. and Valette, B. (1982). Generalized nonlinear inverse problems solved using the least squares criterion. *Revs. Geophys. Space Phys.*, **20**, 219–32.

Thacker, W. C. (1989). The role of the Hessian matrix in fitting models to measurements. *J. Geophys. Res.*, **94**, 6177–96.

Thièbaux, H. J. and Pedder, M. A. (1987). *Spatial Objective Analysis: With Applications in Atmospheric Science*. London: Academic.

Tomczak, M. and Large, D. G. B. (1989). Optimum multiparameter analysis of mixing in the thermocline of the Eastern Indian Ocean. *J. Geophys. Res.*, **94**, 16141–50.

Trefethen, L. N. (1997). Pseudospectra of linear operators. *SIAM Rev.*, **39**, 383–406.

Trefethen, L. N. (1999). Computation of pseudospectra. *Acta Numerica*, **8**, 247–95.

Tziperman, E. and Hecht, A. (1987). A note on the circulation in the eastern Levantine basin by inverse methods. *J. Phys. Oc.*, **18**, 506–18.

Tziperman, E., Thacker, W. C., Long, R. B. and Hwang, S.-M. (1992). Oceanic data analysis using a general circulation model. Part I: Simulations. *J. Phys. Oc.*, **22**, 1434–57.

Ueno, H. and Yasuda, I. (2003). Intermediate water circulation in the North Pacific subarctic and northern subtropical regions. *J. Geophys. Res.*, **108**(C11), Art. No. 3348.

Van Huffel, S. and Vandewalle, J. (1991). *The Total Least Squares Problem. Computational Aspects and Analysis*. Philadelphia: SIAM.

van Laarhoven, P. J. M. and Aarts, E. H. L. (1987). *Simulated Annealing: Theory and Applications*. Dordrecht: Kluwer.

Van Trees, H. L. (2001). *Detection, Estimation and Modulation Theory. Part 1. Detection, Estimation, and Linear Modulation Theory*. New York: Wiley.

von Storch, H., Bruns, T., Fischer-Bruns, I. and Hasselmann, K. (1988). Principal oscillation pattern analysis of the 30-to-60-day oscillation in a general circulation model equatorial troposphere. *J. Geophys. Res.*, **93**, 11022–36.

von Storch, H. and Zwiers, F. W. (1999). *Statistical Analysis in Climate Research*. Cambridge: Cambridge University Press.

Wagner, H. M. (1975). *Principles of Operations Research. With Applications to Managerial Decisions*, 2nd edn. Englewood Cliffs, NJ: Prentice-Hall.

Wahba, G. (1990). *Spline Models for Observational Data*. Philadelphia: SIAM.

Wallace, J. M. (1972). Empirical-orthogonal representation of time series in the frequency domain. Part II: Application to the study of tropical wave disturbances. *J. Appl. Met.*, **11**, 893–900.

Wallace, J. M. and Dickinson, R. E. (1972). Empirical-orthogonal representation of time series in the frequency domain. Part I: Theoretical considerations. *J. Appl. Met.*, **11**, 887–92.

Welander, P. (1983). Determination of the pressure along a closed hydrographic section 1. The ideal case. *J. Phys. Oc.*, **13**, 797–803.

Whittaker, E. and Robinson, G. (1944). *The Calculus of Observations*. Glasgow: Blackie and Sons.

Whitworth III, T., Warren, B. A., Nowlin Jr., W. D. *et al.* (1999). On the deep western-boundary current in the Southwest Pacific Basin. *Prog. Oceanog.*, **43**, 1–54.

Wiggins, R. A. (1972). The general linear inverse problem: Implication of surface waves and free oscillations for earth structure. *Revs. Geophys. Space Phys.*, **10**, 251–85.

Worcester, P. F., Cornuelle, B. D., Dzieciuch, M. A. *et al.* (1999). A test of basin-scale acoustic thermometry using a large-aperture vertical array at 3250-km range in the eastern North Pacific Ocean. *J. Acoust. Soc. Am.*, **105**, 3185–201.

Worcester, P. F. and Spindel, R. C. (2005). North Pacific acoustic laboratory. *J. Acoust. Soc. Am.*, **117**, 1499–510.

Wunsch, C. (1977). Determining the general circulation of the oceans: a preliminary discussion. *Science*, **196**, 871–5.

Wunsch, C. (1978). The North Atlantic general circulation west of 50 W determined by inverse methods. *Revs. Geophys. Space Phys.*, **16**, 583–620.

Wunsch, C. (1984). An eclectic Atlantic Ocean circulation model. Part I: The meridional flux of heat. *J. Phys. Oc.*, **14**, 1712–33.

Wunsch, C. (1985). Can a tracer field be inverted for velocity? *J. Phys. Oc.*, **15**, 1521–31.

Wunsch, C. (1988). Transient tracers as a problem in control theory. *J. Geophys. Res.*, **93**, 8099–110.

Wunsch, C. (1989). Tracer inverse problems. In *Oceanic Circulation Models: Combining Data and Dynamics*, ed. D. L. T. Anderson and J. Willebrand. Dordrecht: Kluwer, pp. 1–77.

Wunsch, C. (1994). Dynamically consistent hydrography and absolute velocity in the eastern North Atlantic Ocean. *J. Geophys. Res.*, **99**, 14071–90.

Wunsch, C. (1996). *The Ocean Circulation Inverse Problem*. Cambridge: Cambridge University Press.

Wunsch, C. and Heimbach, P. (2006). Practical global oceanic state estimation. Submitted for publication.

Wunsch, C. and Minster, J.-F. (1982). Methods for box models and ocean circulation tracers: Mathematical programming and non-linear inverse theory. *J. Geophys. Res.*, **87**, 5647–62.

You, Y. Z. (2002). Quantitative estimate of Antarctic Intermediate Water contributions from the Drake Passage and the southwest Indian Ocean to the South Atlantic. *J. Geophys. Res.*, **107**(C4), Art. No. 3031.

Zhang, H. M. and Hogg, N. G. (1992). Circulation and water mass balance in the Brazil Basin. *J. Mar. Res.*, **50**, 385–420.

Index

Made in the USA
Monee, IL
07 July 2026

56551353R00219